# X-Ray Spectrometry in Electron Beam Instruments

# X-Ray Spectrometry in Electron Beam Instruments

Edited by

## David B. Williams

*Lehigh University*
*Bethlehem, Pennsylvania*

## Joseph I. Goldstein

*University of Massachusetts at Amherst*
*Amherst, Massachusetts*

and

## Dale E. Newbury

*National Institute of Standards and Technology*
*Gaithersburg, Maryland*

Plenum Press • New York and London

Library of Congress Cataloging-in-Publication Data

---

X-ray spectrometry in electron beam instruments / edited by David B.
  Williams, Joseph I. Goldstein, and Dale E. Newbury.
       p.   cm.
    Includes bibliographical references and index.
    ISBN 0-306-44858-0
    1. Electron beams--Instruments.   2. X-ray spectroscopy.
  3. Electron probe microanalysis.   I. Williams, David B. (David
  Bernard), 1949-    . II. Goldstein, Joseph, 1939-    . III. Newbury,
  Dale E.
  QC793.5.E622X14   1994
  543'.08586--dc20                                              94-45877
                                                                   CIP

---

ISBN 0-306-44858-0

To Charles E. (Chuck) Fiori (1938–1992)

# Contributors

*G. F. Bastin* • Laboratory for Solid State Chemistry and Materials Science, University of Technology, Nl5600 MB Eindhoven, The Netherlands

*J. J. Friel* • Princeton GammaTech, Inc., Princeton, New Jersey 08540

*H. J. M. Heijligers* • Laboratory for Solid State Chemistry and Materials Science, University of Technology, Nl5600 MB Eindhoven, The Netherlands

*K. F. J. Heinrich* • National Institute of Standards and Technology (ret.), Present address: Rockville, Maryland 20850

*D. C. Joy* • EM Facility, University of Tennessee, Knoxville, Tennessee 37996-0810, and High Temperature Materials Laboratory, Oak Ridge National Laboratory, Oak Ridge, Tennessee 37831

*B. G. Lowe* • Oxford Instruments, Abingdon, OX14 1TX, United Kingdom

*M. W. Lund* • MOXTEK, Inc., Orem, Utah 84057

*J. J. McCarthy* • Noran Instruments, Inc., Middleton, Wisconsin 53562

*J. R. Michael* • Materials and Process Sciences Center, Electron Microscopy and Metallography Department, Sandia National Laboratories, Albuquerque, New Mexico 87185-0342

*R. B. Mott* • Princeton GammaTech, Inc., Princeton, New Jersey 08540

*R. L. Myklebust* • National Institute of Standards and Technology, Gaithersburg, Maryland 20899.

**D. E. Newbury** • Microanalysis Research Group, National Institute of Standards and Technology, Gaithersburg, Maryland 20899

**A. M. Panin** • Brigham Young University, Provo, Utah 84602.

**S. J. B. Reed** • Department of Earth Sciences, University of Cambridge, Cambridge CB2 3EQ, United Kingdom

**R. Rybka** • Microspec Corporation, Fremont, California 94539

**R. A. Sareen** • Schuster Laboratories, The University, Manchester M13 9PL, United Kingdom

**P. J. Statham** • Oxford Instruments Microanalysis Group, Halifax Road, High Wycombe, Bucks HP12 3SE, United Kingdom

**C. R. Swyt** • National Institutes of Health, Bethesda, Maryland 20892.

**D. B. Williams** • Department of Materials Science and Engineering, Lehigh University, Bethlehem, Pennsylvania 18015-3195

**R. C. Wolf** • Microspec Corporation, 45950 Hotchkiss Street, Fremont, California 94539

**S. M. Zemyan** • Department of Materials Science and Engineering, Lehigh University, Bethlehem, Pennsylvania 18015-3195

# Dedication

This volume is derived from a special symposium on x-ray spectrometry organized by the Microbeam Analysis Society at Loyola Marymount University in Los Angeles as part of the 1993 Annual Conference to celebrate the memory of the late Charles E. "Chuck" Fiori (1938–1992). Chuck Fiori was a highly creative and stimulating colleague who interacted with a wide range of people working in the field of microbeam analysis throughout his highly productive but prematurely shortened career of almost 30 years. His work spanned most of the period of the development and application of electron probe x-ray microanalysis, a field to which he made many significant contributions. His career is remarkable in that he began as a general physical science technician, found his way into electron beam/x-ray spectrometry instrumentation, developed a fascination for the subject, and advanced through much personal effort to reach the level of a world recognized authority in his field. He worked at Scripps Institute, the Smithsonian Institution, the National Bureau of Standards, the National Institutes of Health, and finally returned to what proved to be the completion of his career at the National Bureau of Standards, renamed the National Institute of Standards and Technology. He was an adjunct professor at Lehigh University, where he taught in the short courses on electron microscopy and x-ray microanalysis for the past 18 years. As an author, he published more than 50 journal articles and co-authored three books.

Throughout his career, Chuck gave unstintingly of his time for the common good, serving first as secretary and later as president of the Microbeam Analysis Society (MAS), contributing 12 years of direct service to the Society. Because of his emphasis on the practical aspects of microanalysis and his highly effective presentation style, Chuck was one of the most popular national scientific tour speakers sponsored by MAS to the regional societies. In recognition of his many contributions to the scientific community, the Microbeam Analysis Society has established an annual Fiori Memorial Tour Speaker who will concentrate on topics of practical interest to analysts. Royalties from this publication will support this activity.

Chuck was an extraordinarily gifted person, skilled in both experimental and theoretical work. He had those wonderful mechanical, electrical, and electronic engineering skills, as well as knowledge of the mathematics and physics, that enabled him to operate at every level necessary to carry out complex experiments, from the assembly of the instrument, the electronic signal processing, the computer data manipulation, and the mathematical deconvolution of x-ray spectra, to the electron/x-ray physics for interpretation.

In selecting the invited speakers to present papers at the symposium, the editors sought to provide an overall view of current research in the field of x-ray spectrometry as it relates to the practice of electron probe x-ray microanalysis. In reviewing these papers, it is interesting to note how many of the topics presented at the symposium were areas of active research interest to Chuck at the time of his death. Chuck was always a bundle of creative energy, and he tended to have numerous projects going on in parallel at various stages of development. He was in the process of creating, fortunately in close collaboration with Carol Swyt of NIH and Bob Myklebust of NIST, a comprehensive x-ray spectrometry calculation engine, DeskTop Spectrum Analyzer (DTSA). He planned DTSA as a software workbench to extend his personal research as well as to serve as an incredible resource for the rest of the x-ray spectrometry field. The initial version of DTSA for energy dispersive x-ray spectrometry (EDS) had been successfully released prior to his death. Chuck's last DTSA projects included adaptation of peak deconvolution for the case of wavelength dispersive spectrometry (WDS), application of WDS deconvolution to the problem of measuring trace peaks in the presence of higher intensity peak interferences, a quantitative exploration of the trade-offs between count rate and spectrometer resolution, and the modeling of energy dispersive x-ray spectra from first principles. Chuck was also involved with the development of digital signal processing for EDS.

Chuck Fiori would certainly have been an enthusiastic contributor to the 1993 MAS symposium on x-ray spectrometry. Fortunately for his colleagues, he has left behind for us a legacy of scientific achievement upon which we can continue to build and, through his development of DTSA, he has even provided us the tools with which to labor. We extend our sympathies to his wife, Virginia, and his family, and we shall always remember Chuck for his intelligence, wit, and humanity.

Dale E. Newbury
*Gaithersburg, Maryland*
David B. Williams
*Bethlehem, Pennsylvania*
Joseph I. Goldstein
*Amherst, Massachusetts*

# Foreword

From its early days in the 1950s, the electron microanalyzer has offered two principal ways of obtaining x-ray spectra: wavelength dispersive spectrometry (WDS), which utilizes crystal diffraction, and energy dispersive spectrometry (EDS), in which the x-ray quantum energy is measured directly. In general, WDS offers much better peak separation for complex line spectra, whereas EDS gives a higher collection efficiency and is easier and cheaper to use. Both techniques have undergone major transformations since those early days, from the simple focusing spectrometer and gas proportional counter of the 1950s to the advanced semiconductor detectors and programmable spectrometers of today. Because of these developments, the capabilities and relative merits of EDS and WDS techniques have been a recurring feature of microprobe conferences for nearly 40 years, and this volume brings together the papers presented at the Chuck Fiori Memorial Symposium, held at the Microbeam Analysis Society Meeting of 1993.

Several themes are apparent in this rich and authoritative collection of papers, which have both a historical and an up-to-the-minute dimension. Light element analysis has long been a goal of microprobe analysts since Ray Dolby first detected carbon $K$ radiation with a gas proportional counter in 1960. WDS techniques (using lead stearate films) were not used for this purpose until four years later. Now synthetic multilayers provide the best dispersive elements for quantitative light element analysis—still used in conjunction with a gas counter.

At the other extreme of energy, the Analytical Electron Microscope requires an efficient spectrometer system for x-ray energies up to 100 keV, often with very small count rates. Although the early work was undertaken with WDS spectrometers, these have been largely replaced with silicon EDS detectors for improved efficiency. Germanium detectors are also increasingly used, both on account of their improved absorption of high energy x-rays and their better resolution at all energies.

For general use, however, it is the lithium-drifted silicon detector which is by far the most widely used today. It was first adopted for microprobe analysis from the nuclear field by Kurt Heinrich and others in 1968, and has since undergone major improvements in almost every respect—energy resolution, maximum count rate, and energy range—even complementing the use of multilayer spectrometers for light element analysis. These advances have been made possible by improved techniques of manufacture, both of the counter and the associated pulse processing circuitry, and by a better understanding of the fundamental physics of operation. Much has also been done in the associated software for spectrum processing to make the whole system accessible and friendly to the operator.

With his colleagues at the National Institute of Standards and Technology, Chuck Fiori had worked in almost every topic covered by the Symposium. This volume provides a fitting tribute in his memory, reminding us not only of his versatility, but of the energy and enthusiasm with which he shared his work with others. In both respects he will be remembered by microprobe analysts across the world for a long time to come.

Peter Duncumb

*Cambridge, England*

# Contents

## CHAPTER 4.  **Germanium X-Ray Detectors**

*R. A. Sareen*




## CHAPTER 5.  **Modeling the Energy Dispersive X-Ray Detector**

*D. C. Joy*




## CHAPTER 6.  **The Effect of Detector Dead Layers on Light Element Detection**

*J. J. McCarthy*

**CHAPTER 7.   Energy Dispersive X-Ray Spectrometry in Ultra-high Vacuum  Environments**

*J. R. Michael*



**CHAPTER 8.   Quantifying Benefits of Resolution and Count Rate in EDX  Microanalysis**

*P. J. Statham*



**CHAPTER 9.   Improving EDS Performance with Digital Pulse Processing**

*R. B. Mott and J. J. Friel*

CHAPTER 10.  **A Study of Systematic Errors in Multiple Linear Regression Peak Fitting Using Generated Spectra**

*C. R. Swyt*




CHAPTER 11.  **Artifacts in Energy Dispersive X-Ray Spectrometry in Electron Beam Instruments. Are Things Getting Any Better?**

*D. E. Newbury*




CHAPTER 12.  **Characterizing an Energy Dispersive Spectrometer on an Analytical Electron Microscope**

*S. M. Zemyan and D. B. Williams*

## CHAPTER 13. Wavelength Dispersive Spectrometry: A Review

*S. J. B. Reed*



## CHAPTER 14. Synthetic Multilayer Crystals for EPMA of Ultra-light Elements

*G. F. Bastin and H. J. M. Heijligers*



## CHAPTER 15. A von Hamos-Type Parallel Collection Wavelength Dispersive Spectrometer for Microbeam Analysis

*A. M. Panin and M. W. Lund*



## CHAPTER 16. Fitting Wavelength Dispersive Spectra with the NIST/NIH DTSA Program

*R. L. Myklebust*

**CHAPTER 17. Application of Layered Synthetic Microstructure Crystals
to WDX Microanalysis of Ultra-light Elements**

*R. Rybka and R. C. Wolf*



**CHAPTER 18. An Evaluation of Quantitative Electron Probe Methods**

*K. F. J. Heinrich*

# 1

# The Development of Energy Dispersive Electron Probe Analysis

*K. F. J. Heinrich*

Castaing's first electron probe microanalyzer[1] was a modified electrostatic transmission electron microscope equipped with a curved-crystal x-ray spectrometer. This combination was perfected in later years by the introduction of Langmuir-Blodgett devices[2] and of diffractors consisting of evaporated metal layers,[3] so that the elements of atomic number above 3 could also be observed and analyzed.

Electron probe microanalysis (EPMA) can therefore be used to detect with moderate sensitivity (typically of the order of $10^{-4}$ weight fractions) almost all elements, in a volume at the specimen surface of the order of 1–50 $\mu m^3$. Scanning the focused electron beam over areas up to hundreds of beam diameters provides an additional technique of characterization of a microscopic surface. But the most outstanding characteristic of EPMA, responsible for its wide application, is its potential for quantitation.

Castaing proposed in his thesis that a comparison of the electron-excited x-ray line intensities from the specimen with those obtained from pure elements could be used to determine the mass fractions ("concentrations") of the elements in the specimen. The atomic-number correction[4] provided a factor for introducing the effects of electron deceleration and of backscattering, and a fluorescence correction accounts for the effects of secondary (photon-photon) excitation.[5]

Tracer experiments by Castaing, Descamps, and Henoc[6,7] provided information for several elements on the depth distribution of the generation of primary x rays. This was used, with graphic interpolation, to estimate the absorption losses of x rays, until Philibert developed an algebraic model for the calculation of the absorption correction,[8] modifications of which have been used routinely until the present. The corrections for atomic number ($Z$), primary absorption ($A$), and fluorescence excited by

K. F. J. HEINRICH • National Institue of Standards and Technology (ret.). Present address: Rockville, Maryland 20850

*X-Ray Spectrometry in Electron Beam Instruments*, edited by David Williams, Joseph Goldstein, and Dale Newbury. Plenum Press, New York, 1995.

characteristic lines ($F$) were combined, as multiplicative factors, in a procedure called ZAF correction which, with minor variants, is traditionally used in quantitative electron probe microanalysis.[9]

Duncumb, in Cosslett's group at the Cavendish Laboratory in Cambridge, UK, introduced electronic beam scanning in microanalysis,[10] and with Melford, he demonstrated the usefulness of qualitative x-ray microanalysis, combined with the microscopic procedures based on the detection of secondary electrons in the scanning mode, as a tool in metallurgical research. From then on, microanalysis was usually performed by qualitative investigations of scanned microscopic areas and quantitative local analysis with static beams, or along a linear trajectory.

In the 1960s, the electron probe analyzer had become an indispensable tool in materials research and mineralogy. Yet its application to quantitation was practically limited to the traditional flat polished specimen with normal beam incidence. The quantitative treatment of data obtained with an electron beam inclined with respect to the specimen surface was less certain; special specimen shapes, such as fibers, particles of a micrometer or less in diameter, thin films, and irregular specimen surfaces, could not be handled accurately. The determination of elements of atomic number below 10 was hampered by the lack of accurate mass absorption coefficients, and the handling of specimens of low electrical conductivity offered special problems which have not been totally overcome to this day.[11] The resolution of the method seemed to be definitely limited by the diffusion of the electrons at the energies needed to excite efficiently the x rays of interest, and the low concentrations of elements of interest in most specimens of soft biological tissues, their fragility, and the possible migration or loss of elements in the preparation of such materials greatly limited the use of the microprobe in biological problems. In addition, curved x-ray spectrometers are affected by several shortcomings:

1. Their signal intensity is limited by the small solid angle subtended and by the losses in the diffraction by the crystal.
2. The spectrometer efficiency for a given wavelength is difficult to determine and varies abruptly at the absorption edges of the crystal constituents and of the detector gas. Hence, quantitative analysis is not possible without standards for every element that is measured.
3. Qualitative wavelength scans with crystal spectrometers are very time consuming.
4. The defocusing of the curved-crystal spectrometer with the displacement of the x-ray source severely limits its application to quantitative area scans.

To overcome the limitations in spatial resolution arising from the diffusion of electrons that penetrate the specimen, Cooke and Duncumb built the Electron Microscope Microanalyzer (EMMA),[12] in which electrons of high energy traversed thin foils of specimen, reemerging before they could disperse significantly in the specimen. This instrument was the first analytical electron transmission microscope. However, the success of EMMA was limited by the low sensitivity of the crystal dispersive spectrometer.

The limits in the speed of qualitative scans, the problems of low spectrometer sensitivity, and the effects of defocusing in beam scanning operation could only be overcome by the introduction of energy dispersive spectrometers of adequate resolution and of on-line computers having sufficient speed, programmability, and data storage.

Pulse-height discrimination had been used as an adjunct to wavelength dispersive x-ray analysis with gas-filled proportional detectors in x-ray fluorescence analyis[13] to eliminate interferences due to higher-order crystal reflections. In 1963, Dolby[14] demonstrated that by means of a network of three single-channel analyzers applied to the output of a flow proportional counter he could separate in a scanning electron microprobe the $K$-lines of beryllium, carbon, and oxygen without the need for diffracting devices. Unfortunately, the resolution of such a detector system was too broad to permit a separation of the signals from elements of high atomic number.

Therefore, the problems listed above, which are particularly serious for biological applications, could not be solved until the introduction, in 1968,[15] of the lithium-drifted silicon detector, which permits rapid simultaneous detection of all major constituents of atomic number above 3 or 4. For most elements, its efficiency is high, predictable, and insensitive to minor displacements of the point of beam impact. Therefore, it is applicable to quantitative procedures in combination with beam scanning. The major limitation of the Si(Li) detector, and of similar solid state devices, is the limited wavelength resolution that is an inherent consequence of the detecting process.

The speed with which this detector generates data, and the requirements for its reception, storage, and manipulation, have greatly increased the dependence of microanalysis on the presence of powerful, fast, and easily programmable on-line computers. The lack of data storage and processing mechanisms was for many years a serious limitation even for qualitative area scanning. The usual technique consisted of collecting on a photographic emulsion the light blips produced on the screen by detected photons. The procedure produced coarse area scans, for which both the statistical arrival of the photons and the background disturbed the image quality and border definition. The information registered on the photographic negative could not be further processed; if the result was not satisfactory, the operation had to be repeated. Color scans representing several elements had been demonstrated,[16,17] but were not widely used since their production was quite cumbersome, and the process could not be modified after the registration. These problems were exacerbated when the Si(Li) detectors were used; due to their low energy resolution, the background and its variations with specimen composition and configuration became a serious obstacle both to the imaging of low concentrations and to any attempt of quantitative evaluation.

The full utilization of the photon energy spectrum obtainable with the solid state detector required new techniques which were slow to develop. Initially, the lines of the elements of interest were enclosed within energy "regions of interest," which then were treated in the same manner as the line intensities obtained from crystal spectrometers. Such a procedure did not provide a good background correction and discarded a large proportion of the experimental spectrum. Fiori, however, insisted that the use of the entire spectrum, as well as information simultaneously gathered with crystal spectrometers, would increase the accuracy, sensitivity and speed of analysis. Such a procedure not only requires the full knowledge of relative line and background intensities, but also sophisticated techniques for successive approximations to the composite spectrum. If, on the basis of such an analysis, the energy spectrum can be synthesized and compared with the observed one, false interpretations can be detected swiftly. The DTSA (DeskTop Spectrum Analyzer) procedure developed in 1991 by Fiori, Swyt, and Myklebust[18] along these precepts made full use of the small and fast computers now available at affordable cost, for an efficient qualitative and quantitative evaluation of data produced with both crystal spectrometers and energy dispersive detectors; it also constitutes a significant advance in the evolution of electron probe microanalysis.

To obtain from a theoretical model a spectrum that matches the experimental results from a specimen of unknown composition, an iterative procedure is required in which each iterative step produces an estimate of composition to be used in the next step, until satisfactory matching is achieved. Although simple iterative procedures, usually based on the hyperbolic approximation of Ziebold and Ogilvie,[19] have been used for years, a study of the corresponding literature suggests that the models proposed for the iteration were often inadequate or unnecessarily cumbersome. When the goal of the procedure is to analyze an entire spectrum, the problem becomes even more critical. For this purpose, the DTSA method offers a choice of a multiple linear least square method or the non-linear Sequential Simplex procedure.

Such a program requires, in the first place, an accurate description of the characteristics of the detector process, which are dependent on features such as the resolution, deadtime characteristics, the effects of windows, possible ice deposition on the detector, incomplete charging, and so on. The DTSA program provides input for all relevant known or estimated detector parameters.

Although the program was originally conceived primarily as a tool for the synthesis and interpretation of spectra generated by the lithium-drifted silicon detector, Fiori *et al.* recognized that its potential would be greatly extended if lines generated by crystal spectrometers could also be included in the analytical procedure. Therefore, the artifacts characteristic of wavelength detection, such as higher order reflections and satellite lines observable with such spectrometers, had to be incorporated, in addition to the artifacts arising in the energy dispersive mode, such as escape peaks, two-photon excitation peaks, and the effects of absorption edges in the detector material. It was also recognized that, unlike the energy dispersive peaks, the crystal spectrometer peaks cannot be described accurately by Gaussian distributions; a combination of Gaussian and Lorentzian distributions is used in this case. This aspect of DTSA is still being implemented.

The program of Fiori *et al.* permits a wide range of modes of operation, including the standardless mode. Clearly, if all mechanisms of x-ray photon generation are included, the uncertainties and contradictions among various proposed models for ionization, and x-ray generation and absorption, as well as the uncertainties in the parameters involved, become a serious obstacle. The program allows a wide choice of modes, models, and parameters so that the user can make selections. It is being upgraded constantly as new proposed modalities arise. In the future, this feature can be used to test the accuracy of modes, models, and parameters by sets of analytical results obtained from standard materials of known composition under a variety of experimental conditions. In spite of the lamented and premature death of the first author, this work is being actively pursued, and a full treatment of biological analysis by the Hall method, applicable to analytical electron microscopy, has been incorporated into the program. Techniques for the analysis of particles, thin films, and layered materials will also be incorporated fully.

The synthesis capabilities of such a program are particularly useful in testing the effects of operating parameters, such as operating voltage and choice of lines, and in predicting limits of detection for all observable elements in any matrix; this feature, in turn, suggests the use of the program for automatically setting the operating conditions and data collection sequences for the instrument.

These new tools and techniques will lead to a more exhaustive study of physical parameters, such as relative line intensities, and to general procedures for the unconventional specimen shapes mentioned previously. Unfortunately, the present economic pressures have slowed work in this direction, but the eventual development of an accurate general procedure of quantitative microanalysis with the use of all available detection devices and applicable to diverse specimen configurations is expected.

## REFERENCES


1. R. Castaing, Doctoral Thesis, University of Paris (1951).
2. I. Langmuir, *J. Franklin Inst.* **218**, 153 (1934).
3. J. F. Bastin and H. J. M. Heijligers, in: *Electron Probe Quantitation* (K. F. J. Heinrich and D. E. Newbury, eds.) Plenum Press, NewYork, p. 145 (1991).
4. R. Castaing and J. Descamps, *J. Phys. Radium* **16**, 304 (1955).
5. S. J. B. Reed and J. V. P. Long, *NBS Technical Note* 521, National Bureau of Standards, Washington, DC, p. 317 (1970).
6. R. Castaing and J. Descamps, *J. Phys. Radium* **16**, 304 (1955).
7. R. Castaing and J. Henoc in: *Proceedings, Fourth International Congress on X-Ray Optics and Microanalysis* (R. Castaing, P. Deschamp, and J. Philibert, eds.) Hermann, Paris, p. 120 (1966).
8. J. Philibert, in: *Proceedings, Third International Conference on X-Ray Optics and Microanalysis* (H. H. Pattee, V. E. Cosslett, and A. Engström, eds.) Academic Press, New York, p. 379 (1963).
9. K. F. J. Heinrich and D. E. Newbury, eds., *Electron Probe Quantitation*, Plenum Press, New York (1991).
10. V. E. Cosslet and P. Duncumb, *Nature* **177**, 1172 (1956).
11. G. F. Bastin and H. J. M. Heijligers, in: *Electron Probe Quantitation* (K. F. J. Heinrich and D. E. Newbury, eds.) Plenum, New York, p. 163 (1991).
12. C. J. Cooke and P. Duncumb, in: *Proceedings, Fifth International Congress on X-Ray Optics and Microanalysis* (G. Möllenstedtand and K. H. Gaukler, eds.) Springer, Berlin, p. 245 (1969).
13. K. F. J. Heinrich, *Advances in X-Ray Analysis 3*, Plenum Press, New York, p. 370 (1960).
14. R. M. Dolby, in: *X-Ray Optics and X-Ray Microanalysis* (H. H. Pattee, V. E. Cosslett, and A. Engstöm, eds.) Academic Press, New York, p. 483 (1963).
15. R. Fitzgerald, K. Keil, and K. F. J. Heinrich, *Science* **159**, 528 (1968).
16. P. Duncumb, in: *X-Ray Microscopy and Microradiography* (V. E. Cosslett, A. Engström, and H. H. Pattee, eds.) Academic Press, New York, p. 617 (1957).

17. K. F. J. Heinrich, *Electron Beam X-Ray Microanalysis,* Van Nostrand Reinhold, New York, p. 193 (1981).
18. Standard Reference Data Base 36, available from the Office of Standard Reference Data, National Institute of Standards and Technology, Gaithersburg, MD 20899.
19. T. O. Ziebold and R. E. Ogilvie, *Anal. Chem.* **36**, 322 (1964).

# 2

# Problems and Trends in X-Ray Detector Design for Microanalysis

*B. G. Lowe*

## 2.1. INTRODUCTION

This paper presents some of the problems and trends associated with silicon and germanium x-ray detectors for microanalysis as seen from the manufacturer's point of view. The crystal and field-effect transistor FET comprise a transducer which is sensitive enough to detect charge of the order of 20–30 electron-hole pairs. At the same time, it must be fixed accurately in confined space in a hostile radiation environment and must be maintained at cryogenic temperatures without its magnetic, electrical, and mechanical properties in any way influencing the performance of the microscope to which it is interfaced. Such considerations impose severe constrictions on the design of these detectors, and many compromises must be made. For example, the demand for the low-energy efficiency afforded by a truly windowless (WL) detector is being relaxed more and more in favor of competing considerations such as protection from light, as in the ultra-thin window detector (UTW), and cross contamination and simplicity, as in the atmospheric pressure-supporting window detector (ATW).

## 2.2. GEOMETRICAL CONSIDERATIONS

Spectral artifacts have been reviewed by Newbury.[1] They can result from a number of sources, including the geometrical limitations resulting when an energy dispersive x-ray (EDX) detector is interfaced to the EM column. Figure 2.1 illustrates one of the most difficult problems in the transmission electron microscopy (TEM) environment as an example, namely space restriction. In many applications, the x-ray

B. G. Lowe • Oxford Instruments, Abingdon, OX14 1TX, United Kingdom

*X-Ray Spectrometry in Electron Beam Instruments*, edited by David Williams, Joseph Goldstein, and Dale Newbury. Plenum Press, New York, 1995.

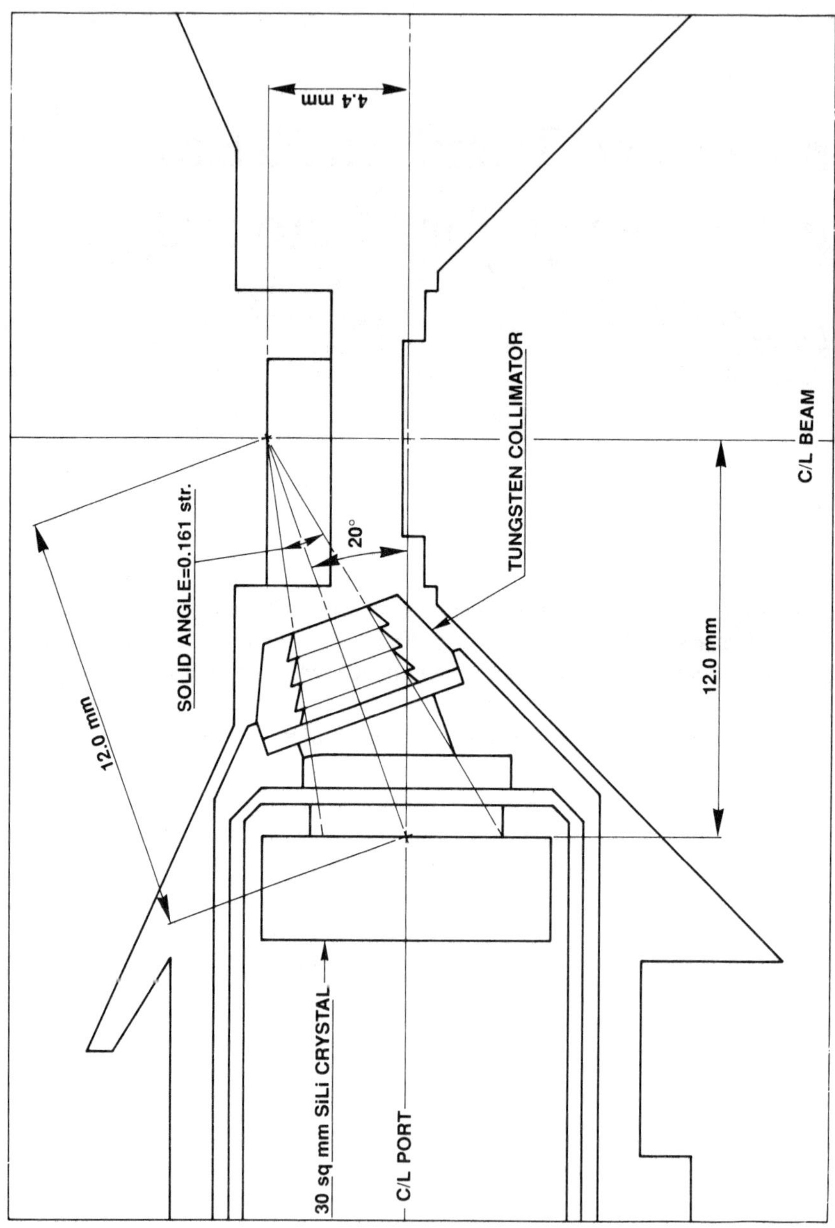


FIGURE 2.1. EDX close geometry and collimation in a typical TEM.

yields are low and the collection solid angle is of paramount importance. A crystal of 30 mm$^2$ active area and 9 mm diameter must be located as close to the specimen as possible, sometimes in a 12 mm diameter detector probe with clearances down to 0.1 mm from the pole pieces. One solution to the solid angle problem is to sum the spectra from two independent detectors in the same TEM.[2] This has the advantage of doubling the collection angle without compromising the energy resolution to any great extent. Any magnetic materials (for example, some tungsten and all nickel alloys) in such close proximity to the beam would cause image astigmatism and shift. It is possible for electromagnetic radiation from the scan coils to interfere with the detector unless adequate shielding is provided. Also, in this position and under certain microscope conditions, the crystal can be exposed to high fluxes of energetic electrons and x-rays. This is particularly true of detectors looking at high take-off angles in the HV TEM. The crystal must be collimated and protected from such stray radiation, imposing further geometrical constrictions. Figure 2.2 shows some of the precautions taken in a high take-off angle TEM detector. Not only will such irradiation impose abnormally high dead times due to the increase in the frequency of the FET charge restoration and pile-up rejection, but in severe cases can cause long-term effects, discussed in more detail below. A recent trend in TEM design has moved away from such high interface ports on the column and toward horizontal detectors with tilted crystals and lower take-off angles. Such a design is shown schematically in Figure 2.3. Space restrictions would make such a design very difficult to achieve for the system in Figure 1. In some cases such a restriction has led detector manufacturers to lap a flat on the crystal to give clearance. This is not a very popular solution with the crystal processors! There are other disadvantages of the tilted crystal from the manufacturers' point of view. The location of the probe in the column becomes very critical. Any inaccuracy in this setting will lead to the crystal seeing x rays from regions other than the beam spot on the sample and there is no line-of-sight adjustment for solid angle. Also, tilted crystals can be more difficult to hold securely against the FET gate-pin, and the increased stray capacitance presented by the crystal back-contact to the probe can lead to higher noise and greater susceptibility to microphony.

Every microscope design and every port on it demands a different geometry detector and interface. Consequently, it has been impossible for detector manufacturers to make models to stock. Most manufacturers have now attempted to rationalize the situation by looking for commonality of parts. For example, many now use a design which allows the probe angle to be adjusted on a finished detector.

Figure 2.4 shows a copper spectrum taken under conditions where the magnetic fields in the vicinity of the specimen were insufficient to take electrons away from the detector. Such high-energy electrons can even penetrate a beryllium window and give rise to the high-energy component of the background, as shown. The effect of interposing a thin aluminum foil in front of the detector is also shown. This is a good diagnostic since it shows the attenuation of the electrons and the bremsstrahlung generated in the foil. Electrons are, of course, a common problem with WL, UTW, and ATW detectors on scanning electron microscopes (SEMs) where the specimen is not in a magnetic field and an electron-trap (usually a pair of ceramic magnets fitted immediately in front of the detector) is necessary. Such a trap must have a special mild-steel yoke to prevent the field from extending to the beam region.

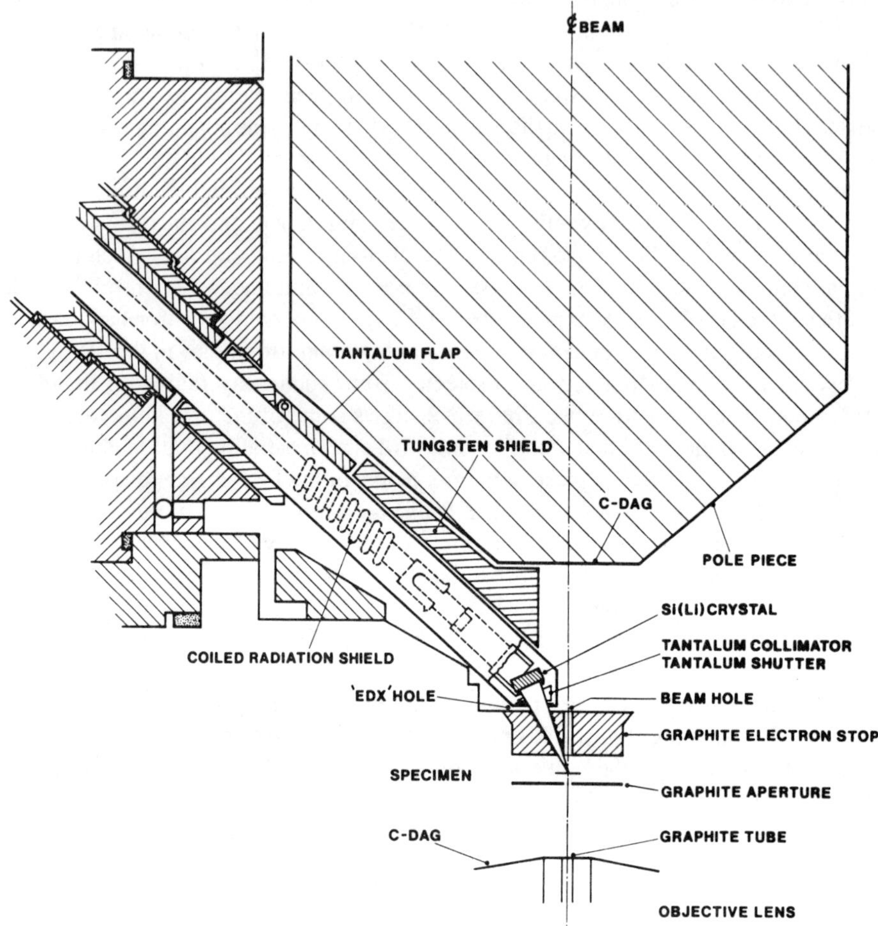



FIGURE 2.2.  Precautions against electrons and hard x rays in a typical high take-off angle TEM detector.

## 2.3. COLLIMATION AND PROTECTION

Compton scattering in the crystal itself becomes important at high energies, and this increases the spectral background at low energies. It is important to have good internal collimation of these high energy x rays to ensure that the first Compton scatter is away from the side walls of the crystal. There is some advantage in having the efficiency roll-off due to the silicon crystal thickness (usually 2–4 mm) and in using germanium where there is less likelihood of the scattered x ray being lost from the crystal. Another effect of this scattering is to put charge into the weak peripheral field regions of the crystal. This can cause charge trapping at the interfaces and consequent field distortion and high leakage current, at least temporarily. These traps can often be discharged by introducing light, removing the detector bias, or annealing to a higher

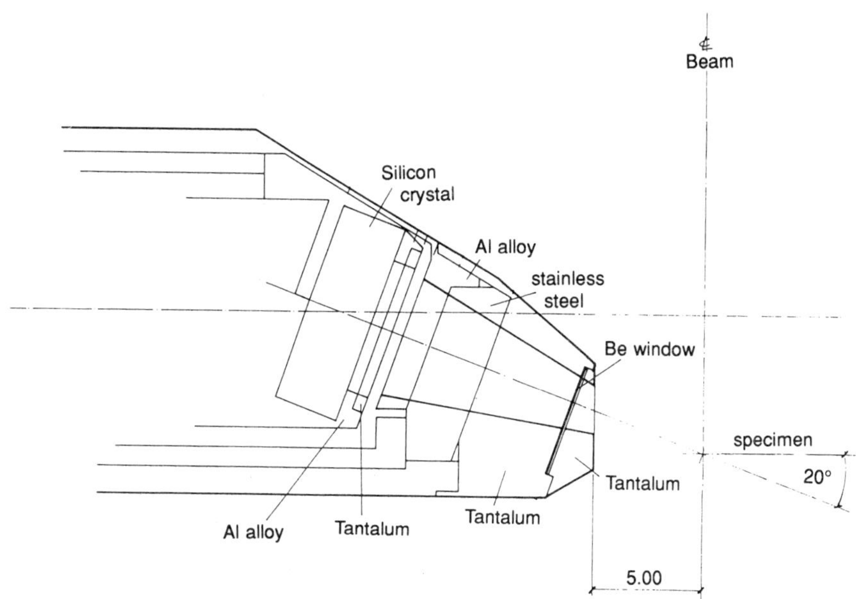



FIGURE 2.3. Horizontal tilted crystal detector.

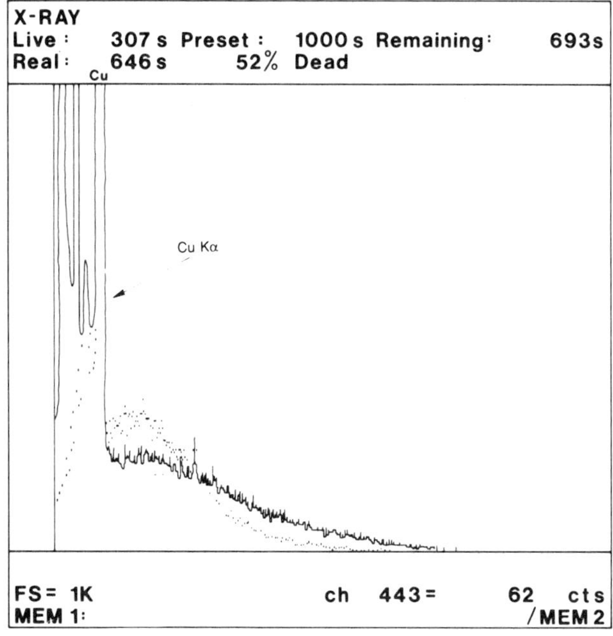



FIGURE 2.4. Spectra of copper grid bar at 100 kV showing the effect of electrons entering the detector. (Dots are with aluminum foil interposed.)

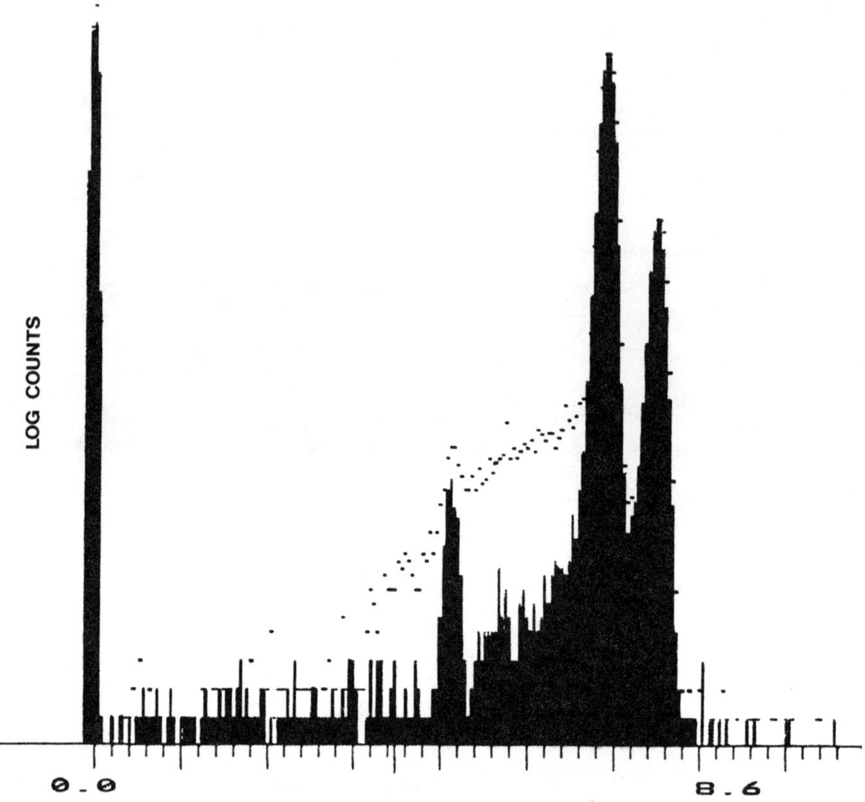



FIGURE 2.5.  Effect of heavy irradiation with processing of the Si(Li) crystal. Dots, standard process crystal; bars, radiation hardened crystal.

temperature. If the detector bias is removed during heavy irradiation, recombination of carriers takes place and charge trapping is reduced. Another partial solution to this problem is to avoid surface oxide by employing special processing techniques. Figure 2.5 shows the effect of irradiation on a poor crystal and on a radiation hardened crystal. Such distortions can be enhanced by x rays entering the crystal edges and by transverse magnetic fields deflecting carriers into the edges.[3] These problems indicate the importance of good collimation (typically 0.5 mm of Ta screened by 0.5 mm Al to stop the Ta $L$-x ray) and shielding (usually many millimeters of Ta). The accurate location of these with respect to the crystal active volume and the specimen is also important. Above 200 keV electron energy there is also the increased possibility of bulk damage to the semiconductor active volume by recoil and photoelectrons, with bulk trapping occurring as a consequence. Heavy irradiation with electrons can also cause charging of various insulated parts of the detector, both externally (such as the epoxy used to attach the beryllium window and the window itself, if it is floating) and internally (such as the boron nitride or Teflon insulators inside the cryostat). In extreme cases of irradiation, the internal dielectrics discharge in the cryostat vacuum causing "self-counting"[4] and occasionally fluorescence of the surrounding materials.

In view of the above mentioned problems, the most appropriate philosophy is to shield and collimate accurately and to physically shutter off the detector during overload conditions whether due to electrons or x rays. Such conditions can occur during low magnification, electron "showering" of the chamber, or when the beam strikes a grid bar; with the very high dead times attendant, no analysis is possible anyway. The shutters can be internal to the detector (see Figure 2.2), triggered automatically by the pulse processor in the overload condition, or can be triggered by the TEM, or both. These "autoshutters" are preferable to "autoretraction" of the detector, which can cause vibration of the column and liquid nitrogen during and after the motion of the detector. Some TEM manufacturers also provide for detector protection by incorporating beam-blanking under certain conditions.

Apart from the necessity of preventing heavy irradiation of the detector, the manufacturer is also responsible for preventing the irradiation of the user. Levels of radiation around the detector external to the column and at its interface are kept below 2 μSv/hr even under the most adverse conditions (such as the full energy beam hitting a heavy metal aperture). Some of the techniques used to ensure protection are shown schematically in Figure 2.6.

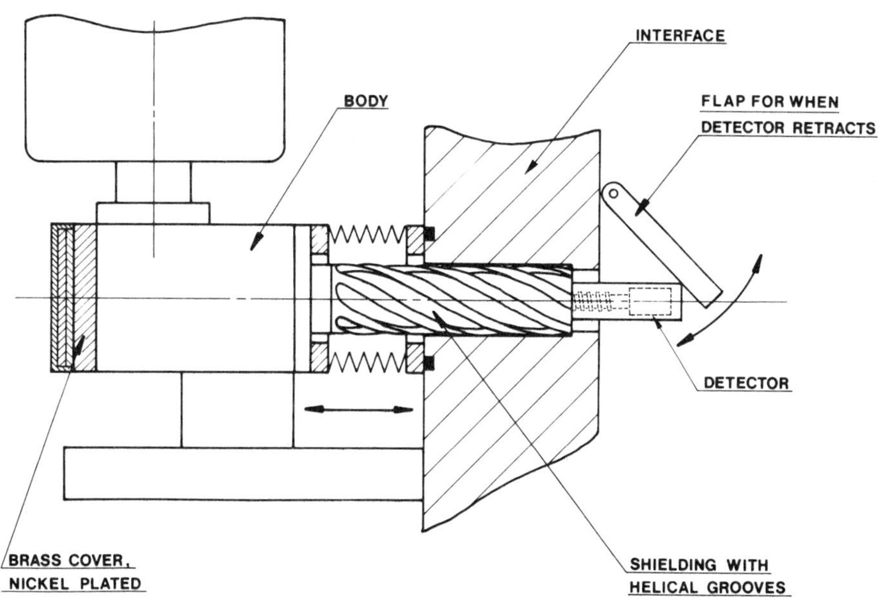


FIGURE 2.6. Schematic diagram illustrating the typical precautions to minimize external radiation in a TEM (detector shown partially retracted)..

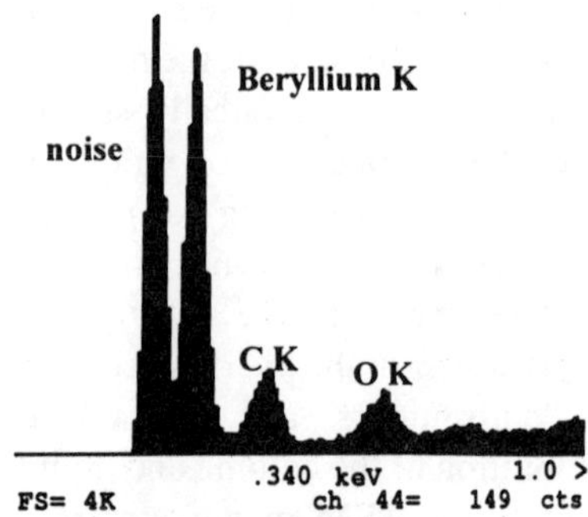


FIGURE 2.7.  X-ray spectrum obtained with a Si(Li) detector from beryllium foil in a SEM at 10 kV.


## 2.4. PROBLEMS OF LOW ENERGY ANALYSIS

In order to obtain state-of-the-art sensitivity to ultra-soft x rays, detectors must be used in the WL or UTW mode. Figure 2.7 illustrates the best low energy sensitivity achieved to date. Such detectors are still in demand on some specialized UHV TEMs, and it is essential that they can be removed from the column and that collimators and UTWs can be inspected or exchanged without breaking the vacuum of either the detector or the TEM chamber. This is achieved by retraction behind in-line gate valves. Such detectors also must be retracted behind a gate valve whenever the specimen is changed and in many cases this is triggered automatically. This precaution is necessary to prevent ice formation on the crystal entrance window,[5,6] which can seriously affect the low energy efficiency due to the presence of the oxygen $K$ absorption edge even to relatively high energies.[7] Ice on the crystal sides may also give rise to spectral artifacts. In any WL or UTW detector the rate of icing depends on the quality of the chamber vacuum, degree of cold-trapping in the chamber, and the care of use. Furthermore, some specimens may also emit water vapor. The effect of icing on the Ni $L/K$ ratio is shown in Figure 2.8. The saturation ratio represents an ice layer of about 0.7 μm. Icing has also been shown to affect Be window and some ATW detectors where small porosities with respect to water vapor become apparent despite the vacuum apparently not being degraded. However, improved pumping techniques have reduced effects due to the outgassing of vacuum dewars, and most manufacturers now offer thermally cyclable detectors and small "cool-on-demand" dewars. However, if ice is to be removed quickly and efficiently without requiring removal of the detector, the liquid nitrogen, or the power, a "conditioner" can prove invaluable. This is an internal heater fitted to some detectors that removes all ice from the crystal and FET package by a process of sublimation. The conditioner is also found occasionally to reduce dielectric noise where this is due to ice. It has also been found recently to be effective in removing carbon (possibly as $CO_2$) from the crystal, but it would not be expected to be very effective against vacuum pump oil contamination. The same authors[8] have also shown that ice and carbon can have effects in the high energy region of spectra,

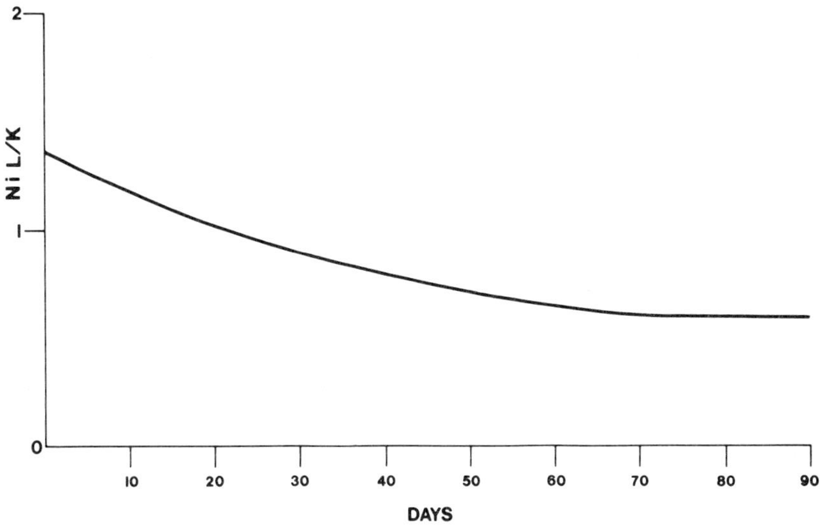


FIGURE 2.8. Effects of ice formation on the Si(Li) crystal on the Ni *L/K* peak ratio as a function of time in a moderate vacuum SEM.

where the layers can be effective in reducing back-scattered electrons from entering the crystal, thus reducing background and dead time.

Apart from potential contamination of a WL or UTW detector, the converse is also possible, i.e., the potential contamination of the microscope by the detector. Despite extreme precautions in the cleaning of internal parts, a buildup of carbon on a specimen by the beam may indicate such contamination. One method of addressing this problem has been to isolate the dewar vacuum from that of the microscope. On UHV instruments the detector is retracted behind a gate valve when the chamber is baked. In cases where extremely clean UHV is required (such as in Auger spectrometers at $10^{-10}$ mb pressure), a truly windowless detector is impractical and a sealed detector which can have its external surfaces baked *in situ* in the chamber is necessary. Such a detector, bakeable to 200°C without removing the liquid nitrogen, is possible with a metal sealed window.[9,10] Similar techniques extending to water cooling of the collimator can be used for applications involving a hot stage.

There has been a trend in recent years away from WL and UTW detectors for low energy analysis on TEMs and UHV microscopes (except for some specialized examples) and toward ATW detectors. This is because of the problems encountered with shared vacuum, as explained above, and also because of the desirability of simpler, more reliable detectors. This has further been encouraged by the improved transmission and reliability of these windows. The ATWs consist of submicron thick amorphous materials (such as boron nitride, silicon nitride, or diamond) or of polymer films. All of these are usually supported by some sort of grid or rib structure. The latter have superior transmission for C-$K\alpha$ energy and below (being largely carbon based) and Be x rays have been detected through them.

## 2.5. HIGH-PURITY GERMANIUM DETECTORS

The properties of high-purity germanium (HPGe) detectors have been reviewed elsewhere.[11] Their introduction for x-ray microanalysis[12] was initially seen as a means of extending the useful range of microanalysis up to 100 keV on HV TEMs,[13] but this trend has not occurred for several reasons. Firstly, the excitation efficiency for the $K$-shell of high atomic number elements by electrons is disappointingly low,[14] so that the greatly improved efficiency for detection at high energies is partially negated. Secondly, there has not been a great requirement for analysis based on $K$-lines in the range 20–100 keV due to the improved resolution of the $L$-lines. Where HPGe has shown itself to be powerful, however, is in its greatly reduced noise and improved resolution at both short and long processing times due to the higher intrinsic gain and the reduced statistical broadening. These benefits greatly outweigh the noise penalty caused by the increased crystal capacitance due to the higher dielectric constant of germanium. The shorter processing times allow much faster analysis than can be achieved by a Si(Li) detector *for the same resolution*. This advantage is seen quite dramatically in the case of elemental mapping where a tenfold improvement in rate can be realized.

Figure 2.9 shows a comparison of jadeite spectra obtained with a Si(Li) detector and a HPGe detector under identical conditions. At low energies (even

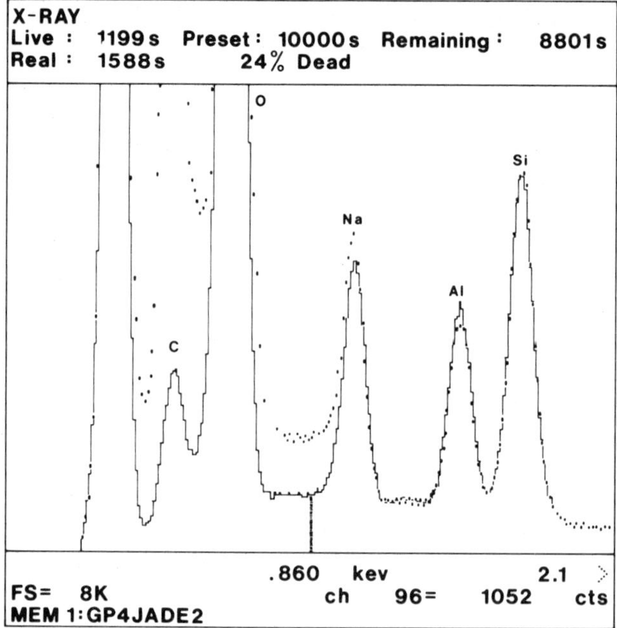


FIGURE 2.9. Jadeite spectra obtained with a HPGe detector (line) and a Si(Li) detector (dots) under identical conditions on a SEM. The carbon peak is due to specimen coating and was variable.

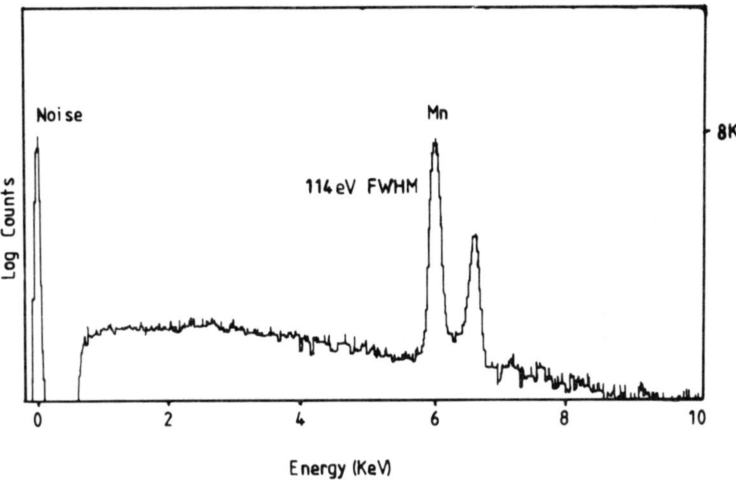


FIGURE 2.10. Manganese spectrum obtained with a HPGe detector on a SEM at 10 kV.

across the Ge *L*-absorption edge just below the Al peak) the efficiencies are very similar, and as the HPGe resolution is superior, the limits of detection are lower. The longer processing times allow much improved resolution (Figure 2.10) even to low energies (Figure 2.11). Furthermore, Figure 2.10 shows the complete absence of any escape peaks (the probability of a Ge *L*-escape peak is very low). Indeed, escape peaks are rarely significant up to beam energies of about 20 keV, and above this they can be effectively subtracted by software. To summarize, the major demand at present is for HPGe detectors on SEMs and, if low energy efficiency is important, for the ATW versions of these detectors.

## 2.6. ELECTRONIC ARTIFACTS AND NOISE

Artifacts can result from inadequate or incorrectly adjusted electronics. For example, noise discriminators must be correctly set to optimize the low energy efficiency and no excess high frequency noise from external sources must be allowed in the pile-up rejection channels. Multiple triggering of pile-up rejectors, for whatever reasons, can also give rise to anomalous loss of efficiency at higher energies, such as reported by Newbury.[1] To overcome such difficulties, manufacturers are now moving toward automated system optimization for the particular detector and its operating environment.

In a well designed, assembled, and installed microanalysis system, the major noise components are due to the detector semiconductor crystal and the first stage of the amplifier, namely the JFET. The former is addressed by using low electrical capacitance geometries to reduce the short processing-time noise and

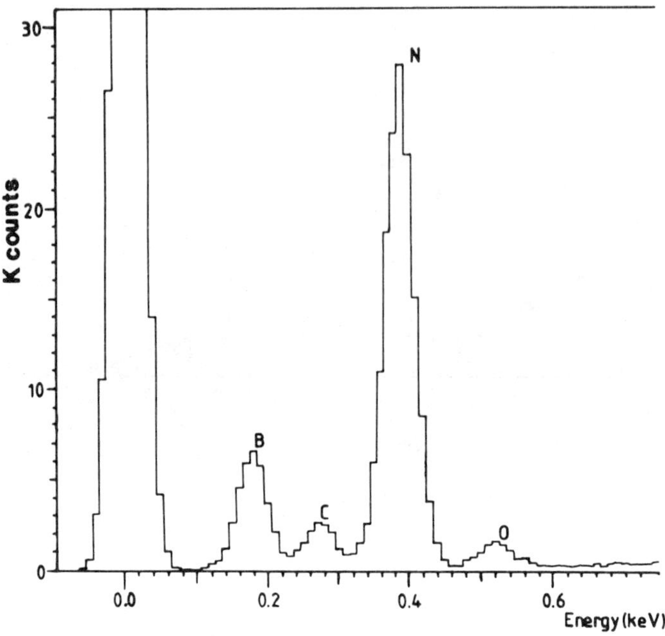


FIGURE 2.11. Boron nitride spectrum obtained with a HPGe detector on a SEM at 10 kV.

refined processing and passivation techniques to reduce the surface leakage currents and so-called residual noise, both of which add to the long processing-time noise. Great advancements have been made in FET designs since it was realized that commercially available FETs (such as the 2N4416) could be greatly improved upon for this application, particularly at high rates.[15] The gate capacitance has also been reduced to about 1 pF, making a much better match to the crystal capacitances. These improvements in design have gone hand in hand with better techniques for charge restoration to avoid saturation at the amplifier. Pulsed optical restoration[16] has been followed by charge injection restoration.[17] The latter devices give restore times less than 1 µs while maintaining noise in the 40 eV full-width-at-half-maximum (FWHM) region. The very efficient restore mechanism has relevance not only to the TEM environment, where very high-energy x rays and back-scattered electrons can lock up a detector for a considerable time, but in any application where very high rates can be used (SEM, XRF, PIXE, etc.). A further advantage of the nonpulsed-optical restore mechanism is that it is no longer a necessity for the detector designer to employ light-tight packages for the FET (which is very close to the detector crystal to reduce stray capacitance). For example, different or thinner low-loss dielectric materials can be used or even eliminated altogether to reduce noise.

## 2.7. SUMMARY

These improvements in detector design have come about by solving the problems presented by the needs of the microscope manufacturers and microanalysis community. The problems have arisen with more widespread uses of WL and ATW detectors, HV TEMs, UHV chambers, hot-stages, etc., and no doubt the list will continue. There have been radical improvements in FET technology, and a choice of detecting semi-conducting material has evolved. Not all of the problems have been solved, but with the continued co-operation between users and EM manufacturers, all should benefit.

## REFERENCES


 1. D. E. Newbury, *Microbeam Analysis* **2**, S180 (1993).
 2. D. B. Williams, J. I. Goldstein, C. E. Lyman, D. W. Ackland, S. von Harrach, and P. J. Statham, *Microbeam Analysis* **2**, S236 (1993).
 3. A. J. Craven, C. P. M. McHardy, and K. A. Pears, *Ultramicroscopy* **28**, 157 (1989).
 4. J.-P. Chavalier, Proc. NSF/CNRS Workshop on Electron Beam Induced Spectroscopies with High Spatial Resolution. Aussois, France, p. 231 (1988).
 5. R. G. Musket, *Nucl. Instrum. Methods* **B15**, 735 (1986).
 6. P. Muller, F. Riehle, E. Tegeler, and B. Wende, *Nucl. Instr. Methods* **A247**, 569 (1986).
 7. D. D. Cohen, *X-Ray Spectrom.* **16**, 237 (1987).
 8. P. Hoverington, G. L'Esperance, E. Baril, and M. Rigaud, *Microbeam Analysis* **2**, 277 (1993).
 9. M. M. El Gomati, C. G. H. Walker, B. G. Lowe, and M. Prutton, *Inst. Phys. Conf. Ser.* **98**, 551 (1990).
10. M. Prutton, C. G. H. Walker, J. C. Greenwood, P. G. Kenny, J. C. Dee, I. R. Barkshire, R. H. Roberts, and M. M. El Gomati, *Surf. and Interface Anal.* **17**, 71 (1991).
11. R. A. Sareen, *Microbeam Analysis* **2**, S170 (1993).
12. C. E. Cox, B. G. Lowe, and R. A. Sareen, *IEEE Trans. Nucl. Sci.* **35**, 28 (1988).
13. B. G. Lowe, *Ultramicroscopy* **28**, 150 (1989).
14. E. B. Steel, *Microbeam Analysis* (A.D. Romig, Jr. and W. F. Chambers, eds.) San Francisco, 439 (1986).
15. K. Kandiah and G. White, *IEEE Trans. Nucl. Sci.* **NS-28**, 613 (1981).
16. F. S. Goulding, J. T. Walton, and D. F. Malone, *Nucl. Inst. Methods* **71**, 273 (1969).
17. T. Nashashibi and G. White, *IEEE Trans. Nucl. Sci.* **37**, 452 (1990).

# 3

# Current Trends in Si(Li) Detector Windows for Light Element Analysis

*M. W. Lund*

## 3.1. INTRODUCTION

Over the last 10 years much of the progress in x-ray spectroscopy has been in the area of light element analysis. In wavelength dispersive spectroscopy, this has resulted from the development of multilayer synthetic crystals. In energy dispersive spectroscopy, this has resulted from the development of new window technologies. Energy dispersive Si(Li) detectors have become more sensitive over the years, until the window can be the limiting factor in light element x-ray analysis. The window allows x-rays to pass and protects the detector from light and gases. It must withstand atmospheric pressure and repeated pressure cycling. Several technologies have been developed for this purpose.[1,2]

An EDX system for light element analysis requires more than just a thin window. The detector, preamp, electronics, and software all contribute to light element performance. I do not intend to discuss these issues in this paper, but it is important to remember that a system for light element analysis should be purchased on its performance as a system, and not just on the basis of the window.

There are basically six different window technologies used for light element analysis. Of these, most are proprietary, meaning that they are available on only one EDX manufacturer's instruments. The two windows that are not proprietary are the MOXTEK polymer window and the diamond window. All manufacturers of Si(Li) detectors have qualified the MOXTEK window, and the same is true of the diamond window. Therefore, these windows can be requested of any EDX manufacturer.

M. W. LUND • MOXTEK, Inc., Orem, Utah 84057

*X-Ray Spectrometry in Electron Beam Instruments*, edited by David Williams, Joseph Goldstein, and Dale Newbury. Plenum Press, New York, 1995.

## 3.2. BERYLLIUM WINDOWS

Beryllium windows are useful down to sodium $K\alpha$. Recently, 5-$\mu$m Be windows have been introduced by at least one vendor, which extends this range down to fluorine, but this material can be used only for the smallest area windows. Table 3.1 gives acceptable clear apertures for 1-atm pressure differential for the standard beryllium foil thicknesses as well as nominal transmission for fluorine, sodium, magnesium, and aluminum. Transmissions have been calculated with the latest version of Optical Constants Grapher,[3] using the 1993 Henke Tables.[4]

Production of thin beryllium foil is difficult because Be is not strictly ductile. The primary mechanism of deformation is basal plane slippage. This causes the basal crystal planes to become oriented in the plane of the sheet while rolling. This alignment produces high strength and ductility in the plane of the foil, but brittleness and low strength perpendicular to the foil surface.[5] For this reason, extreme care must be taken in cleaning Be windows, particularly while the window is supporting an atmosphere of pressure differential. The in-plane strength of cross-rolled Be varies from 380 to 685 N/mm$^2$ (55,000 to 100,000 psi).

Thin foils consist of large crystal grains, with the foil only one to three grains thick, each grain surrounded by beryllium oxide. To form a pinhole-free thin foil, the Be is cross rolled between stainless steel plates at 700–900°C.[6-9] This procedure may introduce impurities into the foil, which rarely has less than 300 ppm of iron regardless of source.

Problems persist in getting leak-tight beryllium in the 5–12-$\mu$m range. Even light-tight 8-$\mu$m window foils have measurable leaks at least half the time. These leaks are dealt with in various ways. Foils with unacceptably high leak rates are discarded. Smaller leaks are sealed with coatings. Typical coating materials in use are vacuum grease, varnish, and parylene. All organic coatings suffer from water vapor and helium diffusion. Detectors using these coatings will gradually degrade due to icing,[10] although on many detectors this may not be evidenced for years. Those detectors used in He environments will degrade due to thermal transfer problems unless there is a He pumping mechanism. It sometimes surprises researchers to find that cleaning contamination off a beryllium window results in higher transmission than the system had when it was new. This increase in transmission is caused by removal of the organic coating used to seal the window and may result in increased air and water leakage.

TABLE 3.1. Thickness, Tolerance, Maximum Clear Aperture, and Transmissions at F, Na, Mg, and Al $K\alpha$'s for Beryllium Foil

| Thickness | Tol | $\phi_{max}$ | $T^F$ | $T^{Na}$ | $T^{Mg}$ | $T^{Al}$ |
|---|---|---|---|---|---|---|
| 5 $\mu$m | 0<br>+3 | 4 mm | 11% | 56% | 73% | 83% |
| 8 $\mu$m | 0<br>+4 | 6 mm | 3% | 40% | 60% | 74% |
| 12 $\mu$m | 0<br>+5 | 8 mm | — | 25% | 46% | 64% |
| 25 $\mu$m | ±3 | 14 mm | — | 6% | 20% | 39% |

There is a more subtle problem of microleaks in the Be foil that can leak water vapor but are not picked up by a He leak detector. The resulting degradation due to icing is most noticeable on the highest quality detectors. A boron coating for Be has been developed to solve this problem.[11] This is a refractory material applied with chemical vapor deposition (CVD) that allows no water vapor diffusion and much reduced helium diffusion. The coating itself is usually 0.5 μm thick on two sides of the foil and has been applied to foils of 5 μm thickness and greater. The coating protects the window from corrosion and allows the window to be cleaned without damaging it or creating toxic waste. Diamond coating of Be has also been attempted to seal Be foil, but no product has been announced yet.

Corrosion of Be is enhanced by several mechanisms. Beryllium may, to some extent, be considered self-protective against atmospheric oxidation. Like aluminum, it forms a passivating oxide that protects it.[12] This protection can be disrupted by several mechanisms. The first involves corrosion by atmospheric moisture in the presence of beryllium carbide.

$$Be_2C + 4H_2O = 2Be\,(OH)_2 + CH_4$$

This usually requires beryllium carbide to be a component of the foil surface. Modern Be foil does not appear to have carbide inclusions, but sputtering of hot carbon onto the window in an electron microscope may cause them. The beryllium hydroxide is the source of the gelatinous material that is sometimes found on Be windows that have been exposed to water. White powder corrosion products are mostly beryllium hydroxide or beryllium oxide.[13]

There are several ions, including chloride, sulfate, and nitrate ions, that will also cause the passive oxide to be disrupted. These ions can come from salt spray, fingerprints, or washing with tap water. Chloride ions are particularly corrosive, especially in hot water solutions. Atmospheric water can slowly hydrolyze the beryllium oxide at the edge of the foil grain boundaries. This allows a chemical path for water vapor to diffuse through an otherwise airtight foil, and is one of the causes of slow icing of detectors.

Helium diffusion is important in x-ray fluorescence analysis in which helium is used to prevent liquids from boiling. We measured helium diffusion on 8-μm beryllium foils and on CVD boron-coated foils. The results are given in Table 3.2. Helium diffuses very fast through polymer films. Rimbert et al.[14] point out that the distance between polyimide chains is larger than the helium atomic diameter (2.7 Å). They measured the helium diffusion rate of 100 μm thick polyimide as $3.5 \times 10^{-8}$ cm$^3$ mm$^{-2}$ sec$^{-1}$ atm$^{-1}$, over 3500 times faster than uncoated Be foil one-tenth as thick. Polymer coated Be is thus expected to have diffusion similar to uncoated Be.


## 3.3. ULTRA-THIN WINDOWS

EDX light element analysis became practical in about 1987 when Kevex, Inc., introduced their Quantum window. Since then the forces of competition have accelerated development in this area, much to the delight of many electron microscopists. There are basically five commercially available window technologies used for light element analysis: windowless, boron nitride, polymer, diamond, and silicon nitride.

Windowless detectors are the ultimate in sensitivity until they are contaminated. When ice and other contaminants form on the detector, the sensitivity drops.[15] A film of 0.5 μm of ice is equivalent in transmission to a typical ultra-thin atmospheric window. The manufacturers of windowless detectors understand these problems very well and have developed conditioning cycles to decontaminate the detector *in situ*. That these procedures work well has been shown by Hovington *et al.*[15] Windowless detectors will always present more difficulties to the end user. Because windowless detectors are sensitive to light, they may also require an aluminum filter to block light that is generated at the sample or leaks into the system from other sources. As the thin window technology has proven itself, windowless detectors have declined in importance.

A schematic diagram of an ultra-thin window is shown in Figure 3.1. It consists of two 150-nm-thick layers of polymer, each with a 20-nm layer of aluminum. The composite film is supported by a grid structure that gives the window strength.

X-ray transmission of the window has three components: transmission of the window material, transmission of the aluminum light-blocking layer, and transmission of the supporting grid. The first two are a function of x ray energy. The last is usually independent of the x ray energy (but see below). These three transmissions are usually multiplied together to give a total effective transmission. Alternately, the supporting grid, because it is wavelength independent at low to medium energies, can be thought of as reducing the geometrical efficiency of the detector. The advantage of this is that variations in film and grid can be measured separately and entered into software separately.

At higher energies the grid material can become transmissive. The silicon grid of the MOXTEK AP1 polymer windows and the diamond windows is nominally 350 μm thick and will transmit 7% at 10 keV and 46% at 15 keV. Thus, the nominally 80% grid increases its effective transmission to 81% at 10 keV and to 88% at 15 keV. This should be taken into account when explaining *K*-to-*L* ratios for heavier elements. The tungsten foil used for polyimide window grids is nominally 25 μm thick and transmits less than 2% at 15 keV.

Figure 3.2 shows the transmission spectra of the various thin window materials, showing the absorption edges. For illustration, all the thicknesses are the same: 3000

TABLE 3.2. Helium Diffusion Rates through 8 μm Beryllium
Foil: Uncoated and Coated with 0.5 μm CVD Boron on Both
Sides

| Sample | Uncoated He leak rate ($cm^{-2}$ $mm^{-1}$ $atm^{-1}$) | Coated He leak rate ($cm^3$ $sec^{-1}$ $atm^{-1}$) |
|---|---|---|
| 1 | $9.7 \cdot 10^{-12}$ | $0.6 \cdot 10^{-12}$ |
| 2 | $9.8 \cdot 10^{-12}$ | $3.0 \cdot 10^{-12}$ |
| 3 | $10.1 \cdot 10^{-12}$ | $1.8 \cdot 10^{-12}$ |
| Average | $9.8 \cdot 10^{-12}$ | $1.8 \cdot 10^{-12}$ |

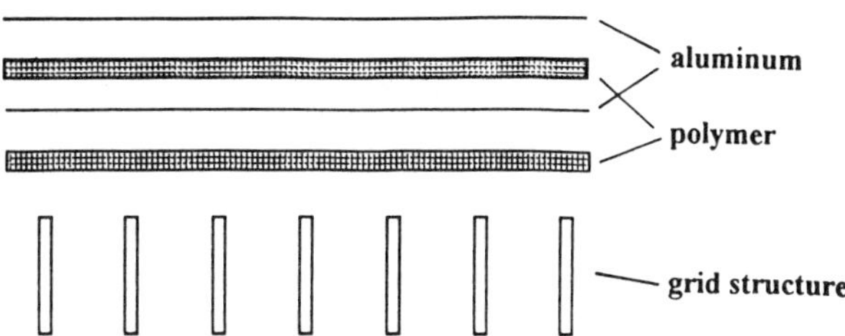


FIGURE 3.1. Schematic diagram of an ultra-thin window structure.

Å. For chemical analysis the structure of this graph is not as important as the value of the transmission at the discrete $K\alpha$ positions. The transmission curves are discontinuous at the absorption edges of each element in the window. If you were researching bremsstrahlung you might want a smooth transmission, which 3000 Å of Be would give you. If you were doing elemental analysis you might want the steps, because the $K\alpha$ emissions are on the high transmission side of each edge.

An interesting effect can be seen by comparing the diamond and the polymer curves. Diamond, consisting of only one element, has higher transmission at the carbon $K$-edge, but on the high energy side of the edge it absorbs very strongly. This edge is still strongly absorbing at the nitrogen line. Beryllium also has this problem and is highly absorbing at the boron and carbon lines. A multielement film, such as boron nitride or a hydrocarbon polymer, does not have the same high peaks, but neither does it have the pronounced valleys, giving a more balanced spectrum.

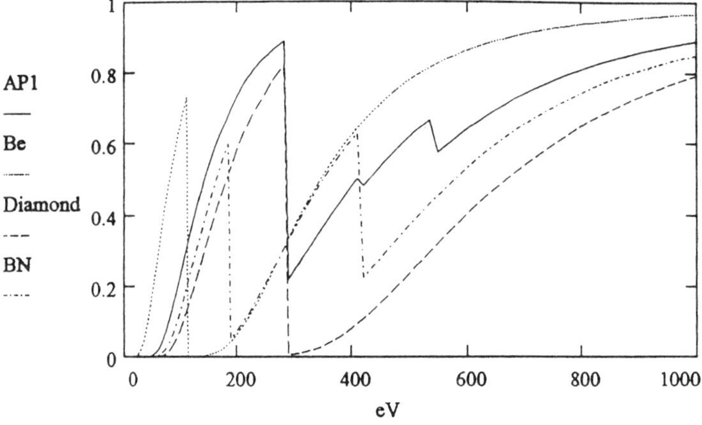


FIGURE 3.2. Spectral transmission of window materials in the soft x ray region. All films are 3000 Å.

Under conditions of high bremsstrahlung background, the window will filter the continuous radiation. A multielement window may form structure in the spectrum that might be interpreted as elemental lines. This phenomenon must be taken into account when measuring low levels of light elements under these conditions.

Table 3.3 lists the currently available window materials with, in most cases, the manufacturer's data on total effective window transmission. The numbers in roman type were supplied by the manufacturer. The numbers in boldface were measured by independent researchers. The numbers in italics were calculated using Optical Constants Grapher. As you can see, from a user's point of view these windows perform similarly. For this table, all the windows were coated with 400 Å of aluminum. In some instances boron nitride is used without an aluminum coating, which increases transmission for boron and carbon $K\alpha$, but leaves the detector sensitive to light.

A simplified model for the minimum detection limit is

$$MDL \propto \frac{\sqrt{I_B}}{I_P\sqrt{t}}$$

where $I_B$ is the background count rate, $I_P$ is the peak count rate, and $t$ is the acquisition time. If we further assume that most of the background is signal induced, due to bremsstrahlung and incomplete charge collection,

$$I_B \approx I_{B0}T, \ I_P = I_{P0}T$$

where $T$ is the window transmission, and $I_{P0}$ and $I_{B0}$ are the peak and background, respectively, without the window. This means the detection limit

$$MDL \propto \frac{\sqrt{B_0}}{P_0\sqrt{Tt}}$$

is proportional to $T^{-1/2}$, and losses in transmission can be made up by a linear increase in acquisition time. Instrumental effects will limit the use of increased acquisition time to compensate for transmission loss, but this still leaves considerable latitude in window transmission. Thus, the difference between 57% and 41% transmission is not significant compared to other factors such as the electron beam current, sample-detector distance, collection time, detector dead layer, detector resolution, charge collection, and system noise. These factors are going to be far more significant than a

TABLE 3.3. Transmission of Available Thin Windows for Si(Li) Detectors

| Window | Area | Be $K\alpha$ | B $K\alpha$ | C $K\alpha$ | N $K\alpha$ | O $K\alpha$ | F $K\alpha$ | Na $K\alpha$ |
|---|---|---|---|---|---|---|---|---|
| AP1 | 100 mm$^2$ | 7% | 24% | 58% | 39% | 52% | 61% | 71% |
| Polyimide | 100 mm$^2$ | 10% | 27% | 61% | 38% | 57% | 75% | 78% |
| BN | 30 mm$^2$ | 9% | 26% | 20% | **36%** | **42%** | 58% | **74%** |
| Si$_3$N$_4$ | 30 mm$^2$ | — | 1% | 15% | 43% | 41% | 61% | 90% |
| Be (5 µm) | 28 mm$^2$ | — | — | — | — | 3% | 18% | 61% |
| Diamond | 28 mm$^2$ | *4%* | *14%* | *36%* | *3%* | *13%* | *29%* | *51%* |

20% difference in window transmission. It would not be wise to buy a system for light element analysis based on window transmission efficiency alone.

Icing has been a problem with ultra-thin windows in the past,[16] but these problems have been solved. Modern ultra-thin windows can be processed in a manner similar to Be windows. A bake-out/pump-down operation hermetically seals the detector. Polymer windows, however, will leak both water vapor and helium if they are not properly coated. To prevent this diffusion, polymer windows are coated with either pure Al or an Al/AlN sandwich.[1,2]

Aluminum is also necessary to block light transmission. It is the most opaque material to visible light, requiring a thickness of only 400 Å for 0.02% transmission. Other light element metals are much more transparent to light, for example, Be has half the attenuation coefficient of Al at $\lambda = 0.5$ µm. Scaling metal film thickness by the absorption of light, an equivalent Be film is less transmissive to light element x rays than an aluminum film.

It is tempting for a manufacturer to reduce the aluminum film thickness in order to show increased transmission for beryllium, boron, and carbon $K\alpha$ x rays. Alternately, since many customers measure samples with cathodoluminescence, or have optical alignment aids in the microscope, more aluminum may be appropriate. Figure 3.3 shows the x-ray transmission for the light elements versus the visible light transmission (at $\lambda = 0.5$ µm), with aluminum thickness as a parametric variable. Typically the aluminum thickness is 400 Å, which allows visible light transmission of about 0.02%. A thickness of 800 Å would give a visible light transmission of $5 \times 10^{-6}$ but eliminate most of the Be and B $K$ x rays.

Ultra-thin windows must be supported by a grid to withstand atmospheric pressure. A schematic window structure is shown in Figure 3.4. It consists of a barrier film and supporting grid. The barrier film can be part of the grid, or it can be attached to the grid. Boron nitride windows are monolithic, with a grid of thicker material built in. Diamond windows

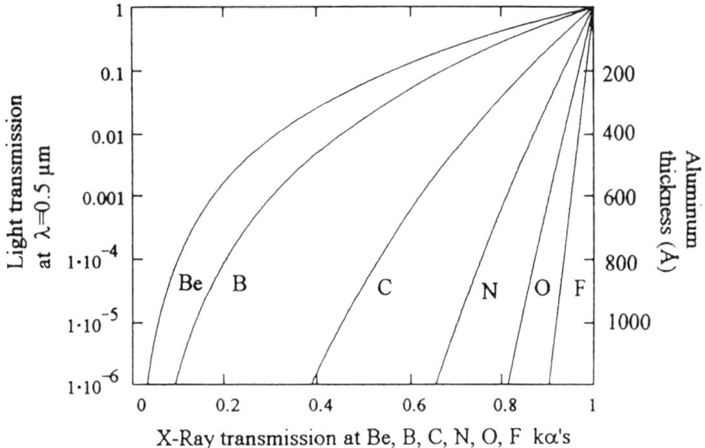


FIGURE 3.3. The transmission of light and x rays through aluminum film.

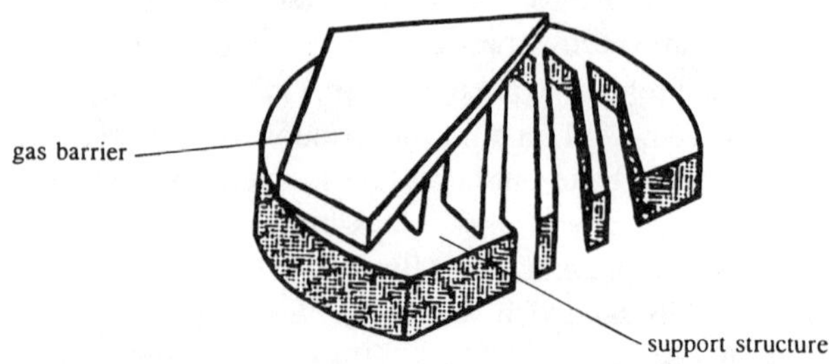


FIGURE 3.4. A schematic of a typical ultra-thin window structure including the gas barrier film and its supporting structure.

have the diamond film deposited onto the silicon before etching the grid, giving a strong chemical bond between the film and the grid. AP1 polymer windows are attached to a micromachined silicon grid[1] after the grid is made. They do not have a strong bond to the grid and therefore cannot withstand any back pressure which would force the film off the grid. Polyimide windows are supported on etched tungsten grids.[2] A typical micromachined silicon support grid has about 78% transmission at all wavelengths.

Silicon grids are made by anisotropic etching of single crystal silicon.[17] Selective etchants will etch silicon (100) planes up to 60 times faster than (111) planes. This allows structures with high aspect ratios to be formed using conventional lithography techniques. An oriented silicon wafer of the desired final thickness is patterned on one side with a grid structure using photoresist. The other side of the wafer is protected with unpatterned photoresist. After developing the photoresist, the structure is placed in the etchant and etched from one side to form ribs 25 $\mu$m wide and 350 $\mu$m deep, separated by 250 $\mu$m. The ribs extend over an area from 10 to 100 mm$^2$. The structure thus formed will fail a burst test at about 6 atm pressure.

Tungsten grids are made by reactive plasma etching.[2] The 25-$\mu$m-thick tungsten foil is coated with an aluminum film. This film is then patterned using standard photolithography and a chemical etch. The tungsten is etched using the aluminum as a mask in a $CH_4/O_2/CHF_3$ plasma. This plasma etches anisotropically due to plasma polymerization of the methyl fluoride, which forms a protective polymer coating on the side walls. The resulting mesh has 50-$\mu$m-wide ribs on a square grid with 0.5-mm period. The tungsten mesh is not strong enough to withstand atmospheric pressure over large areas, so it is supported on a laser machined tungsten grid (called a "strongback") with 2.5-mm spacing.

## 3.4. SPECIAL PURPOSE WINDOWS

### 3.4.1. Thicker Windows

Most manufacturers offer a thicker version of their ultra-thin windows in order to give their customers a choice for oxygen, fluorine, and sodium analysis at a lower price than the standard light element window. There has also been a feeling that the thicker windows are more reliable. This is true of polymer windows under severe use conditions, such as those found in environmental microscopes.

### 3.4.2. Ultra-pure Windows

The sheath-rolling of beryllium between hot stainless steel plates results in a moderate level of contamination of iron, vanadium, and nickel in the 300 ppm range, as well as a detectable level of uranium and thorium. These contaminants are bothersome to the lower limit of detection of these elements. Most of the technologies developed for thin windows use very pure materials, making these windows useful as ultra-pure windows. This makes ultra-thin windows attractive in applications that do not require light element sensitivity.

### 3.4.3. Germanium Detector Windows

High purity germanium detectors with better resolution and stopping power than Si(Li) detectors have been introduced. Because germanium has a smaller band gap than silicon, it is more sensitive to infrared radiation coming from the warm window. For this reason, it is useful to use a very thin, cold window of aluminum between the outer window and the detector.[18] This window does not have to be robust, since it does not support air pressure, but it adds some absorption for beryllium, boron, and carbon x rays.

## 3.5. RELIABILITY

Window failure can be fast or slow. A catastrophic failure is not impossible, even for 12-μm beryllium windows. More subtle failures are light leaks (which will raise the noise floor), very slow gas leaks (which could result in detector icing[19]), and window charging noise. For these reasons, it is important to test your system regularly to identify system degradation.[20]

Ultra-thin windows can be quite reliable. Reported mean time between failure for one manufacturer is over seven years. In certain applications, however, reliability can be poor. The biggest problem is particle impact on the window during microscope venting. This causes "bullet holes" in ultra-thin windows and depends on the gas dynamics inside the microscope. Many electron microscopes do not show this problem, but some models have a high propensity for window damage.

Ultra-thin means ultra-fragile. You cannot touch an ultra-thin window with your finger, bump it with a stage, or clean it with a cotton swab. However, ultra-thin windows can be quite reliable. Reliability data is proprietary, so I can only report on MOXTEK results. One EDX manufacturer who uses MOXTEK windows has well over 1000 ultra-thin window systems in the field and an accumulated mean time before failure of 390 weeks and rising as of May 1993. That is seven and a half years. They report that without thin windows on their environmental microscopes, this number would be much higher. When fitting a new thin window system on an old microscope, it would be a good idea to discuss window reliability with both the microscope and EDX manufacturer.

Environmental electron microscopes can be hard on window reliability, with some microscope models being worse than others. Windows fail on environmental microscopes as a result of exposure to corrosive gases, particles, and liquids. An intense effort is being made by the window, spectrometer, and microscope manufacturers to increase window reliability under these difficult conditions.


## 3.6. FUTURE WINDOW TECHNOLOGIES

It is interesting to speculate what future progress in materials science might contribute to this field. Of course, the ultimate advance would be a vacuum window, which is already with us in the form of windowless detectors. If we limit ourselves to atmospheric windows, lithium, at $\rho = 0.53$ and $Z = 3$, has the lowest density and lowest $Z$ of any material. Even lithium hydride is denser with $\rho = 0.82$. Lithium is neither strong enough nor inert enough to make a good thin foil window. Beryllium makes only brittle ionic compounds with lithium, and the other row 2 elements all form brittle insulators with lithium.

Many lithium alloys are extremely reactive. The lightest element that benefits lithium by alloying is magnesium. Lithium-magnesium alloys are stable in air and might be made into thin foils, though they are probably more reactive than beryllium because they lack the passivating oxide. A typical research alloy, $Li_2Mg_8$, has transmission comparable to equal thicknesses of AP1 or polyimide, up to Mg $K\alpha$, after which it absorbs quite strongly. If the technology of lithium magnesium foils could be perfected, they would have the advantage of having no absorption edges between Be $K\alpha$ and Mg $K\alpha$. They would have the disadvantages of negligible transmission of Be and B $K\alpha$'s and strong absorption at energies higher than the Mg $K$-edge.

Other combinations of row 2 elements have either been used already (BN, diamond, polymers) or cannot be used, since they are gases, water soluble, or unstable. It is possible that polymers with more strength and less density will be developed, but present yield strengths around 300 MPa will be hard to beat. The superpolymers, such as AP1 and polyimide, typically have oxygen, nitrogen, or fluorine, but a pure carbon-hydrogen polymer with high strength and low density could make a significant contribution. In particular, a carbon-hydrogen polymer with a strength of 300 MPa and a density $\rho = 1$ would have slightly better transmission than presently available windows and only one absorption edge. Recent progress in polymer technologies that might benefit windows for light element analysis are ion implantation, which increases the surface hardness,[21] and the processing of liquid crystal polymers into thin films.[22]

As discussed in Section 3.3, small improvements in transmission fail to make a significant contribution to the lower limit of detection. Large transmission improvements are possible only for Be $K$-radiation. It is doubtful, therefore, that another window technology is likely to come along that will overwhelm the current ones.

In conclusion, this is a great time for lovers of carbon, nitrogen, oxygen, fluorine, and sodium. Under intense competition, thin window technology has developed quickly and may be said to be maturing. No material breakthroughs are anticipated. Reliability is very good. No one technology is likely to knock the others out. Future progress in light element analysis will likely come in detector and preamplifier technology, which will give better resolution and lower background.

ACKNOWLEDGMENTS. I would like to thank all the manufacturers of energy dispersive spectrometers for their help in obtaining transmission and lifetime data. I also acknowledge the assistance of my colleagues at MOXTEK: Tracy Andersen who performed the beryllium helium leak tests, as well as Fang Yuan, Raymond Perkins, Mark Larson, Clark Turner, and Madlyn Tanner.

## REFERENCES


1. R. T. Perkins, D. D. Allred, L.V. Knight, and J. M. Thorne, (1990) *Adv. X-Ray Anal.* **33,** 615 (1990).
2. R. Mutikainen, V-P. Viitanen, and S. Nenonen, presented at: *Proceedings of the Conference on Soft X-Rays in the 21st Century* (Provo, UT, 1993) to be published in *J. X-Ray Sci. Technol.*
3. M. W. Lund and P. Moody, Optical Constants Grapher, available from MOXTEK, Inc., Orem, UT.
4. B. L. Henke, E. M. Gullikson, and J. C. Davis, *Atomic Data* **54,** 181 (1993).
5. W. W. Leslie, *Beryllium Science and Technology* Vol. 2, (D. R. Floyd and J. N. Lowe, eds.) Plenum Press, New York, p. 57 (1979).
6. J. J. Truhan and L. M. Wagner, *Nucl. Instrum. Methods Phys. Res.* **176,** 481 (1980).
7. J. Wittenauer, T. G. Nieh, and G. Waychunas, *J. Mater. Sci.* **27,** 2653 (1992).
8. J. C. Bomberger, U.S. patent no. 3,150,436, 1964.
9. V. P. Krivko, P. M. Romanko, L. I. Kolesnik, N. V. Nagnibeda, and Yu. I. Kokovikhin, *Metal Sci. Heat Treat.* **1,** 12 (1991).
10. D. D. Cohen, (1987) *X-Ray Spectrom.* **16** 237 (1987).
11. F. Yuan, D. Allred, I. Rudich, U. S. patent no. 5,226,067, 1993.
12. J. J. Mueller and D. R. Adolphson, in: *Beryllium Science and Technology* Vol. 2 (D. R. Floyd and J. N. Lowe, eds.) Plenum Press, New York, p. 417 (1979).
13. N. F. Gmur, *Nucl. Instrum. Methods Phys. Res.* **A266,** 362 (1988).
14. J. N. Rimbert and O. A. Testard, *Nucl. Instrum. Methods. Phys. Res.* **A251,** 95 (1986).
15. P. Hovington, G. L'Esperance, E. Baril, and M. Riguad, *Microbeam Anal.* **2,** 277 (1993).
16. A. H. Foitzik, J. S. Sears, A. H. Heuer, and N. J. Zaluzec, in: *Proceedings of the 49th Annual Meeting EMSA* (G. W. Bailey, ed.) San Francisco Press, San Francisco, p. 49 752 (1991).
17. W. Kern, *RCA Rev.* **39,** 278 (1978).
18. C. E. Cox, B. G. Lowe, and R. A. Sareen, *IEEE Trans. Nucl. Sci.* **35,** 28 (1988).
19. D. D. Cohen, *X-Ray Spectrom.* **16,** 237 (1987).
20. S. M. Zemyan and D. B. Williams, *Microbeam Anal.* **2,** S182 (1993).
21. E. H. Lee, G. R. Rao, M. B. Lewis, and L. K. Mansur, *Nucl. Instrum. Methods Phys. Res.* **74,** 326 (1993).
22. Data Sheet for AP10 Procon Windows, available from MOXTEK, Inc., Orem, UT, 1994.

# 4

# Germanium X-Ray Detectors

*R. A. Sareen*

## 4.1. INTRODUCTION

Methods of employing semiconductor materials to detect electromagnetic radiation, charged particles, and neutrons have been developed extensively during the last forty years.[1] Within this period, a particular semiconductor has been chosen for a specific application following extensive research and development. For example, the lithium drifted silicon (Si(Li)) detector[2] of a particular shape and size has been the preferred choice for detecting low energy x rays and has helped to promote the science of microanalysis on electron microscopes and other excitation sources. These include x-ray excited fluorescence systems, diffractometers, synchrotrons and particle accelerators. Within the last five years[3,4] it has become clear that germanium is capable of rivaling silicon with certain advantages. This presentation reviews the development of these germanium detectors as devices for high resolution energy dispersive x-ray spectroscopy and compares their performance with silicon.

The development of high purity germanium has progressed to the point where this material is commercially available[5,6] in both *n*-type and *p*-type forms with net impurity concentrations as low as $10^9$ atoms/cm$^3$ and long carrier lifetimes. At this purity level the material, when processed as a diode, can be depleted to great depths (up to cms) with biases below its breakdown voltage. In parallel with materials development, new passivation techniques[7] have been introduced that, together with improved ohmic and rectifying contact technology, give near flat band conditions at the surfaces. The resulting devices can have low leakage currents, low capacity, low $1/f$ noise, and undistorted fields, but the yield of good devices is still low.

The stimulus for the original work on this material was to make large volume, high resolution gamma-ray detectors which were stable when stored at room temperature and, when required, could be annealed at higher temperatures to remove radiation

R. A. SAREEN • Schuster Laboratories, The University, Manchester M13 9PL, United Kingdom

*X-Ray Spectrometry in Electron Beam Instruments*, edited by David Williams, Joseph Goldstein, and Dale Newbury. Plenum Press, New York, 1995.

damage. These devices are used in large numbers in national laboratories in what is often referred to as gamma ball[8] experiments.

As often happens with technology, other areas of measurement can benefit from these developments, and this report is about applying this germanium to the measurement of low energy x rays. The results obtained to date have been made possible by similarly important developments in FETs,[9] pulse processors,[10] cryostats employing materials with low dielectric losses, and exceptionally thin self-supporting atmospheric windows.[11]

Both silicon and germanium can be used to detect low energy x rays, but the activation energy is smaller in Ge (2.96 eV versus 3.86 eV at 77 K) offering the possibility of improved resolutions. The higher atomic number and density of Ge gives an advantage in stopping power, but when used for measuring very low energy x rays (Be $K\alpha$ at 110 eV), Ge must be processed with an exceptionally thin dead layer at the entrance window. The smaller bandgap also demands efficient cooling to suppress the thermally excited carriers (to equal the low leakage of silicon, Ge must cooled a further 25 degrees Kelvin), and the device must be thermally screened from room temperature surfaces (including the entrance window) because of the high sensitivity of Ge to infrared radiation. The higher dielectric constant compared with silicon results in a higher capacity per unit volume, which must be offset by careful geometrical design and the use of the more advanced lower capacity field effect transistors.[12]

These improvements have been incorporated into the latest designs and x-ray spectrometers with high efficiencies in the energy range up to 100 keV, premium resolution (less than 120 eV at 5.9 keV) with full energy peaks of near Gaussian shapes, and good peak to background ratios (typically 10,000:1) are now in use. (There is also the possibility of increased radiation damage resistance and storage at room temperature for long periods without degradation allowing the user to cool when required.)


## 4.2. DETECTION OF X RAYS

The primary function of these detectors is to convert radiation into charge in an efficient manner and to allow this charge to flow to the terminals with minimum loss. For elemental identification we are interested in radiation in the energy range which includes $K$-shell x rays from elements up to the heavy elements, e.g., the $K\alpha$ x-ray of lead at 74.957 keV. The mass absorption equation (Figure 4.1) is replotted for germanium, with silicon as comparison, and it can be seen that several millimeters of Ge are required to absorb up to 100 keV whereas Si thicknesses of tens of millimeters would be required.

Comparing the $K$- and $L$-shell x-ray yields in Figure 4.2 shows the $K$-shell dominates at all energies making it possible to identify elements more efficiently through their $K$ x rays. (*Note*: The preferred route to fill a vacancy in the $K$-shell is via the $L$-shell, although other routes are permitted, with these processes taking place in about $10^{-16}$ s). The extra advantage of the less cluttered $K$-spectra is welcome, but such advantages must be balanced against the increasing difficulty in exciting the $K$-shell of the heavier elements by electron excitation, since the innermost shells are effectively screened from the incident electrons. This limitation is not a problem with x-ray excitation (Figure 4.3 illustrates these differences). The

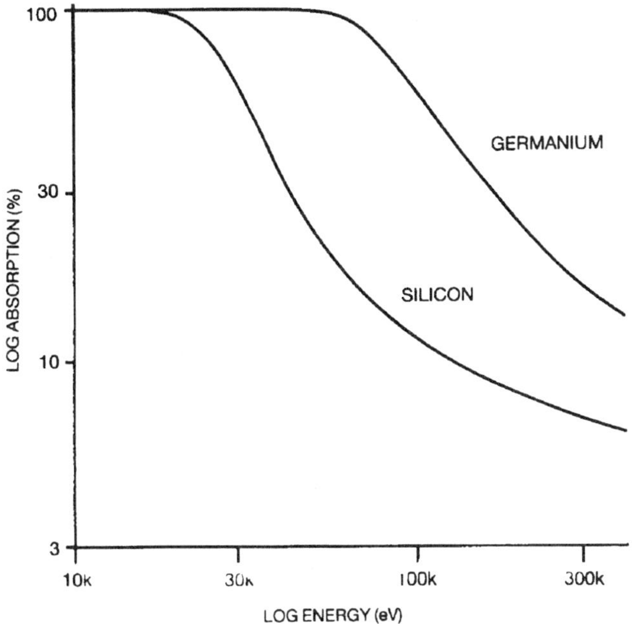


FIGURE 4.1. Absorption in 3-mm slices of silicon and germanium.

energy loss process in the range 0–100 keV is governed by the photoelectric effect at the lower energies and the Compton effect at the upper end of the energy range. Figure 4.4(a,b) graphs these responses for germanium and silicon. A portion of the $Am^{241}$ spectrum shown in Fig. 4.5 indicates how well the Ge detector absorbs the Compton-scattered higher energy (59 keV) photon, whereas in silicon many of these escape, leaving the first Compton interaction appearing in the background. Silicon is a more effective Compton scatterer at low energies. For this reason, a good internal collimator is required to make sure the first interaction is away from the side walls of the detector. This collimation requirement will also apply to Ge since in both devices there is the possibility of weak fields at the boundaries. Lithium drifted devices can be particularly bad because lithium can sometimes accumulate in pockets on the crystal walls where there may be crystal damage. Such regions act as another contact capacitively coupled into the bulk of the detector, producing a range of attenuated signals appearing as "ghost" peaks in the spectrum. These will be eliminated as bad devices by the manufacturers, but similar effects could occur on detectors stored at room temperature for long periods.

## 4.3. CHARGE FLOW

Fundamental to the flow of charge through a semiconductor are parameters like carrier velocity and trapping, and these are implicit in Ramo's Theorem,[13] which

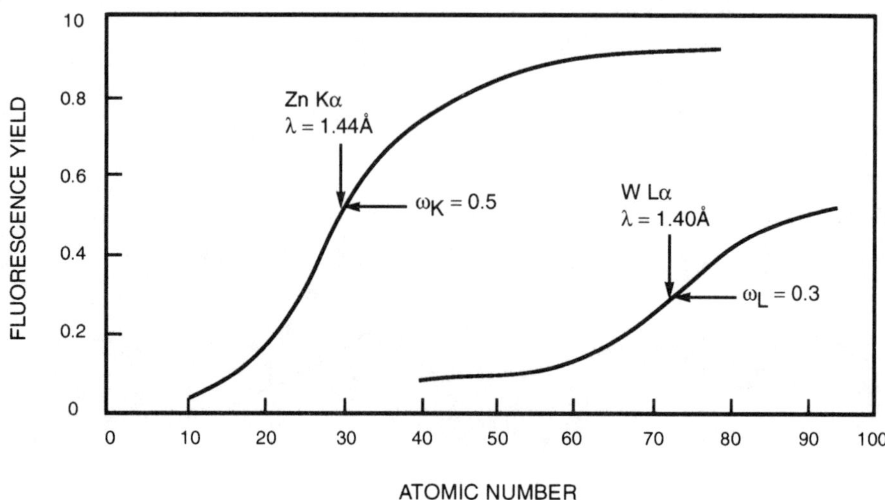


FIGURE 4.2. Fluorescence yield as a function of atomic number. [Published with permission from Ron Jenkins, *Introduction to X-Ray Spectrometry*, John Wiley & Sons (1974)].

states that once produced the charge will flow to the terminals such that the quantity of charge induced in the external circuit is proportional to the sum of the particular charge carrier multiplied by the fraction of the voltage this charge carrier moves through. This is expressed in Fig. 4.6; the contacts are aware of this flow of charge by the process of induction as soon as the charge starts moving. Equally so, if charge is trapped then the measured quantity is reduced to the product of the carrier type and the fraction of the voltage it traveled through prior to being trapped. Charge trapping can occur anywhere in the detector, e.g., at surfaces, contacts, interfacial regions, weak field areas, and in the sensitive volume, and is responsible for backgrounds and distorted peak shapes. As pointed out by Gatti *et al.*[14] the collected charge signal is independent of any fixed space charge in the lattice, the latter only affecting the signal rise time. In principle, a drifted device should perform like a depleted device operating at the same voltage. Having converted the incident radiation into charge, it will then flow into the external circuit and the detector must contribute as little noise as possible to its measurement. In the following sections some of the physical and electrical effects which contribute to noise are discussed.

## 4.4. RESOLUTION

The resolution of a semiconductor detector is the quadratic sum of several terms describing the noise inherent in the detector under the specified conditions at the time of measurement, the statistical spread in the charge collection process, and any charge losses that occur at trapping sites. These are expressed in Figure 4.7 and it can be shown

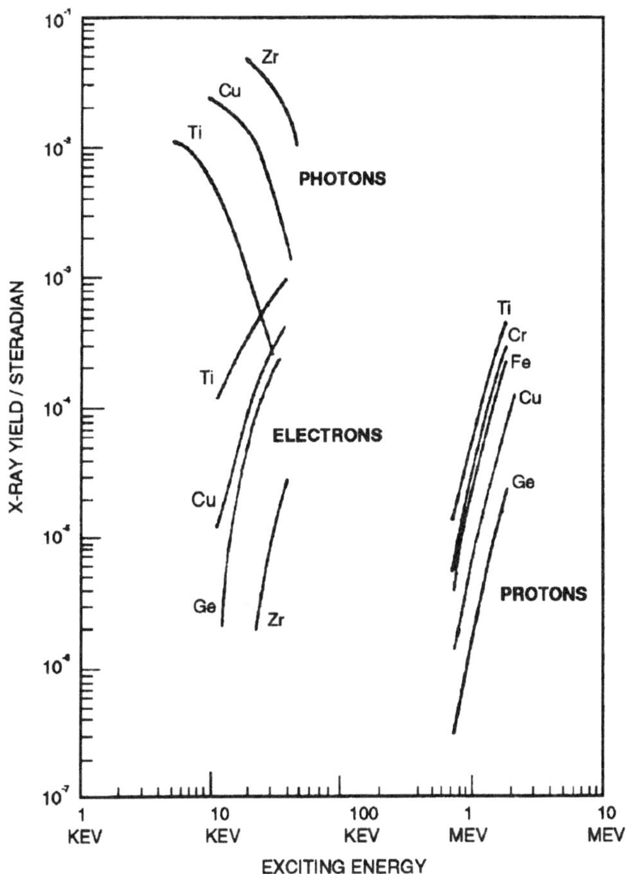



FIGURE 4.3. X-ray yields as a function of exciting nergy for different types of radiation. –, $K$ series. [Published with permission from R. Woldseth, *X-Ray Energy Spectrometry*, Kevex, Inc. (1973)].

that changing from silicon to germanium brings an improvement to the first two terms, provided the physical size of the Ge detector is adjusted to compensate for its higher dielectric constant and hence electrical capacity. It is assumed that trapping is near zero. The best treatment of noise is by Goulding.[15] Here a physical model is used to describe the benefits of various filters in processing the signals from semiconductor detectors and covers noise due to leakage current, the effect of detector capacity, and the contribution from the FET. It is also important to consider the effects of $1/f$ noise (Llacer[16]). (There is now the possibility of additional effective filtering using the new digital filtering[17] techniques appearing in the literature, but the consequences of adaptive filtering and non gaussian peaks need careful consideration.)

The argument showing an improvement in electronic noise when using germanium instead of silicon from two devices with identical capacity is shown schematically in

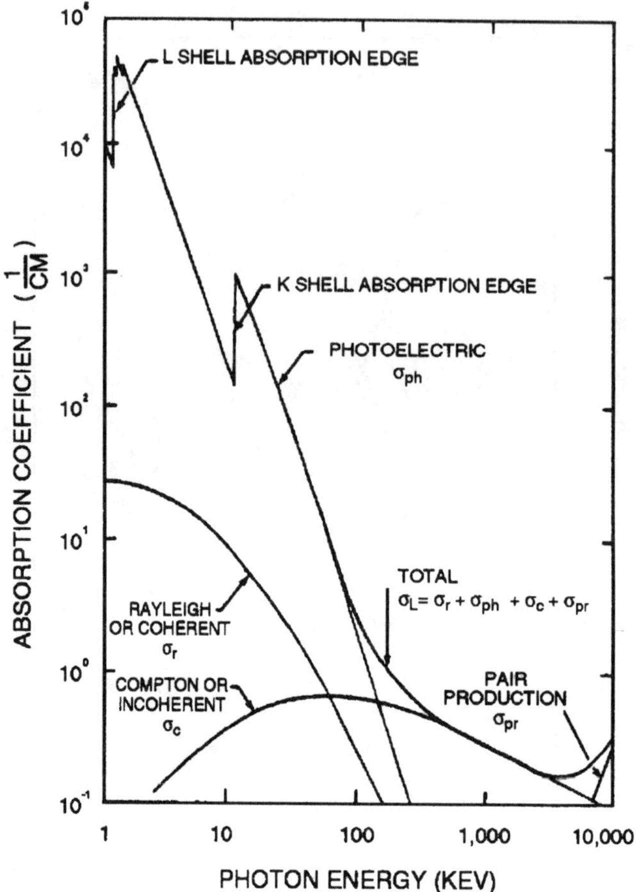


FIGURE 4.4. (a) Energy loss in germanium.

Figure 4.8. The charge from the detecting element is integrated onto the feedback capacitor and amplified to a voltage $V$, producing an event in the multichannel analyzer at channel $N$. If the sensing element is Si then this channel number is determined by the quantity of charge produced during ionization, the fixed gain in the amplifier, and the conversion gain in the analog-to-digital converter (ADC). Provided the feedback capacity remains the same when the sensing element is changed to Ge, then to maintain the peak in the same channel the gain can be reduced by a factor equal to the ratio of the energy per electron-hole pair in Ge and Si (2.96/3.86). In other words, we are able to produce similar size pulses at lower gain settings, which is equivalent to a reduction in noise. Besides improvements in resolution, this reduced noise increases the electronic efficiency in the recognition circuits, enhancing the detection of the very light elements. This advantage is partially offset in Ge by the need for an internal window held at low temperature to reduce Ge increased sensitivity to infrared radiation.

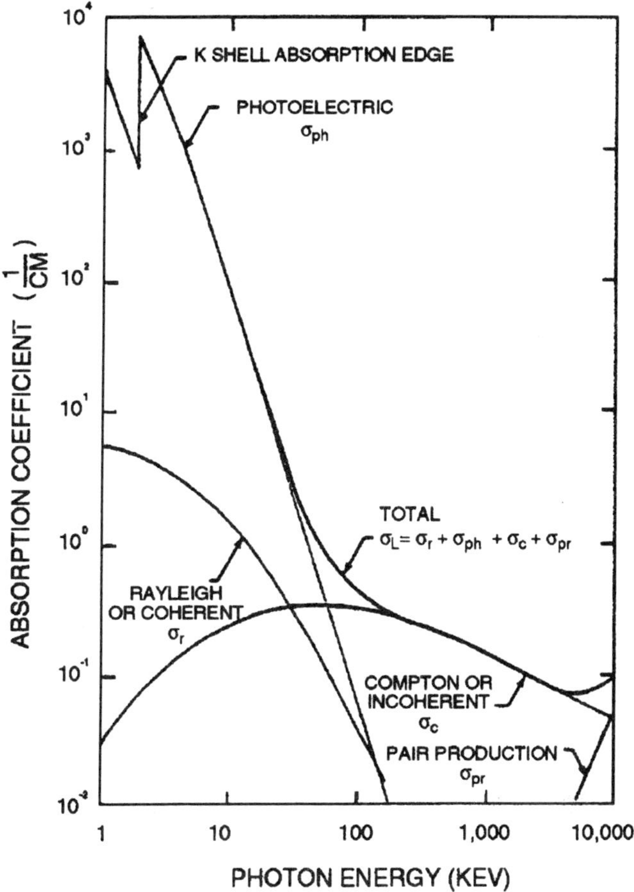


FIGURE 4.4. (b) Energy loss in silicon.

The second term in the resolution formula is also improved because the reduced activation energy in Ge produces more carriers per photon. The statistical spread in the number of carriers ($N$) created is proportional to $1/N^{1/2}$ and thus improves with $N$.

The Fano factor was introduced by Fano[18] in 1947 to explain the difference between the observed resolution and that predicted by calculation. If all the energy appeared as electron-hole pairs where each pair was totally unaware of its neighbors the factor would be zero, but in the energy loss process the pairs produced can interact with each other and with the lattice, producing phonons. Thus, the process is not governed solely by Poisson statistics, and the Fano factor is proportional to the ratio of the square of the standard deviation to the number of pairs produced. The Fano factor in germanium rather surprisingly is similar to that in silicon and values between 0.10 and 0.12 have been measured consistently, although values as low as 0.06 have been reported (but perhaps without regard to peak shape) with a reported variation

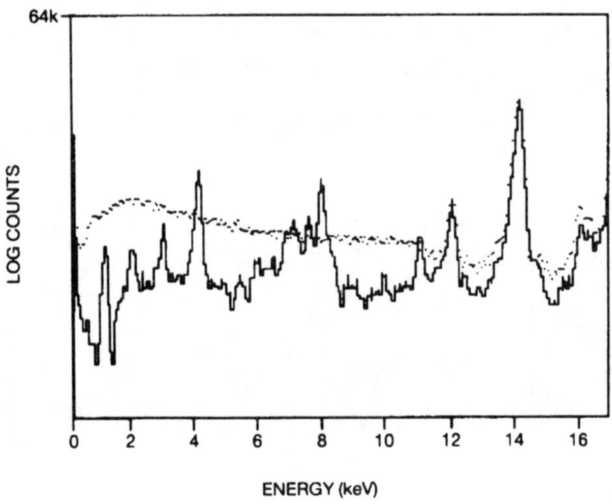


FIGURE 4.5. Low energy region of Am$^{241}$ spectrum. Solid line, HPGe; dotted line, Si(Li).

with energy in both materials.[19] The range of the photoelectron in Ge is smaller, the number of carriers created and their velocities higher and yet the statistical processes in charge collection and phonon generation end up being very similar!

## 4.5. DETECTOR GEOMETRY

The detector is in parallel with the input capacity of the first stage of the measuring circuit. This includes the input FET, the feedback capacitor, and any stray capacity. To achieve the best signal to noise ratio the total capacity has to be minimized, and for a given detector the best FET is the one with the lowest possible noise and an input capacity nearly equal to that of the detector.[20] For low energy x-ray measurements the detectors are physically small to achieve a low capacity (0.5 pfs). Commercial FETs are not available with matching low values. The 2N4416, the main device used by the manufacturers of Si(Li) and small HPGe detectors in these applications, has an input capacity of 3.5 pfs and a cooled voltage noise of 1.5 nV/Hz$^{1/2}$. Recently new FETs have become available with much improved performance: for example, the NJ14 from Interfet,[21] the Pentafet[9] from Oxford Instruments, and the new range from MOXTEK.[11]

The capacity is a function of geometry, dielectric constant, and the processing used to fabricate the detector. The techniques that were developed for silicon[22] have been applied to germanium[23] with some success. However, with further development and new geometries, the possibility exists of a reduction in detector capacity and improved collection fields. The grooved and top hat designs shown in Figure 4.9 were first developed for silicon and enabled the manufacturers to successfully compensate

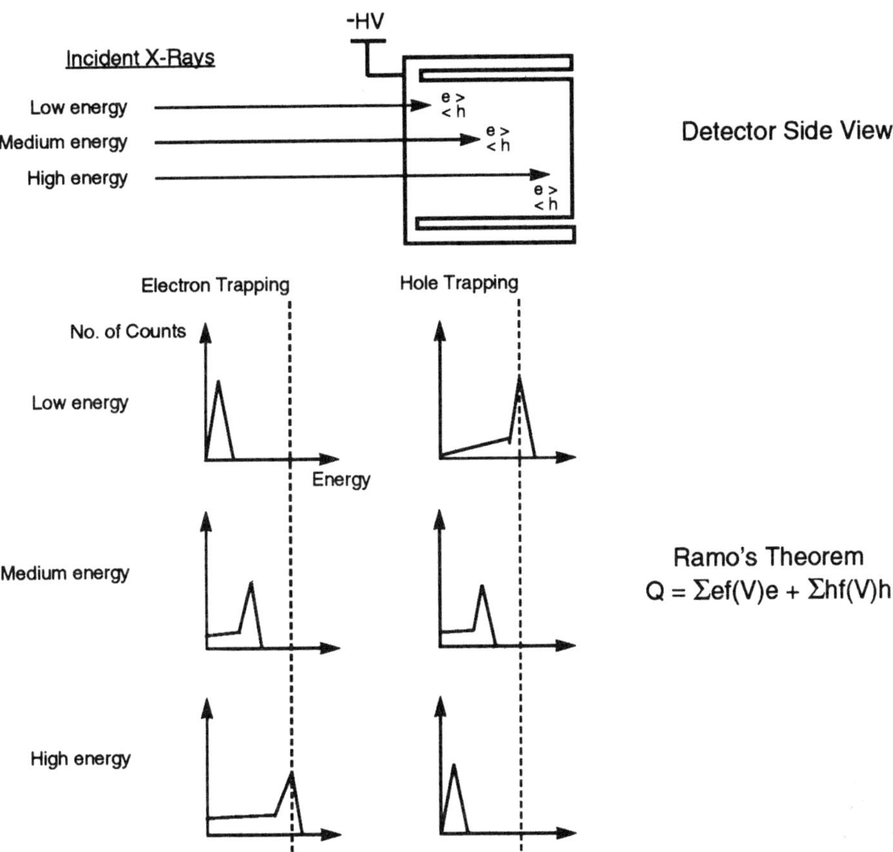



FIGURE 4.6. Signal contributions from different parts of the detector.

for the materials and obtain low leakage currents. Recently Madden[24] introduced a new geometry, illustrated in Figure 4.9, which further reduced the capacity and the volumes containing weak fields resulting in better charge collection and peak to backgrounds. The principal disadvantages of the grooved and top hat designs were the

$$\text{FWHM (ev)} = \text{Sqr.Rt} \left( (\text{FWHM})^2_{\text{Noise}} + (2.35\,\text{sqr.rt}(FEe))^2 + \text{Sigma}^2_T + \text{Sigma}^2_{DE} \right)$$

$F = 0.11$

$e = 2.96$ ev

$E = $ Incident energy in ev

$\text{Sigma}_T = $ spread due to trapping  (assumed zero in a good detector)

$\text{Sigma}_{DE} = $ spread due to drift in the electronics (assumed zero in  a good system)

FIGURE 4.7. Parameters affecting detector resolution.

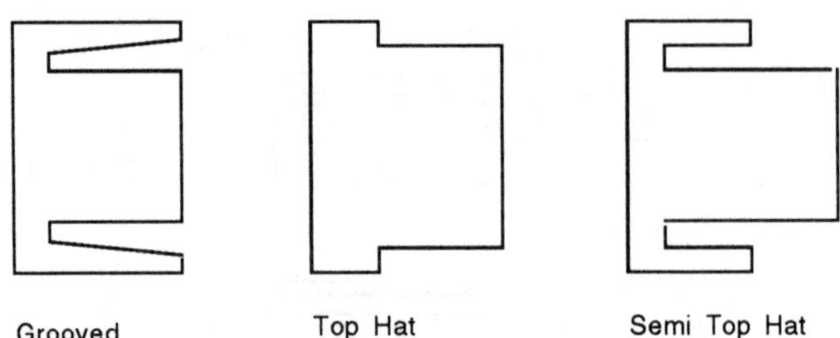



Side View of Standard Detectors supplied by Manufacturers.

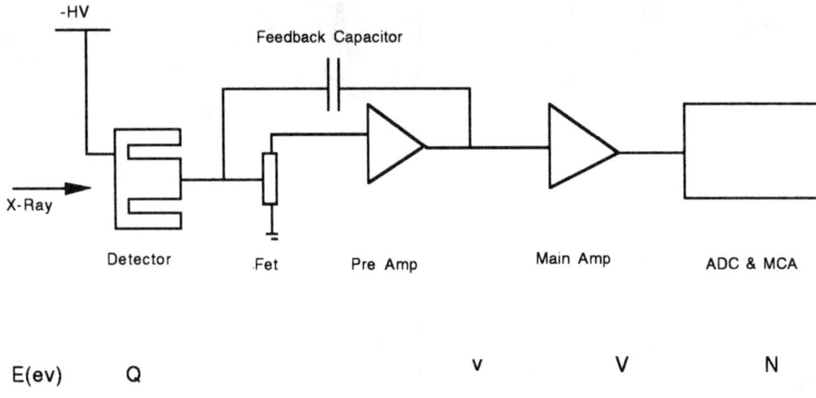



FIGURE 4.8. Input stage.

weak fields produced at the base of the groove and the spreading out of the drifted region as the drifted profile or the depleted volume extended beyond the bottom of the groove. The Madden design overcomes these weaknesses and is a design that could benefit germanium. Another interesting design is shown in 4.10, where the capacity is determined by the diameter of the hole and there are no contact structures in the way of the incident x rays.

With refinements in passivation techniques,[7] a design may now be possible that offers thinner effective dead layers compared with the usual entrance contact structures. This design departs from the normal planar geometry, and with its logarithmic field, may advantageously employ the consequences of Ramo's theorem.

## 4.6. GERMANIUM

During the early development of germanium as a gamma ray[25] detector, the available $p$-type Ge was too low in resistivity to deplete, and like silicon, the electrical impurities were compensated for using the lithium compensation process.[26] Lithium

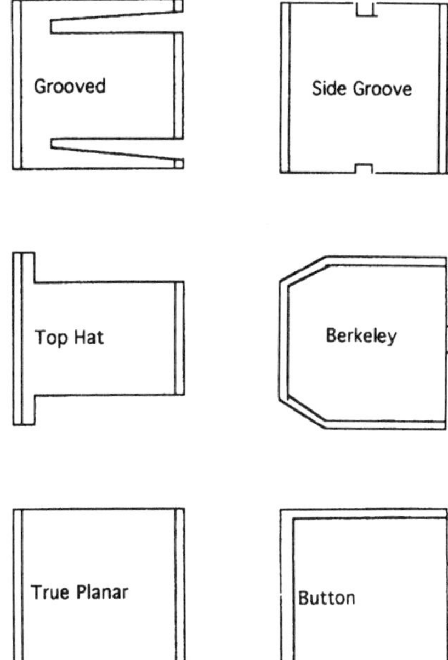



FIGURE 4.9. Different planar detector geometries. All of these shapes are attempts to reduce capacity, leakage current, and weak field regions in the detector. The ideal device is the true planar but it is difficult to process. The "Bullet" is ideal for a large area, low capacity diode. The "Berkley" structure is an excellent approximation to the "Bullet" with more uniform field values throughout the device.

ions were "drifted" through the material and settled interstitially at locations where there was a net negative charge. The most likely place for this was at a trivalent impurity site. The process was successful in both semiconductors, but particularly in the case of Ge the compensated diode was unstable to lithium precipitation at room temperature. This mode it essential for such devices to be stored at low temperatures, a major drawback in their use. Tremendous materials development effort resulted in today's high purity or near intrinsic Ge where compensation by the lithium drifting process is no longer required.

When suitably contacted this "intrinsic" material can be depleted to many centimeters,[27] making a very successful gamma ray detector and, more recently, a low energy x-ray detector. It should be noted that the development as a low energy device was aided by improvements in contact technology that minimized trapping at the entrance window, giving results like those shown in Figure 4.11. These results should be compared with those shown for earlier devices, as in Figure 4.12.[28] One possible disadvantage of having moved to depleted devices from compensated devices was that it was believed that the compensation techniques also compensated for active sites at dislocations. This put great demands on the high purity material to be crystal-lographically excellent as well as low in impurities, and recent results do indicate that this can be realized (Figure 4.11). The technique of depleting as opposed to compensating can be applied to silicon provided very high resistivity material is available. Resistivities of greater than 50 kohm-cm would be required to deplete to several millimeters at achievable voltages. (Using $d = 0.3 \, (\rho V)^{1/2}$ for $p$-type material and with $V = 500$ volts then $d = 1500$ micrometers). The choice in both materials of whether to

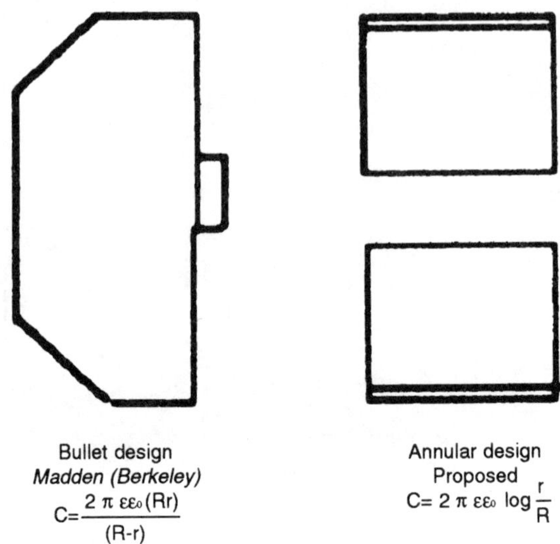



$$C = \frac{2\pi\,\varepsilon\varepsilon_0\,(Rr)}{(R-r)}$$

$$C = 2\pi\,\varepsilon\varepsilon_0\,\log\frac{r}{R}$$

FIGURE 4.10. Recent planar detector geometries.

use $n$-type or $p$-type is governed surprisingly by the availability of FETs, since all the low noise FETs are $n$-channel devices[29] requiring collection of electrons; the FET should be connected to the low capacity side and preferably the low voltage side of the detector. These points all favor $p$-type material, negatively biased with the $n$-channel FET connected to the $n^+$ contact.

## 4.7. RESPONSE TO IRON-55

Examination of a $Fe^{55}$ spectrum taken with a silicon and a germanium detector as shown in Figure 4.13(a,b) reveals the characteristics of the full energy peaks (Mn $K\alpha$ and Mn $K\beta$) and the background. With sufficient statistics, escape peaks and absorption edges from contact materials, entrance windows, and any internal windows will be revealed. All peaks should be Gaussian and well developed above background. Any events at energies greater than the full energy peaks will give a measure of the effectiveness of the pile-up rejector. Solving Poisson's equation in the one dimensional case and using typical values for the detector, it can be shown that the field at the entrance window is similar in both the depleted and compensated devices.

By considering where ionization is taking place in the detector and what carrier is likely to be trapped, one can predict the shape of the background. First consider low energy x rays that can interact in the metal contact, the interfacial layer, or the start of the sensitive part of the detector. Since there is zero electric field in the metal, no information will be recorded unless the ionization electron has sufficient energy to move deeper into the detector. If it moves into the interfacial layer, then there is a

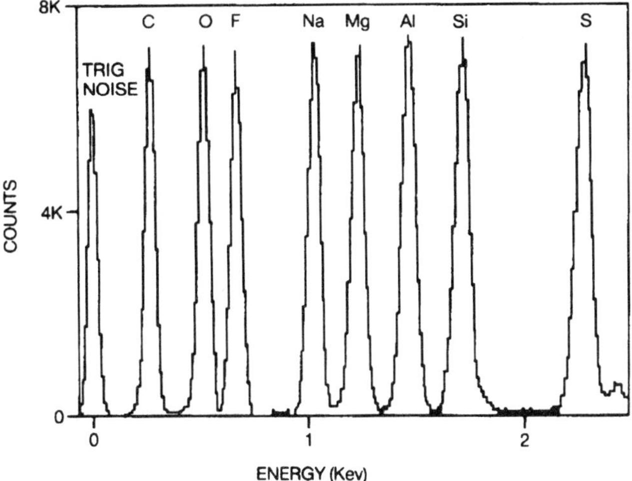


FIGURE 4.11. Results obtained with a recent detector.

possibility it would experience some of the electric field prior to being trapped, contributing to the low energy part of the spectrum (sometimes referred to as the "spur"). If ionization takes place in the interfacial layer, the electrons and holes do have a probability of being collected. The hole will make an infinitesimally small contribution to the signal because it experiences almost none of the applied voltage. The efficiency for collecting the electron will vary from zero at the metal contact to 100% as the electron enters the sensitive volume. The probability of electron trapping is very high in this region since the metal contact and the applied bias (-ve) attract a net positive charge density to this interfacial area. The difference between a good detector and a bad detector can often be attributed to the trapping that takes place in this region. For optimum results, the contact must be thin and have a low sheet resistance. The interfacial layer also must be thin and designed to support an electric

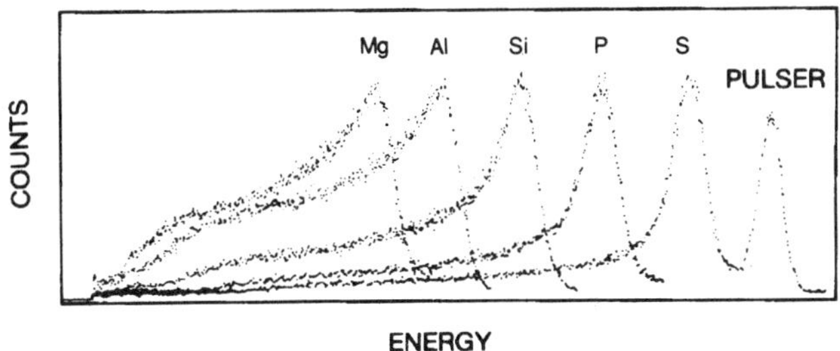


FIGURE 4.12. Results obtained with an early HPGe detector.

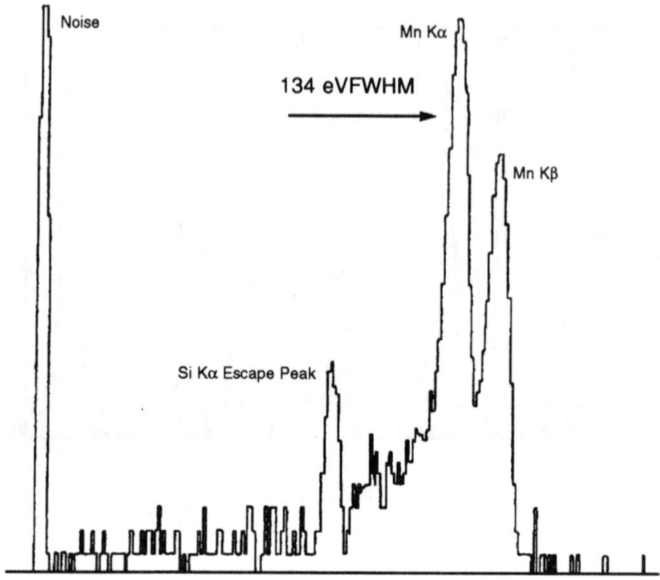



FIGURE 4.13.    (a) Iron-55 spectrum taken from a silicon detector.

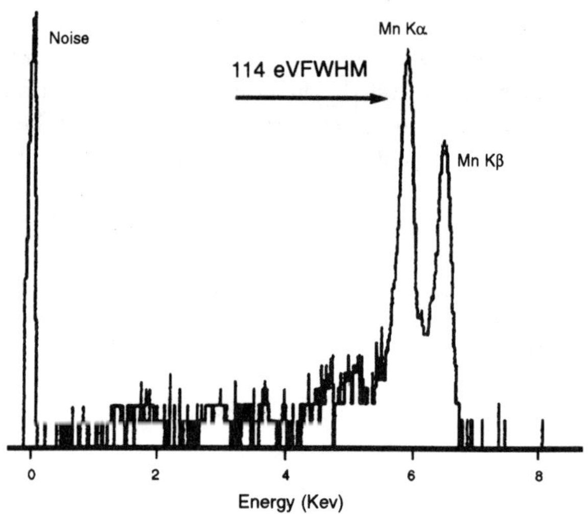



FIGURE 4.13.    (b) Iron-55 spectrum taken from a germanium detector.

field. It is important to avoid a thick oxide and to prevent water vapor or organic material from condensing in this area. If the electrons are trapped in this area, the spur will develop in the background near the zero of energy and the low energy peaks ( Be, B, C, O...) will be broadened and down shifted in energy. Only a fairly bad detector will exhibit this spur.

When ionization takes place in the bulk of the detector and both carriers move through the field to their respective contacts then, on reaching, they will account for an event in the full energy peak. It is important that both carriers are collected. Electron trapping as the electron approaches the rear contact is also possible. For penetrating radiation where ionization has occurred close to the rear contact most of the signal will derive from the hole traveling toward the front contact. Any of these trapping mechanisms that occur after the charge has traveled through the majority of the applied voltage will account for the events that appear in the tail. There are several papers that suggest mathematical models for charge collection in the detector.[30–34] In practice, the reasons for the difference between a very good detector and an average detector are not understood, but it is thought to be a consequence of surface charge densities and electric field values at the entrance window and back contact.

## 4.8. RESPONSE TO HIGHER ENERGY X RAYS

When higher energy x rays enter the detector there is a probability of exciting the germanium $K$-shell so some of the Ge $K$ x rays will escape as shown in Figure 4.14(a,b). This is a possibility whenever x rays of energy greater than 11.1 keV are incident on the detector, and these escape peaks can mask and confuse the characteristic x rays we are trying to measure. This problem was present in silicon from the $K\alpha$ line at 1.74 keV.[35] These escape peaks were only a problem up to about incident energies of 20 keV since these x rays produce Si $K\alpha$ x rays at depths where they will be internally absorbed. These peaks can be effectively removed by software. However, in the case of germanium, there is a probability of a $K$-escape from higher energy incident x rays, with the result that such peaks appear over a wider range of the spectrum starting from the threshold of 11.1 keV. Their energy can be identified, but their removal is complicated by determining the contribution to their intensity the excitation of germanium as a function of energy and by the probability of escape as a function of position in the detector. Also, because these peaks are present over a wide energy range, allowance must be made for the square root dependence of their line width with energy.

## 4.9. TIME EFFECTS AND PEAK SHIFTS

The background measured with a germanium detector system is now comparable with that available from a good Si(Li) detector. In the energy range 0–10 keV it can be seen that the spectrum is free from artifacts and escape peaks. *However, there is the possibility that the background can change with time due to changes at the surfaces of the detector.* Although, this problem is not fully understood, it may be caused by organic vapors or water vapor out-gassing from the cryostat, or may be due to the mechanical conditions of the surfaces or the effects on the surface of the processing or any residual infrared radiation continually incident on the detector. Also, just like in silicon where the carbon peak can be shifted in energy to as low as 230 eV, there can be a shift at aluminium (i.e., just above the germanium $L$-edge). These phenomena

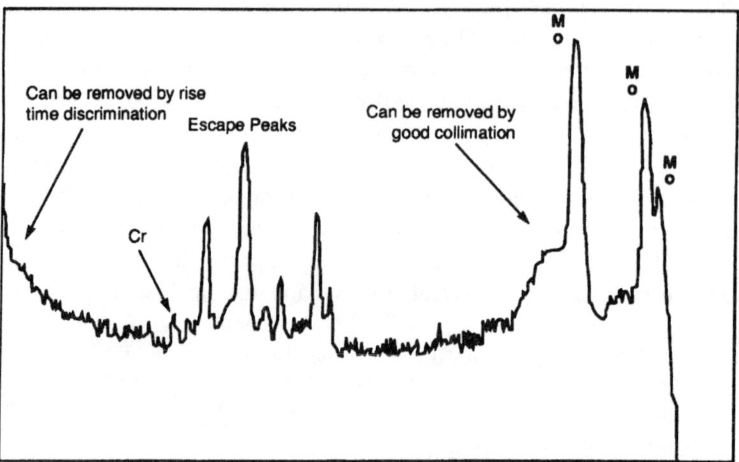


FIGURE 4.14. (a) Molybdenum excited by $Am^{241}$. The general high value of the background comes from scatter of materials external and internal to the cryostat.

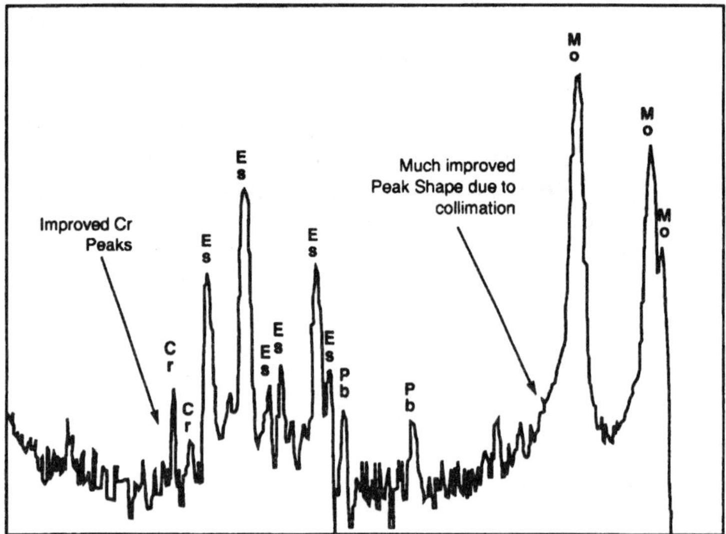


FIGURE 4.14. (b) Molybdenum excited by $Am^{241}$ (Es = escape peaks).

vary from detector to detector and again are not fully understood but are probably related to the thickness of the interfacial layer which is a mixture of semiconductor, oxygen, and contaminants. They correlate with the $L$-edge of the semiconductor. It is most likely a surface phenomenon, evidenced by the fact that the escape peaks (generated from a parent with more range) are always in the expected position. For example, in silicon the phosphorous escape peak is at 275 eV (2.015–1.740) whereas carbon can be downshifted to 260 eV.

(*Note*: The electronic system must have a linear ADC and amplifier, both of which can be checked with a signal generator. However, the very nature of triggering the system with noise to extend the measuring capability to below, say, 100 eV in order to see the Be $K\alpha$ puts a shift to the position of true zero. The reason for this is that when the system is triggered the noise is already instantaneously above the discriminator, and the amplifier will measure the noise at that instant. A better system is to trigger the measurement of a zero peak by a pulse generator which, having tested that no signal is present, will measure the noise, whatever its instantaneous value.[36] Only when true zero is measured and the system has been checked for linearity can the centroid positions of the soft x rays be measured.)

## 4.10. DISCUSSION AND CONCLUSIONS

Intrinsic germanium is a good alternative to silicon for applications involving the detection of low energy x rays. It offers a number of advantages with few disadvantages and the theory and practice is *reasonably* well understood. *However, germanium detectors are presently more difficult to make and may not yet have the long term stability typical of silicon.* This situation would be expected to improve as manufacturers gain more experience. Areas of future research will include a more detailed understanding of $1/f$ noise, including a determination of its contribution from the detector; the FET; and the dielectrics in the vicinity of the detector-to-gate lead connection. Future designs will further improve the peak-to-background and reduce or eliminate the small peak position shifts at low energies, together with removing any time-varying element. These designs may well include a grounded or active guard ring or structures with minimum surface to volume ratio.

Whenever such devices are being used to measure the very soft x rays, there is a concern that monolayers of ice will form on the entrance window of the detector. These can be removed temporarily by internally heating[37] the detector to a temperature where the ice evaporates and recondenses elsewhere in the cryostat. The benefits of resolution are well understood in areas where the intrinsic properties of the detector's excellent peak-to-background can be used as in, for example, x-ray fluorescence. In microanalysis the backgrounds are dominated by bremsstrahlung but resolution can still be an advantage. The bremsstrahlung falls to zero at the low energies, and it is here that the resolution is extremely important. Also, the minimum detection limit for an isolated peak is reported to be proportional to the square root of the resolution divided by the count rate.[38] Rate improvements will come through use of digital signal processing techniques and pulsed operation of the microscope or x-ray tube, etc. Another method of improving the count rate capability of the system is to segment the detector, with each segment connected to its own amplifier. The rate and pile up limitations imposed by one amplifier can thus be shared among the number of amplifiers connected to the individual segments; the final rate is determined by summing the results from all amplifiers. This technique will call for the design of a number of fixed time constant amplifiers on one board.

Germanium has a useful role to play in x-ray measurements, but these devices must be cooled close to liquid nitrogen temperatures. The next few years will see many uses for these detectors. Similarly, recent developments in arrays of silicon pixels integrated with an FET and silicon drift chambers offer the possibility of reasonable performance at temperatures that can be achieved by Peltier coolers and Sterling engines (−60°C). This approach could be extended to germanium. It is also possible that we may see silicon and germanium used together in one sensor.

ACKNOWLEDGEMENTS. I am grateful for discussions and input over the years from the following experts in this field: C. Cox, B. G. Lowe, T. Nashashibi, G. White, F. Goulding, N. Madden, K. Kandiah, R. Henck, D. Gutknecht, and P. Sansingkeow.

## REFERENCES

1. G. F. Knoll, *Radiation Detection and Measurement*, 2nd Ed., John Wiley and Sons, New York (1989).
2. E. M. Pell, *J. Appl. Phys.* **31,** 291 (1960).
3. C. E. Cox, B. C. Lowe, and R. A. Sareen, *IEEE Trans. Nucl. Sci.* **35,** 28 (1988).
4. J. J. McCarthy, M. A. Ales, and D. J. McMillan, in: *Proceedings of the Xth International Congress for Electron Microscopy* (1988).
5. R. N. Hall and T. J. Soltys, *IEEE Trans. Nucl. Sci.* **NS-18,** 160 (1971).
6. Hobokenin, Belgium, and Tennelec, Oak Ridge, Tennessee, Manufacturers of germanium.
7. P. Luke, *et al. IEEE Trans. Nucl. Sci.* **39,** 93 (1992).
8. P. Twin, J. Sharpey-Schafer, P. Nolan, J. Simpson, and P. Butler, Davesbury Report, University of Liverpool, Chesire, England.
9. T. Sashashibi, *Nucl. Instrum. Methods* **A322,** 551 (1992).
10. G. White, *et al.*, Proceedings of Second ISPRA Nuclear Electronics Symposium. Commission of the European Community, Stresa, Italy, p. 153 (1975).
11. M. Lund and R. Perkins, MOXTEK, Provo, Utah.
12. T. Nashashibi and G. White, Internal Report, Link Analytical, Halifax Road, High Wycombe, England (1991).
13. S. Ramo, *Proc. Inst. Radio Eng.* **27,** 584 (1939).
14. Cavalleri, Fabri, Gatti, and Svelto, *Nucl. Instrum. Methods* **21,** 177 (1963).
15. F. S. Goulding, *IEEE Trans. Nucl. Sci.* **1,** 493 (1972).
16. G. Llacer, LBL Report 5585, Berkeley, California (1976).
17. A. Georgiev, *Nucl. Instrum. Methods* **1** (1992).
18. U. Fano, *Phys. Rev.* **70,** 44 (1946); **72,** 26 (1947).
19. R. Zulliger, *IEEE Trans. Nucl. Sci.* **NS-17,** 187 (1970).
20. V. Radeka, *Nucleonics* **23,** 53 (1965).
21. Interfet, Dallas, Texas, manufacturer of FETs.
22. F. S. Goulding, UCRL Report 16231, UCRL, Berkeley, California, 293 (1965).
23. F. S. Goulding, *et al.*, *IEEE Trans. Nucl. Sci.* **NS-19** (1972).
24. N. Madden *et al.*, *IEEE Trans. Nucl. Sci.* **1** (1992).
25. Freck and J. Wakefield, *Nature* **193,** 669 (1962).
26. R. Tavendale, *IEEE Trans. Nucl. Sci.* **NS-11,** 191 (1964).
27. R. Trammell *et al.*, *Nucl. Instrum. Methods* **176,** 595 (1980).
28. M. Slapa *et al.*, *Nucl. Instrum. Methods* **196,** 575 (1982).

29. K. Kalinka and I. Taniguchi, *Nucl. Instrum. Methods* **B75,** 91 (1993).
30. D. Joy, *Rev. Sci. Instrum.* **59,** 83 (1985).
31. J. L. Campbell and J. X. Wang, *X-Ray Spectrom.* **20,** 191 (1991).
32. Heckel and Scholz, *X-Ray Spectrom.* **16,** 181 (1987).
33. Inagaki, Shima, and Maezawa, *Nucl. Instrum. Methods* **B27,** 353 (1987).
34. Bloomfield, Love, and Scott, *X-Ray Spectrom.* **13,** 69 (1984).
35. Reed and Ware, *J. Phys.* **E5,** 148 (1972).
36. White, *IEEE Trans. Nucl. Sci.* **35,** 125 (1988).
37. Lowe and Tyrell, Patent owned by Link Analytical, Halifax Road, High Wycombe, UK (1988).
38. P. Statham and T. Nashashibi, *Microbeam Analysis* (1988).

## SUGGESTED FURTHER READING

1. Reference 1.
2. Delaney and Finch, *Radiation Detectors*, Clarendon Press, Oxford, England (1992).
3. Debertin and Helmer, *Gamma Ray and X-Ray Spectrometry with Semiconductor Detectors*, Elsevier, North Holland, Amsterdam (1988).
4. Tsoulfanides, *Measurement and Detection of Radiation*, Hemisphere Publication Corporation, Washington (1983).

# 5

# Modeling the Energy Dispersive X-Ray Detector

*D. C. Joy*

## 5.1. INTRODUCTION

The use of a solid state detector to record an x-ray spectrum was first demonstrated by Miller and Wilkinson.[1] A second significant milestone was the description of the Si(Li) diode configuration by Pell,[2] which led directly to the first devices recognizable as modern energy dispersive x-ray analyzers (Elad and Nakamura[3]; Fitzgerald *et al.*[4]). In the succeeding 25 years the Si(Li) detector has become the single most widely used piece of analytical instrumentation in the field of electron microscopy. Despite this popularity, however, little attention has been paid to the properties of the detector itself. For most electron beam microanalysts, any discussion has been confined to trying to estimate the absorption of soft x rays in the window protecting the diode and to assessing the fraction of high energy x rays transmitted directly through the diode itself. The influence of the detector diode itself generally has been ignored.

As long as attention is confined to x-ray photons with energies of at least 3–4 keV, this neglect is not a serious matter because the diode acts as a 100% efficient transducer converting a flux of mono-energetic photons into a train of charge pulses which have a predictable Gaussian intensity distribution about some mean value which is linearly related to the x-ray energy (Hubbell[5]). But when photons of lower energy are observed, it is found that the resultant spectrum no longer has peaks that are Gaussian but instead has peaks which display a low energy "tail" whose size and shape may vary significantly with energy. At the same time the calibration of the detector diode is found to become non-linear. Both of these effects indicate the detector is no longer completely efficient at converting the photon energy into a pulse of charge, and

D. C. JOY • EM Facility, University of Tennessee, Knoxville, Tennessee 37996-0810, and High Temperature Materials Laboratory, Oak Ridge, National Laboratory, Oak Ridge, Tennessee 37831

*X-Ray Spectrometry in Electron Beam Instruments*, edited by David Williams, Joseph Goldstein, and Dale Newbury. Plenum Press, New York, 1995.

thus these artifacts are collectively labeled "incomplete charge" phenomena. In the case of electron beam analysis, the presence of the beam-induced continuum reduces the visibility of these effects, but the non-Gaussian peak shapes makes it difficult to measure the integrated intensity under a peak, and the distortion and non-linearity make procedures such as peak overlap deconvolution difficult and unreliable. In analyses where x rays are fluoresced by protons or photons, bremsstrahlung is effectively absent, and these incomplete charge effects become more pronounced leading to a requirement for detailed corrections to the spectrum.

This paper discusses a method for analyzing incomplete charge effects which considers the semiconductor physics of the Si(Li) detector. Effects due to external factors, such as the interaction of the diode with the preamplifier electronics, or to departures from ideality in the geometry of the detector are not considered. It is shown that this model permits both qualitative and quantitative predictions to be made about the behavior of the detector. The results of the investigations also suggest ways in which energy dispersive x-ray detectors might be improved.

## 5.2 THE PHYSICS OF THE DETECTOR DIODE

A variety of different configurations have been used as x-ray detectors, but the most common is that illustrated schematically in Figure 5.1. By a process of diffusion or implantation a junction is formed at some depth $DD$ beneath the entrance surface, creating a $p$-$i$-$n$ diode with three distinct regions (Pell[2]; Wolfgang $et\ al.$[6]; Muller[7]; Shima $et\ al.$[8]). The first region, made from degenerate $n+$ doped material, is provided to facilitate the production of an ohmic back contact to the diode. The second region, comprised of intrinsic silicon compensated by the presence of the drifted lithium, is the active volume of the device which, when the diode is reverse biased, is space charge depleted. The combination of the high resistivity of the intrinsic material and the large applied bias ensures that a large active region is depleted. The $p$-$i$-$n$ geometry allows the active region to be extended to a depth of several millimeters, sufficient to capture all of the charge deposited by photons with energies of 10 keV or greater, although, because of limitations on the size and shape of the diode set by the need to minimize capacitance, the efficiency of charge collection deteriorates toward the outside edges of the active region. A high efficiency of charge collection is vital if the detector is to have linear calibration. If a photon deposits its energy $E_{\text{x-ray}}$ in the active region, then $n_{e-h}$ electron-hole pairs are generated where

$$n_{e-h} = \frac{E_{\text{x-ray}}}{e_{\text{eh}}} \tag{5.1}$$

and $e_{eh}$ is the effective energy required to generate one electron-hole pair, for example 3.6 eV in silicon. If all of these carriers are collected in the depleted region then a voltage pulse

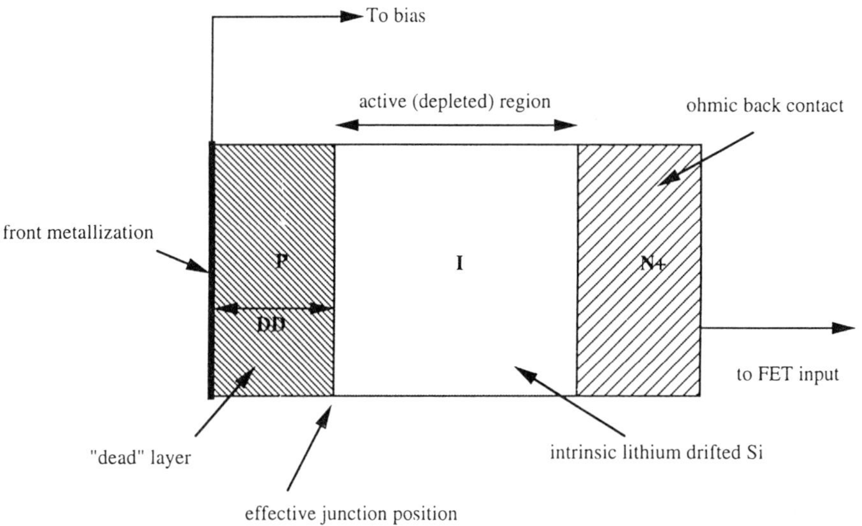


FIGURE 5.1. Schematic diagram of a typical *p-i-n* lithium drifted silicon detector.

$$\Delta V = \frac{n_{\text{e-h}} \cdot e}{C} = E_{\text{x-ray}} \cdot \frac{e}{e_{\text{eh}} \cdot C} \qquad (5.2)$$

(where $C$ is the capacitance of the diode and $e$ is the electronic charge) is sensed by the input field effect transistor of the preamplifier. The magnitude of this pulse is then seen to be directly proportional to the energy of the incoming photon.

The third region is that between the surface of the diode and the junction. Because of the *p*-doping used to form the junction, the resistivity of this region is low and so the space charge depletion region does not extend through to the surface. This region is popularly referred to as the "dead layer," but its action is far from passive and is, in fact, the key to the problem of incomplete charge. The connotation of "dead" as implying a lack of activity is incorrect, because in the region immediately beneath the surface, there are actually two competing effects: firstly, the diffusion of electrons toward the surface where they are lost, and secondly, the drift of electrons in the bias field towards the junction where they will be collected. As shown independently by Caywood et al.[9] and Goulding[10], the consequence of this is that there is a region over which the charge collection efficiency is effectively zero. If $\mu$ is the carrier mobility, $T$ is temperature (K), $V_s$ the saturation drift velocity of the electrons in the drift field, and $k$ is Boltzmann's constant, then the average diffusion distance $s_{\text{diff}}$ traveled by a carrier in some time $\tau$ is

$$s_{\text{diff}} = \left( \frac{kT}{e} \cdot \mu\tau \right)^{1/2} \qquad (5.3)$$

while the corresponding distance travelled $s_{\mathrm{drift}}$ under the influence of the drift field is

$$s_{\mathrm{drift}} = V_s \tau \tag{5.4}$$

Assuming the electrons will not be collected if the diffusion distance exceeds the drift distance, we can then find this "dead layer" thickness $d$ by equating it to the two values

$$d = V_s \tau = \left( \frac{kT}{e} \cdot \mu\tau \right)^{\!\!1/2} \tag{5.5}$$

hence eliminating $\tau$:

$$d = \frac{kT}{e} \cdot \frac{\mu}{V_s} \tag{5.6}$$

Inserting typical values for silicon at 77 K (QRM[11]) of $\mu \approx 4 \times 10^4$ cm$^2$/V and $V_s \approx 10^7$ cm/s gives the thickness of the dead layer as about 0.25 µm, a value close to that usually accepted for commercial Si(Li) detectors. At room temperature, $\mu \approx 10^3$ cm$^2$/V, so the effective dead layer will be only 200 Å or so in thickness, again close to values observed experimentally (Inskeep *et al.*[12]). It is clear that this dead layer is not, as is sometimes suggested, the result of some defect in processing, or of a surface film, or any other materials problem, but is a necessary result of the physics of charge transport within a semiconductor.

A more quantitative analysis of the effect of these competing processes than the order of magnitude estimate derived above can be made by examining the transport of the charge carriers. In the simplest approximation, if the minority carrier diffusion length is $DL$, then the probability $\eta$ of a carrier generated at a depth $z$ beneath the surface reaching the junction at depth $DD$ and so being collected is

$$\eta = \exp\left( -\frac{(DD - z)}{DL} \right) \tag{5.7}$$

However, the competition between drift and diffusion, discussed above, modifies this result. If the diffusion to the surface is characterized by a normalized recombination velocity $s$ (where $s$ is itself a function of the mobility $\mu$ and the temperature $T$), then the effect of surface recombination is to produce an effective minority carrier diffusion length $DL_{\mathrm{eff}}$ where (Jastrzebski *et al.*[13])

$$DL_{\mathrm{eff}} = DL \cdot \left[ \left( 1 - \frac{s}{(s+1)} \right) \cdot \exp\left( -\frac{z}{DL} \right) \right]^{\!\!1/2} \tag{5.8}$$

For a perfect, non-recombining surface, $s = 0$ and $DL_{eff} = DL$ at all depths, but for any other surface, $s > 0$ and thus $DL_{eff} < DL$ and varies with depth. The fractional charge collection efficiency is then

$$\eta = \exp\left(-\frac{(DD - z)}{DL_{eff}}\right) \tag{5.9}$$

As $s$ and $DL$ are varied, the collection efficiency predicted by (5.8) and (5.9) changes greatly. Figure 5.2 plots the collection efficiency $\eta$ as a function of $(z/DD)$ for some typical values of $s$ and $DL$ and for a junction depth $DD = 0.3$ μm. At one extreme, for a good surface (small $s$) and quality material ($DL \gg DD$), the collection is everywhere close to 100% and the effective dead layer thickness is negligible; but for a lossy surface and poor material only charge generated in the immediate vicinity of the junction is collected and the dead layer thickness is of the same order of magnitude as the junction depth. It will be demonstrated below that it is the parameters s and $DL$ that determine the peak profiles and distortions that result from incomplete charge effects.

## 5.3. THE MONTE CARLO SIMULATION

A simple analytical model for the quantum efficiency of a $p$-$i$-$n$ detector has been given by Lee and Sze.[14] However, a more detailed analysis permitting a visualization of the effects of incomplete charge on peak shapes requires the application of a Monte Carlo method (Joy[15]; Gardner et al.[16]; Geretschläger[17]; He et al.[18]; Wang and

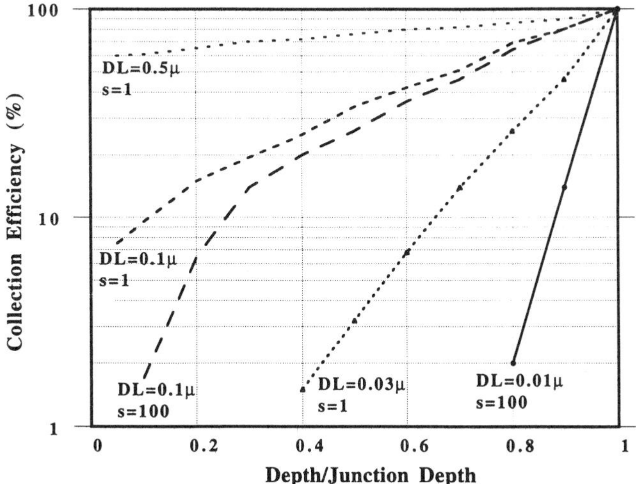


FIGURE 5.2. The charge collection efficiency $\eta$ as a function of the depth of the carrier normalized by the junction depth $DD$ for typical values of surface recombination velocity $s$ and diffusion length $DL$, for a junction depth $DD = 0.3$ μm.

Campbell[19]). Three steps are involved: firstly, the depth distibution of the photoelectrons introduced by the incident x ray photon must be determined; secondly, the subsequent motion of the photoelectrons must be followed; and finally, the charge collected by the diode as the result of these operations must be calculated.

An x-ray photon gives up all of its energy to a photoelectron in a single event. The depth distribution of such events can readily be found since the absorption of x rays is described by Beer's law in the form

$$I(z) = I(0) \cdot \exp\left(-\frac{z}{\lambda}\right) \tag{5.10}$$

where $\lambda$ is a mean free path given (in cm) as $\lambda = 1/(\rho \cdot MA)$, where $MA$ is the mass absorption coefficient ($cm^2/gm$) for the photon energy of interest, and $\rho$ is the density ($gm/cm^3$). Figure 5.3 plots $\lambda$ as a function of photon energy for silicon, showing that $\lambda$ varies from less than 0.25 $\mu m$ at the carbon $K$-line (280 eV) to as much as 14 $\mu m$ just before the onset of the silicon absorption edge at 1.8 keV. In the Monte Carlo simulation, the photon deposits its energy at a depth $z$ below the entrance surface given (Joy[20]) as

$$z = -\lambda \cdot \log (\text{RND}) \tag{5.11}$$

where RND is an equi-distributed random number lying between 0 and 1.

The photoelectron is allowed to leave the generation point in an arbitrary direction with an energy equal to that of the incoming photon and its motion is followed with a standard plural scattering Monte Carlo model (Joy[20]). The range of the electron is calculated using a corrected low energy form of the Bethe stopping power equation (Joy and Luo[21]). Over the energy region of interest (0.2–3 keV), the electron range in silicon varies from about 50 Å to nearly 2000 Å (0.2 $\mu m$) and so is always smaller

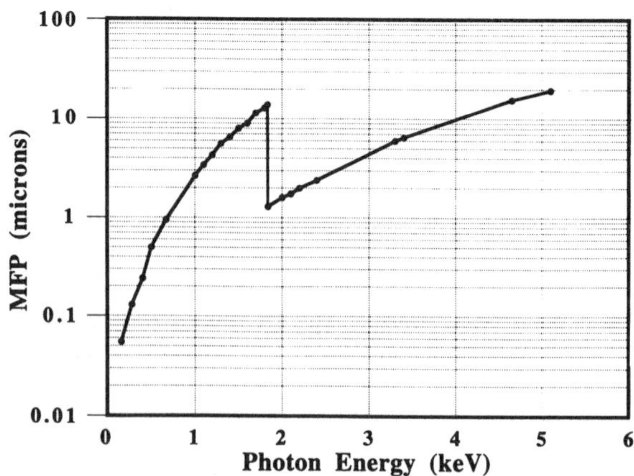

FIGURE 5.3. The variation of the mean free path $\lambda$ for x ray photons in silicon as a function of energy.

than the x ray mean free path. However, although the range is small it cannot be neglected, because as shown above, the collection efficiency of the diode can vary by an order of magnitude when the electron position varies by just a few hundred angstroms. It has also been suggested (Llacer *et al.*[22]; Shima *et al.*[8]) that the emission of photoelectrons from the front face of the detector might be responsible for some of the observed manifestations of the incomplete charge effect.

The Monte Carlo program divides the range into 50 steps of equal length. The energy of the electron at the start of the $k$-th step, $E[k]$, is found using the modified Bethe equation. The energy deposited along the $k$-th step of the trajectory is then ($E[k]$ $- E[k + 1]$) and $n_{e-h}$, the number of electron-hole pairs created as given by Akamatsu *et al.*[23] is

$$n_{e-h} = \frac{E[k] - E[k + 1]}{e_{eh}} \tag{5.12}$$

The fraction $\eta$ of the carriers that are captured and so contribute to the detected charge packet is found by assuming that the charge is deposited at the mid-point of the trajectory step and then applying (5.8) and (5.9) for the appropriate values of $s$ and $DL$. No attempt is made to account for any effect of the drift field on the trajectory of the electron because the field is so small ($< 10^5$ V/cm) that distortions are negligible. The drift velocity of the electrons (and holes) is about 10 m/s, so the time required for complete charge collection is only a few hundred nanoseconds, and it is the summed charge evaluated, as discussed above, that appears across the capacitance of the diode and is measured by the analog to digital converter.

The simulation program, written in Turbo Pascal to run on an MS-DOS computer, typically calculates from 3000 to 10,000 trajectories for each set of experimental conditions. The computation time for such a run, on a typical 386 machine equipped with a math coprocessor, is about three to five minutes. The energy of each output pulse from the diode at the end of a trajectory is recorded and can be displayed on a multichannel analyzer display on the computer. After the simulation is completed, the spectrum is convoluted with a Gaussian function equal in width to the energy resolution of a typical Si(Li) detector, about 140 eV, to facilitate comparison with experimental data. The program also tabulates the magnitude of the incomplete charge, defined as the fraction of collected energy lying more than 15 eV away (in the unconvoluted spectrum) from the channel, expected from the incident photon energy. This definition is more stringent than that normally used but has the merit of being easy to apply. In addition, other quantities such as the calibration error (i.e., the percentage shift between the computed and expected peak positions) and the emitted photoelectron yield are also calculated and displayed.

## 5.4. RESULTS AND DISCUSSION

The first result that can be obtained from this model is the magnitude of the incomplete charge effect itself. If the photoelectron range were zero, then the charge produced by any photon traveling beyond the junction at depth $DD$ would be collected with 100% efficiency, while the charge generated by any photon not reaching the junction would contribute to the incomplete charge tail. In this approximation (Shima *et al.*[8]), the fraction of incomplete charge is simply $[1 - \exp(-DD/\lambda)]$. Figure 5.4 plots some of the experimental data for incomplete charge obtained, for a junction depth of 0.25 μm, from the Monte Carlo model for a variety of values of $DL$ and $s$, together with the analytical result of Shima *et al.*[8] shown as a dotted line. The computed values generally lie close to, but below, the analytical line, although in some cases the discrepancy is as high as an order of magnitude. While the magnitude of incomplete charge is typically below 10% for x-ray lines above the silicon $K$-edge, the magnitude rises rapidly at low energies until at around carbon it is in excess of 60%. The consequences of this will be discussed below.

It is also interesting to note that in the extreme case when the surface recombination is very high ($s > 100$) and the diffusion length $DL$ is small compared to the junction depth, then as shown in Figure 5.2, the "dead layer" extends right up to the junction. In this case photons that do not reach the junction do not contribute any incomplete charge either. Thus a truly dead layer would not cause an incomplete charge effect, but instead, the efficiency of the detector would be lowered by an amount equal to the fraction of photons failing to reach the junction. Incomplete charge effects are thus associated with the presence of dead layers that only extend some of the distance from the surface to the junction.

More information can be obtained by looking at the actual form of the computed x-ray spectrum. Figure 5.5 shows the output data from the program for excitation by

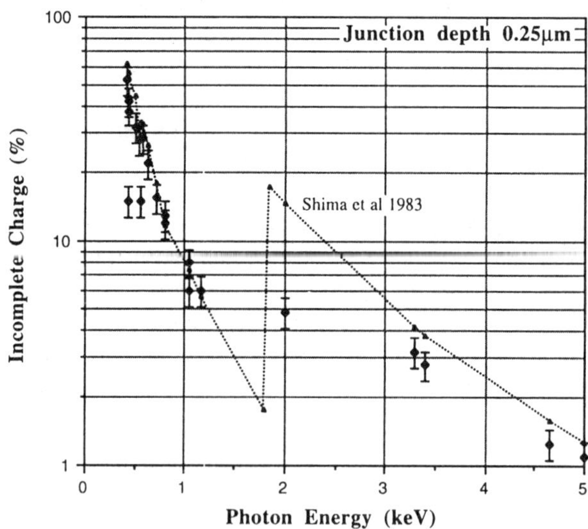


FIGURE 5.4. The computed variation of incomplete charge, for various values of $s$ and $DL$, for a junction depth of 0.25 μm and as a function of photon energy. The analytical result of Shima *et al.* (1983)[8] is also shown for comparison.

Ca $K$-photons and for both a good diode ($s = 0.3$) and a poor diode ($s = 100$). The very wide range of intensities in the spectrum make it necessary to use a logarithmic presentation of the data and also demonstrates why many trajectories must be simulated in order to obtain adequate statistics. Two types of effect can be distinguished on the spectum. Figure 5.6 shows how the value of the minority carrier diffusion length changes the overall shape of the spectrum for a typical set of data for the calcium $K$-line, an arbitrary junction depth of 1 μm, and a surface recombination velocity of $s$ = 1. When the diffusion length is small, the incomplete charge is concentrated at the bottom end of the spectrum, producing a low energy peak similar in appearance to the zero-peak observed with ultra-thin window detectors and usually atttributed to thermal noise in the preamplifier. Between this peak and the actual calcium line the spectrum is flat and low in intensity. In an experimental situation using electron beam excitation this intermediate region of incomplete charge would be submerged in the bremsstrahlung, while the low energy peak would probably be removed by the discriminator. The real extent of incomplete charge is only visible when x-ray excitation is used (e.g., Craven *et al.*[24,25]). As the diffusion length is increased, however, the intensity in the low energy peak is redistributed over the spectrum, gradually taking up the familiar form of a tail, or step, on the low energy side of the characteristic peak.

Figure 5.7 shows how the shape of a peak is affected by the parameters of the semiconductor. Data is shown for the $K$-lines of sodium, oxygen, and carbon, in each case with $DL = 0.5$ μm and with $s = 0.1$ or 100. For the highest energy line, sodium, the effect of incomplete charge is clearly visible as a step on the low energy tail. As the rate of surface recombination becomes higher, this tail increases in magnitude as a fraction of the peak intensity, but the incomplete charge is readily distinguished from the peak. A similar effect is noted for the oxygen line with the incomplete charge becoming a very significant fraction of the total peak intensity in the $s = 100$ case. This

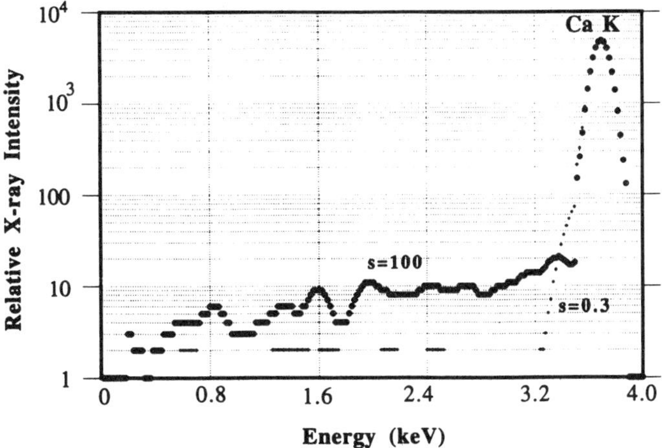


FIGURE 5.5. Typical simulated spectra from the Monte Carlo program for excitation by the calcium $K$-line. Two cases are shown; a good surface ($s = 0.3$) and a poor surface ($s = 100$).

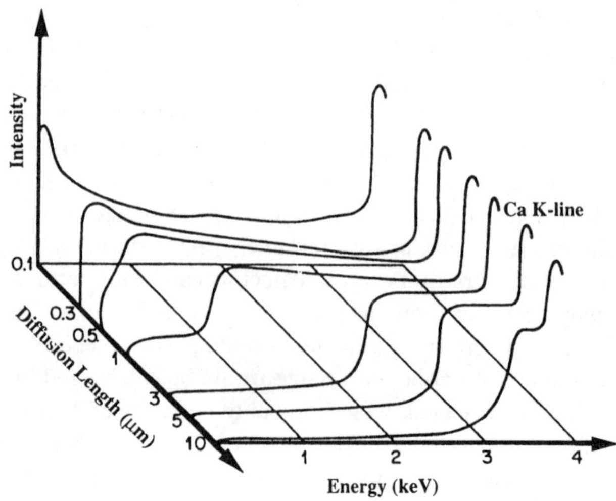



FIGURE 5.6. The variation of general spectrum shape as a function of minority carrier diffusion length. Here the excitation is the calcium $K$-line, the junction depth is 1 μm, and $s = 1$.

is to be expected because the photon mean free path is now much shorter and the incomplete charge is higher. At the lowest energy, that for the carbon $K$-line, the behavior is different. The incomplete charge now merges into the characteristic peak and, for many sets of conditions such as the $DL = 0.5$ μm, $s = 0.1$ case illustrated, the composite peak becomes almost Gaussian in shape again. It can also be seen from the dotted line, which indicates the expected peak position, that the peak position falls below the expected energy value. This anomalous behavior, which has been confirmed experimentally (Statham[26]; Muskett[27]), results from the fact that the photon mean free path (0.13 μm) at this low energy is now smaller than the junction depth, and hence a majority of the photons do not reach the depletion region. The characteristic peak thus disappears leaving in its place a peak that is all incomplete charge, although the fact that the peak appears to be Gaussian in shape disguises its origin. Under these conditions the sensitivity of the detector will be significantly different from the value predicted from the usual models, and care must therefore be taken to employ a suitable standard if quantitative data is required.

A shift in calibration, such as that noted above, is an indication that the peak is mostly or completely incomplete charge. Experimentally, the magnitude of the calibration error is found to be sensitively dependent on the applied bias, because this changes the position of the onset of the depletion layer relative to the surface, and hence the apparent value of the junction depth $DD$. Careful optimization of the bias can usually keep the peaks between boron and nitrogen to within one channel or so of the exact calibration, but some experimentation is essential to get the best performance from a detector in this energy range.

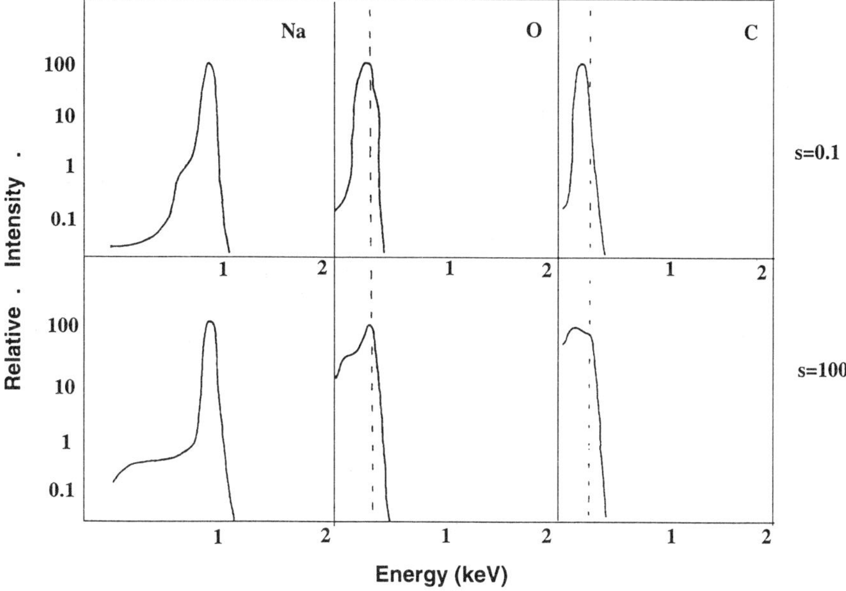

FIGURE 5.7. The computed variation in peak shape and position as a function of $s$ for the Na, O, and C lines for the case of a junction at 0.25 µm and a diffusion length of 0.5 µm.

## 5.5. DISCUSSION AND CONCLUSIONS

This model, although obviously oversimplified, accurately predicts both the major features and the minor details of the spectrum from a Si(Li) detector. In particular it demonstrates that it is the interaction of the three main parameters, the diffusion length, the effective junction depth, and the surface recombination velocity, that determines the behavior of a detector in the low energy region of the spectrum. Because of the sensitivity of the results to the surface recombination velocity, and thus to the condition of the surface itself, it is clear that control of the diode surface should be the key priority during manufacture of these devices.

In order to apply the model quantitatively, it is necessary to find appropriate values for these parameters, which can be done by iteratively comparing model peaks with experimental data, preferably obtained by x-ray fluorescence. Once a set of values has been derived, then the subsequent agreement between experimental and computed spectra is sufficiently good that the model can be used for accurate background removal or corrections to measured data (e.g., Wang and Campbell[19]). Values derived from a number of commercial Si(Li) detectors indicate that the typical junction depth is between 0.2 and 0.25 µm, with $DL$ of the order of 1 µm. The surface recombination velocity is usually in the range $s = 1$–10, although on detectors exhibiting pathological defects, values as high as 100 have been found. This typically arises from the charging, under stray electron impact, of a layer of oxide beneath the gold metallization on the front face of the detector.

The design of Si(Li) detectors for x-ray spectroscopy has evolved slowly and empirically from that of nuclear particle detectors. Although this has resulted in devices with excellent performance in the middle energy range (3 keV and upward), it would seem from the analysis presented here that attempting to cover the full energy range from 0.2 to 30 keV with just one detector requires degrading the performance at one end of the spectrum or the other. For example, the requirements for an efficient detector of 10-keV photons mostly involve obtaining a high depletion depth while minimizing capacitance. The depth of the junction is not critical, and "killing" the front face of the detector by ion implantation will eliminate residual incomplete charge effects with only a minimal effect on sensitivity. On the other hand, a detector designed for low energy use (< 3 keV) needs only a shallow depletion region but requires a perfect (non-recombinant) surface. Since the requirements for these two energy ranges are incompatible with each other, it would seem sensible to consider using two separate detectors, each optimized for the appropriate energy region.

For example, a specialized low energy x-ray detector could be made using a thin (30–100 Å thick) surface silicide Schottky layer. This places the junction at the surface and eliminates the effect of surface recombination. In addition, if built on high resistivity silicon (e.g., 100–1000 $\Omega$-cm) it would be self-depleting to a depth of several micrometers, which is more than sufficient to ensure efficient photon collection in the required energy range. Such a device would be simpler to construct and use than current detectors and should offer enhanced performance because of its lower capacitance and optimized device properties.

ACKNOWLEDGMENTS. The author is grateful to Drs. D. E. Newbury (NIST), S. Davilla (4pi Analysis Inc.), and J. McCarthy (NORAN) for valuable discussions and clarifications. Oak Ridge National Laboratory is managed by Martin Marietta Energy Systems, Inc., under contract DE-AC05-84OR21400 with the U.S. Department of Energy.

## REFERENCES


1. M. M. Miller and R. G. Wilkinson, *Phys. Rev.* **82**, 981 (1951).
2. E. M. Pell, *J. Appl. Phys.* **31**, 291 (1960).
3. E. Elad and M. Nakamura, *Nucl. Instrum. Methods* **41**,161 (1966).
4. R. Fitzgerald, K. Keil, and K. F. J. Heinrich, *Science* **159**, 528 (1968).
5. J. H. Hubbell, *Rev. Sci. Instrum.* **29**, 65 (1958).
6. L. G. Wolfgang, J. M. Abraham, and C. N. Inskeep, *IEEE Trans. Nucl. Sci.* **NS-13**, 1, 30 (1966).
7. J. Muller, *IEEE Trans. Electron Devices,* **ED-25**, 247 (1978).
8. K. Shima, S. Nagai, T. Mikuma, and S. Yasumi, *Nucl. Instrum, Methods* **217**, 515 (1983).
9. J. M. Caywood, C. A. Mead, and J. W. Mayer, *Nucl. Instrum. Methods* **79**, 329 (1970).
10. F. S. Goulding, *Nucl. Inst. Methods* **142**, 213 (1977).
11. Quality Reference Manual, available from: AT&T Bell Laboratories, Murray Hill, New Jersey 07974, p. 125 (1986).

12. C. Inskeep, E. Elad, and R. A. Sareen, *IEEE Trans. Nucl. Sci.* **NS-21**, 379 (1974).

13. L. Jastrzebski, J. Lagowski, and H. C. Gatos, *Appl. Phys. Lett.* **27**, 537 (1975).

14. H.-S. Lee and S. M. Sze, *IEEE Trans. Electron Devices* **ED-17**, 342 (1970).

15. D. C. Joy, *Rev. Sci. Instrum.* **56**, 1772 (1985).

16. R. P. Gardner, A. M. Yacout, J. Zhang, and K. Verghese, *Nucl. Instrum. Methods* **A242**, 399 (1986).

17. M. Geretschläger, *Nucl. Instrum. Methods,* **B28**, 289 (1987).

18. H.-J. He, T.-W. Zhang, R.-C. Shang, and S.-D. Wu, *Nucl. Instrum. Methods* **A272**, 847 (1988).

19. J. X. Wang and J. L. Campbell, *Nucl. Instrum. Methods* **B54**, 499 (1991).

20. D. C. Joy, *Scanning Microsc.* **5**, 329 (1991).

21. D. C. Joy and S. Luo, *Scanning* **11**, 176 (1989).

22. J. Llacer, E. E. Haller, and R. C. Cordi, *IEEE Trans. Nucl. Sci.* **NS-24**, 53 (1977).

23. B. Akamatsu, J. Henoc, and H. Henoc, *J. Appl. Phys.* **52**, 7245 (1981).

24. A. J. Craven, P. F. Adam, and R. Howe, *Inst. Phys. Conf. Ser.* **78**, 189 (1985).

25. A. J. Craven, C. P. McHardy, and W. A. P. Nicholson, *Inst. Phys. Conf. Ser.* **90**, 345 (1987).

26. P. J. Statham, NBS Special Publication 604 (K. F. J. Heinrich, D. E. Newbury, and R. Myklebust, eds.), National Bureau of Standards, Washington, DC, pp. 127–136 (1981).

27. R. G. Muskett, (1981), NBS Special Publication 604 (K. F. J. Heinrich, D. E. Newbury, and R. Myklebust, eds. ), National Bureau of Standards, Washington, DC, pp. 97–108 (1981).

# 6

# The Effect of Detector Dead Layers on Light Element Detection

*J. J. McCarthy*

## 6.1. INTRODUCTION

Detection of x rays below 1 keV is influenced by many factors. These factors include the electronic noise of the system, the material properties of the semiconductor detecting element, absorption by the vacuum window, the high voltage contact metal, and the dead layer associated with the detector front surface.

Progress has been reported in reducing the effect of several of these factors. Reduction of the electronic noise in the energy dispersive spectrometer (EDS) system[1] and the use of new materials for vacuum windows[2] have improved light element detection. The substitution of alternative semiconductor materials, specifically high purity germanium (HPGe), in place of lithium drifted silicon has had some success.[3,4] On the other hand, the subject of the "dead layer" has not been explored in much detail in the recent literature.

As the use of EDS for the detection and quantification of light elements became increasingly commonplace in the mid 1980s, a number of artifacts were reported. The most common artifact observed was that spectral peaks generated by low energy x rays (less than 2 keV) were not Gaussian in shape, but possessed a "tail" on the low energy side of the peak. The size and shape of this tail varied dramatically with energy. Often the peak position was shifted to lower energies than expected, and the amount of the shift increased with lower energy. These effects have been noted by several authors,[5–7] and are attributed to "incomplete charge collection" (ICC) in the dead layer or in a larger region which overlaps the dead layer. Minimizing these artifacts in the spectra

---

J. J. McCarthy • Noran Instruments, Inc., Middleton, Wisconsin 53562

*X-Ray Spectrometry in Electron Beam Instruments*, edited by David Williams, Joseph Goldstein, and Dale Newbury. Plenum Press, New York, 1995.

of light elements requires focusing on understanding and reducing incomplete charge collection in the dead layer.

The impact of these artifacts is to complicate both qualitative and quantitative analysis. The presence of the significant peak tails lowers the observed peak to background ratio and complicates peak overlaps, while the energy shifts make peak identification by KLM markers ambiguous at best. Peak fitting routines (that do not rely on measured peak profiles) must be modified to model the change of the tails as a function of energy and to account for any peak shifts. These requirements make many peak fitting routines impossible to calibrate properly. The only accurate peak fitting approach for light elements may be fitting with collected spectral reference peaks (used in filter fit approaches) or using nonlinear fitting techniques (such as the simplex fit).

Because of the increasing importance of light element microanalysis, it is clearly desirable to minimize the impact of these artifacts. This can be done by improving the charge collection properties of the dead layer region of the detector. To accomplish this, we need to improve our understanding of what causes ICC and develop methods to reduce the ICC in the detector dead layer.

This paper describes ongoing progress toward reducing dead layer effects acomplished in our laboratory and manufacturing process since the early 1990s. We will illustrate the improvement these efforts have made on the spectroscopy of x rays below about 2 keV by comparing the performance of newer detectors to older detectors manufactured in the mid 1980s. We have also used a model of the EDS detector first reported by Joy[8] to simulate the response of the detector to x rays. The comparison of simulation results to experimental data has contributed to our understanding of the physical processes involved in incomplete charge collection and suggests that improved models will lead to further advances.

We have found that the reduction of incomplete charge collection has led to dramatic improvements in light element spectral quality. Further improvements will require a deeper understanding of the sources of incomplete charge collection near the detector surface.

## 6.2. CHARACTERIZATION OF DEAD LAYER EFFECTS

### 6.2.1. Peak Asymmetry and Peak Shifts

When detector vacuum windows with high transmission for softer x-ray lines became available, the use of the EDS detector to identify and even quantify elements between Be and Na quickly became commonplace. Examination of the spectrum of light elements collected by a detector manufactured in the mid-1980s, however, revealed several noticeable spectral artifacts not normally observed for peaks above 2 keV. While these artifacts could vary from detector to detector and manufacturer to manufacturer, they were observed consistently. These effects can be seen in detectors manufactured today, although to a much lesser degree, and may not be obvious in a casual inspection. For the purpose of this discussion, we shall assume that all detectors possess these properties and proceed to describe how to minimize the effects.

The first artifact is the deviation of the peak from a Gaussian shape. Light element peaks below Na have a noticeable low energy "tail," whose size and shape can vary dramatically with photon energy. The second artifact observed is what appears to be a non-linearity in the energy calibration of the detector. The location of the peak centroid may not agree with that predicted by tables of known x-ray energies. Instead, the peaks appear to be consistently shifted to lower energies. This peak shift generally increases with decreasing x-ray energy.

Both of these artifacts suggest that the detector is no longer collecting all of the charge generated by the incoming photons. The effect is known as incomplete charge collection and has been discussed by many authors. The magnitude of these effects is larger for the light elements and the most severe distortions are observed below 1 keV. These low energy photons have a mean free path in the detector that is of the order of the measured dead layer thickness. This suggests that considerable incomplete charge collection occurs near the front surface and within the dead layer of the detector.

The appearance and the magnitude of these artifacts for carbon and oxygen are illustrated in Figure 6.1. Both peaks are quite asymmetric and shifted toward lower energy, as illustrated by the peak labels (at the true energy) and the marker bar and cursor positions. The spectra were collected with the same 30 mm$^2$ Si(Li) detector at the same operating conditions in September 1986. Figure 6.1(a) shows the carbon peak, which exhibits a pronounced shift of 40 eV (10 eV per channel) toward lower energy and the presence of a lower energy tail. In Figure 6.1(b) the peak shift for oxygen is about 20 eV, while the peak tail appears to be even worse than for carbon. The oxygen tail is probably exaggerated by the presence of a small amount of carbon contamination.

### 6.2.2. Detector Dispersion Measurements

Characterizing the effects of incomplete charge collection by measuring data element by element is useful but rather tedious. We would also like to obtain informa-

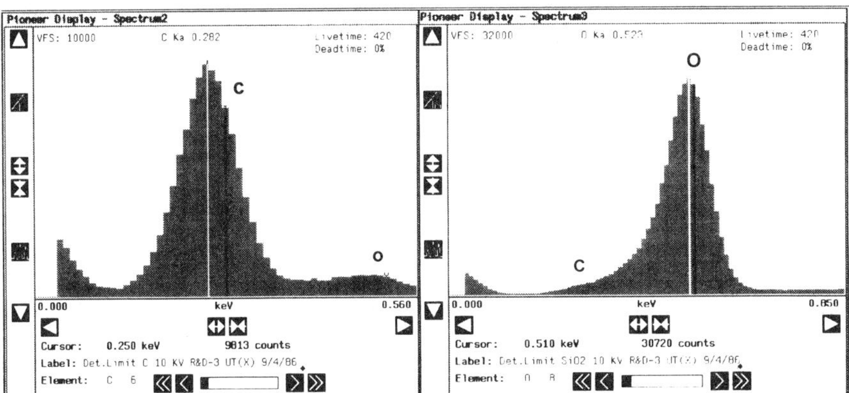


FIGURE 6.1. Example of peak tailing and peak shifts in oxygen and carbon. Spectra were taken from the same 30 mm$^2$ Si(Li) detector. The horizontal expansion and energy per channel are the same in both plots. The dark vertical bar is a KLM peak marker indicating expected peak location. The lighter vertical bar is the cursor placed at the peak channel.

tion on the energy dependence of ICC and visualize that dependence. This goal can be achieved by studying the quantity known as the detector dispersion, $D$.

In the simplest theory, an x-ray photon which enters a detector will produce a Gaussian peak in the energy spectrum with resolution (defined as the FWHM) given by[9]

$$FWHM^2 = N_e^2 + D^2 \qquad (6.1)$$

where $N_e$ is the contribution from electronic noise and $D$, the "dispersion" of the detector, is given by

$$D = \sqrt{2.355^2 \cdot F \cdot E \cdot \varepsilon} \qquad (6.2)$$

$F$ is the Fano factor for the detector material, $\varepsilon$ is the mean number of charge carriers (electron-hole pairs) produced per unit energy, and $E$ is the energy in eV of the x-ray photon. For silicon, the Fano factor is about 0.11, and $\varepsilon$ is 3.78 eV at 77 °K. In practice, we find that the resolution of spectral peaks is not adequately described by this simple equation. The reason for the difference in measured resolution is incomplete charge collection caused by trapping of charge carriers in the bulk material of the detector or by effects at the surfaces of the detector. These effects cause a deviation between measured dispersions and that predicted by (6.1). The advantage of using the dispersion curve for comparison between detectors is that the effects of electronic noise (including the FET and preamplifier, etc.) are removed, leaving only the properties of the Si diode to be studied. The dispersion curve also provides us a method to examine the effects of incomplete charge collection over a wide range of photon energies. This gives us a way to directly compare detectors manufactured in different ways or at different times.

In order to illustrate the deviations from the ideal detector represented by (6.2), we measured the dispersion of various Si(Li) detectors. A typical result is shown in Figure 6.2, where the measured dispersion of an older Si(Li) detector is compared to the prediction of (6.2). This curve was obtained by acquiring the spectrum from a series of pure element standards in an SEM. For each element, the FWHM of the peak of interest was measured from the spectrum with the same program used to determine resolution specifications, which is based on an IEEE standard procedure. The width of the electronic noise peak is automatically computed during the acquisition by a routine in the pulse processor used in our microanalysis system. The dispersion is then calculated from (6.1). The theoretical dispersion is obtained from (6.2) by using the peak energy and a value for the Fano factor which gives the same dispersion value at Mn as the measured value. Significant deviations occur near the Si $K$ absorption edge and at energies below 1 keV, where a prominent "hook" appears in the measured values. The area near the Si $K$-edge is interesting since it implies that one should observe some peak tailing and energy shift for elements just above Si, for example for P or S. The large deviations in the "hook" in the light element region clearly show that, for this older detector, the dead layer is a major source of ICC, which in turn causes spectral artifacts.

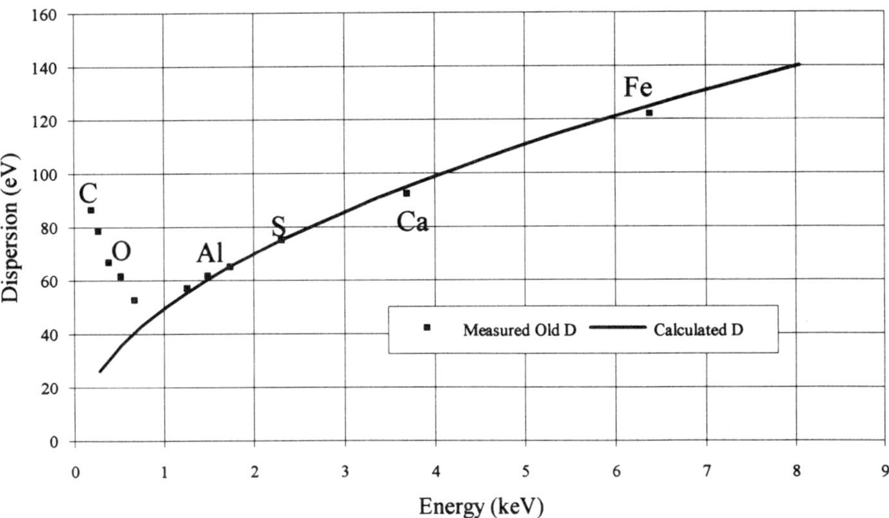


FIGURE 6.2. Measured dispersion of an older Si(Li) detector compared to the prediction of (6.2). Note the large deviations of the measured dispersion from predicted dispersion for C and O.

## 6.3. MINIMIZING THE EFFECT OF THE DEAD LAYER

In our work, we were interested in improving detector performance for light elements by reducing the ICC caused by the dead layer. Since dead layers have been measured in the 0.1–0.2 µm range, efforts were directed at the actions that change the properties of the first few micrometers of the detector. A set of designed experiments were conducted in which the process variables that affect the front surface of the detector were modified. The impact of each experiment was judged by monitoring the peak shifts, peak to background ratios, and dispersion curves for each set of the experimental devices. It is clear from the data presented in Figure 6.2 that any reduction in these values indicates a reduction in the ICC due to the dead layer. The process changes, the measurements performed, and the comparison of results from newer detectors to older detectors are described in the following paragraphs.

### 6.3.1. Designed Experiment

A number of process variables related to detector fabrication were considered to be likely candidates to influence the properties of the dead layer. These processes and the properties impacted are briefly listed in Table 6.1.

This set of experiments was planned to modify the detector fabrication processes and determine what changes would minimize the observed dead layer effects. These experiments identified key variables that affect the dead layer and its charge collection properties. The two most important areas identified in the experiments were improvements in the surface treatments and metal deposition. In order to easily compare the performance of detectors fabricated by different process steps, we will refer to the earlier detectors with poorer performance as "old" detectors and the detectors with improved performance as "new" detectors.

### 6.3.2. Experimental Results

With the results of the designed experiments incorporated into the detector fabrication, the improvements in light element "spectral quality" are quite dramatic. For example, at oxygen the peak shift is reduced by a nearly a factor of 2, the peak to background is improved by a factor of 2, and the dispersion is decreased by nearly 20 eV.

This improvement is illustrated in Figure 6.3, which shows an oxygen spectrum collected on a new Si(Li) detector. In this spectrum the oxygen peak appears within one channel of the correct energy, as indicated by the peak label and the KLM marker. The peak is also very Gaussian in comparison to the example shown in Figure 6.1. The spectrum returns to the baseline background level between the oxygen and carbon peaks. This spectrum is a dramatic improvement our those shown in Figure 6.1.

Another convincing example of the improvement is shown in Figure 6.4, which compares the boron spectrum from an old detector and from a new one. In the spectrum from the new detector, the boron peak is very symmetric and exhibits less than one channel (10 eV) peak shift. The peak to background ratio, as measured by the ratio of the counts in the peak channel to those at the cursor placed below the peak in this spectrum, is about 70:1. In the spectrum from the older detector, the peak is poorly defined. The peak to backgound ratio is only 2:1. Note the large difference in counts in the peak for much different collection times (20 s versus 420 s), implying much improved boron efficiency. The peak to background ratio between the boron peak channel and the lowest background point between the boron peak and the noise peak has been improved by a factor of nearly 30!

The dispersion curve was measured for several new detectors in order to explore the energy dependence of the ICC in these devices. A typical result is shown in Figure 6.5. This figure compares the dispersion curve from a Si(Li) detector made using the new processing methods with the prediction of (6.2). The dispersion data was collected in the same manner as described for the old detector. In the new data, the deviations from the prediction of (6.1) are in the same energy regions as for the old detector, but are much reduced. This suggests that, while the ICC in the dead layer has been

TABLE 6.1. Processes Examined with Designed Experiments

| Process area | Process conditions | Properties impacted |
| --- | --- | --- |
| Surface treatments | Lapping, etching, and passivation | Surface damage, surface states, metal adhesion |
| Oxide layer | Etching, device aging | Surface states and trapping, barrier height |
| Metal deposition | Methods and conditions: time, temperature, and contact thickness | Barrier height and trapping |
| Contact metals | Different metals | Barrier height and trapping |

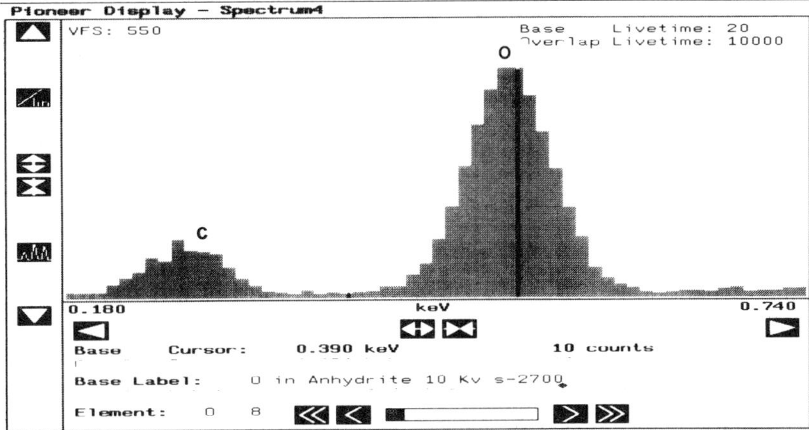

FIGURE 6.3. Spectrum of oxygen and carbon from a new Si(Li) detector. The dark vertical bar is the KLM peak marker; the lighter bar is the cursor placed between peaks. The peak to valley ratio is better than 50:1.

considerably reduced by our new fabrication process, it has not been eliminated completely. The most notable difference in the new data is the absence of the "hook" at low energies exhibited by the older Si(Li) detector dispersion curve.

Finally, in Table 6.2 the light element performance of the old and new Si(Li) detectors for selected light elements is compared. In each case the new detectors exhibit significant improvements over the old. The dispersion values were calculated from measured peak widths by subtracting the electronic noise width according to (6.1).

From these results, it is clear that changing the detector fabrication process has reduced, but not totally eliminated, the incomplete charge collection of the dead layer. This has resulted in a considerable reduction in the spectral artifacts in the spectrum

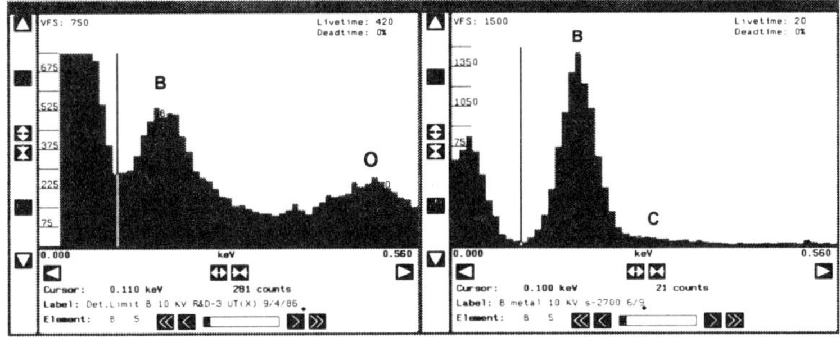

FIGURE 6.4. Comparison of light element performance from a Si(Li) detector from the new process and old process. The specimen is pure boron metal. As in previous figures, the dark vertical bar is the KLM peak marker and the lighter bar is the cursor placed in the valley between the noise peak and the boron peak.

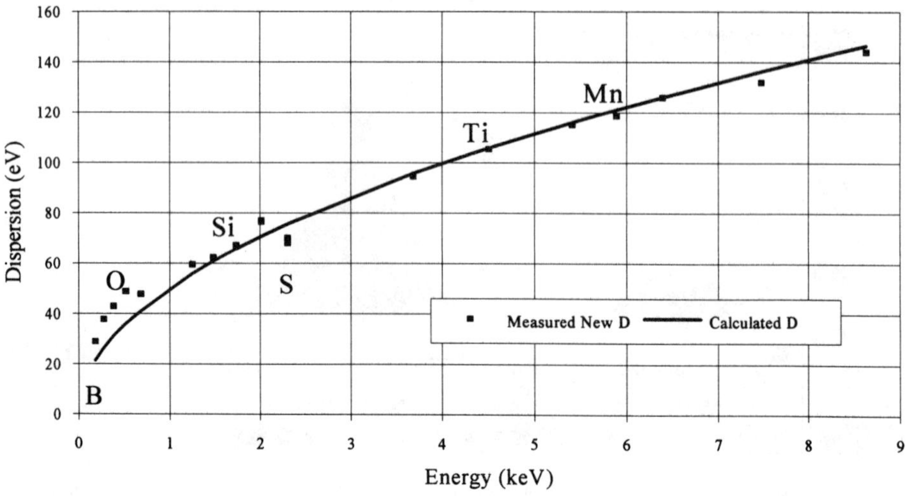

FIGURE 6.5. Comparison of measured dispersion for a Si(Li) detector fabricated using the new process compared with the prediction of (6.1). Note the absence of the "hook" at low energies and the much better agreement to the calculated values.

of the light elements. What actions might lead to further improvements? In the following section a model of the EDS detector, which includes predictions of the contribution of ICC, is compared to the experimental results we have obtained.

## 6.4. DISCUSSION OF EXPERIMENTAL RESULTS

One explanation for the reduction of the ICC effects is that the width of the dead layer has been substantially reduced by the process changes. In order to test this hypothesis, we measured the dead layer thickness on several old and new detectors.

### 6.4.1. Dead Layer Thickness

We have used two methods to measure the dead layer thickness. One common method[10] is to measure the jump in the spectrum which occurs at the Si $K$-absorption edge. The ratio of counts in the spectrum just above the edge to the counts measured

TABLE 6.2. Comparison of Light Element Performance: Old versus New Process Si(Li) Detector

| Element | Peak shift (eV) | | Peak/Background | | Dispersion (eV) | |
|---|---|---|---|---|---|---|
| | Old | New | Old | New | Old | New |
| B | −28.5 | −7.0 | 1.4:1 | 45:1 | 85.3 | 23.5 |
| C | −37.7 | −11.0 | 22:1 | 50:1 | 79.2 | 33.8 |
| O | −17.3 | −11.6 | 21:1 | 50:1 | 63.2 | 45.7 |
| Mn | −0.0 | −0.0 | 1000:1 | 3000:1 | 119.0 | 117.3 |

just below the edge (the "jump ratio") is assumed to be directly proportional to the product of the difference in the mass absorption coefficients on either side of the edge and the thickness of the dead layer. We measured the jump ratio on spectra acquired from a carbon block mounted in a SEM. Spectra were acquired for 2000 seconds at 10 keV with currents that kept the dead time under 20%. Even with these long acquisition times, the variance in the results is due largely to statistical variation in the measured intensities and the fact that the result is the difference of two numbers, both with large uncertainties. The results of this first method are labeled JUMP in Table 6.3.

A second method to measure dead layer thickness employs x-ray fluorescence (XRF) spectra.[7] In this method all of the tailing below the peak in an XRF spectrum is attributed to interactions in the dead layer. It is then possible to calculate the dead layer thickness from the ratio of the integrated counts in the peak tail to the counts in the main Gaussian peak. We used a $^{55}$Fe source to excite bulk targets of sulfur and calcium. The $K\alpha$ peaks were then fitted to a Gaussian and the integrated counts in the peak obtained. The peak counts where subtracted from the spectrum and the integrated counts in the region below the peak obtained. The results of these measurements are also reported in Table 6.3, labeled as XRF.

The most interesting result of these measurements is that the values of the dead layer thickness for the old and new process detector are the same within the errors of the measurements! If this is true, a variation in thickness of the dead layer can not be responsible for the reduction in ICC seen in our light element measurements. This result was unexpected and implies that a change in some other parameter is responsible for the improved incomplete charge collection. In order to explain our results we found it necessary to use a more detailed model of the response of the detector to x rays.

### 6.4.2. A Model of ICC in the Dead Layer

In order to determine what physical properties of the detector dead layer are important in incomplete charge collection for the light elements, it is useful to have a model of the detector. The parameters of this model can then be used to simulate the detector response to low energy x rays and the result compared to experimental measurements. In our study we have adopted a model first proposed by Joy.[8] After a brief review of the Joy model, the results of comparisons of simulated and measured data will be discussed.

In the model proposed by Joy, the detector can be divided into three distinct regions. The first region is the back electrical contact, an $n^+$ region formed from the excess lithium at the surface where the drift process began. This region forms an ohmic

TABLE 6.3. Si(Li) Dead Layer Measurements

| Method | Old process (nm) | New process (nm) |
| --- | --- | --- |
| XRF-CA | 58.3 | 36.6 |
| XRF-S | 76.5 | 84.7 |
| Jump Si | 81.2 | 88.5 |
| Jump Si | — | 109.0 |
| Average values | $76.6 \pm 30.6$ | $79.7 \pm 32.0$ |

contact on the back of the diode. For the purposes of our model, this region is of little importance other than as an electrical contact.

The second region, which comprises the bulk of the detector, is formed by drifting lithium ions through the material under the influence of an electric field. This region is often several millimeters in depth and is characterized by a high resistivity approaching that of intrinsic Si material. For this reason it is often called the $i$ or intrinsic region of the detector. This region forms the "active" volume of the detector. The charge carriers generated in this region by an incident x ray will be collected to produce a charge pulse in the external circuit of the preamplifier.

The third region is between the junction (at a depth $DD$ from the surface) and the front surface of the detector. Because of the doping required to form a junction, and the incomplete nature of the lithium drift process, the resistivity of this area will be considerably lower that in the active region of the detector. Hence the penetration of the collection field is considerably reduced in this region, which reduces the charge that will be collected from this region. In the simplest approximation, minority carriers in this region are assumed to have a finite diffusion length $DL$. This assumption means that charge carriers generated a distance $z$ from the front surface have a probability of diffusing to the junction and being collected that is described by the following equation:

$$n = \exp[-(DD - z)/DL] \qquad (6.3)$$

In a real detector, the properties of the front contact surface influence the charge collection properties in this region, and (6.3) is not an accurate estimate of the fractional charge collected from this region. The presence of trapping centers and electrically active defects results in the recombination of some of the charge carriers. Since these recombined carriers cannot contribute to the total collected charge, the surface conditions can increase the amount of incomplete charge collection. This situation can be described by introducing a parameter $s$, known as the normalized surface recombination velocity, that modifies the diffusion length $DL$ to account for the presence of a "lossy" surface. This defines a new effective minority carrier diffusion length $DE$ given by

$$DE = DL\sqrt{1 - [s/(s + 1)] \exp (-z/DL)} \qquad (6.4)$$

For a perfect surface, $s$ is zero and $DE$ is a constant equal to $DL$. For a less than perfect surface, $s$ is greater than zero, and $DE$ will be less than $DL$ and is a function of $z$. This equation describes the competition between diffusion of charge carriers toward the surface and drift to the junction where charge can be collected. This competition defines a depth in which the efficiency of charge collection is effectively zero. This is the region commonly referred to as the dead layer. We can now rewrite (6.3) in the following form:

$$n = \exp(-DD - z/DE) \qquad (5)$$

which states that the fractional charge collection efficiency, $n$, depends upon the junction depth $DD$, the x-ray range (via $z$), and the properties described by $DL$ and $s$.

Based upon this model of the detector, Joy developed a Monte Carlo program that simulates the pulse height distribution of the detector as a function of the incident x-ray energy and the parameters $DD$, $DL$, and $s$. While a complete description of the Monte Carlo procedure is outside of the scope of this paper, the reader is encouraged to read the original reference to gain an understanding of the simulation method. As shown in the following paragraphs, we have found the comparison of Monte Carlo simulations to measurements on experimental detectors has helped us to better understand ICC in the dead layer.

### 6.4.3. Monte Carlo Simulations of Peak Shapes and Dispersion Curves

The Monte Carlo simulation requires values for the parameters $DD$, $DL$, and $s$ in order to simulate the response of the detector to x rays. We began by using the values of the measured dead layer thickness for $DD$ and nominal values suggested by Joy for $DL$ and $s$. The values of these parameters were refined by a manual iteration procedure defined in the next paragraph.

Best fit values for $DD$, $DL$, and $s$ were found by simulating two peaks for each detector, Ca and O, and comparing the simulated peaks to the measured peaks. These elements were chosen to include both a light element that would show considerable ICC and a heavier element far enough above the Si $K$-edge to minimize ICC. For each set of parameters the simulated peaks were compared to measured peaks. This process was repeated until the simulated peak centroid and peak tail were judged, by eye, to be the "best fit" to the experimental peaks. Using the best fit parameters for Ca, the O peak was simulated next. The parameters were then adjusted to give the best match to both peaks. Finally, the entire dispersion curve was be simulated with these "best fit" parameter values.

The comparison between measured and simulated peak shapes for Ca and O is shown in Figure 6.6 for an old process detector. Note that the Ca $K\beta$ peak was not included in the simulation. The figure shows an overlap of the measured peak (darker colors) with the simulated spectrum (in white). The peak labels are placed in the channel that corresponds to the true peak energy. In general, the simulated peak shift and shape agree well with the measured peaks.

The corresponding results for the new process detector are given in Figure 6.7. As seen in the figure, the peak shape and peak position are well fit in this case, at least as well as for the old process detector.

The dispersion simulation for an older detector is compared to the measured dispersion curve in Figure 6.8. The agreement between measured and simulated dispersion is quite good. The simulation reproduces the shape of the measured dispersion curve reasonably well. What is most encouraging is the area of the light elements, where the prominent "hook" feature is modeled nicely. The model seems to underestimate the dispersion for the lightest elements and over estimate it from about Mg upward.

Figure 6.9 compares the measured dispersion curve to the simulation for a new detector. Again the overall agreement between the data and the simulation is good. As

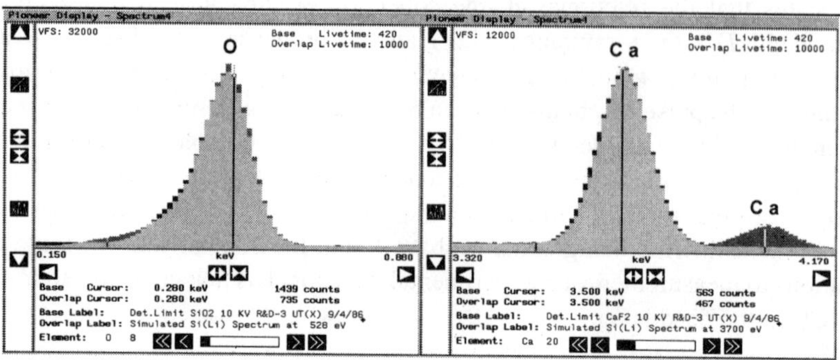

FIGURE 6.6. Comparison of simulated peaks with measured O and Ca peaks for old process detector. The simulated peak (solid light gray) is overlapped with the measured data (darker gray dots and bars). Peak labels and marker bar indicate the expected peak locations.

in the simulation for the old detector, the measured dispersion for the light elements is generally underestimated, while the dispersion for the heavier elements (in this case above S) is overestimated by the simulation. Unlike the estimates for the old detector, the shape of the simulated dispersion for the lightest elements for the new detector differs from the shape of the measured dispersion. A slight increase in the dispersion is indicated by the simulation, while the data actually continues to decline from oxygen to boron. Still, it is impressive to see that the simulation does indeed accurately predict a considerable improvement in the dispersion values for the new detectors versus the old.

Table 6.4 compares the best fit values of *DD*, *DL* and *s* for the old process and new process detectors. The surprising result is that the improvements in ICC did not come about by a reduction in the thickness of the dead layer itself. The Monte Carlo simulation results give dead layer values for the old and new detectors that are nearly

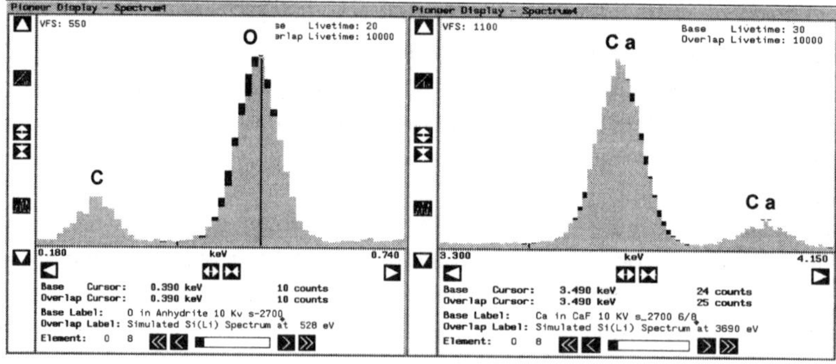

FIGURE 6.7. Comparison of simulated peaks with measured O and Ca peaks for new process detector. The simulated peak (solid light gray) is overlapped with the measured data (darker gray dots and bars). Peak labels indicate the expected peak locations.

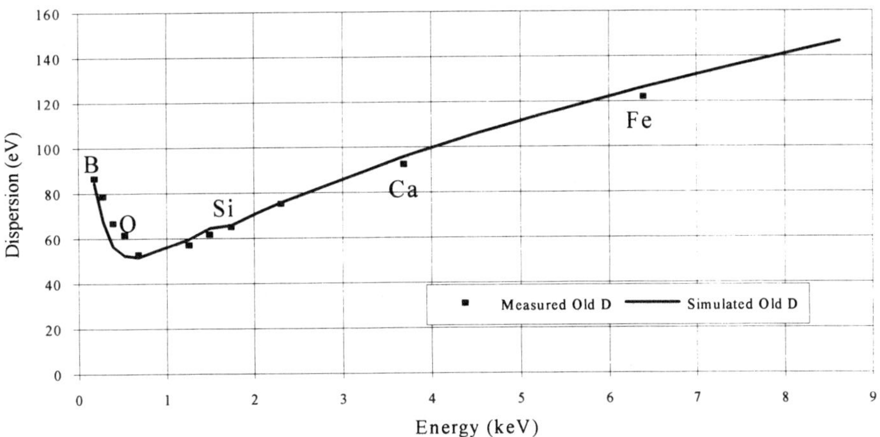


FIGURE 6.8. Comparison of simulated dispersion for an old process detector with the measured dispersion curve.

the same, in agreement with the experimental measurements of the dead layer. The values for the simulation dead layer thickness are also are in agreement with the measured values within the measurement error.

It is clear from this comparison that the major difference between the old process and the new process is a significant reduction in the size of $s$. This implies that the properties of the front surface of the detector may be the most significant factor in the reduction of incomplete charge collection effects in the light element spectrum from the new process detectors. This conclusion is consistent with the results of the designed experiments in which changes in the surface treatments and front contact metal deposition during fabrication were found to produce the most improvement in light element detection.

While the dispersion simulations are in good agreement with the dispersion data, there appears to be a tendency to underestimate the actual dispersion for elements below Na. When the surface recombination is reduced, as in the new detectors, the shape of the simulated dispersion curve more closely matches the heavier elements, but begins to deviate for the light elements. This suggests that our detector model may be overly simple, and when the surface recombination is reduced, other more subtle effects are revealed.

## 6.5. CONCLUSIONS AND RECOMMENDATIONS

The use of the EDS detector for the detection and quantification of the light elements has been increasing for some time. Several artifacts have been routinely observed in the spectra of light elements. The most common of these artifacts are the presence of tails on the low energy side of peaks and noticeable shifts of the peak

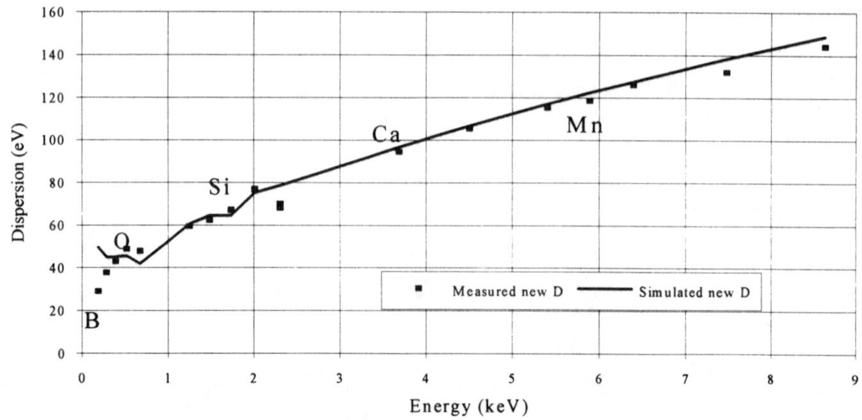


FIGURE 6.9. Comparison of simulated dispersion for a new process detector with the measured dispersion curve. Note the improved agreement between the simulation and measured dispersion for the light elements.

centroid to lower energies. These artifacts complicate qualitative and quantitative analysis and raise detection limits for the light elements.

This paper has presented the results of a set of designed experiments that significantly reduced the incomplete charge collection effects in Si(Li) detectors. This reduction has resulted in originally improved spectral quality for the light elements. Understanding the source of this improvement has proved to be more challenging.

A Monte Carlo simulation of the detector charge collection was used to successfully model the effects of ICC in the dead layer, including peak shapes and peak shifts, and is in general agreement with measured detector dispersions. The primary conclusion derived from the detector simulations is that a sharp reduction in the amount of surface recombination is responsible for the improvements. Another significant conclusion is that the thickness of the dead layer was not changed in the experiments.

Although the agreement is good between experimental spectra and dispersion curves and those simulated here, the need for further refinements is also indicated. The comparison of the dispersion curves for the light elements indicates that the simulations systematically underestimate the observed experimental dispersion. As the surface recombination is reduced, the agreement between measured and simulated dispersion worsens. This lack of agreement indicates that other effects that contribute

TABLE 6.4. Comparison of "Best Fit" Parameters for Monte Carlo Simulations

| Parameter | Old process detector | New process detector |
| --- | --- | --- |
| $DD$: Dead layer thickness (nm) | 110 | 100 |
| $DL$: Carrier diffusion length (mm) | 1.75 | 1.5 |
| $s$: recombination velocity | 35 | 0.1 |

to incomplete charge need to be considered in our model. Some other mechanisms that may contribute include the diffusion of "hot" electrons toward the surface from the junction,[6] and the escape of both photoelectrons and Auger electrons from the intrinsic volume.[4] The addition of these effects to the model may further our understanding and point to additional ways to improve detector performance.

It has been suggested that the values for the parameters $DL$ and $s$ can be obtained from measurements of electron beam induced current (EBIC).[11] Such measurements would help to verify the results of any improved model by providing values for $DL$ and $s$ to be used in the simulations.

Finally, efforts should be made to further reduce the thickness of the dead layer itself, since it was not changed by the processes modified in these experiments. Work with HPGe detectors[3,4] indicates a marked reduction in peak tailing by using implanted contacts to produce thinner dead layers.

ACKNOWLEDGMENTS. The author wishes to thank D. McMillan, M. Meisenhiemer, J. Howard, S. Foote, and the many others at NORAN Instruments who helped construct detectors and collect and analyze this data. D. C. Joy, B. Drummond, and R. Sareen also contributed with helpful comments and conversations.

# REFERENCES

1. P. Statham and T. Nashashibi, in: *Microbeam Analysis* (D. E. Newbury, ed.), San Francisco Press, San Francisco, p. 50 (1988).
2. M. Lund, *Microbeam Anal.* **2**, 5178-5179 (1993).
3. J. J. McCarthy, M. A. Ales, and D. J. McMillan, in: Proceedings of the XIIth International Congress for Electron Microscopy, Vol. 2, (L. D. Peachey and D. B. Williams, eds.) San Francisco Press, San Francisco, pp. 90–91 (1990).
4. J. Llacer, E. E. Haller, and R. C. Cordi, *IEEE Trans. Nucl. Sci.* **NS-24**, 53 (1977).
5. J. M. Caywood, C. A. Mead, and J. W. Mayer, *Nucl. Instrum. Methods* **79**, 329 (1970).
6. K. Shima, S. Nagai, and T. Mikumo, *Nucl. Instrum. Methods* **217**, 515 (1983).
7. J. L. Campbell and J. X. Wang, *X-Ray Spectrom.* **20**, 191 (1991).
8. D. C. Joy, *Rev. Sci. Instrum.* **56** , 1772 (1985).
9. R. Woldseth, *X-ray Energy Spectroscopy,* available from: Kevex Corporation, Burlingame, CA, p. 2.6.
10. R. G. Muskett, NBS Special Publication 604, National Bureau of Standards, Washington, DC, 97.
11. L. G. Wolfgang, J. M. Abraham, and C. N. Inskeep, *IEEE Trans. Nucl. Sci.* **NS-13**, 30 (1966).

# 7

# Energy Dispersive X-Ray Spectrometry in Ultra-high Vacuum Environments

*J. R. Michael*

## 7.1. INTRODUCTION

The addition of an energy dispersive x-ray spectrometer to the ultra-high vacuum (UHV) environment present in many modern electron microscopes should be done under two major constraints. First, the addition of the spectrometer should not compromise the UHV environment in the specimen chamber, and second, the vacuum environment of the electron microscope should not compromise the performance of the spectrometer. Thus, the successful addition of an EDS to a UHV vacuum system requires an understanding of UHV and of how the constituents of the vacuum can affect the performance and the sensitivity of the EDS. Materials used in the construction of the spectrometer as well as the mechanical design of the spectrometer must be compatible with UHV conditions.

The advantages and disadvantages of UHV systems will only be mentioned here briefly but will be fully more discussed later. The major advantage of conducting experiments in UHV is the greatly reduced specimen contamination that allows the study of clean specimens and allows longer x-ray acquisition times for better counting statistics. Window-less ED spectrometers can be used in UHV conditions with a reduced concern for detector contamination, resulting in longer times between detector de-icing treatments. Lack of space in the specimen region is one of the most difficult problems facing EDS and microscope designers. Many non-UHV microscopes attempt to reduce specimen contamination by using liquid nitrogen cooled cold traps very near the specimen, which occupy some of the

J. R. MICHAEL • Materials and Process Sciences Center, Electron Microscopy and Metallography Department, Sandia National Laboratories, Albuquerque, New Mexico 87185–0342

*X-Ray Spectrometry in Electron Beam Instruments*, edited by David Williams, Joseph Goldstein, and Dale Newbury. Plenum Press, New York, 1995.

limited space available around the specimen for other detectors. There is no need to use cold traps near the specimen in UHV systems, thus freeing up valuable space for the addition of other detectors or allowing detectors a closer approach to the specimen for improved acceptance solid angle. The main disadvantage to the use of UHV systems is their increased cost and operational complexity compared to other vacuum systems. There is an additional concern that the clean specimen surfaces made available by UHV may be more easily damaged by sputtering at high accelerating voltages ($\geq 100\,kV$).

This paper will first describe ED spectrometers followed by a discussion of UHV conditions and the gas composition that exists in a UHV system. The design of ED spectrometers for UHV use will then be described, followed by a discussion of spectrometer contamination and its effect on spectrometer performance. Finally, an example of light element microanalysis using windowless EDS in UHV will be discussed.

## 7.2. ENERGY DISPERSIVE SPECTROMETERS

There are many excellent papers in the literature that discuss EDS in detail, and the reader is referred to them for a more complete treatment of the subject.[1–5] Because this paper's aim is to discuss the use of EDS on UHV systems, it is necessary to briefly outline the components of ED spectrometers. The x-ray detector can be based on either lithium drifted silicon (Si(Li)) or high purity germanium (HPGe). The detector is about 10 mm in diameter (typical area 10 or 30 $mm^2$) and is mounted on the end of a metal cold finger so that it can be cooled to a temperature near that of liquid nitrogen. The front face of the detector is usually coated with a thin layer of evaporated gold so that an electrical contact is produced for application of a bias voltage. The detector may be contained in its own vacuum system, maintained by sorption pump materials (zeolites) or small ion pumps, separated from the microscope vacuum by a window of Be or a thin polymer film coated with thin layers of aluminum or other proprietary materials. Alternatively, the detector may be windowless and share the microscope vacuum environment. The x rays to be collected must pass through the window material to reach the detector. Vacuum-tight Be window foils are usually about 7 $\mu$m thick and will nearly totally attenuate x rays with energies below about 1 keV.[4] Also, the window in front of the detector prevents scattered electrons from entering the detector and producing spectral artifacts.[6] Ultra-thin window materials will permit x rays with lower energies to pass. Windowless detectors provide the ultimate in sensitivity to low energy x rays and were installed on microscopes shortly after the development of the Si(Li) detector.[7,8] The relatively poor vacuums achieved in electron columns at that time (early 1970s) resulted in rapid condensation of gases onto the detector surface. This limited the use of windowless x-ray detectors to UHV systems. Also, detector windows are normally made to be opaque to shield the detector from light produced by cathodoluminescent materials. Since windowless detectors do not have light shielding they have difficulty studying materials which cathodoluminesce.[6] High-purity germanium detectors are not usually available in a windowless configuration due to their sensitivity to infrared radiation.[9]

## 7.3. THE VACUUM

Although "vacuum" is a word taken directly from Latin meaning "empty," it is impossible to achieve a space that is truly empty.[10] At the lowest vacuum level attainable through modern pumping systems, there remain hundreds of gas molecules per cubic centimeter. There are now about 19 orders of magnitude of pressures that can be achieved through pumping below the pressure which corresponds to atmosphere.[10]

The degree of vacuum in a system is categorized as low, high, or ultra-high. Low vacuum is usually used to describe pressures that are above 1 Pa ($10^{-2}$ torr). High vacuums are those between $10^{-1}$ and $10^{-5}$ Pa ($10^{-3}$ to $10^{-7}$ torr), and ultra-high vacuums are those with pressures lower than $10^{-5}$ Pa ($10^{-7}$ torr). These limits may seem to be arbitrary, but they are based on the physical concepts of molecular density, mean free path, and the time required to form a monolayer on the surfaces of the vacuum chamber and the specimen. The molecular density is simply the number of molecules of gas per unit volume within the vacuum system. The mean free path for a gas molecule within the vacuum is the average or mean distance that the molecule travels before interacting with other gas molecules within the system. When a freshly cleaved surface is exposed to a vacuum, gas molecules will begin to cover the surface. The time for the surface to be entirely covered with a single layer of gas molecules is referred to as the time to form a monolayer. Better vacuums will require longer times for the monolayer to form, while at lower vacuums monolayer formation may be almost instantaneous. At room temperature and a pressure of one atmosphere, the molecular density is about $2.5 \times 10^{19}$ molecules/cm$^3$, the mean free path for the gas molecules is $7 \times 10^{-6}$ cm, and the time to form a monolayer is about $3 \times 10^{-9}$ s. At a pressure of $10^{-7}$ Pa ($10^{-9}$ torr), the corresponding values are $3.3 \times 10^7$ molecules/cm$^3$, $5.1 \times 10^6$ cm and $2.2 \times 10^3$ s.

By using the concepts of molecular density, mean free path, and time to form a monolayer, we can better understand the pressure levels designated as low, high, and ultra-high.[10] When the number of molecules in the gas phase is much greater than the number of molecules covering the surfaces of the vacuum chamber, the vacuum level is called low. At this vacuum level, vacuum pumps are mainly attempting to rarefy the gas phase. At higher vacuum levels the gas molecules in the system are mainly located on the chamber surfaces, and the mean free path is larger than the dimensions of the vacuum chamber, continued pumping removes molecules from the vacuum system which have become desorbed from a surface. Ultra-high vacuum corresponds to the vacuum level where the time to form a monolayer is longer than the usual time for a particular experiment to be conducted. Thus, in ultra-high vacuum conditions, the specimen surface can be kept clean and is the reason that surface science experiments are nearly always conducted in UHV systems.

There are many sources of gases and vapors within a vacuum system. Real leaks can of course contribute to the gas load within a system, but should not be important in a well maintained system. Backstreaming of pump fluids is one of the more common sources of vapors within the vacuum system. Fortunately, this source of contamination can be readily controlled by properly designed baffles and traps within the system. At UHV, other sources of vapors contribute to the gas load within the system. Virtual leaks also contribute to the gas load. A virtual leak, or internal leak, is a general source of gases arising from inside the vacuum system. Examples of virtual leaks are blind threaded screw holes with no gas relief path from the threaded section once a bolt has been fitted or an internal crack in the vacuum chamber that is continually evolving gas into the system.[11]

Another major source of gases within a vacuum system is desorption of gases adsorbed on to vacuum chamber walls. These gases are most commonly adsorbed when the chamber was last exposed to atmosphere or when gases are introduced to the vacuum system during specimen exchange. As the system pressure is reduced, these adsorbed gas molecules are released from the chamber walls. This is usually the case for water vapor. In order to reach the ultimate vacuum for the system, the adsorbed gases must be desorbed from the chamber walls and pumped from the system. This is accomplished in UHV systems by baking the system at 100°–200°C to assist the desorption process.[11]

The gas composition within the UHV system is important to the successful continued operation of an EDS. At low vacuums the gas composition reflects that of the atmosphere. At higher vacuums the major constituent within the vacuum is water vapor. Much of this water vapor is adsorbed onto the vacuum chamber walls and can be reduced by baking the system. As the vacuum level reaches the ultra-high range, the major gas constituent is hydrogen. At a vacuum level of $10^{-7}$ Pa ($10^{-9}$ torr), the gas consists mostly of hydrogen but also contains partial pressures of water vapor as high as $1.3 \times 10^{-8}$ Pa ($10^{-10}$ torr). In some systems there may exist significant partial pressures of hydrocarbons from cracked pump fluids.[10] Water vapor and hydrocarbons from cracked pump fluids are particularly important because they can condense onto the cold ED detector. When the partial pressure of water vapor in the vacuum system exceeds the vapor pressure that exists in equilibrium with ice at a given temperature, condensation of the water vapor will occur onto the cold surface. The vapor pressure of ice at liquid nitrogen temperature is about $1.3 \times 10^{-20}$ Pa ($10^{-22}$ torr), which is much lower than the partial pressure of water vapor that is achievable in a UHV system.[10] Therefore, water vapor will condense onto the cold detector surface, leading to a buildup of ice on the detector and a decrease in detector efficiency for low energy x rays. Hydrocarbons in the vacuum system will behave in a similar manner.


## 7.4. DETECTOR CONSTRUCTION AND OPERATION FOR ULTRA-HIGH VACUUM SERVICE

The successful addition of an EDS to a UHV system requires careful attention to materials selection and design of detector components. Surfaces of the detector which are exposed to the UHV environment should be made of low out-gassing materials. Different materials have widely varying out-gassing rates, which is the rate at which gas is given off by the material. Aluminum generally has a higher out-gassing rate than stainless steel, which when baked has one of the lowest out-gassing rates for a structural material.[10] Detectors for UHV service should be constructed of stainless steel where possible. Certain polymers and epoxies that are used at low to high vacuum are not suitable for UHV systems due to their relatively high out-gassing rates. It is extremely important that all parts exposed to the UHV be cleaned and pre-baked by the manufacturer. One manufacturer of EDS systems bakes detector assemblies under vacuum for six days.

The attainment and maintenance of UHV requires the vacuum system to be baked periodically to 100–200°C for 10–24 hours while at vacuum. Either the spectrometer must be designed to withstand these temperatures, or the mechanical interface of the spectrometer to the microscope must allow the detector to be retracted from the baked UHV region. Some detectors have built-in water cooling passages to cool the detector during baking. The mechanical motion of detectors interfaced to lower vacuum systems has usually been achieved by the use of sliding O-ring arrangements. This is not acceptable for UHV spectrometers where the movement of the detector should be sealed by a stainless steel bellows with no sliding O-rings since these are sources for contamination due to the vacuum grease required.

An important consideration with windowless or thin window detectors that are not isolated from the microscope vacuum is the maintenance of the vacuum within the detector when the detector is retracted and sealed off from the microscope vacuum system. The vacuum in the retracted detector is maintained by sorption material within the detector cryostat at about $10^{-5}$ Pa ($10^{-7}$ torr). The sorption material (usually a zeolite) has an extremely large surface area, and when it is cooled, it captures gas molecules from the environment. However, due to the large surface area of this material, it is impossible to pump it to UHV conditions. When the detector is inserted into the UHV environment, the vacuum in the detector cryostat must be isolated from the microscope vacuum or it will not be possible to achieve UHV conditions. The isolation is usually achieved with a vacuum separator valve, which is closed manually immediately before inserting the detector into UHV.[6] The disadvantages of window-less operation can be avoided by utilizing sealed Be-window detectors at the expense of light element detectability. UHV compatible Be windowed detectors are now available and are not designed to be movable but can be baked to 200°C in the UHV environment without destroying the detector.[6] Energy dispersive spectrometers with ultra-thin windows are also suitable for UHV service, but these detectors must be retracted from the vacuum system during microscope baking.

The mechanical interface of the spectrometer to the microscope column also requires careful attention by the spectrometer designer. It is not desirable for the spectrometer to be grounded to the metalwork of the microscope column. The isolation of the spectrometer improves its performance and electronic stability. Electrical isolation is relatively easy to achieve in non-UHV instruments due to the use of polymeric O-rings between the spectrometer and the electron column and the addition of a Teflon sheet between the flanges for electrical isolation. The attaching bolts must be electrically isolated through the use of one non-metallic flange or insulating bushings in the threaded bolt holes. Ultra-high vacuum systems require the use of copper gaskets and knife edge seals in place of O-rings, and these provide no electrical isolation. Early spectrometers fitted to UHV systems were not electrically isolated and performed well. However, as the spectral resolution of the spectrometers has improved, it has become necessary to isolate the detectors from the microscope metalwork. Electrical isolation is achieved by using copper gasket seals in all of the connections except for the one that is the farthest from the microscope vacuum which is a viton O-ring as described above, to provide electrical isolation.

The rigid attachment of the spectrometer to the microscope column requires that the spectrometer be accurately aligned through design of the mechanical interface, since the copper seals permit little movement of the detector. The detector must be equipped with a gate valve behind which the spectrometer can be retracted when the microscope is vented to atmosphere or the specimen is exchanged. Ideally, the system should be equipped with two gate valves so that the spectrometer can be removed from the system without venting the microscope vacuum. Also, the spectrometer-microscope interface must permit large collection solid angles to be achieved without sacrificing UHV compatibility.

Certain operational procedures should be strictly followed when windowless detectors are used in any level of vacuum. The detector should always be retracted behind an isolation valve whenever specimens are exchanged or vacuum valves operated where there is any danger of vacuum loss. This is also true when sublimation pumps are operated due to the pressure increase associated with the initial firing of the pump and the light that is produced. The spectrometer should not be exposed to specimens that out-gas since the detector is the lowest temperature surface near the specimen and condensation of gases on the detector surface will result. It is important that liquid nitrogen cold traps be kept cold at all times when the detector is inserted into the microscope column. Finally, it is of extreme importance that the isolation valve is opened before the detector is inserted into the microscope vacuum system. Windowless and some UTW detectors are sensitive to light, which can degrade the spectrometer performance. Some vacuum gauges, in particular ion gauges, emit light during operation and should not be used during data acquisition.


## 7.5. SPECTROMETER CONTAMINATION

Careful attention to the materials used in and the design of an EDS for UHV service can result in a spectrometer that does not compromise the cleanliness of the UHV system. It is also important to consider the effect of the gas composition in the vacuum system on the spectrometer. Sealed detectors rely on the cryo-pumping action of the zeolite in the cryostat to maintain the spectrometer at a vacuum of $10^{-5}$ Pa ($10^{-7}$ torr), while windowless detectors rely on the microscope vacuum system when the detector is in use. A sealed detector is no guarantee against contamination of the spectrometer. Microscopic leaks can occur through the thin Be window and various vacuum seals in the cryostat resulting in a degradation of the detector cryostat vacuum with time.[12,13] The contamination rate for windowless detectors will depend entirely on the quality of the vacuum within the microscope.[14]

One of the advantages of windowless detectors is the increased efficiency for detection of light elements.[15,16] Figure 7.1 shows calculated detector efficiencies for windowless and Be-window (7 μm thick) spectrometers. It is readily apparent that the efficiency of the windowless detector is higher than the Be-window detector up to x-ray energies of about 3 keV. The discontinuities in the efficiency are due to absorption edges for the Au contact layer and the Si in the detector. The efficiency for a UTW is intermediate to the two curves shown, with the specific efficiency dependent upon the window material and Al coating thickness. The combination of a windowless

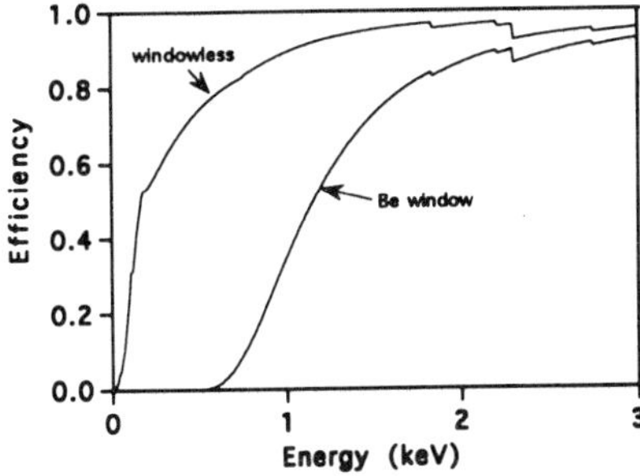


FIGURE 7.1. Calculated detector efficiency as a function of energy for windowless and 7-μm Be-window detectors.

EDS and UHV vacuum system results in the best sensitivity for low energy x rays and a much reduced risk of contaminating the detector compared to use in poorer vacuums.

As was discussed previously, the partial pressures of gases within the vacuum system will determine the rate at which monolayers form on exposed clean surfaces, like the front face of the detector. The rate at which gases adsorb onto surfaces is dependent upon the gas pressure within the system and the temperature of the surface. The number of molecules adsorbed on to a surface from the gas phase is quite low under UHV conditions at room temperature and increases significantly with increasing pressure and decreasing temperature. The increased rate of gas adsorption onto surfaces is related to the molecular sticking coefficient. The sticking coefficient is the probability of an incident molecule remaining on the surface and can range from about 0.05 to 1.0 depending on the particular molecule, the surface composition, surface condition, and the surface temperature.[10,11] Lower surface temperatures result in larger sticking coefficients and a greatly increased rate of molecule accumulation on the surface. Thus, condensable gases, like water vapor, will begin to accumulate on the liquid nitrogen cooled detector when it is exposed to the vacuum. The development of room temperature detectors would greatly reduce the amount of condensate that accumulates on the detector surface and the resulting decrease in sensitivity.

The two important condensable gases in a typical vacuum system are water vapor and hydrocarbons. The accumulation of molecules on the detector surface is not without consequences since this layer greatly reduces the detector sensitivity. As was mentioned previously, windowless x-ray spectrometers were installed on microscopes soon after the development of the Si(Li) detector. The relatively poor vacuum levels achieved in the electron columns at that time resulted in rapid contamination of the detector and thus a rapid decrease in detector efficiency.[7,8]

Water vapor is present in most vacuum systems and is easily condensed onto the cooled detector surface. Ice layers have been noted on sealed Be-window detectors as well as on windowless spectrometers. The contamination of sealed detectors may be attributed to leaks through the thin Be window and around the dewar pumping port, which cause the cryostat vacuum to deteriorate with time. Ice layers on sealed detectors have been measured up to 40 μm in thickness in an extreme case.[12,13] Ice layers of 1

                                                                                      

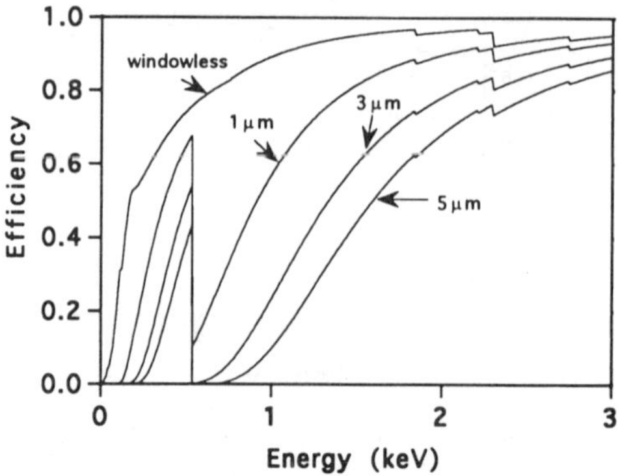


FIGURE 7.2. Calculated windowless detector efficiency for ice layers of 1, 3, and 5 μm compared with an uncontaminated detector.

μm and 3 μm have been measured on windowless detectors installed on UHV systems after one year of use.[17] The differences in ice layer thickness reflect the inherent variability in vacuum system operation.

The efficiency of detection of low energy x rays is greatly reduced by the ice layers on the detector surface. This reduction in efficiency is mainly due to the absorption of the x rays by oxygen in the ice layer. Figure 7.2 is a plot of windowless detector efficiency for ice layer thicknesses of 1, 3, and 5 μm compared to a clean detector surface. The presence of the oxygen x-ray absorption edge is apparent as the discontinuity at 0.53 keV along with the large decrease in detector efficiency above this edge. Figure 7.2 also shows that the detector efficiency above about 3 keV is not significantly reduced by the layer of ice. The detector efficiency can be described by the equation:

$$\varepsilon = \exp(-0.447 x_{ice} E_x^{-2.92})  \tag{7.1}$$

where $\varepsilon$ is the effective detector efficiency relative to ice formation, $x_{ice}$ is the ice layer thickness in μm, and $E_x$ is the x-ray energy in keV for any line greater than 1 keV.[12] An example of the ice layer thickness accumulation along with the detector efficiency for Al $K$ x rays calculated from (7.1) is shown in Figure 7.3 for a sealed Be-window x-ray detector.[12] It is apparent that a large ice layer thickness can, in an extreme case, accumulate over the lifetime of a spectrometer and severely reduce the detector's sensitivity to low energy x rays.

For windowless detectors in UHV, the accumulation of condensed carbonaceous layers on the detector surface may also occur, with an accompanying reduction in detector efficiency. Figure 7.4 shows the effect of 1.0 μm of ice and various amounts of carbon contamination buildup on the detector surface. The discontinuities at 0.28 keV and 0.53 keV are due to the carbon and oxygen $K$-absorption edges. It is apparent that relatively small carbon contamination layer thicknesses are sufficient to severely degrade the spectrometer's performance for low energy x-ray detection. The buildup

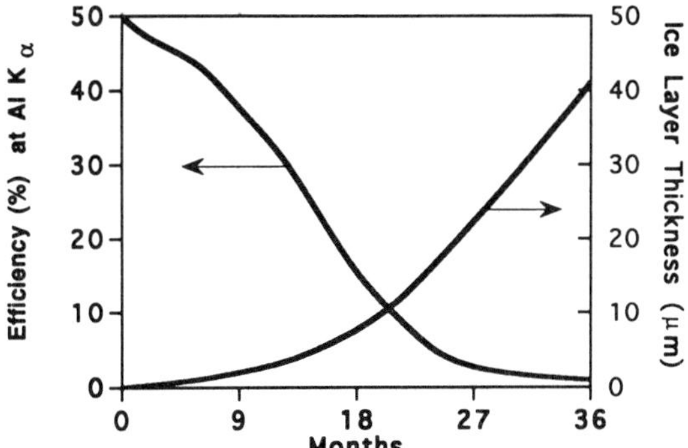


FIGURE 7.3. Ice layer thickness and experimentally measured detector efficiency for Al $K\alpha$ x rays for a particularly poorly sealed Be-window detector.[12]

of carbonaceous layers on the detector surface can be avoided by not exposing the detector to the vacuum system when the cold traps are warm or by using a UHV system that is ion pumped instead of oil-diffusion pumped.

It is possible to remove the contamination layers from the detector surface by heating the detector while it is exposed to the UHV environment. Detector heating is effective in removing ice layers from the detector, but may not be as effective in removing carbonaceous material. Heating of the detector may be accomplished through the use of detector heaters (often called detector conditioners) or by removing the liquid nitrogen from the detector dewar and allowing the detector to warm up. Detector warm-up using a heater for the detector only is obviously the more desirable solution. Detector conditioning circuits are available on some of the newest EDS systems and are operated by a simple switch that removes the high voltage from the detector and then heats the detector. When the detector conditioning cycle is over, the

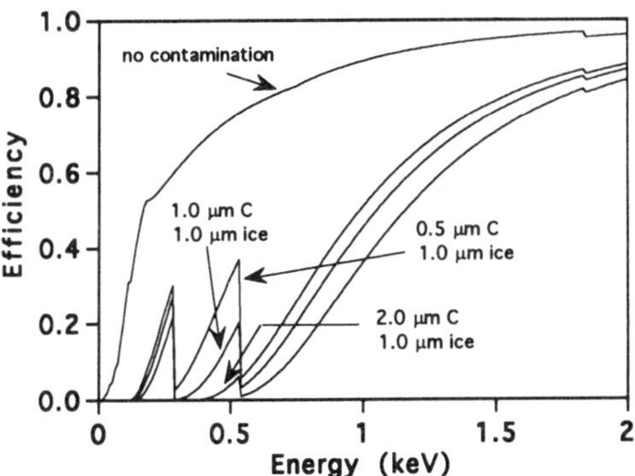


FIGURE 7.4. Calculated windowless detector efficiency for 1 μm ice and various thicknesses of carbon contamination buildup on the detector surface. Note the effect of the carbon edge at 0.28 keV on the detector efficiency.

heater is turned off and the detector cools quickly to the operating temperature and is ready for use. A typical cycle takes about three hours. It is only necessary to raise the detector temperature to the point where the vapor pressure of the ice exceeds the partial pressure of water vapor within the vacuum system. Higher temperatures result in faster sublimation of the ice layer. Alternatively, for detectors not equipped with a conditioning circuit, it is possible to warm the detector by first removing the high voltage from the detector and emptying the dewar of liquid nitrogen. The dewar is then filled with boiling water to heat the detector. It is necessary to pump on the detector and the cryostat dewar while the hot water is in the dewar. This technique removes water from the dewar molecular sieve material as well as removing ice layers from the detector surface. The use of this technique should be approved by the detector manufacturer as damage to the detector is possible.

## 7.6. DETECTOR BENCHMARKING

The routine use of windowless x-ray detectors requires periodic testing or benchmarking of the detector to monitor the performance of the detector and the buildup of contamination layers on the detector surface. Performance monitoring is essential for all detectors when low energy x rays are used for quantitative analysis due to the reduced sensitivity caused by detector contamination. There are a number of possible specimens that can be used for detector benchmarking, but in general the specimens should have low energy and high energy peaks that appear in known ratios. For SEM work, a commonly used specimen is Ni, where there are low-energy $L$-lines and high-energy $K$-lines and the ratio of the intensity of the $L$- and $K$-lines are measured. For analytical electron microscopy (AEM) thin specimens, the $L$-to-$K$ ratio will vary with specimen thickness, specimen tilt, detector window material, and detector contamination layer thickness, but for an electron excited spectrum from a thin specimen, the $L$-lines should be as large or larger than the $K$-lines. For example, Figure 7.5 demonstrates the use of the Ni $L/K$ ratio from bulk Ni to assess windowless detector performance on a SEM. This experimental data shows the Ni $L/K$ ratio as a function of time. The discontinuity in the plot at 100 days indicates that the detector was conditioned to remove contamination; after this the detector performance, as indicated by the $L/K$ ratio, returns to the uncontaminated condition.[6]

An ideal specimen for windowless detector benchmarking on an AEM is a thin film of NiO, which will show the $L/K$ ratio for Ni, but also can be used to determine the sensitivity of the detector to oxygen. Thin NiO films can be prepared from bulk using ion milling or can be produced by oxidizing evaporated Ni thin films deposited on rock salt. The oxidized thin film is then floated off the substrate and collected on suitable specimen supports. A typical spectrum of thin NiO obtained with a clean detector is shown in Figure 7.6. Note that the $L/K$ ratio is greater than 1 and the presence of the obvious oxygen peak. The relative heights of the oxygen $K$-and Ni $K$-and $L$-peaks can be used to assess the condition of the detector. Figure 7.7 shows a comparison of spectra obtained from a thin NiO specimen at 100 kV before and after heating the detector to 80°C where the spectra are scaled so that the Ni $K$-peaks are the same height. The increase in sensitivity after the detector was heated is quite marked. The decreased

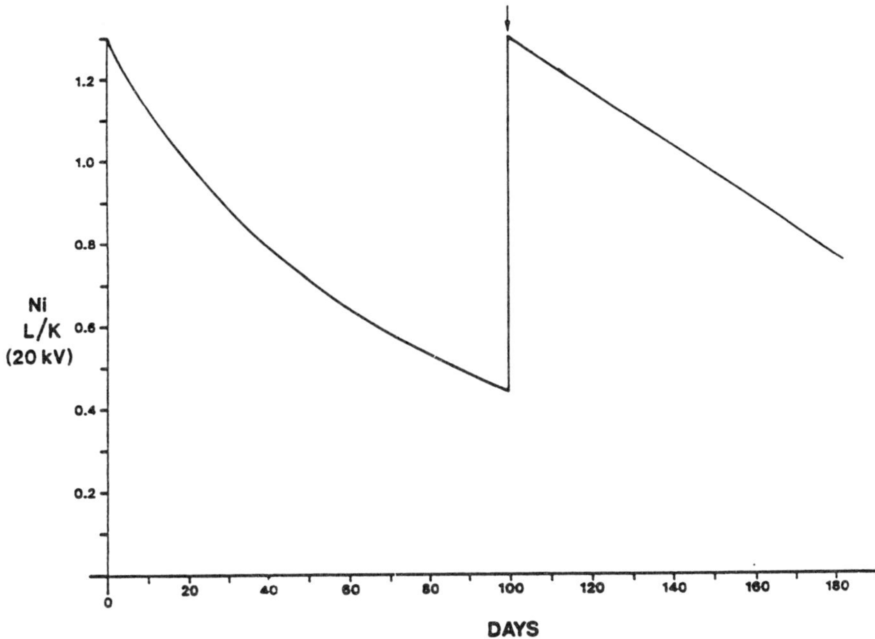



FIGURE 7.5. Plot of the $L/K$ ratio obtained from a bulk Ni specimen as a function of time. Note that at 100 days the windowless detector was conditioned (warmed up), and the $L/K$ ratio returned to the value at the start of the plot.

sensitivity of Ni $L$ indicates that there is ice on the detector surface as demonstrated by the detector efficiency data shown in Figure 7.2. However, the decreased sensitivity to oxygen can not be explained by ice alone. Figure 7.8 shows calculated EDS spectra of NiO for detectors with clean and contaminated detector surfaces. The reduction in the Ni $L$ intensity is mainly due to the reduction in sensitivity due to the oxygen contamination layer, which does not as strongly affect the sensitivity of the detector at the energy of the oxygen peak. The detector sensitivity as a function of energy with 1.0 μm of ice and various thicknesses of carbon contamination is shown in Figure 7.4. The carbon layers result in additional x-ray absorption for x-ray energies below 0.8 keV. In particular, the carbon $K$-edge at 0.29 keV can greatly decrease the intensity of the O peak at 0.525 keV. Figure 7.8 shows calculated spectra for NiO as they would be collected by a detector with no ice or carbon layers and with 2.0 μm of ice and 0.6 μm of carbon. When these calculated spectra are compared with the experimental spectra shown in Figure 7.7, the agreement is quite good. It should be noted that this degree of ice layer formation occurred over the time period of one year. Thus, the NiO specimen is useful in assessing the cleanliness of windowless or UTW detectors because the reduction in the Ni $L$-line indicates the presence of ice and the reduction in the O x-ray signal indicates the presence of carbonaceous contamination. A procedure for modeling the decrease in detector efficiency with C and ice buildup has been developed that uses the C and O absorption edges visible in spectra obtained with contaminated windowless or UTW detectors.[18]

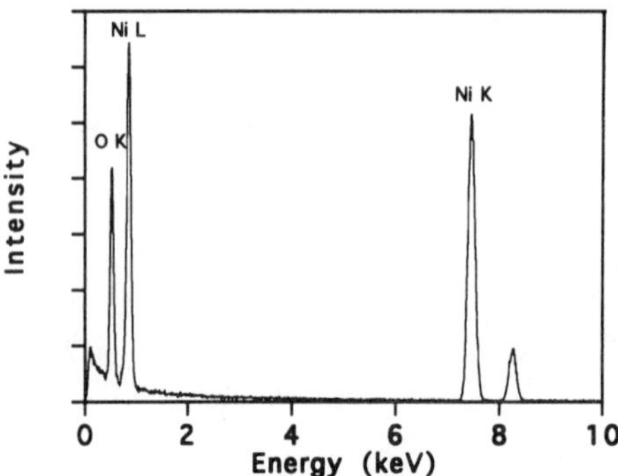

FIGURE 7.6. Windowless EDS spectrum obtained from a thin NiO specimen prepared as described in the text. Note the $L/K$ ratio is greater than 1.

The spectra shown in Figure 7.7 demonstrate the importance of windowless detector benchmarking. The buildup of contamination layers can severely reduce the detector's sensitivity to light elements, and any conclusions drawn from spectra could be misleading. Therefore, when quantitative analysis of light elements is performed, it is important to determine the detector cleanliness prior to the collection of experimental data. It is appropriate to obtain spectra from the NiO standard at least once a month and immediately prior to any light element analysis. The monitoring of the $L/K$ ratio for Ni can be used to determine when detector conditioning is needed. With experience, it may be possible to determine an appropriate schedule for detector conditioning or warm-up to insure that the detector surface is free of surface films thick enough to significantly reduce the detectors sensitivity for low energy x rays.

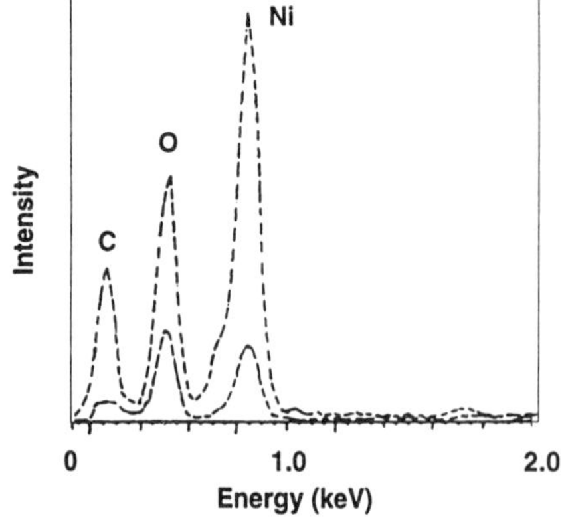

FIGURE 7.7. Comparison of spectra obtained from a thin NiO specimen before and after warming the detector to remove ice contamination. The upper curve was obtained after the detector was warmed. The $L/K$ ratio before warm-up is much less than 1, while after warm-up the ratio is greater than 1. The spectra have been normalized so that the Ni $K$-peaks are the same intensity.

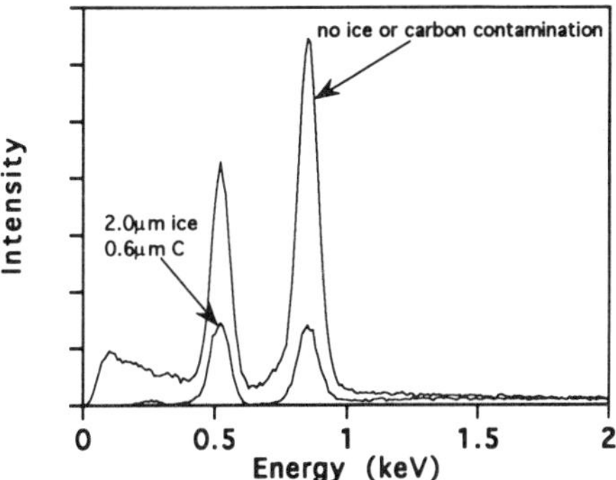



FIGURE 7.8. Calculated NiO spectra for a clean detector and for one contaminated by 2.0 μm of ice and 0.6 μm of carbon.


## 7.7. EXAMPLE APPLICATION OF QUANTITATIVE EDS IN ULTRA-HIGH VACUUM

The use of windowless EDS in UHV microscopes requires more attention to operational details than windowed detectors in lower vacuum environments. However, the use of windowless EDS in UHV microscopes allows light element quantitative analysis to be performed on a routine basis, provided appropriate precautions are taken.[19–21] There continues to be a steady stream of papers published on the use of EDS for light element quantitative analysis in the AEM.[15,16,23–25]

In a properly maintained UHV AEM, it is possible to analyze contamination-free specimens. One must always be aware that atomic sputtering may occur in UHV conditions due to the presence of the very clean specimen surfaces.[26] The major difficulties in performing quantitative thin film microanalysis of light elements are the corrections required for absorption of the low energy light element x rays in the specimen.[27] The explicit corrections for x-ray absorption effects may be difficult to apply since the absorption coefficients are not well known in the case of light elements, but are required for determination of $k_{ab}$-factors for quantitative analysis using the ratio technique.[28,29] A method called the extrapolation technique has been developed which allows determination of $k_{ab}$ at zero foil thickness.[30,31] A recent study has demonstrated the applicability of this technique to light element analysis of thin specimens of AlN in the AEM.[32]

The extrapolation technique requires that the $k_{ab}$ factors be determined for a range of specimen thicknesses, followed by plotting log of $k_{ab}$ versus relative specimen thickness. This relationship should be a straight line with the $y$-intercept yielding the absorption-free $k_{ab}$ value. Normalized x-ray intensity of an element present at constant concentration can be used instead of an explicit measurement of the specimen thickness, which significantly eases the experimental difficulties associated with specimen

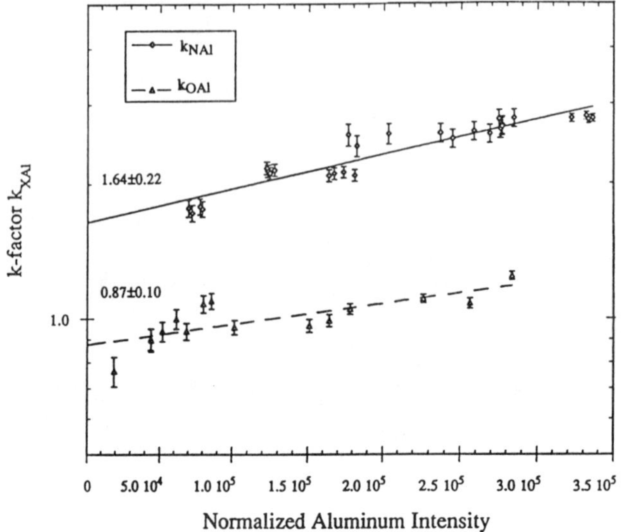



FIGURE 7.9. A log-linear plot of the measured $k_{NAl}$ and $k_{OAl}$ as a function of normalized Al counts. The normalized Al counts are a measure of the specimen thickness. A linear least-squares line was fitted to the uncorrected (raw) data and extrapolated back to zero thickness to provide an *absorption-free k-factor.*[32]

thickness measurement. This technique has been applied to the microanalysis of inversion domain boundaries (IDB) in AlN. Qualitative EDS analysis of IDBs has shown them to be enriched in oxygen. Quantitative analysis of the IDBs was conducted to determine the extent of oxygen enrichment and to aid in determining an interface model for this type of defect.

The use of EDS in UHV was important for this study because surface contamination of the specimen by carbonaceous material would severely limit the detectability of N and O, and due to the length of the study, it was important that the detector sensitivity not be altered by condensed contamination layers from the vacuum system. Prior to the collection of any experimental data, the detector was warmed up and benchmarked with a NiO specimen, as discussed previously. In order to maximize the sensitivity to light elements it is important to use a clean detector. If it is not possible to clean the detector prior to the analysis, or if the analyses were obtained at different times in the life of the detector, a procedure to correct the experimental spectra for the effects of detector contamination can be used.[18] Figure 7.9, which is a plot of $k_{XAl}$ as a function of the normalized Al count rate, demonstrates the extrapolation technique for determining absorption-free k-factors appropriate for use in quantitative analysis using the ratio technique. The k-factors determined by the extrapolation technique were used in conjunction with the x-ray absorption correction to experimentally measure the oxygen, nitrogen, and aluminum concentrations across an IDB in AlN. Figure 7.10 (a,b) shows the results of quantitative thin film analysis, including x-ray absorption corrections, across an IDB in 110 nm thick AlN. Figure 7.10(a) shows the O levels across the IDB, demonstrating that the IDB is indeed quite enriched in O and that the width of the enrichment is quite narrow. Figure 7.10(b) shows the complete quantita-

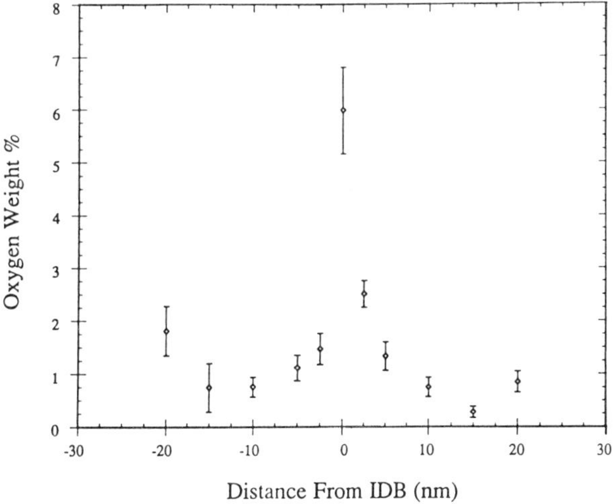

FIGURE 7.10. Quantitative O, N and Al, analyses, corrected for x-ray absorption, across a planar IDB in 110 nm thick AlN. (a) Oxygen analysis across the IDB at high spatial resolution.

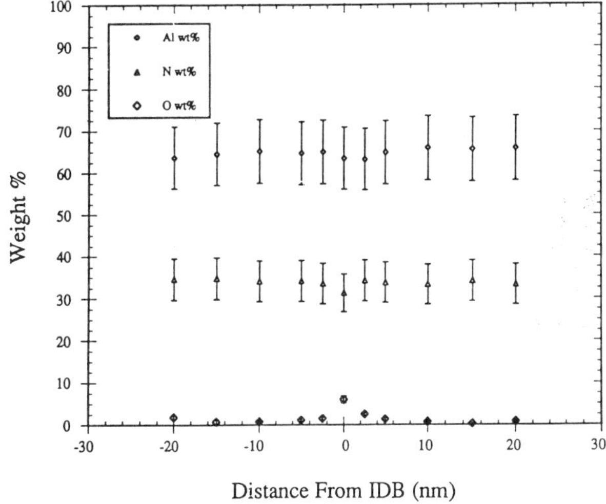


FIGURE 7.10. (b) Complete analysis across the IDB demonstrating the accuracy of the analysis of the AlN matrix.[32]

tive analysis, using the ratio technique, across the IDB. The accuracy of this technique is demonstrated by comparing the experimentally determined matrix composition shown in Figure 7.10(b) (64.8 wt% Al, 34.0 wt% N) to the composition of an ideally stoichiometric AlN matrix (65.2 wt% Al, 34.8 wt% N), which indicates that the value determined for $k_{NAl}$ is accurate.

This study demonstrates that quantitative analysis of light elements is possible with a windowless EDS in an UHV AEM, but can be difficult and time consuming. As with any rigorous quantitative analysis in an electron column instrument, accurate results are only obtained after careful attention is paid to the experimental details. Ultra-high vacuum ensures that the detector does not become rapidly contaminated and that the specimen does not become contaminated by surface layers, which would interfere with the quantitative microanalysis. However, as has been discussed previously, it is important to benchmark the detector before and after conducting quantitative analysis of light elements in order to show that no significant changes in detector sensitivity have occurred.

## 7.8. CONCLUSIONS

The addition of EDS to a UHV system can be accomplished with no compromise of detector performance, or of the UHV system performance, provided the detector designers and manufacturers take into account the special requirements of UHV. The detector designers must use materials that are UHV compatible and must understand the techniques used to achieve UHV. Once the proper interfacing of the detector to the UHV system has been achieved, it is up to the operator to maintain the performance of the system by implementing the following procedures:

1.  The microscope cold traps should always be kept cold when the detector is exposed to the vacuum to reduce the condensation of vacuum system constituents, particularly water vapor, onto the cold detector surface. The detector should always be isolated from the microscope vacuum whenever it is not in use.
2.  Care should be taken in the types of specimens that are introduced into the vacuum system. The detector should not be exposed to specimens which outgas and are not compatible with UHV.
3.  The performance of the detector must be monitored on a regular basis to determine when a warm-up cycle is needed to remove contamination from the detector surface. Benchmarking of the detector is a good quality control tool and will quickly identify detector problems.

When all of the above conditions are met on a routine basis, it is possible to reliably obtain quantitative analysis of light elements with UTW or windowless detectors interfaced to UHV microscopes.

ACKNOWLEDGMENTS. The careful review of this manuscript by C. R. Hills, G. C. Nelson, and A. D. Romig is gratefully acknowledged. This work was supported by the U. S. Department of Energy under contract #DE-AC04–94AL85000.

## REFERENCES

1.  D. A. Gedcke, *X-Ray Spectrom.* **1**, 129 (1972).

2. C. E. Fiori and D. E. Newbury, in: *Scanning Electron Microscopy/1978* (O. Johari, ed) SEM Inc., Chicago, p. 401 (1978).

3. C. E. Fiori, in: *Microbeam Analysis—1979* (D. E. Newbury, ed) San Francisco Press, San Francisco, p. 361 (1979).

4. P. J. Statham, *J. Microsc.* **123**, 1 (1981).

5. K. F. J. Heinrich, *Energy Dispersive X-Ray Spectrometry*, NBS Special Publication 604, National Bureau of Standards, Washington, DC, p. 1 (1981).

6. B. G. Lowe, *Ultramicroscopy* **28**, 150 (1989).

7. R. G. Musket, *Energy Dispersive X-Ray Spectrometry*, NBS Special Publication 604, National Bureau of Standards, Washingon, DC, p. 97 (1981).

8. J. C. Russ and A. O. Sandborg, *Energy Dispersive X-Ray Spectrometry*, NBS Special Publication 604, National Bureau of Standards, Washington, DC, p. 71 (1981).

9. C. E. Cox, B. G. Lowe, and R. A. Sareen, *IEEE Trans. Nucl. Sci.* **35**, 28 (1987).

10. A. Roth, *Vacuum Technology*, North-Holland, New York (1976).

11. J. F. O'Hanlon, *A User's Guide to Vacuum Technology*, John Wiley and Sons, New York (1980).

12. D. D. Cohen, *Nucl. Instrum. Methods* **193**, 15 (1982).

13. D. D. Cohen, *X-Ray Spectrom.* **16**, 237 (1987).

14. G. L'Esperance, G. Bottom, and M. Caron, in: *Microbeam Analysis—1990* (J. R. Michael and P. Ingram, eds.) San Francisco Press, San Franciso, p. 284 (1990).

15. N. Zaluzec and R. Holton, in: *Analytical Electron Microscopy—1984* (D. B. Williams and D. C. Joy, San Francisco Press, San Francisco, p. 353 (1984).

16. C. E. Lyman, D. B. Williams, and J. I. Goldstein, *Ultramicroscopy* **28**, 137 (1989).

18. P. Hovington, G. L'Esperance, E. Baril, and M. Riguad, *Microbeam Anal.* **2**, 277 (1993).

19. D. J. Bloomfield, G. Love, and V. D. Scott, *X-Ray Spectrom.* **13**, 69 (1984).

20. D. J. Bloomfield, G. Love, and V. D. Scott, *X-Ray Spectrom.* **14**, 8 (1985).

21. D. J. Bloomfield, G. Love, and V. D. Scott, *X-ray Spectrom.* **14**, 139 (1985).

22. A. J. Garratt-Reed, T. Thorvaldsson, and J. B. Vander Sande, in: *Analytical Electron Microscopy—1984* (D. B. Williams and D. C. Joy, eds.) San Francisco Press, San Francisco, p. 345 (1984).

23. P. J. Goodhew, P. M. Budd, and D. Chescoe, in: *Analytical Electron Microscopy—1984* (D. B. Williams and D. C. Joy, eds.) San Francisco Press, San Francisco, p. 337 (1984).

24. L. E. Thomas, *Ultramicroscopy* **18**, 173 (1985).

25. J. R. Michael and E. T. Stephenson, *Metall. Trans. A.* **19A**, 953 (1988).

26. C. R. Bradley and N. J. Zaluzec, *Ultramicroscopy* **28**, 335 (1989).

27. R. W. Glitz, M. R. Notis, D. B. Williams, and J. I. Goldstein, in: *Microbeam Analysis—1981* (R. H. Geiss, ed.) San Francisco Press, San Francisco, p. 309 (1981).

28. G. Cliff and G. W. Lorimer, *J. Microsc.* **103**, 203 (1975).

29. J. I. Goldstein, J. L. Costley, G. W. Lorimer, and S. J. B. Reed, *Scanning Electron Microscopy/1977* (O. Johari, ed.) SEM Inc., Chicago, p. 315 (1977).

30. Z. Horita, T. Sano, and M. Nemoto, *J. Microsc.* **143**, 215 (1986).

31. Z. Horita, T. Sano, and M. Nemoto, *Ultramicroscopy* **21**, 271 (1987).

32. A. D. Westwood, J. R. Michael, and M. R. Notis, *J. Microsc.* **167**, 287 (1992).

# 8

# Quantifying Benefits of Resolution and Count Rate in EDX Microanalysis

*P. J. Statham*

## 8.1. INTRODUCTION

A student of energy dispersive x-ray microanalysis soon learns that it is better to have a spectrometer with improved resolution because it is easier to recognize individual characteristic lines from different elements. However, a spectrum obtained in a few seconds can appear "ragged," and random distortion of peak shapes can confuse the distinction between single peaks and overlapping multiplets. Thus, the student learns that it is also important to obtain good counting statistics by increasing beam current or by extending the acquisition time so that the spectrum is not too ragged in appearance.

The student faces a dilemma when the EDX system has an adjustable process time switch. At long process times, resolution is better but electronic processing limits the acquisition rate. At short process times, much higher maximum count rates can be achieved but at the expense of broader peaks. This makes it difficult to choose the best configuration. Different EDX systems usually offer different resolution and count rate capabilities and different alternatives for process time settings. Given the types of work that may be encountered, how can one decide if the choice of system is critical and which option will offer the most valuable performance benefits? It may even be necessary to decide between different technologies for x-ray spectrometry. For example, would a device that offered an excellent resolution of only 7 eV but could only acquire at 60 cps be more useful than a device that could acquire at 10,000 cps but gave much poorer resolution of 150 eV?

P. J. STATHAM • Oxford Instruments Microanalysis Group, High Wycombe, Bucks HP12 3SE, United Kingdom

*X-Ray Spectrometry in Electron Beam Instruments*, edited by David Williams, Joseph Goldstein, and Dale Newbury. Plenum Press, New York, 1995.

Although there are always practical issues to consider when answering these questions, the ultimate limit of accuracy can be considered from a purely statistical viewpoint. This paper introduces new figures-of-merit, or quality factors, which can be used to quantify attainable benefits in terms of analysis speed for various situations. These show that, for simple spectra with minimal peak overlap, increased count rate gives the same proportional benefit as improved resolution. At the other extreme, automatic peak identification requires several overlapping peaks to be deconvolved, and a mere 10% improvement in resolution can improve speed of identification by as much as a factor of 2.6 in count rate.

## 8.2. THE CHALLENGE FOR SPECTRUM PROCESSING

### 8.2.1. Pileup and Distortion at High Count Rates

There are a variety of spectrometers for detecting x-ray photons. Since photons arrive randomly with a Poisson distribution in time, there is always a chance that two photons will arrive within the resolving time of the transducer or associated electronics; such pileup is the root cause of those artifacts and deficiencies that are made worse by increased count rate.

In some devices such as bolometers, the transducer itself presents the fundamental limit to resolving time, and time constants of tens of milliseconds may be involved. For a Si(Li) x-ray detector, the charge collection time in the crystal is only about 50 ns, but there is a high level of noise in the output signal. Since it is necessary to average the output to reduce noise, the effective resolving time is governed more by signal-to-noise considerations than by transit time in the transducer. The resolving time thus depends on the time constants chosen for pulse shaping. Whereas photons of a few keV in energy can be detected reliably in 100 ns, shaping times of the order of 10 $\mu$s are typically required to reduce noise to the level where very low energy photons ($<$ 300 eV) can be detected.

If photons of energy $E1$ and $E2$ do arrive within the resolving time, this constitutes a pileup event that normally gives a contribution to either a sum peak in the recorded spectrum at energy $E1 + E2$ or a pileup continuum extending below this energy. Furthermore, at high count rates, residual voltages in pulse processing electronics can accumulate and give rise to peak shift, broadening, and distortion.

With a suitably designed electronic pileup inspector, pile-up continuum can be avoided and live time correction techniques can be used to compensate for any rejected events. Though reduced in size, sum peaks still remain because of the finite resolving time of the pileup inspector, which is likely to be worse at lower energies.[1] Advanced electronic techniques such as time-variant pulse processing and digital signal processing can be used to minimize count-rate induced distortion. In principle, spectrum processing software can be written to correct for residual sum peaks[1] and provide some compensation for peak shift and broadening, although this is often only practical for a restricted range of analyses.

Although high data rates inevitably improve statistical precision, any system should be checked carefully to make sure any electronic or software corrections for pileup and spectral distortions are effective. The figures-of-merit developed in this paper apply only if there are no systematic errors introduced at the count rates used for analysis.

### 8.2.2. Spectral Background

Even with a perfect electronic system, the transducer response function will still limit measurement accuracy for characteristic line intensities. A Gaussian response function is appropriate for a wide range of transducers and will be assumed throughout this paper. Furthermore, the term "resolution" will be used to refer to the FWHM of a spectral peak recorded from a monochromatic x-ray line.

For electron-excited x-ray spectra, bremsstrahlung background must be subtracted to get accurate peak areas. For a typical Si(Li) spectrometer, the background height is roughly equivalent to a peak of around 1% concentration by weight, so a 10% error in background estimation will give about a 0.1% error in weight percent. At low energies, the absorption edge lies close to the peak energy but the absorption step in the background is usually only a small fraction of the peak height. Therefore, provided there are only a few major elements in the sample so that peaks do not overlap, a straightforward linear interpolation between background points at either side of each peak usually gives sufficient relative accuracy.[2] With better resolution, accuracy increases because the peak-to-background ratio is higher and background points for interpolation can be closer together. Indeed, for resolutions below 10 eV, this simplistic approach can be used for most analyses, but for EDX spectrometers with resolutions in excess of 100 eV, peak overlap is commonplace and more sophisticated background subtraction techniques are required.

### 8.2.3. Peak Overlap

Where peaks overlap, background correction is more difficult because the background has to be interpolated over the much larger energy range spanning the group of overlapping peaks.[2] If the underlying background is curved, background modelling may be necessary to predict the shape. With improved resolution, there will be more peak-free regions available for background fitting, and interpolation will be more accurate over the smaller energy range covered by peaks. Alternative background correction approaches such as iterative peak stripping or digital filtering and fitting[2–4] can be used provided resolution is such that the background is almost linear over the range of a single peak. Clearly, improved resolution makes all types of background correction more effective. This must be taken into account when comparing theoretical performance of systems with vastly different resolutions; accuracy in a poor resolution system may be limited more by systematic errors in background correction than by random counting statistics.

Peak overlap itself can be corrected by several techniques.[2] If an element gives rise to a multiplet of characteristic lines with fixed relative intensity, Peak stripping can be used. For example, if a Cr $K\alpha$ peak is free of overlap but a Cr $K\beta$ peak overlaps a nearby Mn $K\alpha$ peak, the Cr $K\beta$ peak can be stripped in proportion to the area of the Cr $K\alpha$ peak. In this case, the accuracy of overlap correction depends only on accurate determination of the Cr $K\alpha$ peak area. However, this rather *ad hoc* technique is only useful in specific applications, and more sophisticated techniques, such as least squares fitting to a sum of functions, or peak profiles, are invariably required for general analysis of spectra. The shorthand term "deconvolution" is often used to refer to this fitting procedure because the originally sharp, well separated, characteristic emission lines are convolved by the spectrometer response function to produce overlapping peaks and the fitting procedure recovers the intensities of the original lines.

### 8.2.4. Automatic Element Identification

The most extreme test of any system is to identify what elements are present using the spectrum from a completely unknown sample. There are roughly 10 $K$, 25 $L$, and 33 $M$ peak series, one for each element, that might appear in the energy range from 0.5 to 3 keV. Even if the major lines of each series were uniformly spaced, this would imply an average separation of less than 40 eV; thus most peak overlaps are severe for systems with resolutions of the order of 100 eV. Some examples of commonly occurring element combinations that present clusters of severely overlapping peaks are Al$K$ + Br$L$, Si$K$ + Ta$M$ + W$M$, and S$K$ + Mo$L$ + Pb$M$.

For automatic element identification, all elemental peaks that might contribute to a particular overlapping cluster must be included in the fit to find out which are present and which are below the statistical limit of detection. It has been suggested that this is not always necessary for those elements that generate more than one x-ray line series within the observed spectrum. For example, Br$L$ lines appear around 1.5 keV but Br$K$ appears near 12 keV, and Mo$L$ appears at 2.3 keV but Mo$K$ appears at 17.5 keV. The idea is that if the spectrum is examined in the region of the high energy line, then if the line is absent, the element can be eliminated from consideration and thus ease the problem of deconvolution of the low energy peaks. Unfortunately, for routine SEM or microprobe work, this is not nearly as useful as it first appears, because if the electron column is operated at 20 kV, the overvoltage for the higher energy lines is usually insufficient to produce a detectable peak unless the element is present in high concentration. In particular, Mo$K$ is not even excited at 20 kV, and if 10 kV is used to reduce absorption corrections or improve the spatial resolution of x-ray analysis, Br$K$ will not be excited.

Another approach is to use prior knowledge to eliminate "impossibles" from consideration. Of course, this precludes those elements from ever being detected, but this information can be useful in reducing the number of lines to deconvolve; for example the Na$K$ + Zn$L$ + Pm$M$ + Sm$M$ situation can be simplified to Na$K$ + Zn$L$ if the Pm and Sm are known not to be present. Nevertheless, even with the commonly occurring elements, there are still many multioverlaps that are not resolved by a spectrometer with 100 eV resolution, e.g., P$K$ + Zr$L$ + Ir$M$ + Pt$M$ + Au$M$. Therefore, for practical automatic identification, it is essential to be able to deconvolve severely overlapping peaks. For a resolution of 100 eV FWHM, "severe overlap" refers to separations less than 50 eV, and there will be typically up to five peaks to evaluate within a cluster, i.e., 1 $K$, 2 $L$, and 2 $M$ peaks.

## 8.3. STANDARD DEVIATION, PRECISION, AND MINIMUM DETECTABLE MASS FRACTION

Even if it was possible to overcome or correct for the artifacts and deficiencies described in Section 8.2.1 and have an exact method for subtracting spectral background, there would still be random statistical scatter on the results. This scatter originates from photon counting statistics where, for a spectrum channel expected to record $N$ counts, the standard deviation of actual measurements will be $N^{0.5}$. Each technique for spectrum processing (background subtraction and overlap correction) will use different combinations of the raw channel counts. If two different algorithms are used to process a series of spectra obtained from the same point on a sample under identical conditions, the magnitude of fluctuation in results is likely to be different for each algorithm, even if the average result for each measured element is the same. Rather than apply an algorithm individually to several spectra and measure the scatter, a single spectrum can be used to estimate the standard deviation at each spectrum channel and the addition-of-errors rule used to obtain the standard deviation in each derived result.[5,6] Thus, two algorithms can be compared to see which one combines the raw spectral data in a more optimal way to reduce standard deviation.

To be certain that differences in results are due to real differences in concentration, they must be larger than any likely statistical fluctuation; if the statistical standard deviation is less, smaller differences can be detected with certainty so the measurement is more precise. In practice, beam current fluctuations and other instrumental variations may increase the scatter in results. Furthermore, if the sample is inhomogeneous, the real differences in concentration may be interpreted erroneously as due to statistical scatter or instrumental variation. Thus, the predicted statistical standard deviation is always less than or equal to the observed standard deviation in a set of results and therefore determines the best possible precision.

When an element is measured but is not present in a sample, repeated analyses will give both positive and negative results with a statistical scatter about zero. The standard deviation, st.dev, for this case determines the likely maximum extent of such fluctuations: there is only a 1.4% chance that a statistical fluctuation will cause the result to exceed 3 st.dev. In other words, if a result does exceed 3 st.dev, there is a 98.6% probability that a non-zero concentration of the element is present. The concentration equivalent to 3 st.dev is often referred to as the detection limit or minimum detectable mass fraction MMF.[6] Of course, it is only appropriate to use the st.dev predicted from statistics to calculate the MMF if the background subtraction gives errors that are considerably smaller than statistical fluctuations.

If knowledge of the MMF is critical, it is imperative to know what algorithm will be used for the real analysis. If we fit a single peak and background to a series of spectra, the fluctuation in results will be less than if we fit two or more peaks and a background to the same set of data[6]; furthermore, the statistical standard deviation for peak area increases with the degree of overlap of the fitted peaks. If this is not obvious at first, consider the area of a single peak which best fits a particular spectrum: if the possibility of a second peak is considered, the combination of peaks that best fits the same spectrum is likely to involve a different area value for the first peak; including the second peak clearly introduces uncertainty into the result for the first peak. In general, we can say that the more peaks we try to fit to a given set of data, then the larger will be the statistical standard deviation in any one result. Algorithms are designed to give results that are close to true peak areas; if we perform a series of analyses on the same sample, results will vary at random because of statistics, but if the averaged result shows any difference or bias, this is referred to as systematic error. It is only sensible to calculate the MMF once we have found an algorithm which gives a small enough systematic error. For example, if we are trying to measure a very small peak and there is a large neighboring peak, then the algorithm must account for both peaks. In this case, the standard MMF calculated for a single peak on a background will be unduly optimistic. The more complex algorithm that deals with overlap will inevitably give a higher MMF, but this will be more realistic because this is how the spectrum will have to be processed in practice.

Adjusting an algorithm to reduce systematic error often exaggerates statistical fluctuations. For example, to get a precise estimate of background intensity it is important to average over a region containing as many channels as possible. If background is to be subtracted by interpolating between regions at either side of a peak, these regions must be kept narrow if they are not to be influenced by curvature or other peaks in the vicinity. Thus, narrower background regions reduce the systematic error but will not give the lowest possible detection limits. Any systematic error introduced by mathematical treatment of the spectrum must always be substantially less than the standard deviation and so more accurate techniques for background correction are required when standard deviations are smaller. Much experimental time could be wasted if an estimate of the MMF was wildly optimistic. It is important to be aware that the magnitude of standard deviation, and hence the MMF, depend on the method of spectrum processing. For a given method, the standard deviation obviously will be influenced by spectrometer resolution and number of counts in the spectrum.

In summary, statistical effects cause results to vary from analysis to analysis, and this variation is represented by the standard deviation. Standard deviation, which governs precision of measurement for peak areas and the MMF for trace elements, will be different for different processing algorithms. An algorithm must be chosen which will give acceptable accuracy for the particular form of spectrum. Once the most appropriate method has been found, we can consider how a change in system resolution would alter the standard deviation obtained by this method.

## 8.4. HOW DOES RESOLUTION AFFECT STANDARD DEVIATION?

### 8.4.1. Basic Assumptions

The influence of resolution on statistical precision has been addressed in various earlier publications,[5–8] but the assumptions made and the various conclusions reached are not obviously connected. Therefore, simulations have been performed to corroborate some of the earlier findings and to develop simple formulae that can be applied in a wide range of practical situations.

There is an infinite variety of x-ray spectra and spectrum processing methods. In some situations a technique may give large systematic errors in peak areas, and if so, there is no benefit to improving statistical precision. However, if we developed the perfect algorithm that gave the best possible accuracy, then further improvements in precision would always be worthwhile. With the perfect algorithm, we could study how standard deviation in a result would change if we altered system resolution. With any processing method, whenever the resolution is altered, various parameters need to be optimized to get the best possible results. For example, the width of a peak integration region might always be adjusted to be close to FWHM or the width of fitted Gaussian peak profiles adjusted to match the FWHM of spectral peaks. Provided this optimization is always carried out, when the resolution is changed it is likely that the relative change in standard deviation will be similar for all methods that are close to perfect.

If we had a perfect model for the background and fitted it to all available background points in the spectrum, then for all practical purposes we could predict the exact background level beneath any peak. When we subtracted the correct background, statistical channel-to-channel fluctuations would remain. If we had a perfectly stable, well characterized spectrometer, we could use exactly the right functions to fit peaks by a weighted least-squares technique that would then give the most likely result for peak areas (see Appendix). This procedure would then be perfect because it would use data and prior knowledge to the fullest extent to maximize precision. If we are considering peaks within a narrow energy region, the precise form of the background is not critical, so we can simulate this ideal case by superimposing one or more pure Gaussian peaks on a flat background. Under these conditions, calculation of standard deviations for fitted peak areas is quite straightforward and is detailed in the Appendix.

### 8.4.2. Peak on Background with no Overlap

Figure 8.1 shows the parameters used in the simulation. If we used a multichannel analyzer (MCA) and acquired a real spectrum for a set time, the number of counts in the spectrum would stay the same, whatever the system resolution or channel width. The background level, $b$, would be equal to (background counts per eV) × (eV per channel) and the area, $A$, for the characteristic peak would stay constant. For the peak,

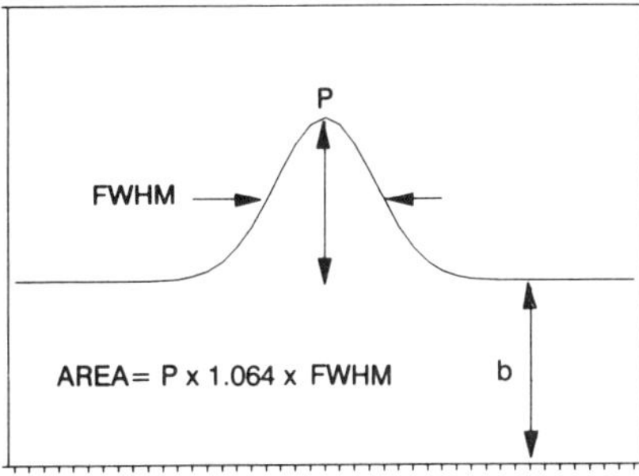


FIGURE 8.1. Single peak on a background: peak height, $P$, is given by 0.94 $A$/fwhm where $A$ is peak area.

a Gaussian shape is used where FWHM in channels = (resolution FWHM in eV)/(eV per channel). Peak height, $p$, is therefore close to (0.94 $A$/ FWHM).

Provided the background is negligible, i.e., $p \gg b$ , then the standard deviation in the fitted area is close to $A^{0.5}$, as we would expect from Poisson counting statistics. Of more practical interest is the situation when $p \ll b$, which corresponds to a very small peak on a high background. In this case, the standard deviation is given by $(1.505 \cdot \text{FWHM} \cdot b)^{0.5}$. A similar result has been calculated by Ryder[5]; although he used a non-weighted least-squares calculation, the results are identical if the spectrum counts are uniform. Thus, when background is dominant

$$\text{st.dev} = const. \text{ FWHM}^{0.5} \tag{8.1}$$

and the same relationship applies to detection limit or MMF. (Note: "*const.*" will be used throughout this paper to mean an arbitrary constant. In each equation, the value of the constant will certainly be different. Since "*const.*" is arbitrary, the units for FWHM are not important. However, if comparisons are to be made to cancel out "*const.*" the same units for FWHM must be used in each case.) This equation is also appropriate for x-ray mapping when the energy window is set equal to the FWHM; st.dev then quantifies the intensity fluctuation due to statistics that affects visibility for areas of low concentration. The relationship holds for any value of resolution fwhm provided $p/b \ll 1$ and there are sufficient channels in the spectrum (i.e., FWHM > 3 channels). If FWHM is reduced by 10%, then (8.1) shows that st.dev will be reduced by 5%.

### 8.4.3. Two Overlapping Peaks on a Background

Figure 8.2 shows the parameters for the two-peak simulations. Peaks of areas $A_1$ and $A_2$, (peak heights $p_1$, $p_2$) of the same FWHM and separated by "*sep*" are placed on a background level $b$, and ideal Gaussian profiles are fitted by weighted least squares to the data. The standard deviations that would be obtained in practice for the fitted peak areas are calculated for a range of FWHM values. (As a consistency check,

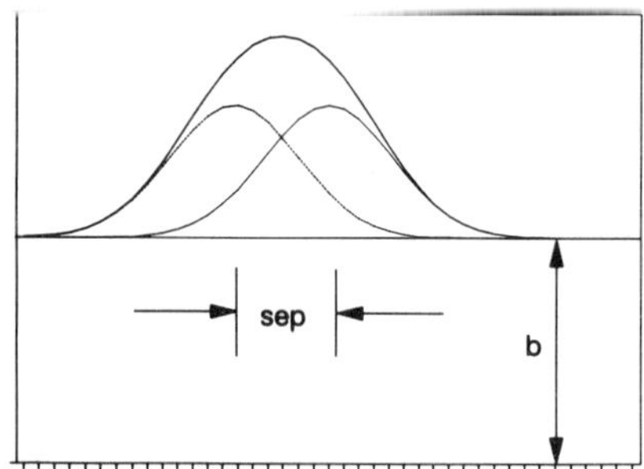


FIGURE 8.2. Two overlapping peaks on background: FWHM is the same but areas are $A_1$ and $A_2$ and *sep* is the separation between centroids.

a large value of *sep* (i.e., no overlap) was used to confirm that predicted standard deviations were identical to those for a single peak on a background. Furthermore, provided FWHM > 3 channels, the results were checked to be independent of the choice of eV per channel.)

It is convenient to determine what power law best expresses the variation of standard deviation with resolution under specific conditions. To do this, the calculated standard deviation values for a range of FWHM values between plus and minus 10% are fitted to a formula of the form

$$\text{st.dev} = const. \ \text{FWHM}^n \tag{8.2}$$

to find the best fit for the exponent $n$. The constant, *const,* will depend on many factors but is not relevant to the current discussion. The power law is appropriate because in certain limiting conditions it is exactly applicable.[5,6] Moreover, (8.2) appears to fit the observed results to better than 1% relative for most situations over $a \pm 10\%$ change in FWHM so it provides a useful parameterization of the results. For small changes in resolution, the power law provides an easy method of seeing the influence on st.dev: a 1% change in resolution produces $n\%$ change in st.dev.

For the two-peak simulations, values of standard deviation for the smaller of the two peaks have been calculated and fitted to (8. 2) as described to find the appropriate constant $n$. The results are summarized in Table 8.1, which shows the exponent $n$ for various regimes of interest.

If peaks are dominant, in situations of severe overlap (*sep*/FWHM < 0.1),

$$\text{st.dev} = const. \ \text{FWHM} \tag{8.3}$$

This equation corroborates the results of Fiori and Swyt for their simulations of overlapping Pb$M$ + S$K$ peaks.[8] It also agrees with Ryder's calculations for non-weighted least squares fitting.[5] The formula is applicable to the results for either peak, irrespective of their relative heights. At the other extreme, where peaks hardly overlap, then st.dev = $A_1^{0.5}$ (or $A_2^{0.5}$), which is independent of FWHM, so $n = 0$ in the Table 8.1. In between, for intermediate overlap, st.dev may depend more strongly on FWHM than at either extreme. This is because, when a small peak is close to a very large neighbor, the st.dev of the fitted result is affected by the skirt of the neighbor, which is like a local background. For $A_1{:}A_2 = 1{:}1000$ and *sep*/FWHM = 1, $n$ is about 2.5 although this is unlikely to represent a case of practical interest. For a more likely situation where $A_1{:}A_2 = 1{:}100$ and *sep*/FWHM = 1, $n$ is about 1.5. Thus, Table 8.1 shows that $n$ can take on values between 0 and 2.5 in marginal overlap situations. (The precision of the large peak in these extreme situations is close to $A_2^{0.5}$ so is insensitive to FWHM changes.)

If background is dominant, in the severe overlap situation,

$$\text{st.dev} = const. \ \text{FWHM}^{1.5} \tag{8.4}$$

a relationship which can also be confirmed analytically.[5,6] This formula is appropriate when estimating detection limits for trace elements. As the peaks become separated, the dependence on FWHM falls off smoothly toward the no-overlap case, so $n$ takes on values between 0.5 and 1.5 in the region of intermediate overlap in Table 8.1.

### 8.4.4. Three Overlapping Peaks on a Background

For more than two peaks, it is difficult to make generalizations across the variety of combinations of separations and relative peak heights. However, the symmetric case (Figure 8.3) where a peak of interest of area $A_1$ is overlapped on both sides by peaks of area $A_2$ at energy separation *sep*, can be analyzed in much the same way and gives a feel for how the number of peaks affects the dependence on FWHM.

If peaks are dominant, in the severe overlap situation, the formula

$$\text{st.dev.} = const. \ \text{FWHM}^2 \tag{8.5}$$

fits the simulated results, irrespective of the relative areas of the peaks. As separation increases, the sensitivity to FWHM depends greatly on the relative heights of the peaks. For example, if $A_1{:}A_2 = 1{:}100$ and $(sep/\text{FWHM}) = 1.4$, $n = 2.7$, whereas if $A_1{:}A_2 = 1{:}1000$ , $n = 4.7$. The dependence on FWHM is clearly greater than in the two-peak case; this is to be expected because the local background for the peak in the middle is affected by changes in the tails of two neighboring peaks rather than just one.

If background is dominant, and overlap is severe,

$$\text{st.dev.} = const. \ \text{FWHM}^{2.5} \tag{8.6}$$

fits the simulated results. As separation increases, the exponent drops away from 2.5 and eventually reaches 0.5 when the peaks are well separated, as for the single peak case. For example, at $(sep/\text{FWHM}) = 0.5$, $n = 2.2$ and at $(sep/\text{FWHM}) = 1$, $n = 0.9$ . This situation corresponds to the case of a small $K$-peak with a pair of $L$- or $M$-peaks nearby (e.g., Si$K$ straddled by Ta$M$ and W$M$), where peak fitting is required to determine which elements are present.

### 8.4.5. Five Overlapping Peaks on a Background

As described in Section 8.2.4 above, if we consider all chemical elements there are many situations of extreme overlap which involve five peaks (e.g., 1 $K$, 2 $L$ and 2

TABLE 8.1. Dependence of Standard Deviation on Resolution for the Smaller of Two Peaks Fitted to Spectrum Data. Separation, *sep*, FWHM and background *b* are as defined in Figure 8.2. $n$ is the exponent for the equation *st. dev = const.* $\text{FWHM}^n$

| | Separated, $sep/\text{FWHM} > 2$ | Some overlap | Severe overlap, $sep/\text{FWHM} < 0.1$ |
|---|---|---|---|
| Background dominant, $p/b \ll 1$ | $n = 0.5$ | $0.5 < n < 1.5$ | $n = 1.5$ |
| Peaks dominant, $p/b \gg 1$ | $n = 0$ | $0 < n < 2.5$ also varies with relative heights | $n = 1$ |

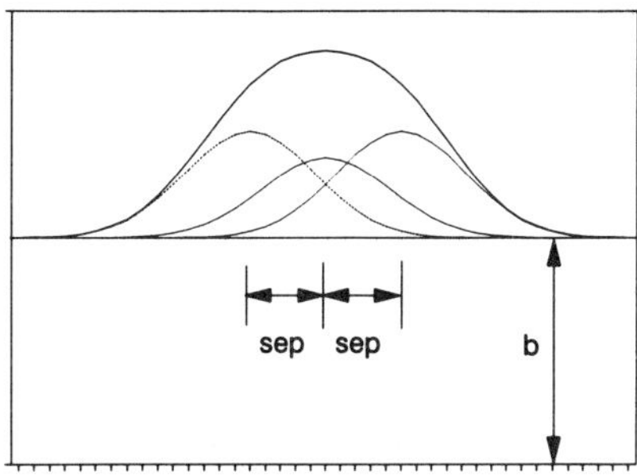

FIGURE 8.3. Peak with overlapping neighbours: FWHM and separation, *sep*, between peaks are the same. The central peak of interest has area $A_1$ and the outer peaks have area $A_2$. For the five peak simulation described in the text, further peaks are added either side.

$M$), so the last set of simulations is for five equally separated peaks on a high background. No simulation was performed for the peak-dominant case since this was thought to be less relevant to the aims of the current study. For the background-dominant case, when overlap is severe,

$$\text{st.dev} = const.\ \text{FWHM}^{4.5} \tag{8.7}$$

When overlap is less severe, the dependence on resolution is less (i.e., smaller exponent, $n$). Equation (8.7) corresponds to determining which elements are present and which are below the limit of detection when an overlapping cluster appears (e.g., P$K$ + Zr$L$ + Ir$M$ + Pt$M$ + Au$M$). It is possible that when a peak is detected that could be a multiplet, the results for all five potential contributing elements fall below the limit of detection so no unique identification is possible! Thus, for qualitative analysis where we simply want to determine if an element is present or not in the spectrum from an unknown sample, a small change in FWHM can have a dramatic effect on the performance of automatic peak identification software. Equation (8.7) suggests that where five peaks are involved, st.dev and hence detection limit would be improved by about 40% for a mere 10% improvement in FWHM. To put this in context, a factor of 2 increase in number of counts would improve detection limits by about 30%.

Although no formal analytical proof has been attempted, the results so far suggest the following generalization for cases of *extreme* overlap when there are m overlapping peaks:

$$\text{st.dev.} = const.\ \text{FWHM}^{(m-0.5)} \tag{8.8}$$

for the background-dominant case, and

$$\text{st.dev.} = const.\ \text{FWHM}^{(m-1)} \tag{8.9}$$

where background is negligible.

### 8.4.6. A Specific Example: SRM-168

SRM-168 is a standard alloy that contains 0.8% Si, 0.95% Ta and 3.95% W, among other elements. A spectrum was acquired at 15 kV excitation using a Si(Li) EDX spectrometer to observe the relative heights of peak and background. By taking account of approximate relative excitations of Si$K$, Ta$M$ and W$M$ under these conditions, the realistic simulation described in Table 8.2 was set up to demonstrate the effect of change of FWHM on ultimate attainable precision. The predicted standard deviations for two different resolutions were converted to equivalent weight percent of the elements. The FWHM for peaks at this energy (1.74 keV) in the real experiment was 97 eV. The simulation demonstrates that if the FWHM were to be degraded by 13% to 110 eV, then the standard deviation for Si would increase by almost 40%, that for Ta would increase by 33%, and for W by 24%. For comparison, note that (8.6) predicts that a 13% increase in FWHM would increase standard deviation by 36%. This confirms that the three-peak generalization, (8.6), does predict the expected level of change quite well for practical cases.

### 8.4.7. General Observations

Reducing FWHM improves precision and detection limits. Resolution becomes increasingly important as

1.  Peaks get closer together.
2.  Peaks get smaller compared to background or tails of nearby peaks.
3.  The number of peaks that need to be deconvolved increases.

## 8.5. FIGURES-OF-MERIT FOR ANALYSIS SPEED

### 8.5.1. Definition of Analysis Speed

All of the simulations and discussion about the influence of resolution on precision given above have been for spectra with the same number of counts. For any

TABLE 8.2. Results of Simulation for 15 kV Excitation SRM-168 Spectrum. Background of 60 counts/eV at 10 eV/$ch$ was used, giving a background of 600 counts per channel. Only a single major peak in each series was fitted

|  | Si$K$ | Ta$M$ | W$M$ |
|---|---|---|---|
| Centroid, eV | 1740 | 1710 | 1775 |
| Area, counts | 3600 | 1900 | 8200 |
| Weight % | 0.8 | 1.0 | 4.0 |
|  |  |  |  |
| *st. dev.* wt% |  |  |  |
| @ 97eV FWHM: | 0.13 | 0.18 | 0.17 |
| @ 110eV FWHM: | 0.18 | 0.24 | 0.21 |

given result, the standard deviation will also depend on $N^{-0.5}$, where $N$ is the number of counts in the spectrum. $N$ can be increased by acquiring data for a longer time, $T$, or by boosting acquisition rate, $R_a$ . Here, the most basic definition of acquisition rate, accumulated counts per *real* second, $R_a = N/T$, is implied rather than a live time-corrected rate or timing based on live seconds, because the *real* time it takes to complete an analysis is most important.

After pressing the "start" button, the analyst may be watching a spectrum display and waiting until enough counts are accumulated so that the peaks are smooth enough to identify. Alternatively, if a map is being acquired, successive image scans may be summed until enough counts are recorded so that a feature appears out of the "snow" due to counting statistics. A spectrum may be repeatedly processed during acquisition to determine if a trace amount of an element is present; as soon as it is above the detection limit, the acquisition can be terminated. The spectrum may have to be matched to one of a series of alternatives, but enough counts must be recorded to remove any uncertainty as to which provides the best match. These examples illustrate that the time required for an analysis is the time it takes to accumulate enough counts to achieve a prescribed statistical precision (i.e., in terms of st.dev or detection limit, for example).

If we take the standard deviation, st.dev, predicted for any of the situations above [(8.1)–(8.9)] where the total counts in the spectrum are not variable, then if we consider the effect of increased total counts,

$$\text{precision} = \textit{const.} \text{ (st.dev.)} \cdot (R_a \cdot T)^{-0.5} \tag{8.10}$$

Therefore, if we fix the precision , the time to reach this precision is

$$T = \textit{const.} \text{ (st.dev)}^2 / R_a \tag{8.11}$$

(*Note: "const."* is not the same constant as in (8.10).)

If we now define "analysis *speed*" as the number of analyses we can perform in a given time, then

$$\textit{speed} \text{ in analyses per second} = \textit{const.} R_a / \text{ (st.dev)}^2 \tag{8.12}$$

As we might expect, *speed* depends on count rate, $R_a$, but it also depends on what method is used to process the data since this influences the st.dev in any derived result. If time has an associated cost, then a system configured to give twice the *speed* will halve the cost of analysis; therefore, *speed* is a quantitative benchmark for comparison that is a direct indicator of practical benefit. If count rate and resolution are the only variables to consider when comparing two configurations or two different spectrometers, we need a figure-of-merit, or quality factor, derived from $R_a$ and FWHM that is directly proportional to *speed* of analysis. As shown in Section 8.4 above, the dependence of st.dev of a calculated result on FWHM depends on both the background magnitude and the degree of peak overlap, so it is necessary to consider a series of quality factors, where each factor applies to a different situation.

### 8.5.2. Detection Limits in Simple Spectra: $Q_1$

If the average local background can be estimated accurately and there are no overlaps that need to be deconvolved by fitting, then (8.1) is appropriate. By using (8.1) to substitute for st.dev in (8.12) and ignoring arbitrary constants, we obtain the following figure-of-merit proportional to speed:

$$Q_1 = R_a \,/\, \text{FWHM} \tag{8.13}$$

This quality factor, which can be applied over any range of FWHM where peak overlap is not significant, is primarily of use when the limit of detection of an element is of interest. It is also appropriate for x-ray mapping where, to maximize contrast, an energy band of interest is defined spanning roughly the FWHM of a peak, and counts falling in that band are integrated for each image pixel. Any units can be used for either $R_a$ or FWHM but, of course, the same units must be used when making comparisons.

### 8.5.3. Precision with Two Unresolved Peaks: $Q_2$

If there are two elements present in major concentration and there is severe peak overlap that can only be resolved by least-squares fitting, for example, then (8.3) is appropriate and the corresponding figure-of-merit is

$$Q_2 = R_a \,/\, \text{FWHM}^2 \tag{8.14}$$

This quality factor can only be used for comparisons over a range of FWHM because, as FWHM gets very small, the peaks become separate and the benefits are better described by $Q_1$. It is appropriate for specific applications where the precision in determining a major element concentration is of interest. Typical examples of two elements where the major lines are closely spaced ($< 50\,\text{eV}$) are $SK + MoL$, $SK + PbM$, $NbL + HgM$, $PK + ZrL$, $PK + PtM$, $ZrL + PtM$, $YL + OsM$, $SiK + WM$, $SiK + TaL$, $RbL + TaM$, $AlK + BrL$, $NaK + ZnL$, $FK + FeL$, $OK + VL$, and $NK + ScL$.

### 8.5.4. Detection Limit with Two Unresolved Peaks: $Q_3$

There are many situations where detection limits for trace elements are of interest, but overlap from a nearby peak is so severe that least-squares fitting is required to resolve the constituents. In this case, where background is dominant, (8.4) applies and the corresponding figure-of-merit is

$$Q_3 = R_a \,/\, \text{FWHM}^3 \tag{8.15}$$

Again, this quality factor is only appropriate for comparisons between cases where FWHM is large enough to cause severe overlap.

### 8.5.5. Precision for Three Unresolved Peaks: $Q_4$

Below about 3 keV energy, there is an increasing likelihood for the major lines of three elements to be close enough together to give an unresolved cluster. Notable examples involving commonly occurring elements are $SK + MoL + PbM$, $NbL + AuM + HgM$, $PK + ZrL + PtM$, $YL + OsM + IrM$, $SiK + TaM + WM$, $NaK + ZnL + GaL$, and $OK + VL + CrL$. If all elements are present in major concentrations so that peaks are well above background level but are still unresolved, then (8.5) describes the best attainable standard deviation and the corresponding figure-of-merit is

$$Q_4 = R_a \, / \, \text{FWHM}^4 \tag{8.16}$$

This quality factor is appropriate where three large elemental peaks overlap and precision of concentration measurement is of concern. Again, it is only applicable over a range of FWHM values for which peaks severely overlap.

### 8.5.6. Detection Limit for Three Unresolved Peaks: $Q_5$

If the same elements as in section 8.4.5 are expected to be present, but are at very low concentrations so that background is dominant, then (8.6) will show how the detection limit varies with FWHM and the corresponding figure-of-merit is

$$Q_5 = R_a \, / \, \text{FWHM}^5 \tag{8.17}$$

This quality factor is appropriate when detecting a trace element if it is possible that two of the others in the triplets mentioned above are also present. As before, comparisons using $Q_5$ are only valid provided the peaks are unresolved at both values of FWHM.

### 8.5.7. Automatic Peak Identification: $Q_6$

For fully automatic peak identification, it is necessary to deconvolve all unresolved multiplets into constituent elemental peaks. Although in principle it is possible to use $K/L$ or $L/M$ ratios to strip peaks, the high energy line of such a pair may be considerably smaller or perhaps not even excited if the incident beam voltage is too low. If an element produces a peak series, such as $K_\alpha + K_\beta$ with known relative intensities, and one of these peaks is well resolved from other peaks in the spectrum, it can be used both to detect whether the element is present and to strip the remaining peaks in the series from the spectrum. However, if the element is close to the limit of detection, only the peak from the major line will be significant and subsidiary lines are then not useful for either stripping *or* detection. Therefore, if the peaks from major lines of elements severely overlap, and peaks are close to the background level, then it is necessary to use some technique like least-squares fitting to resolve which elements are present. It is not difficult to find groups of five elements that have closely spaced ($< 50$ eV ) major lines (e.g., $PK + ZrL + IrM + PtM + AuM$ or $SiK + RbL + SrL + TaM + WM$) that would be unresolved at 100 eV resolution. For five unresolved peaks close to background level, (8.7) applies and the corresponding figure-of-merit is

$$Q_6 = R_a / \text{FWHM}^9 \qquad (8.18)$$

This quality factor is suitable for comparing how fast a system can achieve a unique identification of a given element when it is present in low concentrations. With limited prior knowledge, most elements in the periodic table must be considered as possible candidates, so if a peak appears in the spectrum close to any of the above example lines, all five peaks must be included in a least-squares fit or equivalent deconvolution to resolve the multiplet, as discussed in Section 8.4.5.

For x-ray energies around 2 keV, $Q_6$ is appropriate over a range of FWHM from about 80 eV to 120 eV . Below 80 eV FWHM, the overlap is not so severe for such quintuplets, so a smaller exponent than 9 would be more applicable. Above 120 eV, six candidates, e.g., $PK + YL + ZrL + OsM + IrM + PtM$, or even eight candidates, e.g., $SK + NbL + MoL + TcL + HgM + TlM + PbM + BiM$, may need to be resolved if a peak appears close to 2.0 keV or 2.3 keV. In this case, (8.8) suggests that a suitable figure-of-merit may be $R_a/\text{FWHM}^{11}$ or even $R_a/\text{FWHM}^{15}$. This argument could be extended to even higher values of FWHM, where even more peaks must be considered as possible candidates and the dependence on FWHM becomes even more dramatic. However, there reaches a point when detection limits are so high that automatic peak identification is practically impossible. In this case, the spectrometer can only be used if prior knowledge of the sample is available to restrict the range of possible elements.

## 8.6. THE RESOLUTION/COUNT RATE TRADE-OFF

Sample excitation, detector collection solid angle, and window transmission govern the count rate presented to the transducer. For an EDX spectrometer with a 10 mm$^2$ or 30 mm$^2$ detecting crystal, it is comparatively easy to collect more than $10^5$ photons per second from a bulk specimen in SEM. For a thin specimen in a TEM/STEM, the photon rate is often less than 1000 cps, although much higher rates can be achieved using field emission guns and detectors with high solid angles,[9] With other transducers, the physical size of the sensing element may be much smaller and limit the available photon input rate. However, even if we are able to increase photon input rate without limit, the maximum possible acquisition rate will usually be limited in some way. For example, in a bolometer, which measures the heat deposited by an individual photon, thermal time constants are currently in the range 1–10 ms.[10] In a solid-state EDX spectrometer, the maximum acquisition rate is determined by the electronics because of the need to spend time reducing the noise contribution to each photon measurement. Basically, if more time is spent measuring each signal step, noise fluctuations are averaged out more effectively and resolution is improved. However, this also increases the chance that another photon arrives before the measurement is complete, whereupon the measurement will have to be aborted and a count will not be recorded in the spectrum. Therefore, there is a connection between resolution and count rate capability, and the remainder of this section will concentrate on this aspect of a typical EDX system.

The acquisition rate, $R_a$, is always less than the input rate of photons striking an EDX detector. If SEM beam current is gradually raised, $R_a$ increases to a maximum, $R_{max}$, and then actually begins to decrease. At the maximum, $R_{max}$ is typically 30% of the input photon rate. Meanwhile, provided the electronics are well designed and artifacts can be removed (see Section 8.2.1 above), resolution (e.g., FWHM at 5.9 keV) remains virtually constant, so the best statistical precision can be obtained by adjusting SEM beam current for maximum acquisition rate. On most EDX spectrometers, a multiway switch, sometimes computer controlled, allows the electronic time constant or process time, PT, to be altered by the operator. If a short PT is chosen, the maximum acquisition rate is higher, but resolution will be worse than at longer PT settings. For example, Figure 8.4 summarizes typical performance of three different EDX spectrometers, A, B, and C, in terms of resolution and corresponding maximum acquisition rate values as published in manufacturers' literature (data summarized in Table 8.3). Note that these curves do not represent the change in FWHM with count rate; they merely summarize the trade-off between resolution FWHM and maximum achievable acquisition rate.

EDX detector "resolution" specifications invariably refer to the best resolution that can be achieved, typically at low rate, 1000 cps, with a high PT setting. However, the quoted "maximum acquisition rate" for a spectrometer refers to the minimum PT setting. Thus, spectrometer B in Figure 8.4 might be quoted as "133 eV resolution, maximum acquisition rate 30 kcps"; unfortunately, such a brief specification does not reveal that 133 eV resolution cannot be obtained at the same PT that permits 30 kcps acquisition rate! To add to the confusion, an identical system might be specified as "133 eV resolution, 100 kcps throughput capability" where "throughput" refers to input photon rate rather than the more pertinent acquisition rate!

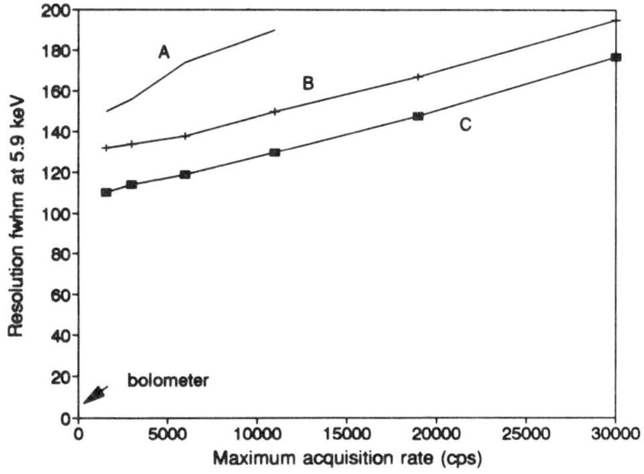


FIGURE 8.4. Resolution FWHM in eV at 5.9 keV at maximum rate. Data for systems A, B, C, and the bolometer are given in Table 8.3.

All EDX systems have circuitry that attempts to compensate for losses by generating a "live time clock" that runs slower than real time. If a spectrum is acquired for a set number of live time seconds then the same number of counts will be recorded as for loss-free counting with the same number of real seconds; of course, the experiment will take longer than it would with the perfect spectrometer. Since many EDX systems display live time rather than real seconds on the video display, or display an estimate of the input photon rate, it is not always straightforward to find out the true acquisition rate into the spectrum memory. To avoid any doubt, it is always wise to use a conventional stopwatch to do timings when checking the true acquisition rate, $R_a$, into the recorded spectrum.

Thus, in an EDX spectrometer, it is usually possible to choose PT either to get the best resolution and limit the maximum acquisition rate or to get the maximum acquisition rate at the expense of degraded resolution. If speed of analysis is important, the figures-of-merit or quality factors above can help to choose the optimum configuration.

## 8.7. EXAMPLES OF COMPARISONS OF SYSTEM PERFORMANCE

Here are the steps to follow to apply the figures-of-merit to an analytical problem:

1. Find a consistent measure of acquisition rate, $Ra$, which is proportional to the total counts per second that will actually be recorded in the spectrum in the

TABLE 8.3. Example Data Used for Text and Figures. Performance data for real systems A, B, and C were found in manufacturer's literature. The values for a hypothetical bolometer are based on the best results reported in ref. 10

| | Electronic process time settings | | | | | |
|---|---|---|---|---|---|---|
| | PT1 | PT2 | PT3 | PT4 | PT5 | PT6 |
| SYSTEM A (Si(Li)) | | | | | | |
| Max acquisition rate, kcps | 11.1 | 6.9 | 4.8 | 3.3 | 2.1 | NA |
| fwhm(5.9keV) @ max rate | 190 | 178 | 170 | 156 | 150 | NA |
| fwhm(2keV) @ max rate | 162 | 148 | 138 | 121 | 113 | NA |
| SYSTEM B (Si(Li)) | | | | | | |
| Max acquisition rate, kcps | 30 | 18.3 | 11.1 | 6.2 | 3.3 | 1.7 |
| fwhm(5.9keV) @ max rate | 195 | 167 | 150 | 138 | 134 | 132 |
| fwhm(2 keV) @ max rate | 168 | 135 | 113 | 97 | 91 | 88 |
| SYSTEM C (HPGe) | | | | | | |
| Max acquistion rate, kcps | 30 | 19 | 11 | 6 | 3 | 1.6 |
| fwhm(5.9keV) @ max rate | 177 | 148 | 130 | 119 | 114 | 110 |
| fwhm(2 keV) @ max rate | 157 | 123 | 101 | 86 | 79 | 73 |
| BOLOMETER (hypothetical) | | | | | | |
| Max acquisition rate, kcps | 0.06 | | | | | |
| fwhm(5.9keV) @ max rate | 7.3 | | | | | |
| fwhm(2 keV) @ max rate | 7.3 | | | | | |

region to be studied. This must take into account excitation of the specimen, collection solid angle, transmission efficiency of the detector window, and count losses in either transducer or electronics.

2. Establish the resolution FWHM for the peaks in the energy region to be studied.

3. Choose the appropriate quality factor $Q_1$–$Q_6$ according to the particular class of problem which is critical to the application.

4. Evaluate the quality factor for the different configurations to establish the relative benefits, the configuration with the highest $Q$ value being the most advantageous.

The word "consistent" in item above 1 means if we are comparing totally different detectors, then for a fair comparison we must use a measure of acquisition rate that includes all the counts contributing to the overlapping characteristic peaks and background over the same energy band of interest in both cases. This is because one detector may have a superior collection solid angle and count rate capability but quite a different detection efficiency at various energies, so it would not be consistent to simply equate $R_a$ to the total spectrum acquisition rate without making some correction for the different efficiencies. Also, the same take-off angle and kV must be used for sample excitation, for example. It is easier if we are comparing different configurations of the same system because the detection efficiency versus energy curve and solid angle for collection will stay the same; $R_a$ can then be the total spectrum acquisition rate because this will be proportional to the count rate in any chosen energy band. Item 2 above is also very important because resolution usually varies with energy and the count rate/resolution trade-off can be quite different at different energies. (Note that the units for FWHM are not critical but the same units must be used when making comparisons.)

In Figure 8.4 a point corresponding to the current state of the art in bolometers is included on the graph. The resolution achievement of 7.3 eV FWHM would make it the transducer of choice for all x-ray analysis were it not for the very low count rate capability of 60 cps; but is the low count rate that important if the resolution is so good? The classic way of assessing sensitivity for an analytical technique is to calculate the minimum mass fraction, MMF, which can be detected under practical conditions. The "standard" formula is often quoted in a form similar to this:

$$\mathrm{MMF} \sim (P \cdot P/B \cdot T)^{-0.5} \qquad\qquad (8.19)$$

(see for example Goldstein and Williams[11]). This relationship was derived in an early paper by Zeibold[12] and applied in particular to a Bragg crystal spectrometer (WDX) where the same effective acceptance band was used for peak and background measurements and $P \gg B$. Here, $P$ is peak count rate and $P/B$ is the peak-to-background ratio for a spectrum from a pure element recorded for an analysis time $T$. For an EDX spectrometer, the formula is still applicable provided the same energy band is used for integrating counts for background and peak. The optimum detection limit is obtained by choosing an energy band of about 1.2 FWHM[13] which always intercepts a constant fraction of the total peak area. Since the integrated background will then be proportional to FWHM, we see that (8.19) is basically equivalent to (8.1). Therefore, calculation of $Q_1$ is equivalent to the traditional approach for assessing sensitivity.

In Figure 8.5, $Q_1$ has been plotted for all the systems summarized in Figure 8.4. Using the maximum total acquisition rate for $R_a$ and the resolution at 5.9 keV for FWHM, and assuming that the same detection efficiency curve applies to the bolometer, then $Q_1$ is only about 8 for the bolometer and much higher values are achieved by all the EDX spectrometers. In particular, we see that system C (HPGe) operating at a PT setting that gives an acquisition rate of about 11 kcps and resolution of 133 eV gives $Q_1$ of over 80; this means system C would be able to analyze samples to the same MMF level ten times faster than the bolometer. In all of the EDX spectrometers shown, when PT is switched to smaller values to increase count rate capability, the reduction in precision due to peak broadening is more than compensated for by the increased count rate. Indeed, the provision of a PT setting allowing 30 kcps acquisition rate for system C gives it a $Q_1$ rating of about 170, which is about three times better than the best that can be achieved for any configuration of system A; in other words, system C could be configured to achieve analyses with the same MMF three times faster than could ever be achieved with system A.

It would be easy to conclude at this point that it is always worth trading resolution for rate. However, note that $Q_1$ and most traditional calculations for MMF are only applicable if no overlapping peaks have to be resolved. This may be true for x-ray mapping or simple x-ray spectra with only marginal overlap problems, but as soon as overlap has to be resolved, resolution becomes much more important. In Figure 8.6, published data from the same EDX systems has been used to plot $Q_3$ values; these indicate the speed of achieving a given MMF when two unresolved peaks have to be deconvolved by least-squares fitting, for example. Total acquisition rate has been used for $R_a$, but the resolution at 2 keV has been used for FWHM, rather than the resolution

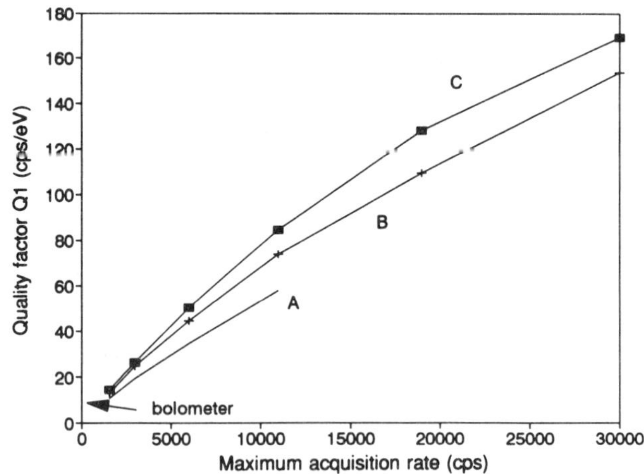


FIGURE 8.5. Peak detection at 5.9 keV: speed factor $Q_1$ plotted for the systems described in Table 8.3 using FWHM for 5.9 keV and maximum acquisition rate for $R_a$.

at 5.9 keV, because severe overlap is much more likely at the lower energy (see discussion in section 8.5). As pointed out above, relative $Q_3$ values are valid in the range of FWHM where overlap is still severe. For peaks separated by 30 eV (e.g., Si$K$ + Ta$M$), overlap is severe for FWHM = 157 eV, and even at the best resolution of 73 eV, peaks are still not resolved; use of $Q_3$ to compare speed at these two extremes gives results only about 10% different from the exact calculation. Neither system B nor system C has any PT configuration that gives FWHM as low as 73 eV FWHM at 2 keV, so it is realistic to use $Q_3$ for all of the configurations shown in Figure 8.6. For system A, increased count rate still offers improved speed of analysis, despite the degradation in FWHM but there are no more PT settings available to allow acquisition rates to exceed 11 kcps, so the highest speed corresponds to $Q_3$ = 0.0026. System C achieves a $Q_3$ value of 0.011 at an acquisition rate of 11 kcps, so it will reach a given level of sensitivity more than four times faster than at the optimum conditions for system A. However, in contrast to the single peak case, system C cannot gain any more benefit by moving to higher count rate capability because of the increased importance of FWHM for resolving peak overlap. Even though the best resolution of system C for 11 kcps acquisition rate is only 16% better than that for system B, system C achieves a $Q_3$ value 40% greater than system B, so it will be 40% faster at completing analyses to the same precision.

As soon as FWHM reaches the level where peaks are resolved, further improvements are much less advantageous, and the formula for $Q_3$ is no longer appropriate for comparison. For this reason, the bolometer has not been included on this graph. However, if we consider the specific case of two peaks separated by 30 eV, then results of simulation show that a bolometer with 7.3 eV resolution at 60 cps reaches a given MMF with the same speed as system C (73 eV FWHM at 1.6 kcps). Although the severe count rate limitation of the bolometer is still an impediment, the benefits of improved resolution are clearly much more useful when there are peaks to be resolved.

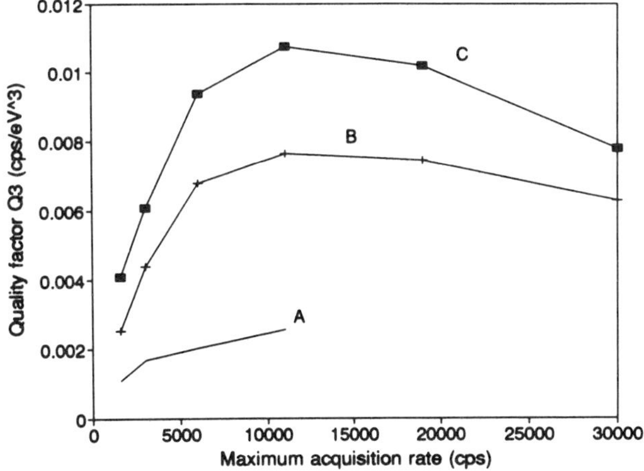


FIGURE 8.6. Detection at 2 keV with overlap: speed factor $Q_3$ plotted for systems in Table 8.3 using FWHM for 2 keV and maximum acquisition rate for $R_a$.

If we now consider the case of automatic peak identification, then $Q_6$ benchmarks the speed for a system to achieve a given level of selectivity in correctly identifying elements present. In Figure 8.7, the FWHM at 2 keV has again been used since this is the most likely region in which to see five or more unresolved major peaks (see Section 8.4), and $Q_6$ has been plotted for the same set of EDX systems. Again, it is worth questioning whether the simple formula for $Q_6$ can be applied over the range of FWHM values for all the systems shown and whether the specific assumptions given in Section 8.5.5 are appropriate for real spectra where peak separations are not all uniform. To test this, a specific set of calculations were made for the following case: $SiK(1740\,eV) + RbL(1694\,eV) + SrL(1806\,eV) + TaM(1710\,eV) + WM(1775\,eV)$. The ratio of $Q_6$ for FWHM = 101 eV relative to FWHM = 157 eV is only 7% higher than with the exact calculation and $Q_6$ for FWHM = 73 eV is 20% higher than the correct speed relative to that for FWHM = 157 eV. Thus, although $Q_6$ is not an exact measure of performance, it does seem to give a good guide for this class of problem, and Figure 8.7 does describe the general trade-off adequately. In most cases, configurations with improved resolution are superior to any with increased count rate. The optimum speed to achieve a given MMF or selectivity in peak recognition with system C ( PT6, 1.6 kcps, 73 eV FWHM) is 50 times greater than at its least optimal configuration (PT1, 30 kcps, 157 eV FWHM). Furthermore, this configuration is more than three times faster than the best that can be achieved for system B and 50 times faster than the best for system A. Clearly, if truly automatic peak identification is required, even a small improvement in resolution can offer substantial speed advantages. Although $Q_6$ is not appropriate when FWHM is small enough to resolve peaks, the relative speed for the hypothetical bolometer (7.3 eV FWHM, 60 cps) calculated for the same specific example ($SiK + RbL + SrL + TaM + WM$) shows that it would achieve the same MMF or selectivity 15 times faster than the best that system C could achieve, even though it would be operating at a count rate 25 times smaller! Although this may be difficult to

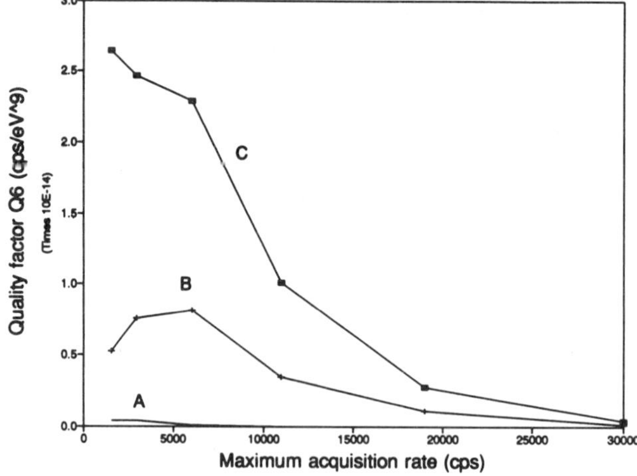


FIGURE 8.7. Identifying peaks at 2 keV: speed factor $Q_6$ plotted for systems in Table 8.3 using FWHM for 2 keV and maximum acquisition rate for $R_a$.

believe at first sight, the intuitive explanation is that, at 7.3 eV, the peaks are totally resolved so that, although we may have to wait longer for counts to accumulate, when a count appears in a given channel, we can be sure from which element it arises.


## 8.8. CONCLUSIONS

It is not the FWHM of a $MnK_\alpha$ peak that is important to the analyst but it is whether a given job can be done and how long it will take. On the basis that cost is usually proportional to time, speed of achieving a given MMF for example is more appropriate than comparing MMFs. Thus, speed of analysis is a good basis for choice between different technologies or different configurations of the same instrument.

Conventional MMF calculations neglect the effect of peak overlap, but a full calculation of statistical precision to cover every algorithm and type of spectrum is clearly impractical. Nevertheless, it is useful to have some simple rules of thumb when planning experiments. The series of figures-of-merit or quality factors, $Q_1$–$Q_6$, described in this paper cover many cases of practical interest and are directly related to speed of analysis in these situations.

Both EDX resolution and spectral intensity vary with energy so $Q_1$–$Q_6$ are not universal values and depend on definition of $R_a$ and FWHM. However, provided the chosen definition of $R_a$ refers to a true acquisition rate and FWHM refers to peaks in the energy range of interest, these quality factors can be used for comparing speeds. For $Q_2$–$Q_6$, comparisons are only valid over a range of FWHM that maintains the stated overlap conditions, but $Q_1$ can be used for any FWHM because no overlap is involved. If there is an opportunity to alter either the fwhm or the acquisition rate in the energy region of interest, the analyst should calculate the appropriate $Q_n$ to decide which combination will be fastest to achieve the desired MMF.

Count rates may be limited by beam current, collection solid angle, or window transmission efficiency. In this case, calculation of the appropriate $Q_n$ can still be useful in quantifying just how beneficial a small change in resolution would be. In the extreme case of automatic element identification, when it is essential to rely on spectrum processing to decide the presence or absence of elements, the conventional resolution specification of FWHM = 5.9 keV does not give a realistic picture of comparative benefits. For example, the standard specification for systems A, B, and C would be 150 eV, 132 eV, and 110 eV, respectively, but the relative analysis speeds for deconvolution in the critical region near 2 keV (Figure 8.7) would be in the proportion 1:15:50! Of course, these huge differences in performance can only be obtained if peak fitting software is designed to make best use of the available data.

If we look to the future and consider what new technologies may improve on the ubiquitous EDX, superconducting tunneling junctions (STJs) and bolometers offer the exciting promise of an order of magnitude improvement in resolution. Unfortunately, physical limitations in these transducers currently limit available count rates to at most a few 10s of counts per second; on the basis of $Q_1$, this makes their performance considerably poorer for x-ray mapping for example. $Q_3$ suggests that even for severe overlap of two peaks, a good EDX can achieve higher precision because of its much higher acquisition rate capability. However, when the spectrum itself is the only indicator of elemental content, a 7.3 eV device that could count at 60 cps would theoretically be much more effective than the best EDX available today, and would certainly appear to be the best choice for fundamental spectroscopy. In practice, instrumental artifacts including systems background must be eliminated and efficiency at low energies improved before such theoretical performance can be exploited, so these new devices still have some way to go to match the versatility and speed offered by Si(Li) and HPGe EDX spectrometers.

For the range of typical semiconductor EDX systems summarized by Figure 8.4, we have the following guidelines for analysis:

1.  Whenever peak overlap is not a problem, Figure 8.5 makes a compelling argument for pursuing higher acquisition rates . For example, high rate is very useful for x-ray mapping. In a specialized application such as silicate mineral analysis, where overlap corrections are usually small, high rate can be particularly advantageous provided practical restrictions of pileup and count-rate-dependent distortions are well understood.[14]

2.  As soon as it becomes necessary to resolve extreme overlaps, resolution is more valuable in the trade-off with rate. It is therefore critical to consider the FWHM for the particular peaks of interest because this may vary with count rate differently than the standard FWHM at 5.9 keV. Figure 8.6 suggests that even with the best EDX system shown, there is little advantage in going much beyond 10 kcps acquisition rate (or 30 kcps photon input rate at the detector) when two close peaks must be resolved.

3.  When automatic peak identification is required or clusters of peaks such as $SiK + WM + TaM$ are involved, resolution is of paramount importance. In this case, Figure 8.7 demonstrates that any significant degradation in FWHM is too high a price to pay for increased count rate capability and most systems should be operated at the best resolution possible, albeit at the highest count rate possible for the configuration.

# REFERENCES


1.  P. Statham, *X-Ray Spectrom.* **6**, 94 (1977).
2.  P. Statham, *X-Ray Spectrom.* **5**, 16 (1976).
3.  F. H. Schamber, in: *Proceedings of the Symposium on X-Ray Fluorescence Analysis of Environmental Samples* (T. Dzubay, ed.) Ann Arbor Science, Ann Arbor, MI, pp. 241–257 (1977).
4.  P. J. Statham, *Anal. Chem.* **49**, 2149 (1977).
5.  P. L. Ryder, in: *Scanning Electron Microscopy/1977*, ITT Research Institute, Chicago, Illinois, pp. 273–280 (1977).
6.  P. J. Statham, in: *Microbeam Analysis—1982* (K. F. J. Heinrich, ed.) San Francisco Press, San Francisco, pp. 1–4 (1982).
7.  P. J. Statham and T. Nashashibi, in: Microbeam Analysis–1988 (D. E. Newbury, ed.) San Francisco Press, San Francisco, pp. 50–54 (1988).
8.  C. E. Fiori and C. R. Swyt, *Microbeam Anal.* **1**, 89 (1992).

9. C. E. Lyman, J. J. Goldstein, D. B. Williams, D. W. Ackland, S. von Harrach, A. W. Nichols, and P. J. Statham, *Microbeam Anal.* **2**, S234 (1993).

10. D. McCammon, W. Cui, M. Juda, J. Morgenthaler, J. Zhang, R. L. Kelley, S. S. Holt, G. M. Madejski, S. H. Moseley, and A. E. Azymkowiak, *Nucl. Instrum Methods* **A326**, 157 (1993).

11. J. I. Goldstein and D. B. Williams, *Microbeam Anal.* **1**, 29 (1992).

12. T. O. Zeibold, *Anal. Chem.* **39**, 858 (1967).

13. S. J. B. Reed, *Electron Microprobe Analysis*, Second Ed., Cambridge University Press, Cambridge, UK, p. 166 (1993).

14. S. J. B. Reed, *Microbeam Analysis—1990* (J. R. Michael and Peter Ingram, eds.) San Francisco Press, San Francisco, pp. 181–184 (1990).

15. G. G. Guest, *Numerical Methods of Curve Fitting*, Cambridge University Press, Cambridge, UK, p. 260 (1961).

## APPENDIX

*Calculation of Standard Deviation in the Simulations*

For a digitized x-ray spectrum represented by a histogram of counts falling in contiguous energy bins of equal width, the counts $y_i$ in each channel $i$ are statistically independent and are distributed according to a Poisson distribution with variance close to $y_i$, provided $y_i \gg 1$. If it is known that a series of Gaussian peaks, $g_{mi}$, are superimposed on a known background, $b$, then to find the best estimates of area for the individual peaks which have the minimum variance,[15] we need the set of $C_m$ which minimises the expression:

$$\sum_i \left( y_i - b - \sum_m C_m \cdot g_{mi} \right)^2 \cdot W_i$$

where $g_{mi}$ is the value of the $m$-th peak profile at channel $i$, normalized to unit total area, $C_m$ is the area of the mth peak, and the statistical weighting, $W_i$, is the reciprocal of the variance in the original data point $i$. For the current purpose it is adequate to set $W_i = 1/y_i$. Differentiating the above expression with respect to each $C_m$ and equating the result to zero to find the minimum gives the "normal" equations in matrix notation as

$$[\mathbf{A}] \times [\mathbf{C}] = [\mathbf{B}]$$

where

$$A_{mn} = \sum_i g_{mi} \cdot g_{ni} \cdot W_i$$

$$B_m = \sum_i g_{mi} \cdot (y_i - b) \cdot W_i$$

The solution for the parameters $C_m$ is given by

$$[\mathbf{C}] = [\mathbf{A}^{-1}] \times [\mathbf{B}]$$

where $[A^{-1}]$ is the inverse of matrix $[A]$. The standard deviation in the fitted peak area $C_m$ is then given by

$$\text{st. dev.}_m = (1/A^{-1}{}_{mm})^{0.5}$$

where $A^{-1}{}_{mm}$ is the $m$th element on the diagonal of the inverse matrix.

# 9

# Improving EDS Performance with Digital Pulse Processing

*R. B. Mott and J. J. Friel*


ABSTRACT

Direct digitization of the preamplifier output of a Si(Li) detector, with all subsequent pulse processing performed digitally, enhances throughput versus resolution performance. Sensitivity, resolution, and especially pileup rejection for photons below 1 keV are also improved. Adaptive pulse shaping allows both low dead time operation with minimal loss of resolution and automatic adjustment to a wide range of beam current without requiring manual selection of processing time.

An overview of the operating principles of a digital pulse processor will be presented and compared with conventional approaches. Examples of spectra taken with the same spectrometer under the same conditions, changing only the pulse processors, will illustrate the advantages of adaptive digital processing.


## 9.1. INTRODUCTION

Each successive amplification and filtering stage of the EDS spectrometer processing chain marginally degrades the original signal generated by an incident photon. All of the information in the 0–10 V shaped pulse that feeds the analog-to-digital converter (ADC) in a conventional system is also contained in the step signal of a few tens of millivolts emerging from the charge-sensitive preamplifier.

It is intuitively attractive to convert the signal into a digital data stream as early as possible, for a number of reasons. First, once in digital form, all further manipulation of the signal is perfectly linear and noise-free. Second, digital filtering is precise and


R. B. Mott and J. J. Friel • Princeton Gamma-Tech, Inc., Princeton, New Jersey 08540



*X-Ray Spectrometry in Electron Beam Instruments*, edited by David Williams, Joseph Goldstein, and Dale Newbury. Plenum Press, New York, 1995.

reproducible compared to the imperfect tolerances and parasitic non-ideal behaviors that are unavoidable with analog components. Third, digital filtering function shapes are not constrained to those realizable with physical components. Within the limitations imposed by quantizing the signal into a finite number of bits at a finite sampling speed, filter shapes can approach the mathematically ideal. Finally, the perfect linearity of digital shaping offers the possibility of adapting the shaping time on a pulse-by-pulse basis, giving the best possible energy measurement for each photon.

The idea of using digital signal processing (DSP) at the preamplifier output is not new. The earliest functioning system of this type of which the authors are aware was constructed by Henriecus Koeman in 1973, working at Philips Research Laboratories. This seminal work, published in a series of three papers in 1975, described the theory and detailed design of a digital system, noted the crucial idea of adaptive processing, and demonstrated impressively high count rates even by today's standards.[1–3] Unfortunately, the best resolution obtained at low count rates was 200 eV at 5.9 keV, much worse than the 145 eV commercially available at the time for conventional systems.[4]

More recently, a group in Hungary led by Lakatos built a system very similar in concept, also with adaptive shaping. The work was published in 1990, so their claim to have built the "world's first real-time fully digital high resolution, very high throughput rate signal processing and analyzing system for semiconductor x-ray spectrometry" was a little overenthusiastic.[5]

Seventeen years of improvements in Si(Li) crystal processing, digitizing technology, and high-speed computational circuitry led to an improvement in best resolution to 160 eV for the Lakatos design, still far short of the 133 eV specified in recent product literature for several analog systems.

A different approach has been taken by a group in Germany, who have built a digital pulse processing system designed for gamma-ray energies using a tail-pulse resistive feedback preamplifier design.[6] Instead of adapting the shaping time pulse by pulse, the overlapped tail pulses are deconvolved in the time domain before shaping. Good results have been published for gamma-ray spectra, but the higher noise level of resistive feedback is not desirable for measuring low energy x rays.

The work described in this paper began with the idea of matching analog resolution at low count rates while improving resolution at higher rates. Commercial EDS resolution specifications are now within 10–15 eV of the theoretical dispersion limits of the semiconductor used (Si or Ge) for middle energies such as 5.9 keV. However, these resolutions are obtained only at such long shaping times that count rates are often below 1 kcps and acquisition for many minutes may be required for adequate counting statistics.

Nearly a quarter of the papers presented in the previous special publication on EDS were concerned with algorithms and statistical considerations for the deconvolution of overlapped peaks.[7–11] These techniques have since become well established. In view of this, some recent papers have suggested that for practical acquisition times it is often reasonable to use count rates well above the traditional few kcps, on the grounds that the resulting improvement in counting statistics outweighs the loss of energy resolution due to shorter pulse shaping times, even for close overlaps such as Pb$M$ with S $K$ lines.[12,13]

It became clear early in the project that the problem of pileup rejection for low energy lines had to be addressed. Current pileup rejectors function well at high energies, distinguishing pairs of photons separated by fractions of a microsecond, but rapidly become less effective at energies below 1 keV. Pileup of two carbon photons at 277 eV may not be detected unless they are separated by 15 μs.[14]

With the availability of new detector window materials which have usable transmission down to B (185 eV) or even Be (109 eV), yet can withstand atmospheric pressure without the mechanical complexity and vulnerability to contamination of windowless configurations, light element EDS has become much more widespread. As Statham points out, the inability to do adequate pileup detection for light elements adversely affects the quantitation of higher energy lines as well, limiting the usable count rate severely.[15] Even at modest count rates (500 cps in the low energy peak), undetected pileup with soft photons will cause counts to be shifted out of the higher energy major peaks into a pileup continuum shelf a few hundred eV above each peak.[15] As we will see later, the detectable interval for a low energy photon following a high energy one may be worse than for two low energy photons.

## 9.2. SYSTEM ARCHITECTURE

The differences between digital pulse shaping and analog pulse shaping are shown in the simplified block diagrams of Figure 9.1. The signal emerging from the charge-sensitive preamplifier is a rising stairstep with occasional resets to the lower operating limit. The mechanism by which the staircase waveform arises has been described many times and will not be repeated here.[16] In analog pulse processors, the

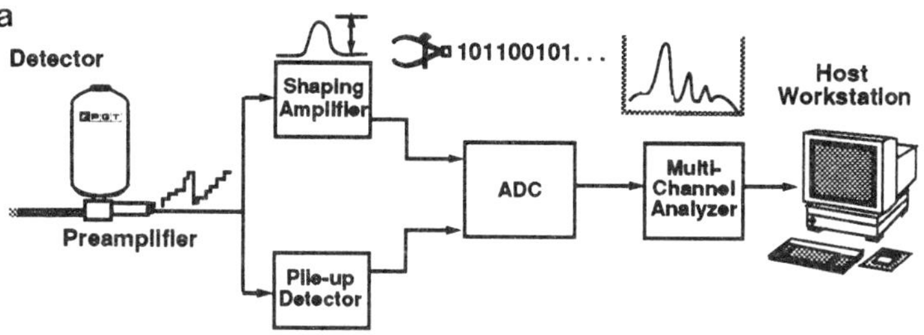


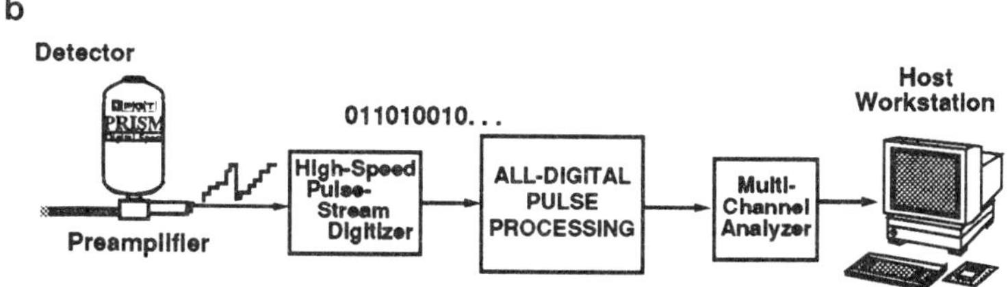


FIGURE 9.1. (a) Traditional analog pulse processing. (b) Digital pulse processing.

signal is then split between a shaping amplifier and one or more pileup detection channels. The height of the shaped pulse is measured by an ADC. The shaping amplifier usually has several processing time constants selectable by the operator, while the pileup detection channels have fixed time constants chosen by the manufacturer as a trade-off between resolving time and threshold energy. If a pileup channel's time constant is shortened, improving time resolution, the threshold energy must be raised to avoid false triggering by noise.

While the optimum trade-off is a continuous function of energy, as a practical matter analog designers must divide the energy scale into a few ranges and choose time constants so as to give noise-free detection of the lowest energy in each range. Therefore, in analog systems, resolving time is not optimum over most of the energy scale. This effect is most noticeable at very low energies, where a small decrease in threshold energy requires a large increase in processing time.

For each pileup channel, the best (shortest) resolving time is obtained for two photons just energetic enough to be detected by that channel. As photon energy increases, resolving time degrades because the shaped pulse remains above the detection threshold longer, until the detection threshold is reached for another pileup channel with a shorter time constant. In general, two nearly coincident photons will have differing energies. The resolving time will be determined by the fastest pileup channel which can detect the lower energy photon, but it will also depend on the arrival order. If a low energy photon arrives just before a high energy one, the faster pileup channel responds later than the slower channel, so the pileup can be detected. However, if the high energy photon arrives first, the slow pileup signal may not have recovered from its strong response to that photon. The trailing low energy photon will then be missed.

In Figure 9.2, the three waveforms show (a) the preamplifier signal with a high energy photon closely followed by one of lower energy; (b) the output of the fast pileup rejector, which sees the first photon but misses the second; and (c) the final shaped

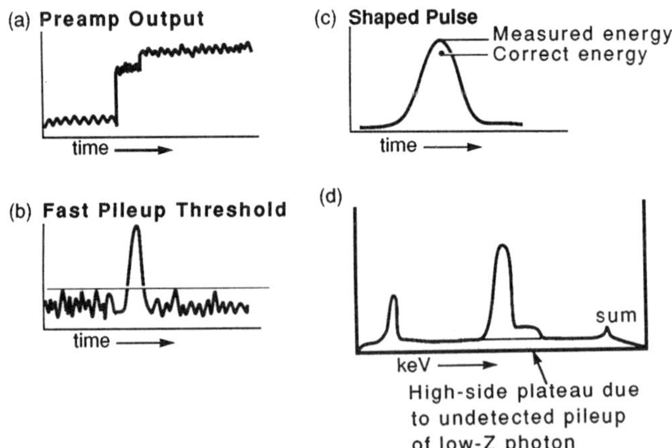


FIGURE 9.2. A trailing low energy photon is the worst case for analog pileup detection.

pulse, which is too high in energy. The spectrum of Figure 9.2(d) shows these undetected pileups spreading into a broad shelf above the higher energy emission peak, as described by Statham.[15]

Digital pileup rejection using pattern recognition techniques on the staircase waveform does not depend on the arrival order of photons. It is also possible to get much closer to the optimum resolving time over a wider range of energies at the low end of the energy scale.

Analog linear amplifiers require baseline restoration, a source of additional noise and complexity that is eliminated by all-digital processing. As described by Knoll, the AC coupling in the first differentiator stage of a shaping amplifier causes variable baseline shifts which must be compensated for by an active restoration circuit; otherwise the height of the shaped pulses will not represent the photons' energies accurately.[17]

Digital processing is not constrained by the physics of coupling capacitors. Figure 9.3 plots one reset cycle of the digitized staircase waveform. The process of digital filtering boils down to taking the difference of a properly weighted average of the signal level before and after each step response due to an incident photon. Therefore, a DC offset which shifts the entire digitized waveform up or down has no effect on the measured energy.

For analog system designers, ADC energy linearity is always cause for concern. Since an incident photon of a given energy is always shaped into the same pulse height by the analog linear amplifier, the centroid of a characteristic peak always falls in the same ADC conversion bin for analog systems. If the ADC's bin widths are not uniform, distortion of the peak shapes will result. Differential nonlinearity is the term for these local variations in bin width, while integral nonlinearity measures error over the full-scale range of the ADC. Both types of error are important in spectroscopy.

Thirty years ago, the sliding scale averager was introduced as a way to reduce the effects of differential nonlinearity.[18] In this technique, a high precision digital-to-analog converter (DAC) generates an offset signal which is added to the pulse height before it reaches the ADC, shifting the result to a different conversion bin. The digital

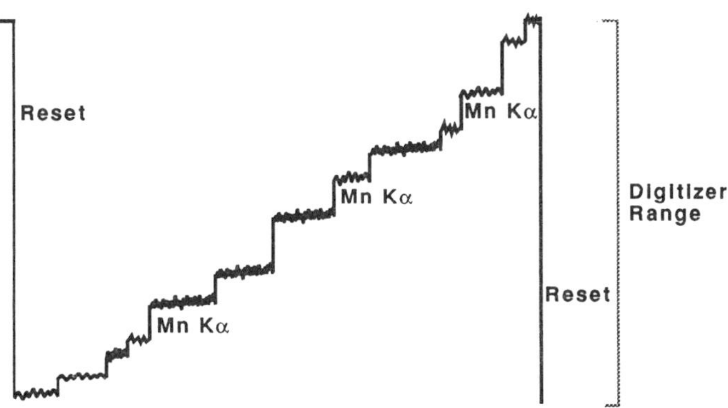


FIGURE 9.3. Digitizing entire preamplifier waveform.

offset value is then subtracted from the converted result to recover the correct digital pulse height. Sliding scale averaging improves differential linearity by a factor approximately proportional to the number of channels in the slider DAC, at the cost of inducing fractional channel centroid shifts in channels near a channel with a large bin-width error. These secondary errors also decline with slider width.[19]

In Figure 9.3, we see that since the entire staircase waveform is digitized, a photon step of a given energy may appear anywhere in the digitizer's range. This is equivalent to a sliding scale averager with zero DAC noise and almost as many slider channels as the ADC itself. Therefore, energy linearity of the digital pulse processor is close to ideal.

While very little additional linearity error is introduced by the digital pulse processor, this does not imply perfect linearity of calibration for low energy peaks that are absorbed close to the front face of the detector. The model described by Joy shows that such low energy lines may be shifted down in energy from their theoretical positions through recombination of some fraction of the electron-hole pairs, depending on various physical parameters of the detector.[20] No pulse processing technique can measure charge that is never collected! However, digital pulse processing does eliminate the shift to lower energies that may occur due to failure of an analog linear amplifier's baseline restorer at high count rates.

Digital pulse processing also eliminates the extra dead time imposed by analog-to-digital conversion after pulse shaping and the time required for the linear amplifier output to return to baseline. At all but the highest count rates, the latter time is dominant. The total dead time for purposes of pileup rejection can be surprisingly long compared to the rise time of the pulse shaping, which determines resolution. Goulding and Landis tabulated the total dead time as a ratio to rise time for a number of shaping functions.[21] Four common shaping functions are given in Figure 9.4, with their noise figure-of-merit and the dead-time-to-rise-time ratio.

The best case dead time is three times longer than the rise time for a triangle shaper. The point made by Goulding is that a pulse cannot be measured if the interval to the previous pulse is less than the total processing time, nor if the next one occurs within the rise time. The Kandiah/Harwell gated integrator looks like it has a very fast return to baseline after shaping, but in fact a "protect time" equal to the processing time is required for the switching noise to settle out.[22] A digital processor can make better use of the available interval between photon events for shaping, thus having a longer effective integration time for a given throughput as well as a near-ideal shape. This will be explained in more detail shortly.


## 9.3. DIGITIZER CONSIDERATIONS

The performance of the waveform-digitizing ADC is obviously critical to overall system performance. While the maximum available bit depths and sampling speeds improve every year, we can identify some general design considerations which should remain valid.

The minimum bit depth is determined by the need to have the peak-to-peak noise excursions in the staircase waveform span a sufficient number of digitizer bins to avoid

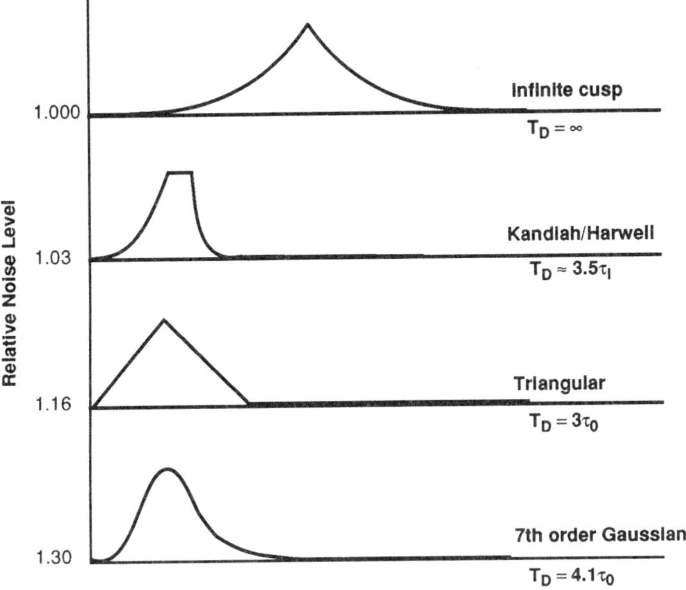


FIGURE 9.4. Various pulse shapes and their noise levels.

quantization artifacts. The concept is the same as the sliding scale averaging discussed above, but the natural noise of the signal provides the bin dithering, instead of an artificially added offset. The reader may have gathered from previous discussion that digitizer linearity is relatively unimportant. On the contrary, it is crucial, but poor linearity manifests itself as a loss of resolution rather than uneven channel widths in the spectrum.

At the low signal levels provided by the preamplifier, the noise performance of the digitizer's input sample-and-hold circuit is also more important than for a pulse-height ADC.

Finally, with respect to sampling rate, since signal-to-noise ratio (SNR) improves with frequency over a wide bandwidth, performance improves with sampling rate–provided nonlinearity and input noise remain constant. Sampling rate also puts a lower bound on resolving time for pileup detection. In practice, faster converters tend toward worse noise and linearity, and the optimum trade-off among these parameters has yet to be reduced to a neat set of equations. Preamplifier response characteristics also influence digitizer design choices. Considerable experimentation is required for any particular system configuration.

## 9.4. BASICS OF DIGITAL SIGNAL PROCESSING SHAPING

The fundamental operation of high speed digital signal processing is time domain convolution. Each sample of the staircase waveform, and a certain number of samples preceding and following it, are multiplied by the corresponding values of a digital

| Data Input Samples→ Time↓ | $S_0$ | $S_1$ | $S_2$ | $S_3$ | $S_4$ | $S_5$ | $S_6$ | $\cdots$ | $S_n \to S_t$ |
|---|---|---|---|---|---|---|---|---|---|
| $t_2$ | $W_0$ | $W_1$ | $W_2$ | $W_3$ | $W_4$ | | | Convolution Weights | |
| $t_3$ | | $W_0$ | $W_1$ | $W_2$ | $W_3$ | $W_4$ | | | |
| $t_4$ | | | $W_0$ | $W_1$ | $W_2$ | $W_3$ | $W_4$ | | |
| . | | | | | | | | | |
| . | | | | | | | | | |

$$C_2 = S_0W_0 + S_1W_1 + S_2W_2 + S_3W_3 + S_4W_4$$

$$C_3 = S_1W_0 + S_2W_1 + S_3W_2 + S_4W_3 + S_5W_4$$

$$C_4 = S_2W_0 + S_3W_1 + S_4W_2 + S_5W_3 + S_6W_4$$

**Convolution Output:** $\quad C_2\, C_3\, C_4 \ldots C_t$

FIGURE 9.5. Convolution with a finite impulse response filter.

weighting function. The products are then summed and scaled to produce one output sample of the convolution. The weighting function is shifted by one sample relative to the data stream and the process repeated to generate the next output sample, as illustrated in Figure 9.5. This type of filter is also known as a transversal filter or finite impulse response (FIR) filter, so called because no data sample outside the weighting function interval can influence the result. By contrast, exponential decay with theoretically infinite memory is characteristic of analog shaping.

This type of digital filtering is familiar to the microanalysis community in another context. If the channels of a spectrum are regarded as "samples" in energy rather than time, the mechanics of the process are identical to the top hat filtering used in filter/fit spectrum background removal, although the weighting functions are quite different. As a useful consequence of FIR filter behavior, there is no need for a long linear-amplifier recovery period after a reset. Only the settling time of the preamplifier itself need be considered.

In contrast to analog processing, it is more convenient to think of the digital processing time as being centered on the rising step of the photon event in the digital data stream as shown in Figure 9.6. A simple digital filtering function is a series of -1 weights followed by a series of +1 weights, as shown in Figure 9.6(a). Time increases to the right, so the reader should envision new data samples appearing at the right edge of the figure and marching right to left like a ticker tape.

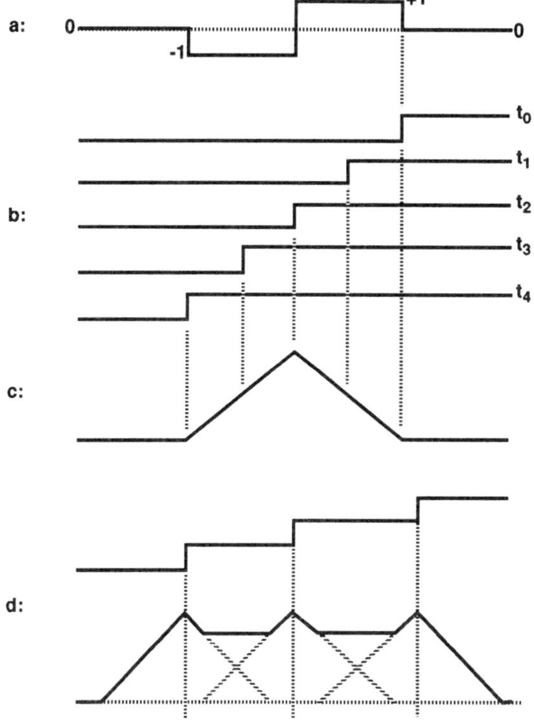

FIGURE 9.6. Digital filter weighting function and resulting waveforms.

Figure 9.6(b) shows the progression of a single ideal noise-free photon step sweeping across the weighting function to produce the triangle output of Figure 9.6(c). The maximum response occurs when the step crosses the center of the weighting function. This weighting function is a digital realization of the triangle shaper of Figure 9.4.

Figure 9.6(d) gives the output response to a series of three photon steps of equal height, spaced at equal intervals just greater than the rise time. The individual triangle responses to the steps are shown by dashed lines. A traditional analog system can only measure the first of these events, whereas a digital system correctly measures all three. When the weighting function is centered on each event, the others are outside the weighting function interval and do not distort the energy measurement. Thus, the FIR property of digital filtering means that the ratio of total dead time to rise time can approach 2, a 33% improvement over the best case shown in Figure 9.4. A recent paper by Jordanov and Knoll also makes use of this property to reduce dead time in a hybrid analog/digital approach. They use a conventional analog prefilter, but replace the gated integrator stage with a digital filter.[23]

The best filter shape for noise performance has been known for many years to be the infinite cusp, and all other shapes are rated relative to the cusp. The various noise sources and their effects are well known.[24] Goulding favors the symmetrical triangle for high-rate work, while other authors have discussed modifications to the cusp shape to account for finite processing times and the effects of $1/f$ noise at longer shaping times.[21,25,26] The beauty of digital shaping lies in its flexibility; not only can the desired theoretical shape be modeled with great accuracy, but the shape is simply a table of numbers in memory chips, not fixed into hardware components. Different shapes can be used for different operating regimes.

Furthermore, the best shaping for a particular spectrometer depends on measurable characteristics of the semiconductor crystal, FET, and preamplifier. In principle, one could customize the filtering to match each individual instrument. We have not yet measured enough systems to report whether such matching yields significant improvement, but the potential is intriguing. If software can be devised to estimate the appropriate parameters from the spectrometer's behavior over a range of count rates, shaping times, and peak energies, it may be feasible to update the shaping periodically for units on-site in the field.

## 9.5. ADAPTIVE SHAPING

The major benefit of digital processing comes from adaptive shaping. If one could build an analog adaptive processor, it would behave as shown in Figure 9.7. The top trace shows a few photon steps of varying amplitudes and spacings arriving at the pulse processor. Traditional fixed-time shaping, shown in the middle trace, would only measure one of the five photons. Adaptive shaping, shown on the bottom trace, would adjust the shaping time for each pulse to avoid overlap with the following pulse, thus measuring all five. Digitally, changing processing times is no problem. There is no

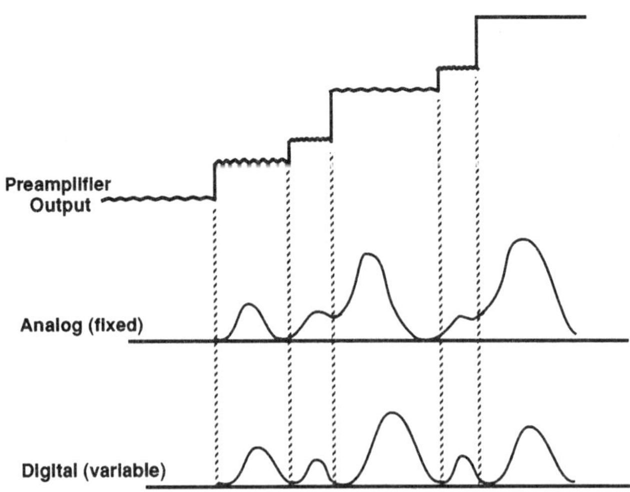


FIGURE 9.7. Adaptive shaping behavior.

physical switching to generate noise, and perfect gain matching between shaping times is simply a matter of scaling.

To do this, of course, it is necessary to know when the next photon is going to arrive. Pulse anticipators are among the trickier circuits to get right, so as an alternative, the digital data stream is stored in a memory buffer until either the next photon appears or enough time has elapsed that no further resolution improvement would result from waiting longer. In principle, analog delay lines could accomplish the same function, but the long delays required for high-resolution shaping would be likely to result in noticeable loss of SNR.

Figure 9.8 compares the throughput count rate versus resolution performance for analog Kandiah/Harwell gated integrator shaping and adaptive digital shaping at 50% dead time. The data points for the analog system for Figure 9.8 were taken from Figure 1 of Fiori and Swyt's two-page abstract, in which they are actually drawn as points.[27] It is traditional to plot count rate versus resolution by drawing a curve through these points, as was done in Fiori and Swyt's full paper and elsewhere.[12,15]

However, this does not correctly represent the operating curve of an analog processor. Given a limited set of selectable processing times fixed in hardware, only the discrete data points correspond to operation at 50% dead time. The discontinuous near-vertical lines in Figure 9.8 are more accurate representations of the set of possible operating conditions. Dead time declines as the beam current is lowered, but resolution improves very little. This is why analog system resolutions for high rates are typically quoted at high dead times. If the same throughput can be obtained with a longer processing time by accepting a higher percentage of dead time, one would not ordinarily use a shorter processing time with worse resolution. However, operation at low dead time is attractive for other reasons, as discussed in the following section.

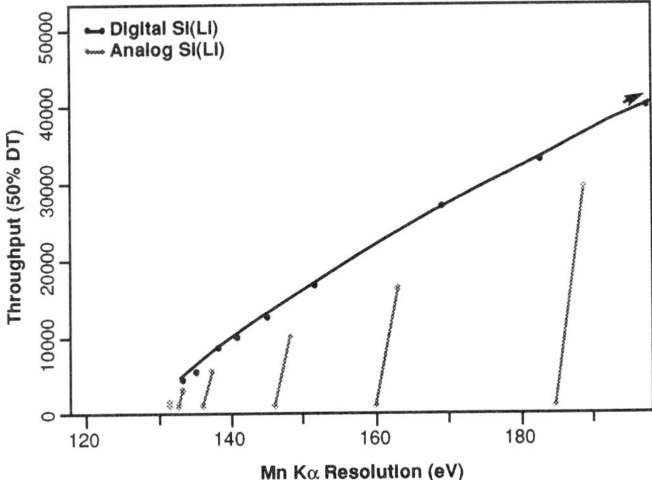


FIGURE 9.8. Operating curves for adaptive digital shaping and Kandiah/Harwell shaping with fixed switchable time constants.

For digital processors, a continuous operating curve is possible at any desired dead time. Figure 9.8 shows that adaptive digital processing improves count rate slightly (15–25%) at the 50% dead time points of the analog system, but its advantage is much greater anywhere else in the middle and upper range of throughput. At the bottom of the throughput curve, nearly all photons are measured with the longest useful shaping given 1/$f$ noise constraints, so adaptive processing has less effect.

As an aside, it is important not to become confused over the terminology of count rates. We use the word "throughput" in its ordinary meaning of counts recorded in the spectrum per real second, and the term "incident count rate" to refer to counts arriving at the detector's active area per real second. "Throughput" has sometimes been abused to mean counts recorded per live second, which is of course numerically the same as the incident count rate per real second if the live time corrector is working properly. Mistakes are easy to make, even for experts. The figures equivalent to our Figure 9.8 in Fiori and Swyt's abstract and full paper disagree; the abstract is correct.[12,27]

To understand how adaptive shaping works, it is necessary to consider the interval distribution of the arriving photons and the effect of shaping time on resolution. Figure 9.9 gives the resolution of fixed, nonadaptive digital shaping as a function of total processing time at constant incident count rates of 1 and 10 kcps. For any given shaping time, there is some loss of resolution with higher incident count rate, but only a few eV. This is consistent with the near-vertical nature of the fixed shaping lines in Figure 9.8. However, below the "knee" of the curve at around 40 µs, resolution begins to worsen rapidly.

After a photon event, the cumulative probability that a second photon arrives within any given time interval $t$ is given by

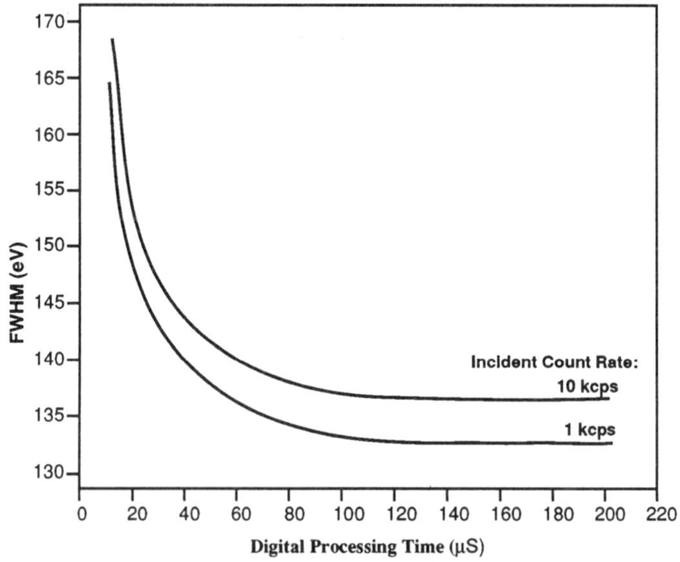


FIGURE 9.9. Resolution as a function of processing time at a fixed incident count rate.

$$1 - e^{(-nt)} \tag{9.1}$$

where $n$ is the incident count rate.[28] This corresponds to the percentage of dead time, which is plotted against $t$ for various incident count rates in Figure 9.10. This is the dead time due to arrival statistics only. At very high rates, other sources of dead time, such as preamplifier reset overhead, also become significant, reducing the available processing time still further if the desired throughput is to be maintained.

With fixed shaping, the shaping time for all photons is determined by the intersection of one of the family of curves exemplified in Figure 9.10 with a horizontal line at the appropriate dead time. The resolution can then be determined using the corresponding curve of Figure 9.9. All photon intervals below this shaping time result in pileup and therefore rejection. However, very few intervals exactly match the shaping time. In fact, for 50% dead time, the median interval of the photons that are not rejected is twice the shaping time, which can be seen by drawing a line at the 75% mark on Figure 9.10. With adaptive shaping, each photon is measured at its actual interval; for the 20-kcps curve in Figure 9.10, half of the photons would be processed for at least 70 μs, as compared with 35 μs with fixed shaping.

The ratio between the median shaping time and the minimum shaping time depends only on dead time and increases with lower dead time as follows. The live time fraction is 1 minus the dead time, and the median interval by definition reduces the live time by half. Leaving $n$ and $t$ defined as in (9.1), and defining $m$ as the median interval and $L$ as the live time fraction

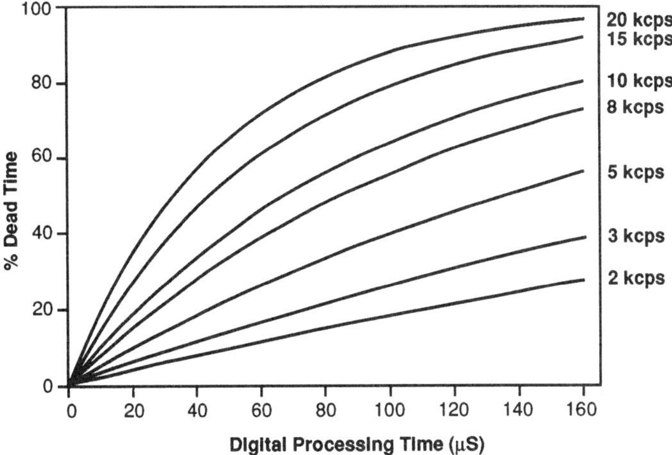


FIGURE 9.10. Dead time due to photon arrival statistics as a function of total processing time for several incident count rates.

$$L = e^{-(nt)}$$

$$0.5L = e^{-(nm)}$$

$$m/t = \ln(0.5L)/\ln L \tag{9.2}$$

For example, at 20% dead time, $m/t$ becomes $\ln(0.4)/\ln(0.8)$, or 4.1. At 20 kcps incident count rate, the minimum interval must decrease to 11 µs to reach 80% live time, but the median interval $m$ would be 45 µs. With analog shaping, both the minimum and median processing times would be 11 µs, and resolution would be degraded.

Note that this analysis is not changed by centering the processing interval on the photon steps as discussed above. Since the interval probability distribution is exponential and successive intervals are independent, the probability of two successive intervals of one-half $t$ is the same as that of one interval of $t$.

R. A. Sareen has pointed out that the peak shape that results from adaptive shaping is not strictly Gaussian, but a more complex shape which is the sum of a series of Gaussians with the same centroid but varying resolutions.[29] The resulting peak will have positive kurtosis, i.e., it will be slightly higher, slightly narrower, and have wider skirts than a Gaussian. A peak with exaggerated positive kurtosis is superimposed on a Gaussian for comparison in Figure 9.11.

From the point of view of minimum detectability and resolving overlaps, peaks with positive kurtosis are actually preferable to Gaussians, in the first case because the higher center channel improves peak-to-background ratio and in the second because

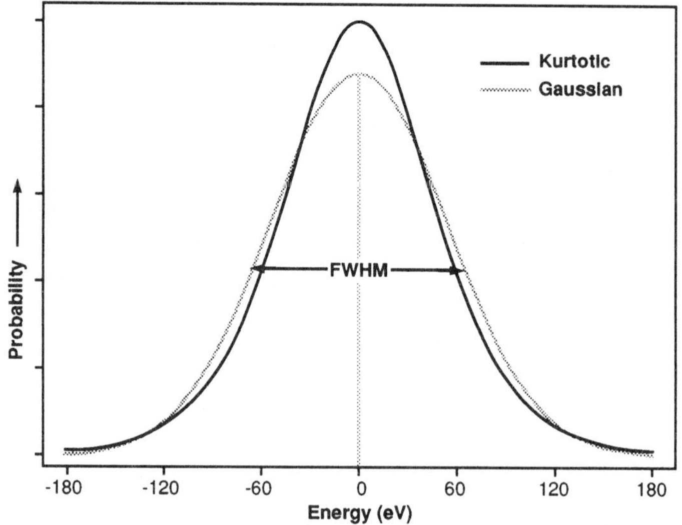


FIGURE 9.11. 130 eV FWHM Gaussian peak versus peak shape with exaggerated positive kurtosis.

the more rapid fall-off away from the center channel reduces the overlap factor. Furthermore, the degree of kurtosis can be predicted from the count rate and dead time, so the standard least-squares fitting deconvolution techniques could be readily adapted to use non-Gaussian peak shapes.

However, it is not necessary to consider these measures except at the very highest count rates, because the range of resolutions is not that large. Figure 9.12(a) shows a pair of generated Gaussian peaks, one with a resolution of 130 eV and the other 160 eV, with the 160-eV peak scaled down by a factor of 3. Figure 9.12(b) shows a weighted sum in which 75% of the peak intensity comes from the 130-eV peak and 25% from the 160-eV peak. Referring to Figures 9.9 and 9.10, note that this results in higher kurtosis than that of a real peak, in which the resolution distribution would vary smoothly.

Figure 9.12(b) also shows a Gaussian peak fitted to the summed peak. Its resolution is 137.3 eV, very close to the weighted average of the parent peaks. They look indistinguishable, but if the vertical scale is expanded, as in Figure 9.12(c) it can be seen that the skirts are indeed slightly wider in the summed peak. The peak channel has 1400 counts. No individual channel differs by more than five counts. Because the summed peak's slightly higher center and slightly faster fall-off tend to offset each other, the difference in the integral of a window covering the FWHM of the peak is less than 0.03 %. Leptokurtophobia (fear of skinny peaks) should not be allowed to cost the analyst any sleep.

## 9.6. LOW DEAD TIME OPERATION

As discussed previously, with fixed shaping there is little incentive to use low dead times except at low count rates. Digital adaptive shaping changes this picture dramatically at what must now be regarded as moderate throughput rates, up to around 15 kcps.

Figure 9.13 is the equivalent of Figure 9.8, but for a dead time of 10%. The resolution of the digital curve is actually slightly worse for the same throughput as in Figure 9.8, but the advantage over fixed shaping is so much greater because the ratio of logarithms in (9.2) works here to the detriment of analog systems. If the shaping time is held fixed, all that can change to reduce the dead time is the incident count rate $n$. In reducing the dead time from 50% to 10%, the incident count rate declines by a factor of $\ln(0.5)/\ln(0.9)$, or 6.6. Therefore, the throughput declines by 6.6 multiplied by the live time ratio, 0.5/0.9, or a factor of 3.65.

Furthermore, with adaptive shaping, the loss of resolution caused by allowing more (but less accurately measured) photons into the spectrum is partially offset by moving to a higher-resolution curve in Figure 9.9. This is due to the lower incident count rate required for the same throughput at 10% dead time.

All this would be interesting, but not compelling, if it were always possible to increase the beam current to obtain whatever count rate is required to use high dead times. However, many SEMs, field emitters in particular, are not designed first and foremost as electron probe x-ray microanalyzers. It may not be all that easy to generate 30–40 kcps at the detector. The beam-limiting aperture may need to be increased

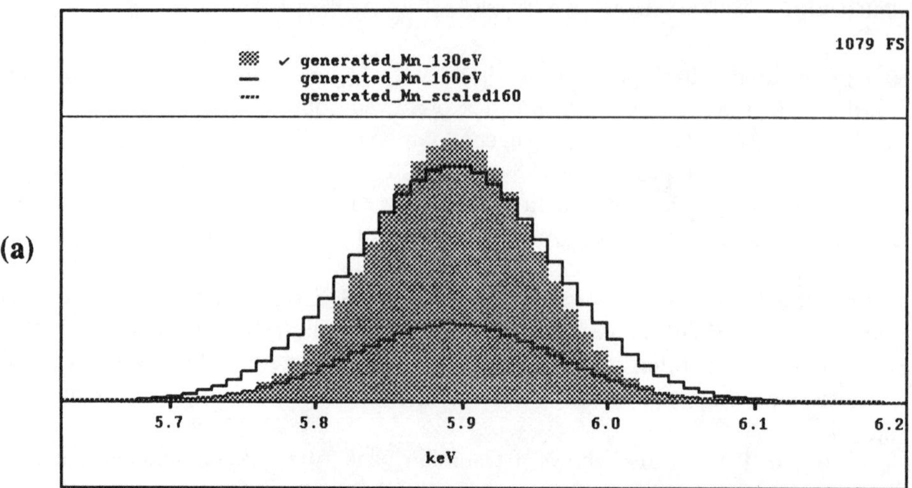

1079 FS
generated_Mn_130eV
generated_Mn_160eV
generated_Mn_scaled160
5.7
5.8
5.9
6.0
6.1
6.2
keV
(a)


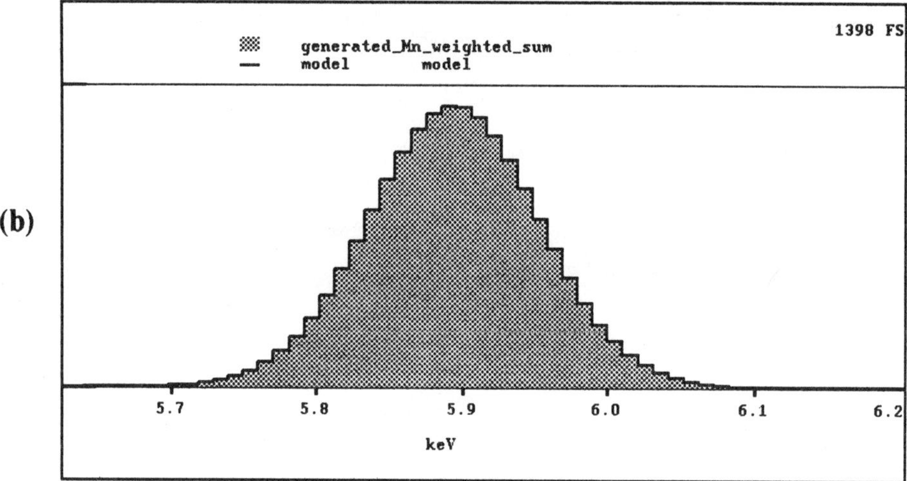

1398 FS
generated_Mn_weighted_sum
model        model
5.7
5.8
5.9
6.0
6.1
6.2
keV
(b)


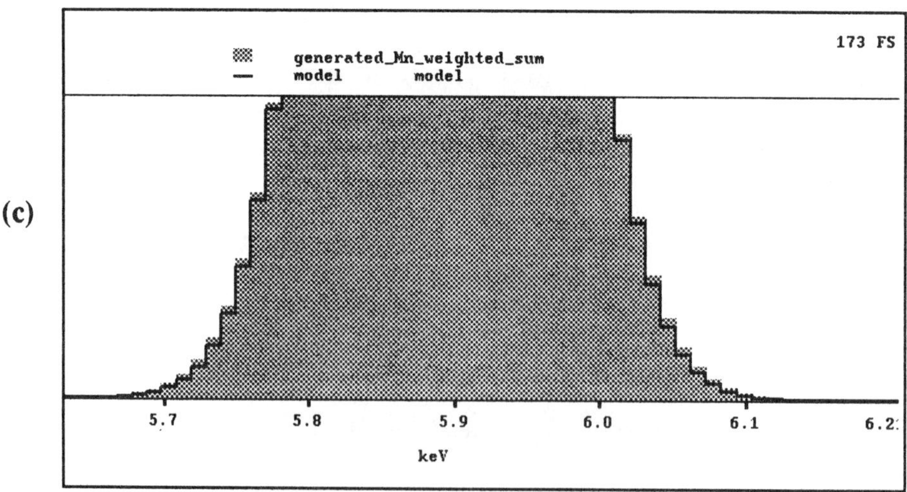

173 FS
generated_Mn_weighted_sum
model        model
5.7
5.8
5.9
6.0
6.1
6.2
keV
(c)

beyond the optimum size for imaging, forcing the analyst to change apertures frequently.[30]

If the incident count rate is limited, throughput can only be improved by lowering the dead time. Consider high spatial resolution x-ray mapping for low concentrations of light elements. The acquisition time required for good visual contrast in the map depends not only on throughput but also on energy resolution, which influences the peak-to-background ratio. Increasing throughput by selecting a fixed short processing time is self-defeating for light elements, because the resolution degradation with shaping time is so much worse as a percentage of peak width at low energies. A processing time which degrades Mn $K\alpha$ resolution from 133 eV to 163 eV (23%) will degrade C $K\alpha$ from 63 eV to 113 eV (79%), according to (9.3) which is presented in the following section on light element performance. Adaptive digital shaping can achieve the same improvement in throughput with a much smaller resolution penalty.

Even if the necessary count rate for high dead time operation is available, there are other reasons why low dead time is attractive. The sample may be sensitive to beam damage; 10% dead time operation cuts the total electron dose almost in half for a given level of statistical precision in the spectrum. In addition less-than-perfect vacuum systems and specimen outgassing lead to a buildup of carbon contamination, playing havoc with attempts to analyze carbon and introducing a peak which overlaps the boron and nitrogen peaks and strongly absorbs the oxygen peak. Not all SEMs are equipped with cold traps or air jets to control this problem.[31] Accumulated contamination is a function of beam current and real acquisition time. Either can be improved by using a lower dead time.

Low dead time operation also reduces spectrum artifacts. The sum peak intensity as a fraction of the parent peak is given by (9.1), which is approximated by $nt$ for small values of $t$.[28] Reducing the dead time from 50% to 10% cuts the incident count rate by a factor of 1.8, thereby decreasing all sum peak intensities by the same amount. In the presence of light-element lines, this is especially important. Digital pile-up discrimination time for C $K\alpha$ photons is under 4 $\mu$s, roughly four times better than the 15 $\mu$s reported previously.[14] Combined with the factor of 1.8 due to operation at 10% versus 50% dead time, this results in sum peaks of C $K\alpha$ with any other line which are over seven times less intense for the same throughput.

Low energy photons are no longer a "nuisance" such that an "EDX spectrometer performs better when there is an absorbing window to stop photons that are too small to be seen by the pileup inspector."[15] With digital low energy pileup rejection and adaptive shaping at low dead times, light element analysis becomes routine at respectably high throughput.

## 9.7. LIGHT ELEMENT PERFORMANCE

The SEM in our research and development area on which most of this work was done is subject to quite rapid contamination buildup. Early in the project, during the

FIGURE 9.12. Gaussian peak fitted against the weighted sum of two peaks with differing resolutions.

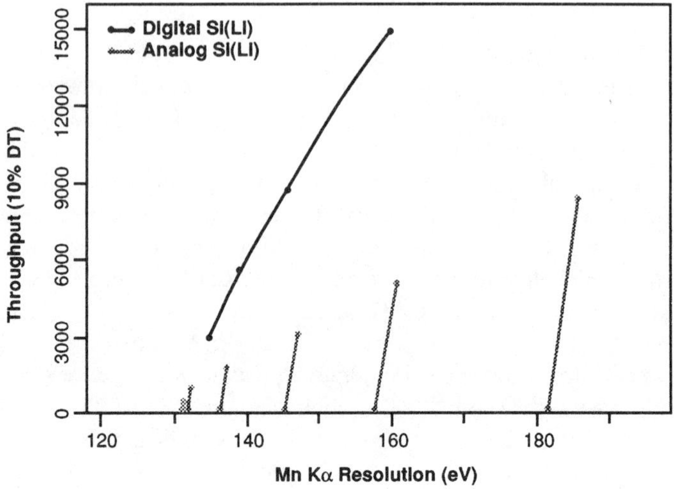


FIGURE 9.13. Operating curves comparable to Figure 9.8, but for 10% instead of 50% dead time.

first comparisons between digital and analog spectra, we observed higher than expected carbon peaks using digital processing, but wrote it off to contamination— until we collected an oxide spectrum and found more intense oxygen as well. It turns out that an unexpected benefit of direct digital shaping is improved sensitivity to low energy x rays.

At low photon energies, the preamplifier signal step height becomes smaller relative to the peak-to-peak noise excursions. As discussed by Statham, traditional processing electronics can fail to detect a low energy photon if its arrival is coincident with a negative noise spike.[32] This causes the apparent rise time of the photon step to be much slower. The result is a lower response from the initial differentiation stage of an analog pulse processor, which may then fail to exceed the detection threshold, as illustrated in Figure 9.14.

Statham refers to this phenomenon as "pulse detection efficiency," which varies from 0 to 100 percent over a surprisingly wide range around the nominal energy threshold of the analyzer. The range is influenced by the effective noise line width of the peak detection circuit. In a gated integrator system, the critical noise line width is that of the circuit which triggers the integrator switching, not that of the final shaper.

In the same way that direct digital photon detection does a better job of detecting low energy pileup and distinguishing real events from noise, it is also better at finding real events buried in noise. This is a serendipitous finding, because light element photons are already subject to heavy absorption in getting out of the sample, through the detector window, if any, and through the crystal dead layer. Recovering photons that were previously "absorbed" in the electronics lowers the threshold of detectability for these elements.

Figure 9.15 shows two spectra taken under identical conditions, with the same spectrometer, of a sample of boron carbide ($B_4C$). The preamplifier output signal was

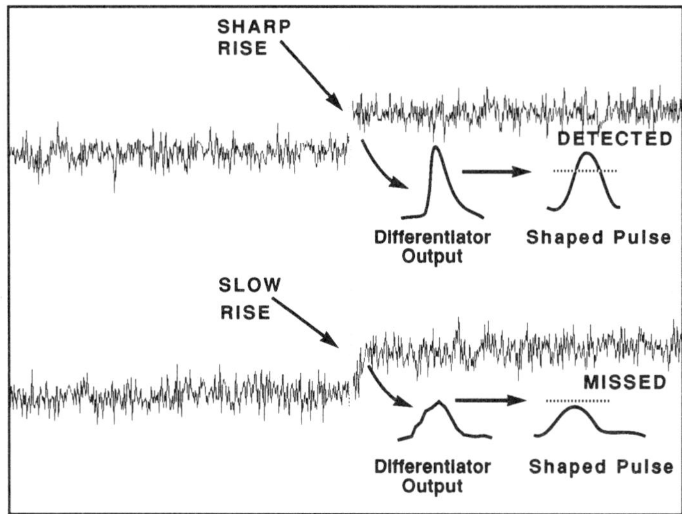


FIGURE 9.14. Two low energy photon traces with differing rise times, showing the resulting output of the initial differentiation stage and the final pulse with its detection threshold for analog shaping.

simply switched between the analog and digital pulse processors. The analog processor for all spectra presented here is a commercial seventh-order semi-Gaussian shaping amplifier, operated at a 12-μs shaping time.

The spectra were collected for 100 s live time at a low incident rate of around 600 cps, to minimize dead time and thereby eliminate differences in pileup performance as a factor. An accelerating voltage of 10 kV was chosen for optimum peak-to-background ratio. The peak channel intensity is about 50% higher for the spectrum with digital pulse processing, but not all of that gain is due to sensitivity. Resolution is also visibly better.

In order to separate the effects of resolution and sensitivity, a best-fit sum of two Gaussians was computed for each spectrum after subtracting the backgrounds and stripping out the noise peaks. The processed spectra and the fitted models appear in Figure 9.16. For digital processing, the fitted B peak has a resolution of 55.4 eV and the fitted C peak, 63.3 eV. For analog processing, the corresponding resolutions are 70.5 eV and 80.0 eV. The area of the fitted B peak was 20% greater for digital processing, which represents the gain from improved electronic sensitivity.

The carbon intensity is 32% greater, but because of contamination buildup it is difficult to quantify any gain in sensitivity. The slightly lower intensity for the small oxygen peak in the second (digital) spectrum is not significant and is probably due to some artifact of beam exposure. Other spectra for which oxygen is a major component show an increase in oxygen sensitivity for digital processing, as we shall see shortly.

Light element peak resolutions show improvment for two reasons. First, since it comes very close to the ideal cusp shape, digital shaping has a much lower noise index than semi-Gaussian shaping, as shown in Figure 9.4. The standard model for peak

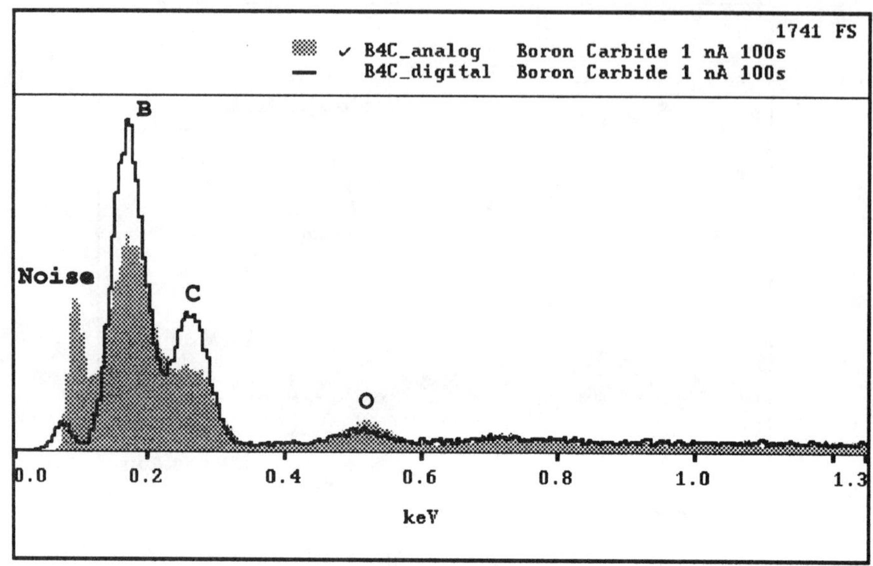


FIGURE 9.15. Boron carbide spectra taken under identical conditions with the same spectrometer at low count rate, one by digital processing and one by analog processing.

resolution at energies above a few keV adds the noise line width and a term proportional to energy in quadrature:

$$FWHM^2 = N^2 + cE \qquad (9.3)$$

where FWHM is the measured peak resolution, $N$ is the noise line width, $E$ is the peak energy in eV, and $c$ is a constant determined by the properties of the detector material. The exact value of $c$ depends on whose estimate of the Fano factor is used; Statham proposes 2.45.[15] At low energies, the noise term dominates, so small variations in $c$ make little difference.

The noise indices of Figure 9.4 relate the noise line widths for various shapers to the ideal cusp index of 1. If we use (9.3) with $c = 2.45$ to compute $N$ from the boron and carbon peak resolutions for analog and digital pulse processing, we get the following values:

$$N = 51.1 \text{ eV (boron by digital shaping)}$$

$$N = 67.2 \text{ eV (boron by analog shaping)}$$

$$N = 57.7 \text{ eV (carbon by digital shaping)}$$

$$N = 75.6 \text{ eV (carbon by analog shaping)}$$

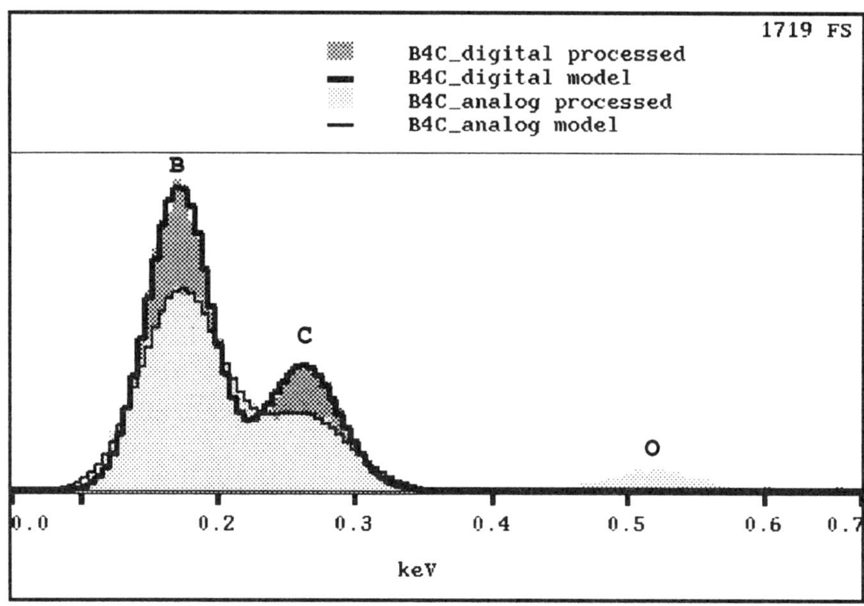


FIGURE 9.16. Modeling of the boron and carbon peaks of the spectra in Figure 9.15 by least-squares fitted Gaussians.

The ratios of analog to digital noise line widths are in excellent agreement (1.315 for boron, 1.31 for carbon), and a nearly exact match for the expected ratio of 1.30 for a seventh-order Gaussian compared to an ideal cusp. This demonstrates that the digital pulse processor approaches optimum noise performance very closely.

The similar discrepancy for the computed values of $N$ between carbon and boron in both spectra is intriguing. The $N$ value of 58 eV derived from the carbon resolution with digital shaping is consistent with higher energy lines. If the boron peaks were artificially narrowed due to a high peak detection threshold cutting into the peak, one would expect visibly non-Gaussian shape and therefore a poor fit using two Gaussians. In fact, the lower tail of the boron peak does seem slightly suppressed in both cases, but not enough to account for a resolution difference exceeding 12%. Also, it seems unlikely that a systematic error due to threshold settings in two shapers would match well enough by chance to allow better than 1% agreement in the digital-to-analog ratios for the two elements. We suspect detector physics at work. It would be interesting to test this behavior against the computer model developed by Joy.[20] S. A. S. Audet has suggested the possibility that the Fano factor or the average number of electron hole pairs generated per unit energy, or both, may not remain constant at these shallow depths near the surface of the crystal.[33]

The second reason why digital shaping gives superior light element resolution is that it is less affected by incomplete charge collection (ICC). Except for electron–hole pair recombination at the surface, much of what is commonly called incomplete charge collection is actually slow collection of a portion of the charge, which is trapped temporarily by defects in the semiconductor material. In Joy's model of a detector, this distinction is quite properly ignored since the initial differentiator will not respond to any charge released on a time scale many times slower than the normal few hundred nanoseconds.[20] However, digital pulse processing can be made relatively insensitive to variations in the rise time of the preamplifier signal.

As discussed by McCarthy and Misenheimer, this trapping is most likely to happen for low energy photons absorbed near the "dead layer" at the front face of the detector, causing resolution to degrade markedly from its theoretical value.[34] Their solution is to modify the detector crystal processing to minimize incomplete charge collection. With digital pulse shaping, it is possible to reduce the effects electronically in addition to whatever can be done with crystal processing.

The spectra in Figure 9.17 neatly capture most of the advantages of digital pulse processing in a single picture. The specimen is Mg metal with an oxidized surface layer. Several spectra were collected from a small area to show that the oxide layer was locally uniform, i.e., the Mg/O peak ratio did not vary significantly. The accelerating voltage was 10 kV. Note that the Mg + C sum peak is a close overlap for Al (44 eV separation), and the Mg + O sum peak is a close overlap for Si (37 eV).

The incident count rate was 12.5 kcps. The throughput rate for digital pulse processing was more than double that for analog (8.5 kcps versus 4.1 kcps), while the resolution was 10 eV better at Mg and 12.5 eV better at C. The noise line width derived from (9.3) for a resolution at Mg of 86.7 eV is 66.7 eV. The measured C resolution of 71.5 eV is a near-perfect match for the expected resolution of 71.6 eV according to (9.3).

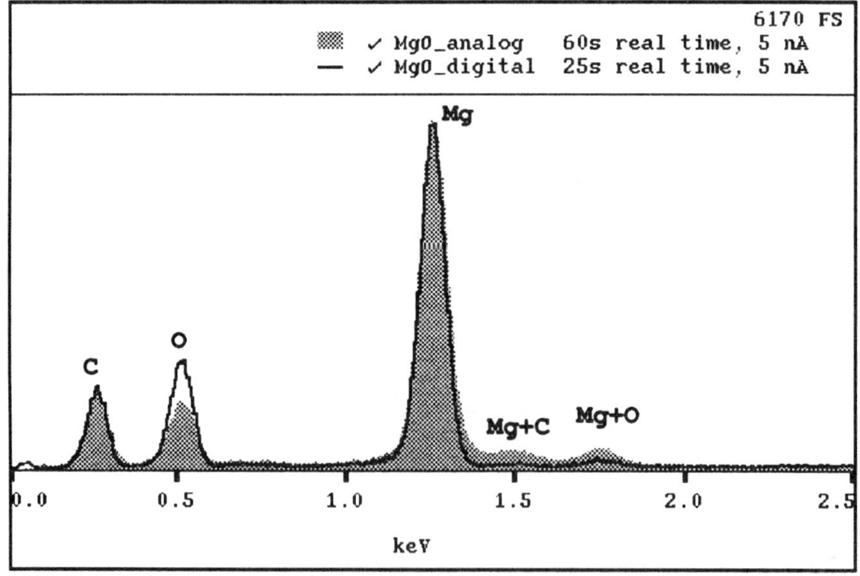


FIGURE 9.17. Spectra of an oxide layer on Mg metal by analog shaping and digital shaping. (line).

The intensities were measured by the areas of fitted Gaussians as before. The analog spectrum was collected for 60 s of real time, and the digital spectrum was collected until the Mg peak heights matched, which took 25 s. Because of its better resolution, the Mg peak's intensity for the digital spectrum was 12% lower than for the analog spectrum, but the O intensity was 50% higher. The O/Mg ratio was therefore 70% greater under digital processing.

The similar carbon intensity is puzzling at first, until one realizes that the analog spectrum was collected for more than twice as long, causing more contamination buildup. The sample contains no carbon; all the carbon intensity is due to coating (whether intentional or not). The beam was moved a few micrometers between spectra, so they started from the same low initial carbon load. The digital sensitivity improvement and higher analog contamination balance each other.

It is easy to see the dramatically reduced Mg + C and Mg + O sum peaks. What is less obvious, unless the vertical scale is increased as in Figure 9.18, is the reduction of background between the O and Mg peaks. While the Mg peak intensity is 12% lower for the digital spectrum, the background is nearly 33% lower. Some of this effect is due to reduced pileup continuum from O, and some is from reduced ICC tailing from Mg. The net result is once again improved detectability of trace elements in this energy range.

Note also that the reduction in the Mg + C sum peak is much greater than the reduction in the Mg + O sum peak. This is because sum peak height is proportional to the incident count rates in the two contributing peaks, and the carbon count rate is not constant due to contamination buildup. The change in count rate is nearly linear with time for the first few minutes of acquisition.[31] Thus, the average carbon count rate in

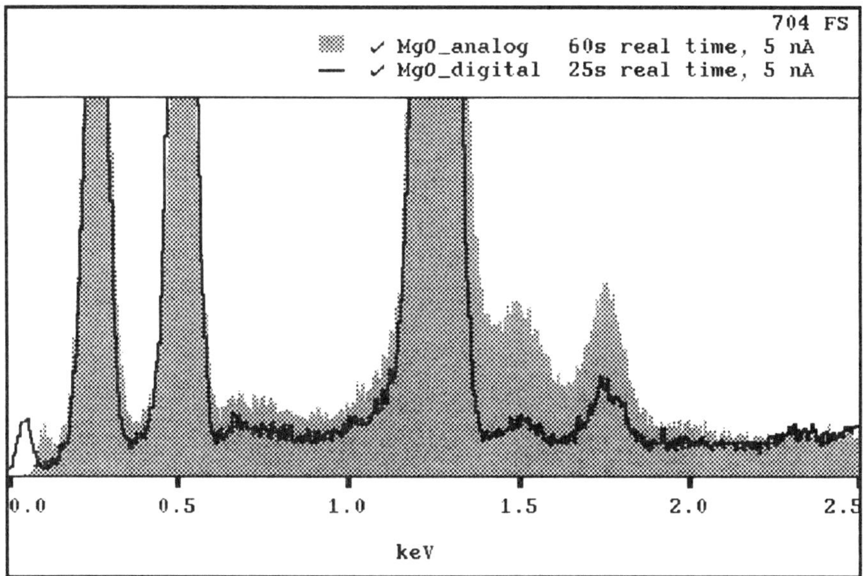


FIGURE 9.18. Note lower sum peaks and lower background between O and Mg peaks.

the digital spectrum (and therefore the sum peak height) is lower by approximately the 2.4 ratio of the real acquisition times (60 s versus 25 s), even before considering the greater effectiveness of digital pileup detection. This is a practical demonstration of the value of operating at lower dead time, as discussed in the previous section.

## 9.8. MEASURES OF LIGHT ELEMENT PERFORMANCE

A number of rules of thumb have been proposed for getting a quick fix on light element performance of a spectrometer, most notably resolution at fluorine and the "peak-to-valley" (P/V) ratio.[14,15] The P/V ratio is computed by measuring the height of the lowest energy peak in the spectrum and comparing that height to the minimum of the valley separating the peak from noise. We argue that these measures are at best not very informative, at worst misleading, and no substitute for actual measurements on all peaks of interest.

Why would one choose an inconvenient element like fluorine as a benchmark for light elements? By current standards, it's not all that low in energy. Fluorides are quite toxic and also tend to glow brightly under bombardment by electrons, making measurement with a windowless or thin-window detector difficult. An answer emerges if we return to McCarthy and Misenheimer's recent paper, which reveals that potential problems with ICC near the dead layer begin in earnest at the oxygen peak.[34] That is, fluorine is the lightest element unlikely to be affected by dead layer problems. It is therefore exactly the wrong peak to use in trying to uncover them.

Preferably, a fitted Gaussian should be displayed superimposed on each light element peak to measure resolution and check for shape, although the ratio of full width at tenth maximum (FWTM) to FWHM is also an acceptable shape measure. According to Goldstein *et al.*, this should be not more than 1.9 for good peak shape.[35] Although this may not be achievable for all light elements, a specification for this parameter should be made for every element of interest when evaluating a spectrometer system.

While FWHM alone is best measured by fitting a Gaussian, the FWTM/FWHM ratio must be measured by interpolating down the sides of the peak for both values. Gaussian fitting and interpolation will give the same answer for FWHM on a good peak, provided the counting statistics are adequate, although fitting a Gaussian is much more reproducible with fewer counts. However, for a peak with tailing, the best-fit Gaussian may give a wider FWHM than interpolation, making the FWTM/FWHM ratio better than it should be. By the same token, beware of interpolated FWHM measurements with inadequate peak heights. The maximum channel in the peak is then likely to be noticeably greater than the height of a fitted Gaussian, giving an artificially low FWHM.

The P/V ratio can be misleading because there are two types of "noise" peak. The natural noise peak depends on the threshold setting for the peak-detecting circuit and results from false triggering due to noise in the preamplifier signal. This peak will have a centroid somewhat greater than zero, due to the energy in the noise spike which triggered the measurement. Lowering the threshold will increase the count rate in the noise peak.

Some systems, in particular those using some variant of the Kandiah/Harwell processor, generate an artificial peak by triggering the energy measurement circuit in the absence of a photon event. This is called a "zero strobe" peak and is used to stabilize offset drift.[22] By definition, this peak has its centroid at 0 eV, and its count rate can be set independent of energy threshold. The P/V ratio will obviously differ substantially if the centroid of the noise peak varies by 50 or 60 eV.

Statham makes the point that measurement of P/V is only valid if the noise peak height is equal to or greater than the peak height for the lowest element of interest.[15] Otherwise, a high peak-detecting threshold may be cutting off the tail of the peak, making it artificially narrow.

Of course, if the only noise peak visible is the artificial zero peak, there is no direct indication of any possible threshold effect other than the shape of the element peak. The zero peak height has nothing to do with threshold setting. If the threshold has been set incorrectly, the FWTM/FWHM ratio may actually be better than that for a pure Gaussian (1.82).

While a zero peak is not required for digital processing, it is also not difficult to modify the hardware to generate one. Figure 9.19 shows two spectra from a sample containing a high concentration of boron, one with only the zero peak and one with both the zero peak and the natural noise peak. Note that the left edge of the display starts at −100 eV rather than zero. The resolution of the zero peak, measured by fitting a Gaussian, is 56 eV. The P/V ratio against the natural noise peak is 13:1, whereas the ratio against the zero peak alone is 140:1. Because the statistics at the bottom of the valley are so poor, it might be better to consider the valley count to be the average of the three lowest channels; this still results in a P/V ratio over 100:1. The tail of the boron peak is not truncated, as can be seen clearly from the shaded spectrum in Figure 9.19.

It is interesting that, although the threshold was lowered by nearly 40 eV in order to match the natural noise peak height to the boron peak height, the boron intensities are roughly the same. This shows that, at least down to boron, the digital peak detection algorithms are fairly insensitive to threshold changes, in contrast to the strong "pulse detection efficiency" effect with threshold cited by Statham.[32] In other words, when the threshold is raised to suppress the natural noise peak, boron sensitivity is not lost.


## 9.9. NATURAL NOISE PEAK BEHAVIOR

Figure 9.19 points out another interesting feature of digital pulse processing. The natural noise peak in Figure 9.19, whose centroid is just under 70 eV, is very different from what one would expect from an analog processor. Most obvious is its width, which is 29 eV—half the width of the zero peak, which is pure electronic noise! It should not be possible to generate a peak narrower than the zero peak, but there it is.

Furthermore, unlike the sharp cutoff on the low energy side associated with conventional thresholding, the natural noise peak from digital processing is a good Gaussian, albeit narrow.

Our current hypothesis (which fits the observations, but which we have not yet worked out in full mathematical detail) is that the natural noise peak is a Gaussian

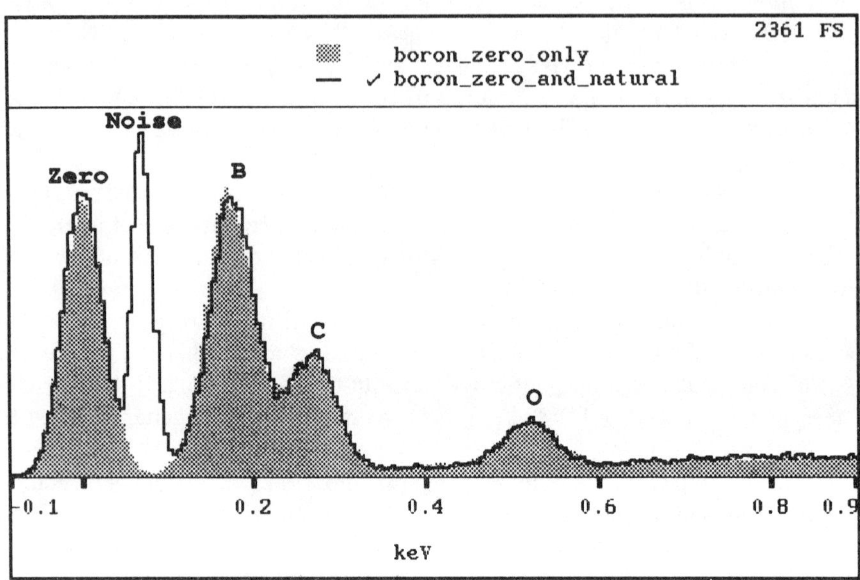


FIGURE 9.19. Boron P/V ratio for zero peak versus natural noise peak.

distribution centered on the threshold energy with the same width as the zero peak. However, it is then scaled down in energy, which causes it to appear narrower.

The effective gain of a digital convolution depends on the shape of the input pulse and on the sum of the weights in the weighting function. For the same sum of weights, the convolution output will be higher for a step function in the preamplifier signal than for a tail pulse of the same height. A step is characteristic of a real photon event, whereas impulse or delta noise is the cause of false recognition and will have tail-pulse characteristics. Thus, the processing system has different gains for real photons and noise events. This is one reason why the digital system gets such extraordinary P/V ratios measured against the zero peak. The noise counts that might otherwise contribute to the valley are shifted down in energy. What's left is true background, which is exceedingly small.

This leads directly to another interesting behavior that is potentially useful at least for qualitative analysis. The spectra in Figure 9.20 are similar to these in Figure 9.19, but the sample is oxidized Be foil with the usual carbon contamination. The spectrum with only the zero peak was taken in the maximum resolution operating mode. Its P/V ratio against the zero noise peak is 15:1. The spectrum with the natural noise peak was taken in a higher-rate, lower-resolution operating mode of the system in which one would not expect to be able to detect Be; however, because the count rate was low (1 kcps), the actual resolution using adaptive shaping is quite similar for the carbon and oxygen peaks. Only the photon detection settings are different. When the threshold is lowered to allow the natural noise peak to emerge, a Be peak appears shifted up from its expected energy and narrower than it should be, indicating the low-energy side of the peak has been truncated by threshold effects.

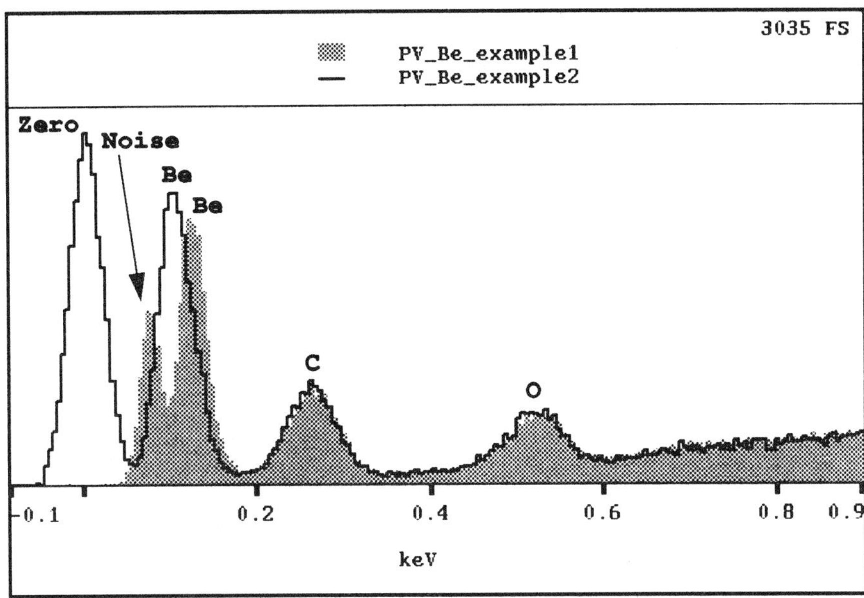


FIGURE 9.20. Beryllium detection in higher rate (shaded) and best resolution (line) photon detection.

The really interesting question is, how can we see a Be peak when the detection threshold is supposedly above the Be emission line energy? Our explanation for this runs as follows: even though a Be photon itself cannot trip the threshold, its presence increases the probability that a positive noise excursion in the vicinity will cross the detection threshold. Once the measurement circuit is engaged, the gain difference will cause Be photons coincident with noise to split from pure noise events. The Be peak will thus be shifted up in energy, but it will be distinct from the noise peak. Sensitivity will suffer because a Be photon must be coincident with a positive noise excursion to be counted. However, the system can be operated at higher count rates than are available in the best resolution mode without completely losing the ability to detect Be.

## 9.10. EXPERIMENTS WITH HIGH PURITY GERMANIUM

Figure 9.21 shows a spectrum collected with a HPGe crystal instead of silicon. Previous authors have pointed out the possible problems with worse ICC and higher noise for HPGe detectors for light element detection.[15,36,37] Previously published boron peaks have been separated from noise, but not to the degree routinely obtained with silicon detectors.[36,37] The spectrum in Figure 9.21 looks essentially no different from Figure 9.19, except the P/V against the natural noise is better, at 30:1. The separation of boron from carbon is noticeably better. The resolutions of the B and C peaks by Gaussian fitting are 56 and 58 eV respectively, with no truncation of the B

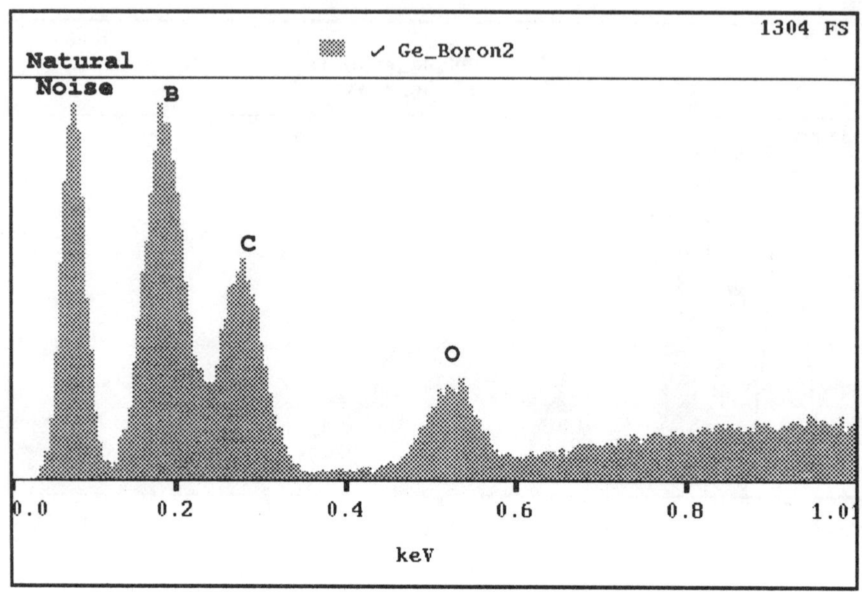


FIGURE 9.21. Spectrum containing boron, taken with a HPGe detector using digital pulse processing.

tail visible. With digital processing, there is no longer any reason to be concerned about the low energy performance of HPGe detectors.


## 9.11. CONCLUSIONS

Adaptive digital pulse processing has been shown to provide better resolution for a given count rate than the fixed shaping times of analog processors. Improved low-end resolution and pileup rejection have been demonstrated, and digital processing has proved able to detect light element photons which are missed entirely by the trigger circuits of analog systems, providing increased sensitivity.

Operation at moderately high count rates (up to 10–15 kcps) with low dead times (less than 15%), which is not possible with fixed shaping time systems without severe degradation of resolution, has been proposed as an effective way to reduce beam dose and sum peak intensities for a given level of spectrum statistics. Alternatively, if the available count rate is limited, low dead time operation allows good statistics to be obtained in less time for a small sacrifice in resolution.

In the future, we believe that most x-ray pulse processors will follow some variant of this basic technology. There are no remaining technical advantages to analog pulse processing over digital pulse processing; the only debate is what form of digital processing is best. Traditional analog processors have reached the limits of their performance, much like the last of the clipper ships at the dawn of the age of steam—intricate, elegant, but ultimately unable to keep pace.

The analogy is good with respect to digital processors as well. The entire field of digital signal processing is young, let alone the application of that body of knowledge to EDS. Any detailed prediction of the future capabilities of digital processors would probably look quaint when the next edition of this volume is written.

One obvious point is that evolutionary pressure will now fall on the front end digitizing components. We are still a long way from the ideal digitizer with infinite speed, infinite bit precision, perfect linearity, and zero cost. There is also clearly more to be gained in the area of low end pileup rejection. Given enough computing power, it may be possible to reach resolving times of a few microseconds down to 100 eV by treating pileup detection as a pattern recognition problem rather than a filtering problem.

Furthermore, digital processing technology opens the possibility of tuning the pulse processor to match the characteristics of a particular spectrometer (crystal, FET, and preamplifier) to a degree which would be completely impractical for analog systems. One can even envisage automated software tools for doing such matching in the field, without the need for specialized equipment.

ACKNOWLEDGMENTS. While no brief mention can do justice to their efforts, the authors would like to acknowledge with gratitude the contributions of Charles Waldman, project mathemagician with an astounding capacity for prying useful information from the most abstruse papers; John "Toby" Wallmark and Margie Mott, who did yeoman work writing the software to control all this highly programmable hardware; Marie Kuszewski, whose artistry with desktop publishing has greatly enhanced this and many other papers; Frank Cerenzia, ADC-master and relentless killer of spurious noise; Dan Ungar and Rich Geddes, the digital designers who had their boards working less than 24 hours after power was first applied; Dan Ungar again, for his work on low-energy pileup detection; Paul Zhu, for his early work on soft photon detection algorithms; Arto Niemelä, who won the nickel, for his help with the preamplifier and noise studies; Sarah Audet, for suggesting that empirical constants may not be; Sarah Audet, Tom Gagliardi, and Jay Patel, for their excellent work on the detector crystals; David Gittens, for ignoring our telling him what he couldn't do; and John I. H. Patterson, for taking the risk. Finally, further thanks are due to Toby Wallmark, who put us on the right track with the noise peak width puzzle by suggesting that a scaling phenomenon was involved.

## REFERENCES


1. H. Koeman, *Nucl. Instrum. Methods* **123**, 161 (1975).
2. H. Koeman, *Nucl. Instrum. Methods* **123**, 169 (1975).
3. H. Koeman, *Nucl. Instrum. Methods* **123**, 181 (1975).
4. E. Lifshin, and M. F. Ciccarelli, in: *Scanning Electron Microscopy 1973*, IITRI Chicago, pp. 89–96 (1973).
5. T. Lakatos, *Nucl. Instrum. Methods Phys. Res.* **B47**, 307 (1990).
6. A. Georgiev and W. Gast, *IEEE Trans. Nucl. Sci.* **40**, 772 (1993).

7. P. L. Ryder, NBS Special Publication 604 (K. F. J. Heinrich, D. E. Newbury, and R. L. Myklebust, eds). National Bureau of Standards, Washington, DC, pp. 177–192 (1981).

8. F. H. Schamber, NBS Special Publication 604 (K. F. J. Heinrich, D. E. Newbury, and R. L. Myklebust, eds.) National Bureau of Standards, Washington, DC, pp. 193–232 (1981).

9. C. E. Fiori et. al., NBS Special Publication 604 (K.F. J. Heinrich, D. E. Newbury, and R. L. Myklebust, eds.) National Bureau of Standards, Washington, DC, pp. 233–272 (1981).

10. J. J. McCarthy and F. H. Schamber, NBS Special Publication 604 (K. F. J. Heinrich, D. E. Newbury, and R. L. Myklebust, eds.) National Bureau of Standards, Washington, DC, pp. 273–296 (1981).

11. J. C. Russ, NBS Special Publication 604 (K. F. J. Heinrich, D. E. Newbury, and R. L. Myklebust, eds.) National Bureau of Standards, Washington, DC., pp. 297–314 (1981).

12. C. E. Fiori and C. R. Swyt, *Microbeam Anal.* **1**, 89 (1992).

13. S. J. B. Reed, *Microbeam Analysis—1990* (J. R. Michael and Peter Ingram, eds.) San Francisco Press, San Francisco, 181–184 (1990).

14. P. J. Statham and T. Nashashibi, in: *Microbeam Analysis—1988* (D. E. Newbury, ed.) San Francisco Press, San Francisco, pp. 50–54 (1988).

15. P. J. Statham, *Inst. Phys. Conf. Ser.* **119**, 425 (1991).

16. J. I. Goldstein, D. E. Newbury, P. Echlin, D. C. Joy, A. D. Romig, Jr., C. E. Lyman, C. Fiori, and E. Lifshin, in: *Scanning Electron Microscopy and X-Ray Microanalysis*, Second Ed., Plenum Press, New York, pp. 296–298 (1992).

17. G. F. Knoll, *Radiation Detection and Measurement*, Second Ed., John Wiley and Sons, New York, pp. 575–577 (1989).

18. C. Cottini, E. Gatti, and V. Svelto, *Nucl. Instrum. Methods* **24**, 241 (1963).

19. T. L. Mayhugh and J. F. Pierce, *IEEE Trans. Nucl. Sci.* **NS-29**, 587 (1982).

20. D. C. Joy, *Rev. Sci. Instrum.* **56**, 1772 (1985).

21. F. S. Goulding and D. A. Landis, *IEEE Trans. Nucl. Sci.* **NS-29**, 1125 (1982).

22. K. Kandiah, A. J. Smith, and G. White, *IEEE Trans. Nucl. Sci.* **NS-22**, 2058 (1975).

23. V. Jordanov and G. F. Knoll, *IEEE Trans. Nucl. Sci.* **40**, 764 (1993).

24. F. S. Goulding, *Nucl. Instrum. Methods* **100**, 493 (1972).

25. J. Llacer, *Nucl. Instrum. Methods* **130**, 565 (1975).

26. E. Gatti and M. Sampietro, *Nucl. Instrum. Methods Phys. Res.* **A287**, 513 (1990).

27. C. E. Fiori and C. R. Swyt, in: Proceedings of the 50th Annual Meeting of the Electron Microscopy Society of America (G. W. Bailey, J. Bentley, and J. A. Small, eds.) pp. 1770–1771 (1992).

28. V. D. Scot and G. Love, *Quantitative Electron-Probe Microanalysis*, Ellis Horwood Ltd., Chichester, United Kingdom, pp. 95–96 (1983).

29. R. A. Sareen, personal communication (1993).

30. J. I. Goldstein, D. E. Newbury, P. Echlin, D. C. Joy, A. D. Romig, Jr., C. E. Lyman, C. Fiori, and E. Lifshin, *Scanning Electron Microscopy and X-Ray Microanalysis*, Second ed., Plenum Press, New York, pp. 59–60 (1992).

31. *Ibid.*, pp. 514–515.

32. P. J. Statham, NBS Special Publication 604 (K. F. J. Heinrich, D. E. Newbury, and R. L. Myklebust, eds.) National Bureau of Standards, Washington, DC, pp. 127–139 (1981).

33. S. A. S. Audet, personal communication (1993).

34. J. J. McCarthy and M. Misenheimer, *Microbeam Anal.* **2**, S178 (1993).

35. J. I. Goldstein, D. E. Newbury, P. Echlin, D. C. Joy, A. D. Roming, Jr., C. E. Lyman, C. Fiori, and E. Lifshin, in: *Scanning Electron Microscopy and X-Ray Microanalysis*, Second Ed., Plenum Press, New York, p. 337 (1992).
36. J. J. McCarthy, M. W. Ales, and D. J. McMillan, in: *Microbeam Analysis—1990*, (J. R. Michael and Peter Ingram, eds.), San Francisco Press, San Francisco, pp. 79–84 (1990).
37. C. E. Cox, B. G. Lowe, and R. A. Sareen, *IEEE Trans. Nucl. Sci.* **35**, 28 (1988).

# 10

# A Study of Systematic Errors in Multiple Linear Regression Peak Fitting Using Generated Spectra

*C. R. Swyt*

## 10.1. INTRODUCTION: ML FITTING

Quantitative analysis using electron, x-ray, or ion beam excited x rays depends on accurate measurement of characteristic peak intensities in the spectrum. Multiple-linear-least-squares (ML) procedures are routinely used to extract peak areas even from spectra with severe peak overlaps. In an ML fitting procedure, a factor is found that scales a reference distribution for each peak or family of peaks from each element of interest contributing to the specimen spectrum. The area of the reference is known, so the fitting factor may be used to calculate the area of the fitted peak. Most conveniently, the reference distributions are segments of spectra from pure element specimens and thus are assumed to include all spectrum features associated with a particular family of lines as well as any spectral artifacts peculiar to the spectrometer.[1] For electron excited x rays, some implementations of ML peak fitting utilize reference distributions that are background-free.[2,3] Others fit distributions that include the continuum background and assume that a correction can be made for the reference background contribution in the fitted area.[4] Continuum can be suppressed before the fitting procedure by digitally filtering both the spectrum and references with a "top-hat" filter of dimensions chosen to preferentially pass only higher frequency features.[5,6] For spectra from bulk specimens, these high frequency features include not only characteristic peaks but also abrupt decreases in continuum due to absorption edges that may introduce an error in calculating fitted peak areas. In fact, as pointed out by Statham,[6] any region of the continuum with a change in curvature will leave

C. R. SWYT • National Institutes of Health, Bethesda, Maryland 20892.

*X-Ray Spectrometry in Electron Beam Instruments*, edited by David Williams, Joseph Goldstein, and Dale Newbury. Plenum Press, New York, 1995.

a residual in the filtered spectrum that will introduce an error in the fitted area of a peak in that region. For example, Kitazawa *et al.*[7] demonstrated a systematic error for Na in spectra from a thin specimen acquired with a beryllium window detector. This error arises from the change in curvature of the continuum near the Na $K\alpha$ energy due to strong absorption of the low energy x rays by the detector.

It is generally assumed that any systematic errors in fitted peak areas will cancel in quantitation procedures that utilize the ratio of measured specimen peak areas to those extracted with the same ML procedure from standard spectra. It has also been assumed that measurement errors for a particular peak due to background shape or spectrometer resolution or calibration can be estimated by fitting for the peak when it is not present in a spectrum, a "null experiment." Statham used this approach to estimate the error as a function of energy separation from a single Gaussian peak with a small shift in energy calibration compared to the reference.[6] The error was obtained from the fit of a second reference over the range of energy separation. He pointed out that the error would be greatest for a small peak overlapped by a much larger one. Kitazawa *et al* estimated the error for Na due to continuum curvature by fitting for the Na peak in spectra from a specimen with no Na.[7] In the same paper they also estimated the fitting errors for a peak with overlap by a much larger peak due to small differences in spectrum calibration and detector resolution between the references and the spectrum for the case of Ca overlapped by K. They showed that these errors may be reduced by also fitting derivatives of the reference for the larger peak: the first derivative for differences in calibration and the second derivative for differences in the resolution. They recommended this derivative method for any two overlapping Gaussians and demonstrated its efficacy by a null experiment: fitting for Ca in a large number of spectra acquired from $KHCO_3$.

In fact, null experiments using real data have been the only way for most analysts to obtain some estimate of systematic errors in ML fitting of spectra. If the source of the error is not specimen-dependent, as in the Na and Ca examples, the estimate may be quite good. However, can a null experiment estimate systematic errors due to features in the digitally filtered spectrum such as absorption edges? In the discussion that follows we will explore the answer to this question by examining the systematic errors that are produced by absorption edges in ML fitting of spectra generated for bulk specimens. With a spectrum generated for a composition of interest, it should be possible to determine the total effect of all systematic errors on fitted peak areas. We will compare fitting errors from references with and without background and compare these to the results of a null experiment. We intend only to illustrate the use of generated spectra to estimate such errors, not to provide estimates for specific cases.


## 10.2. SPECTRUM GENERATION

We employ the NIST/NIH program Desktop Spectrum Analyzer and X ray Data Base[2] to generate spectra and perform ML fitting of peaks. The program provides a means of inputting specimen composition and varying all experimental parameters to generate a spectrum from first principles which matches quite well with a spectrum acquired from a real specimen.[8] The peak areas in generated spectra are known and can be compared to areas measured by the ML fitting routine. Since the spectra are free of counting noise, it is not necessary to analyze a large number of spectra to separate systematic errors from those due to counting statistics. The details of the experimental parameters or particular cross sections used in the calculations are not germane to the discussion and will depend on the experiment the analyst wishes to simulate. The digital filter as implemented in the DTSA ML fitting routine has zero area and width that depends on the detector resolution and is optimized for the peak widths being fit. Spectra were generated for a detector resolution of 140 eV FWHM at the energy of Mn $K\alpha$. We mention this because the errors considered here will vary with the top-hat filter used. It is worth noting that the calculation of mass absorption coefficients in DTSA is based on an algorithm developed by Kurt Heinrich.[9] The examples we will consider are, of course, for specific compositions, but they are representative of the systematic errors due to continuum.

## 10.3. EXAMPLES

### 10.3.1. A Major Constituent and Its Absorption Edge

Consider first a spectrum generated for a bulk sample of pure P. The spectrum is the linear combination of the characteristic peaks and continuum shown in Figure 10.1(a). The respective contributions of the peaks and continuum to the distribution to be fit after digitally filtering the spectrum are displayed in Figure 10.1(b). Clearly the continuum is not completely suppressed, and the P absorption edge produces a peak-like contribution. Since the digital filter is a linear function, one would expect that a fit to the spectrum with a background-subtracted P reference could be corrected for the continuum by subtracting the result of a fit of the P reference to the continuum. However, a residual error remains after this correction. This error arises from the fact that, in the digitally filtered spectrum, neither the data points being fit nor the variances of the fitted data are independent. The weighting function usually used in ML fitting is obtained from the variance of the data. For the digitally filtered data, the variance of each value is calculated from the variances of all the channel values contributing to the digital filter at that channel. It is obvious that the fit to the continuum alone does not have the same weighting function as in the fit to the spectrum. We refer the reader to the paper by Statham for a mathematical treatment of this subject.[6] The percent errors for the fitted P areas are given in Table 10.1, which also shows the errors for ML fitting with the same references but a weighting function of 1. As expected, for this case there is essentially no residual error after the fit to the continuum is subtracted from the peak area obtained with the background-subtracted reference. For a reference with background, the correction to the fitted area is simply a multiplication by the ratio of reference peak area to total area of the reference. There is no difference with weighting in the systematic error for the fits using references with background, because the continuum affects the digitally filtered reference and spectrum in the same way; the weighting factor is absorbed into the scaling factor for the reference. The systematic errors for the fit to a single peak are certainly negligible, but they give some sense of how the continuum and weighting affect the fits through the digital filter.

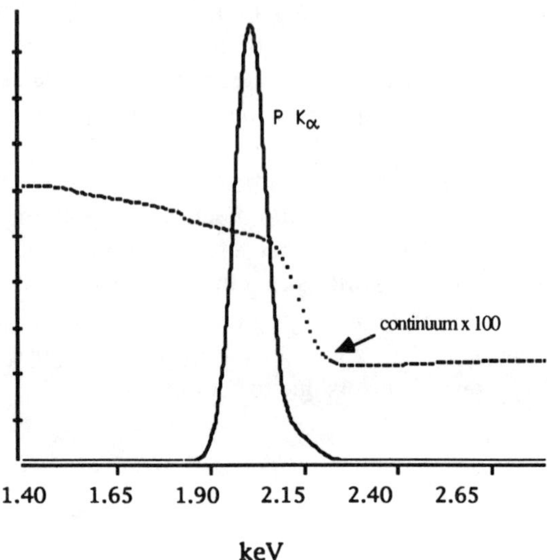



FIGURE 10.1. (a) Gaussian and continuum multiplied by 100 generated for pure P.

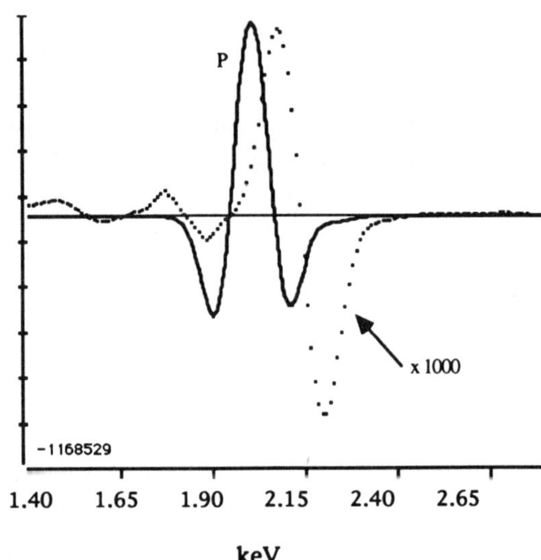



FIGURE 10.1. (b) Gaussian and continuum multiplied by 100 generated for pure P digitally filtered by a zero-area top-hat function of width 20 channels.

## 10.3.2. A Minor Constituent Near a Major Peak

Let us now consider the case of 1% P in bulk Si. The generated distributions of unfiltered and filtered peaks and continuum for this case are shown in Figures 10.2(a) and Figure 10.2(b) respectively. The Si absorption edge leaves a contribution in the digitally filtered spectrum that is almost as large as that from the P peak and that overlaps the digitally filtered P peak. Table 10.2 shows the errors in fitted areas before and after the same corrections described for Table 10.1. For the background- subtracted references, the error from the filtered continuum residual alone is in the opposite sense from the continuum effect on weighting. The fit to the continuum alone overestimates the correction for the fitted P area. Because the contribution of P to the continuum is so small, the errors due to weighting from fits of references with background are almost the same. However, the uncorrected area obtained with background-subtracted references and the normal weighting is better than even the corrected area obtained with references with background.

### 10.3.3. M Family–K Family Interference

Fitting of spectra from bulk PbS provides an example of errors that occur when one element has a number of lines overlapping the absorption edge of another element. The generated continuum and Gaussians for this case are shown in Figure 10.3. The major Pb *M*-peaks in the spectrum occur right across the S absorption edge. The errors in the fitted peak areas are presented in Table 10.3. Once again, the uncorrected areas obtained with weighted background-subtracted references are slightly better than the corrected areas obtained using references with background; both sets of errors, however, are impressively small. The error for the peak represented by Pb would increase rapidly as the concentration of the other element, and consequently the filtered edge area would increase.

## 10.4. A NULL EXPERIMENT

Let us compare the errors found for the example of 1% P in Si to the predictions of a null experiment. We fit for Si and P in a spectrum generated for pure Si and calculate the percent error represented by each fitted value for the P area relative to the true area of P from the generated spectrum for 1% P in Si. As might be expected from the pure P example, fitting with references with background predicts no systematic error for P: areas of zero counts for P are reported. The fit of unweighted background-subtracted references yields a fitted P area that overestimates the error due to the absorption edge residual: 8.5% The weighted fit with background-subtracted references underestimates the total systematic error by a factor of 2, reporting an error of -0.7%.

TABLE 10.1. Errors in ML-fitted P Peak Areas in a Generated Pure P Spectrum. Values expressed as percent of true P area

| References | Weighted | Corrected | Weight = 1 | Corrected |
|---|---|---|---|---|
| With background | 0.73% | 0.00% | 0.73% | 0.00% |
| Without background | 0.07 | 0.04 | 0.02 | 0.00 |

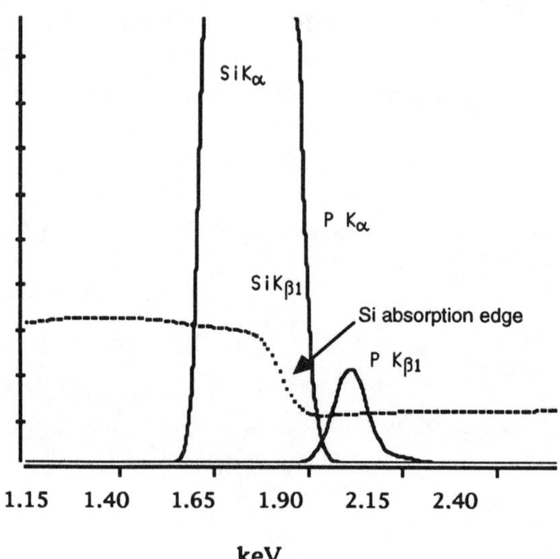



FIGURE 10.2. (a) Gaussians and continuum generated for 1% P in a bulk Si matrix.

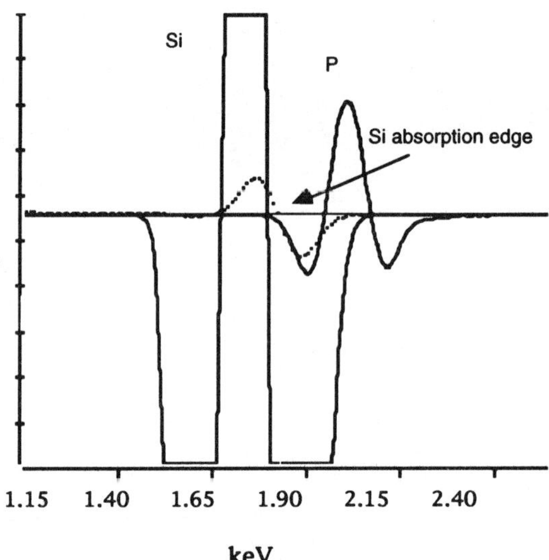



FIGURE 10.2. (b) Gaussians and continuum generated for 1% P in a bulk Si matrix filtered by a zero-area top-hat function of width 20 channels.

## 10.5. CONCLUSIONS

In these examples the corrections for continuum contributions to reference areas are exact because the continuum distribution is accessible. In practice, this is not the case, so errors determined with generated spectra for fitting by references with background must be regarded as minimum values. If generation parameters, including

TABLE 10.2. Errors in ML-Fitted P Peak Areas in a Generated Spectrum for 1% Weight Fraction P in Bulk Si. Values expressed as percent of true P area

| References | Weighted | Corrected | Weight = 1 | Corrected |
|---|---|---|---|---|
| With background | 3.0% | 2.3% | 3.1% | 2.3% |
| Without background | −1.4 | −1.8 | 7.5 | −0.9 |

mass absorption coefficients, can be chosen such that generated spectra match real pure element spectra in the regions to be used as references, references obtained by subtracting the calculated continuum should be used in preference to those with background. Background-subtracted references may be expected to produce smaller systematic errors than references with background, even with a correction to each fitted area for continuum contribution to the reference. For small peaks near the absorption edge of a much larger peak, this advantage may be an important factor in the accuracy of an analysis. References with background have one other complication: each reference must extend over the entire region of the spectrum that is being fit, or the difference in continuum contribution between the digitally filtered reference and data will increase the systematic error. Therefore, in order to accommodate spectra with large regions of overlapped peaks, very large spectrum segments must be stored as references.

The fact that the systematic errors discussed here are dependent on composition has two consequences for accurate quantitative analysis. First, null experiments should not be expected to give very good estimates of the systematic error magnitude, as we showed. Second, systematic errors will not cancel in quantitation by ratio with standards unless the standards are very close in composition to the unknown. There-

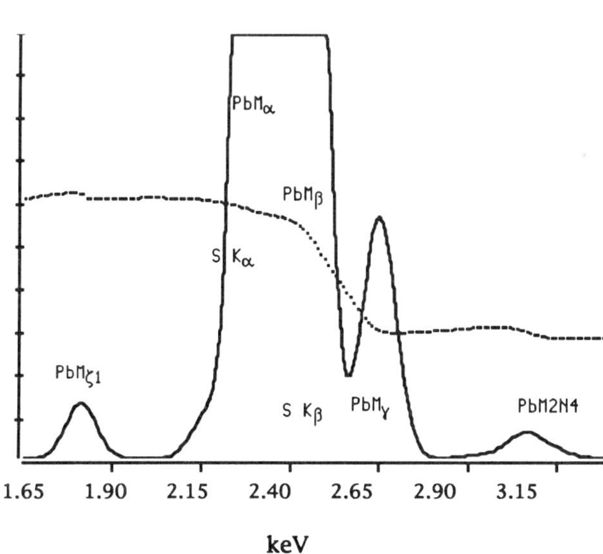


FIGURE 10.3. Gaussians (solid line) and continuum (dashed line) generated for bulk PbS.

TABLE 10.3. Errors in ML-Fitted Peak Areas in a Generated Spectrum for Bulk PbS. Values expressed as percent of true peak area

| References | Element | Weighted | Corrected | Weight = 1 | Corrected |
|---|---|---|---|---|---|
| With background | Pb | 65.7% | 0.42% | 65.9% | 0.53% |
| Without background | Pb | 0.14 | -0.40 | 0.71 | 0.12 |
| With background | S | 5.78 | 0.89 | 5.75 | 0.86 |
| Without background | S | −0.80 | 0.14 | −1.03 | −0.12 |

fore, when an analysis has stringent requirements for accuracy, the analyst may find it advantageous to use generated spectra to examine the effect on minor elements of the systematic ML fitting errors. All systematic errors that arise from the background are greatly reduced with analytical techniques such as x-ray fluorescence and PIXE, which have no continuum radiation.

## REFERENCES


1. F. H. Schamber, in: *Proceedings of the 8th National Conference on Electron Probe Analysis*, Louisiana, Paper 85 (1973).
2. P. J. Statham, *X-Ray Spectrom.* **5**, 16 (1976).
3. DTSA (SRD-38), available from: National Institute of Standards and Technology, Gaithersburg, MD.
4. H. Schuman, A. V. Somlyo, and A. P. Somlyo, *Ultramicroscopy* **1**, 317 (1976).
5. F. H. Schamber, in: *X-Ray Fluorescence Analysis of Environmental Samples* (T. G. Dzubay, ed.) Ann Arbor, MI, p. 241 (1977).
6. P. J. Statham, *Anal. Chem.* **49**, 2149 (1977).
7. T. Kitazawa, H. Schuman, and A. P. Somlyo, *Ultramicroscopy* **11**, 251 (1983).
8. C. E. Fiori, and C. R. Swyt, in: *Microbeam Analysis—1989* (R. Linton, ed.) San Francisco, p. 236 (1989).
9. K. F. J. Heinrich, in: *Proceedings of the 11th International Congress on X-ray Optics and Microanalysis* (J. D. Brown and R. H. Packwood, eds.) Graphics Services, University of Western Ontario, Ontario, p. 67 (1987).

# 11

# Artifacts in Energy Dispersive X-Ray Spectrometry in Electron Beam Instruments. Are Things Getting Any Better?

*D. E. Newbury*


## ABSTRACT

Energy dispersive x-ray spectrometry is subject to artifacts of the measurement process that distort the recorded spectrum from the true x-ray spectrum that is generated in the specimen. Artifacts can be classified as physically inevitable or experimentally reducible or avoidable. Physically inevitable artifacts include absorption in window and detector materials, peak broadening, incomplete charge collection, distortion on the low energy side of a peak, and silicon escape peaks. Artifacts that are reducible/avoidable include stray radiation, direct entry of electrons into the detector, dead time, pulse pileup, microphony/electromagnetic interference, and ground loops. First-principles spectrum generation provides a powerful tool to assess the presence of artifacts and their impact on qualitative and quantitative applications of EDS.


## 11.1. INTRODUCTION

When I look back over the many technical interactions it was my privilege to have with the late Chuck Fiori, I realize that I learned a great deal more from him than


D. E. NEWBURY • Microanalysis Research Group, National Institute of Standards and Technology, Gaithersburg, Maryland 20899



*X-Ray Spectrometry in Electron Beam Instruments*, edited by David Williams, Joseph Goldstein, and Dale Newbury. Plenum Press, New York, 1995.

he probably gained from me. In this regard, one of the most educational experiences for me was my participation in the preparation of the manuscript *Artifacts Observed in Energy Dispersive X-ray Spectrometry in the Scanning Electron Microscopy* (*sic*, the manuscript was submitted to the journal with the last word in the title given as "Microscope").[1] This paper presented a detailed consideration of the artifacts observed in energy dispersive x-ray spectrometry as performed in electron beam instruments from the point of view of the total x-ray detection and measurement process. The assembly of this catalog was a natural development resulting from his intense interest in the use of energy dispersive x-ray spectrometry for practical analysis. From the first description of the application of EDS to electron beam columns, as described by Fitzgerald, Keil, and Heinrich,[2] Chuck was fascinated with the potential of EDS to revolutionize the practice of electron probe microanalysis, and he made many significant contributions to help bring about that revolution. An issue of keen interest to him was the veracity of the x-ray spectrum that was observed by EDS. Energy dispersive x-ray spectrometry offered the tantalizing possibility that the analyst could view the entire x-ray energy range excited by the incident electron beam on a time scale commensurate with human patience, say about 100 s. Such a comprehensive view was not easily accessible with the wavelength dispersive x-ray spectrometer (WDS) at the time the EDS was developed. Even with modern computer-aided spectrometer control, it is not really practical with WDS to view the complete spectrum at all locations sampled by the beam while operating under practical analysis conditions. Moreover, the EDS presented the spectrum in a form that hinted at the "true" x-ray spectrum as emitted from the specimen. The peak signature for the various families of characteristic x rays generated in the specimen could be directly observed superimposed on the background.

Indeed, so much information was contained in the EDS display that an entirely new approach for electron-excited x-ray microanalysis could be developed. With EDS, the complete spectrum was available *from every location* measured on the specimen. This made possible a new analysis strategy in which qualitative analysis could be (and should be!) performed at every location sampled on the specimen. The ultimate limit to the utility of this new strategy depended on the extent to which the EDS spectrum actually represented the "true" x-ray spectrum. Chuck considered any deviation from that true form introduced by the EDS detection process to be an artifact. To develop a comprehensive view of the x-ray spectrum, he also examined the physical artifacts that arise because of electron and x-ray interactions with the sample before the spectrum reached the EDS, so that the modifications imposed by the EDS measurement process could be appreciated.

## 11.2. SCOPE

Many significant technical developments in energy dispersive x-ray spectrometry have occurred during the 15 years that have passed since the original *Artifacts* paper.[1] Chief among these developments are:

1. Resolution improvements to the lithium-drifted silicon Si(Li) detector from approximately 165 eV FWHM measured at Mn $K\alpha$ (5890 eV) to 140 eV or less;
2. Computer-based multichannel analyzers with sufficient calculating speed and memory capacity to provide extensive "real time" support for analysis, including x-ray database information, peak identification, peak deconvolution, and quantitative matrix correction;
3. A variety of window materials that permit passage of low energy x rays in the range 0.1–1 keV with useful efficiency; and
4. The intrinsic germanium detector with improved resolution and greatly increased detection efficiency for x-ray energies above 15 keV, which is particularly useful in analytical electron microscopy.

In this paper, artifacts observed in spectra recorded with Si(Li) detectors of recent (less than three years) construction will be examined. Germanium detectors will be described elsewhere in this volume. Energy dispersive x-ray spectral artifacts can be divided conceptually into two classes: (1) those that are physically inevitable and (2) those that are experimentally reducible or preventable. To appreciate the role of artifacts, we shall first review the ideal x-ray spectrum, making use of the first principles spectrum simulation features of Desk Top Spectrum Analyzer (DTSA), Chuck Fiori's legacy to the x-ray spectrometry field (with great contributions from Carol Swyt and Bob Myklebust).[3,4] DTSA provides detailed simulation of the x-ray spectrum at all stages of its generation, propagation, and measurement to serve as a reference point against which experimental spectra can be compared.

## 11.3. THE IDEAL X-RAY SPECTRUM

Desk Top Spectrum Analyzer enables us to calculate the ideal x-ray spectrum as generated in the specimen and then follow the changes in the spectrum through each stage to the final measurement. Figure 11.1(a) shows the final spectrum calculated (solid trace) for a beam energy of 20 keV impinging on a thick target of NIST SRM-470 (K-411 glass). The certified composition of K-411 is listed in Table 11.1. Figure 11.1(a) also shows an experimentally measured spectrum (dots) superimposed on the calculated spectrum. The degree of correspondence between the theoretical and experimental spectra can be better judged in the residuals obtained when the theoretical spectrum (scaled to the experimental spectrum at the peak of Ca $K\alpha$ to compensate for inexact

TABLE 11.1.  Composition of NIST Glass K-411

| Element | Mass fraction | Atom fraction |
|---|---|---|
| O | 0.424 | 0.603 |
| Mg | 0.0885 | 0.0829 |
| Si | 0.254 | 0.206 |
| Ca | 0.111 | 0.0628 |
| Fe | 0.112 | 0.0457 |

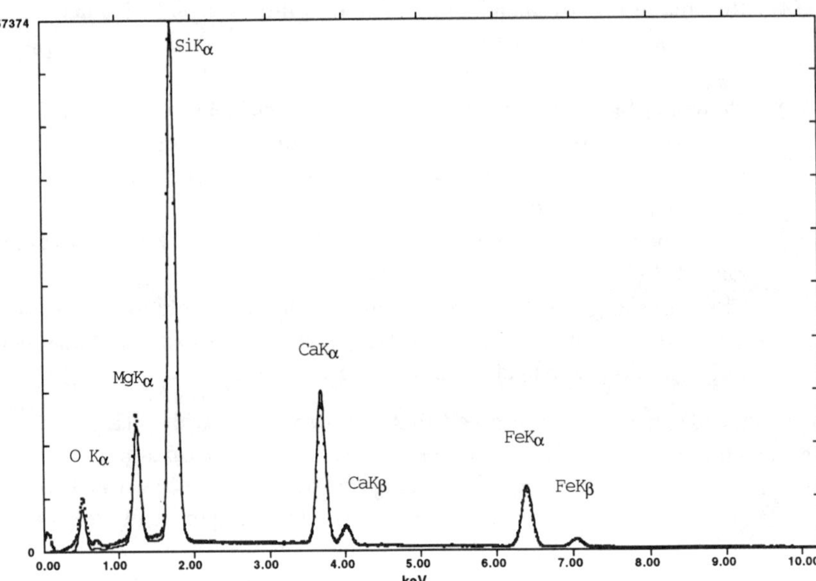

FIGURE 11.1.    (a) X-ray spectrum as calculated (solid trace) by DTSA (Fabre $K$-cross section[5] Bethe $L$-cross section with constants due to Fiori,[6] and the bremsstrahlung cross section from Small, *et al.*[7]) and as measured (dots) for NIST SRM glass K-411; beam energy = 20 keV. The spectrum has been scaled to match the calcium $K\alpha$ peak intensities to compensate for inexact knowledge of the detector solid angle.

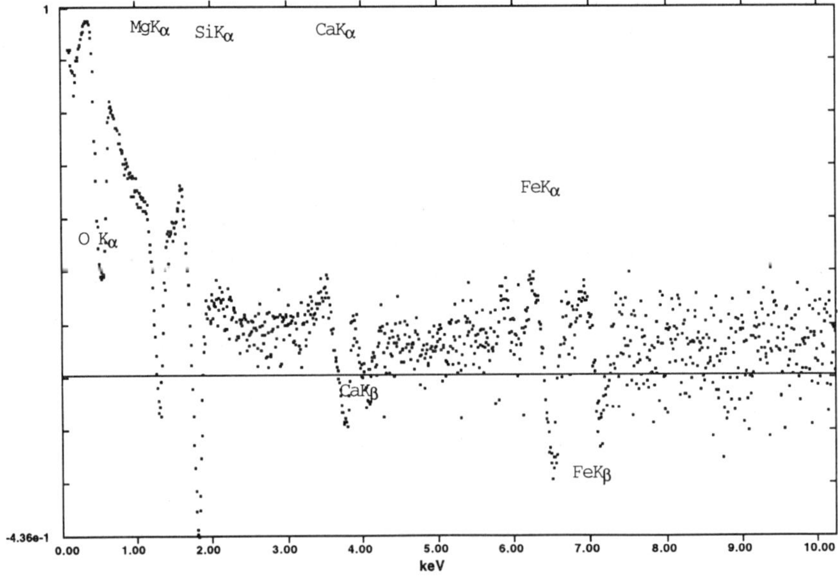

FIGURE 11.1.    (b) Residuals following subtraction of the theoretical spectrum from the measured spectrum and then ratioed to the measured spectrum. Bipolar plot with a range from –0.98 to +0.98.

knowledge of experimental parameters, particularly the detector solid angle) is subtracted from the experimental spectrum in Figure 11.1(b) and then plotted as a ratio of the residuals to the experimental spectrum. The deviations revealed by the scaled residuals are generally small, except at very low x-ray energy where the efficiency of the detector is low. This level of correspondence between the "final" theoretical spectrum, which incorporates all the physics of electron and x-ray interactions with the specimen and detector, with experimental spectrum encourages confidence in the views of the spectrum calculated at other points in the generation, emission, and detection process where it is not possible to make a direct comparison with reality.

### 11.3.1. The X-ray Spectrum as Generated Inside the Target

Figure 11.2(a) shows the calculated spectrum as generated in the K-411 target for an incident beam energy of 20 keV. The Fabre $K$-shell ionization cross section[5] and the Bethe $L$-shell cross section,[6] with constants as suggested by Fiori,[4] were used for the characteristic intensities, and the bremsstrahlung cross section of Small et al.[7] was selected for the background. The widths of the bins of the histogram presentation of the EDS spectrum are 10 eV, and the natural line widths of the characteristic peaks are less than 10 eV, so the characteristic peaks are shown as "stick" plots. The peak-to-background (i.e., the characteristic peak intensity divided by the background in a single 10-eV-wide bin at the peak energy) is approximately 500:1 for the silicon $K\alpha$ peak, so in order to see the shape of the background at low and high energies, the vertical and horizontal scales have been expanded in Figure 11.2(b) and 11.2(c). The distinctive background shape at both extremes of the energy range is dominated by the $(E_0 - E_v) / E_v$ dependence of the bremsstrahlung, where $E_0$ is the incident beam energy and $E_v$ is the bremsstrahlung energy.

### 11.3.2. Specimen Absorption

The x-ray spectrum is generated over a range of depth in the specimen, with the range of generation increasing as the x-ray energy decreases. Characteristic x rays are emitted isotropically, and although bremsstrahlung generation is dependent on the instantaneous electron trajectory, the randomizing effect of elastic scattering on the electron trajectories effectively results in isotropic emission for the bremsstrahlung as well. To reach the spectrometer, the generated x rays must pass through the specimen, while photoelectric absorption and inelastic scattering occur along this path. Figure 11.3(a) shows the K-411 spectrum after it has propagated through the specimen along a path defined by an EDS spectrometer take-off angle of 40° above the surface (that is, 50° to the specimen surface normal), the so-called "emitted" x-ray spectrum. The influence of self-absorption is most evident at the low energy end of the spectrum, as shown in Figure 11.3(b), which compares the generated and emitted spectra. For this particular selection of beam energy and composition, the loss of O $K\alpha$ due to specimen self-absorption is 85% of the original generated intensity. Comparison of the generated and emitted x-ray bremsstrahlung provides a dramatic illustration of the effect of specimen self-absorption throughout the low energy range. Above approximately 4

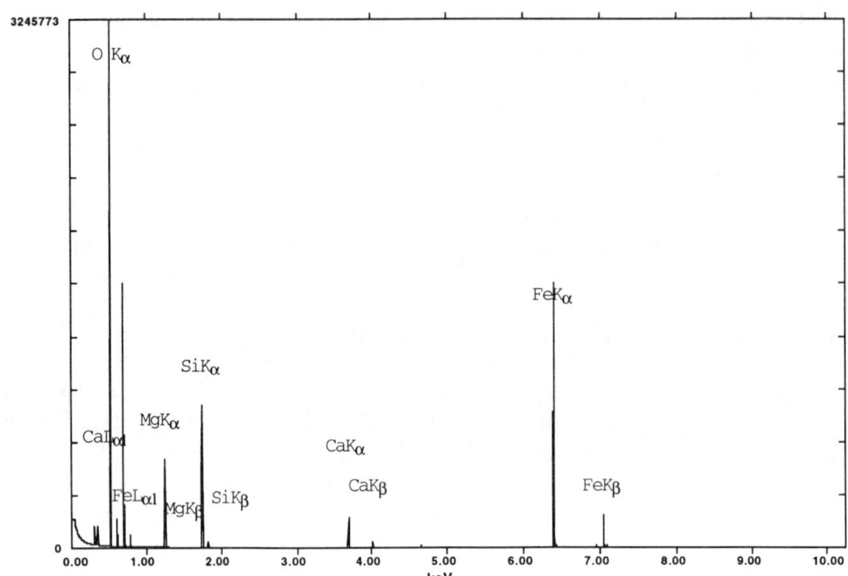

FIGURE 11.2.   (a) X-ray spectrum as generated within the target for SRM K-411 glass, beam energy = 20 keV.

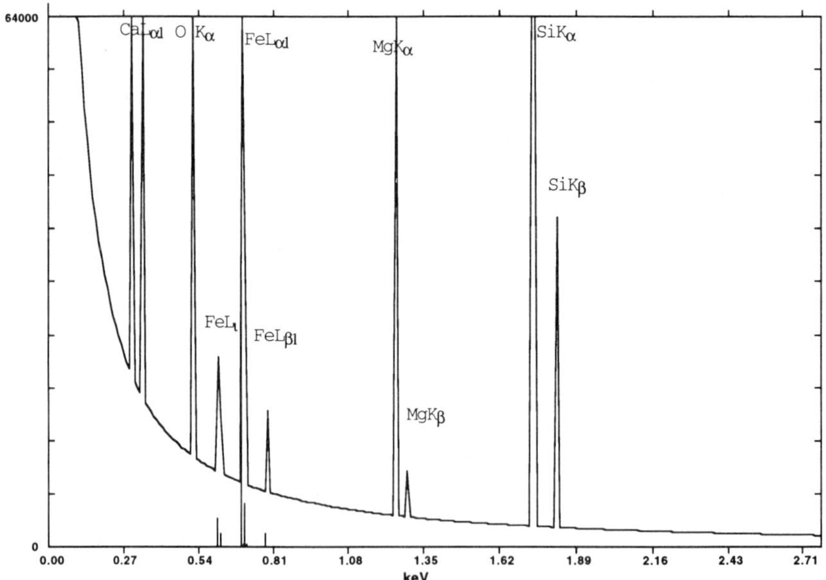

FIGURE 11.2.   (b) Vertical and horizontal scale expansion for the low energy range of the spectrum.

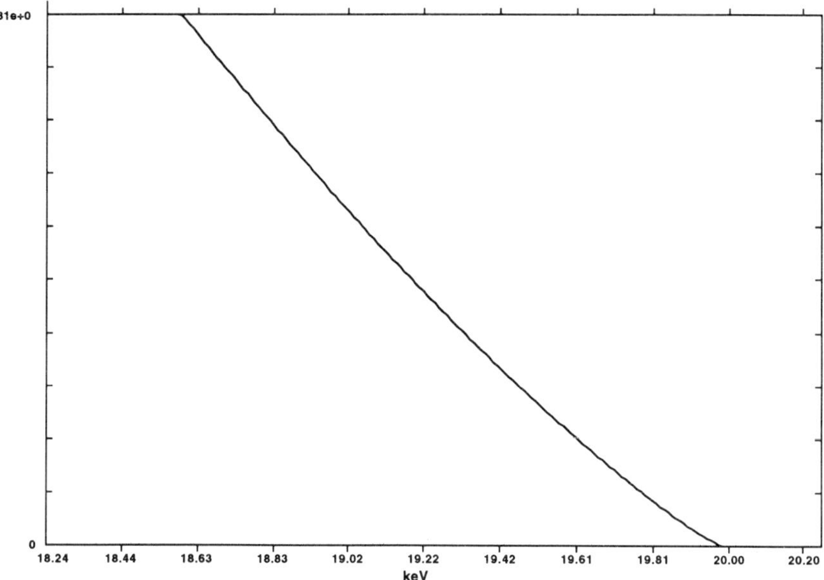

FIGURE 11.2.    (c) Vertical and horizontal scale expansion for the high energy range of the spectrum, showing the Duane-Hunt x-ray cutoff energy.

keV (the exact value depends on the incident beam energy and the specimen composition), the loss of x-ray intensity due to self-absorption becomes negligible.

The spectrum as emitted from the specimen, shown in Figure 3 (a,b), is clearly already substantially modified from that generated by the interaction of the beam with the specimen. Any further modification beyond this point, however, is due to the artifacts of the EDS detection process.

## 11.4. PHYSICALLY INEVITABLE EDS ARTIFACTS

### 11.4.1. Window Absorption

The structure of an EDS detector is illustrated schematically in Figure 11.4. The first hurdle the x rays must pass in the detection process is that of the detector window. The vast majority of EDS spectrometers in existence make use of a window material to isolate and protect the detector from the vacuum environment of the instrument. The detector assembly is cooled near liquid nitrogen temperature to reduce thermal noise. The cooled detector is therefore capable of acting as a cryopump for high vapor pressure substances, such as water and volatile hydrocarbons, that frequently form the major background gases in the microscope chamber, and which may be continually emitted from the specimen, especially under electron bombardment. Only detectors on true ultra-high vacuum chambers (pressure $< 10^{-8}$ Pa) can afford to operate in the

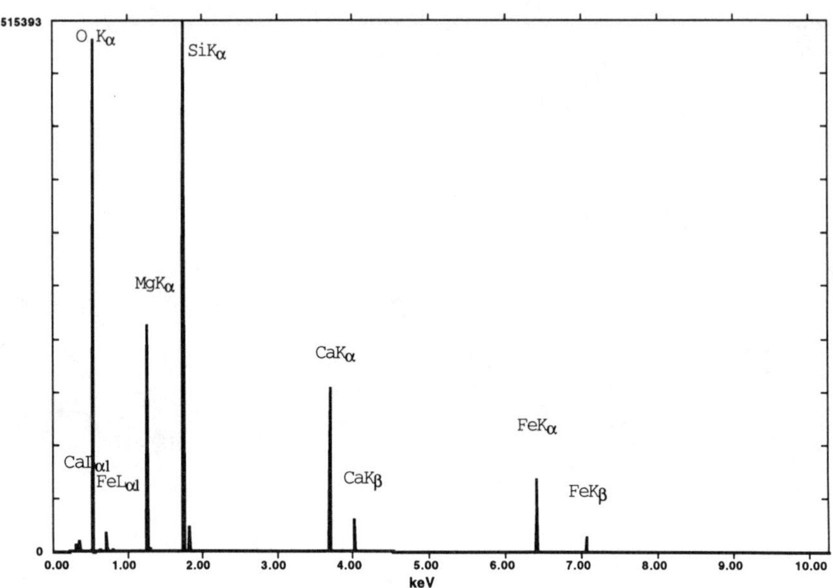


FIGURE 11.3.   (a) "Emitted" K-411 x-ray spectrum after the x rays have propagated through the specimen along a path defined by an EDS spectrometer take-off angle of 40° above the specimen surface.

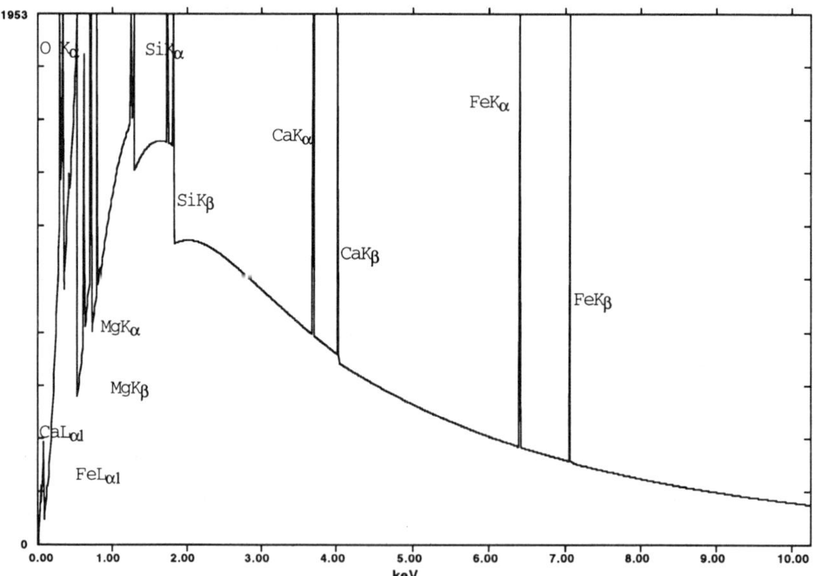


FIGURE 11.3.   (b) Expansion of the vertical scale, showing the loss of intensity due to self-absorption in the specimen at low x-ray energy.

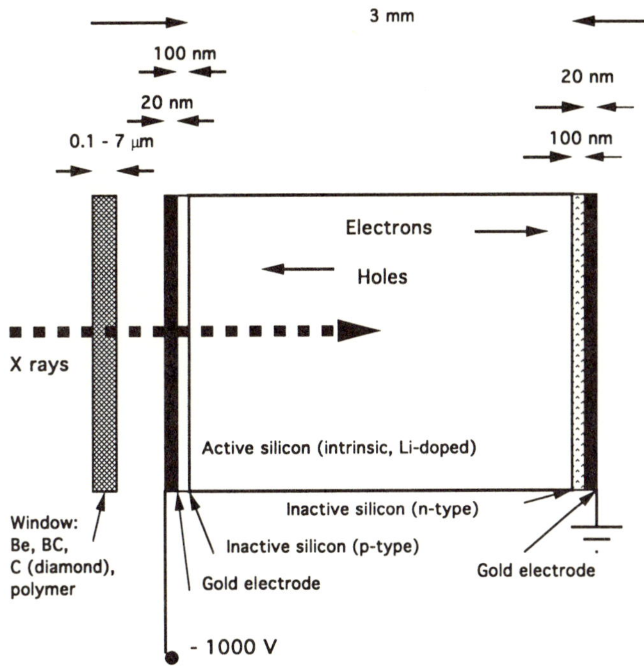


FIGURE 11.4. Schematic illustration of an energy dispersive x-ray spectrometer.

truly windowless mode, and even in such "clean vacuum" systems, the detector is still vulnerable to contamination from the specimen. When our original artifacts paper was written in 1978, the only window material available was beryllium foil, and many detectors equipped with Be windows are still in use. The K-411 spectrum as modified by absorption during passage through a 7.5-μm-thick Be foil window is shown in Figure 11.5(a) (thick trace), compared to the reference emitted spectrum (thin trace). Severe attenuation of the low energy range of the spectrum below 3 keV is evident, and the spectrum is essentially lost below approximately 0.75 keV, which prevents the detection of important low atomic number elements, including carbon, nitrogen, and oxygen.

A major development in EDS in recent years has been the evolution of ultra-thin windows with thicknesses less than 1 μm. The window materials include diamond, boron nitride, and polymeric mixtures, the compositions of which remain proprietary. Generally, the windows are coated with a reflective metal coating (aluminum) to protect the detector from visible light. An example of the low energy performance of one of these window materials, 0.4 μm of diamond with a coating of 0.06 μm of aluminum, is illustrated in Figure 11.5(b), as compared to the performance for the standard beryllium window. Transmission curves for several of the window materials are plotted for the low energy range in Figure 11.6. The sharp discontinuities in these efficiency curves mark the absorption edges for the principal elemental constituents.

D. E. NEWBURY

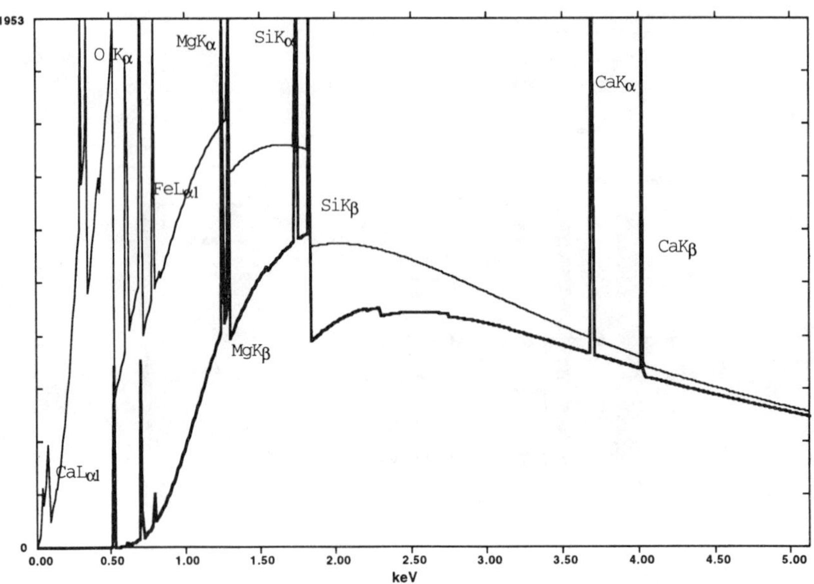


FIGURE 11.5. (a) X-ray spectrum after the x rays have propagated through a 7.5-µm-thick window of beryllium (thick trace) compared to spectrum as emitted from the specimen (thin trace).

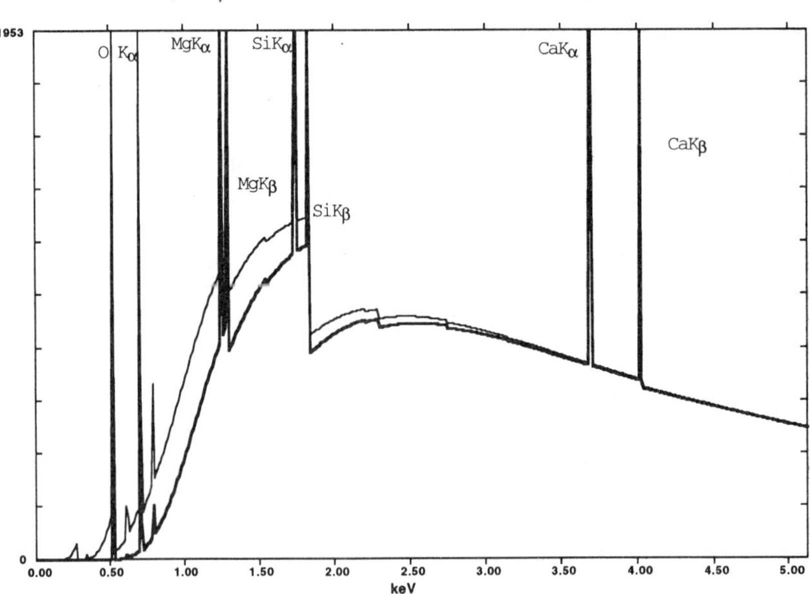


FIGURE 11.5. (b) Comparison of x-ray propagation through a 7.5-µm beryllium window (thick trace) and a 0.4-µm diamond window (with 0.06-µm aluminum coating) (thin trace).

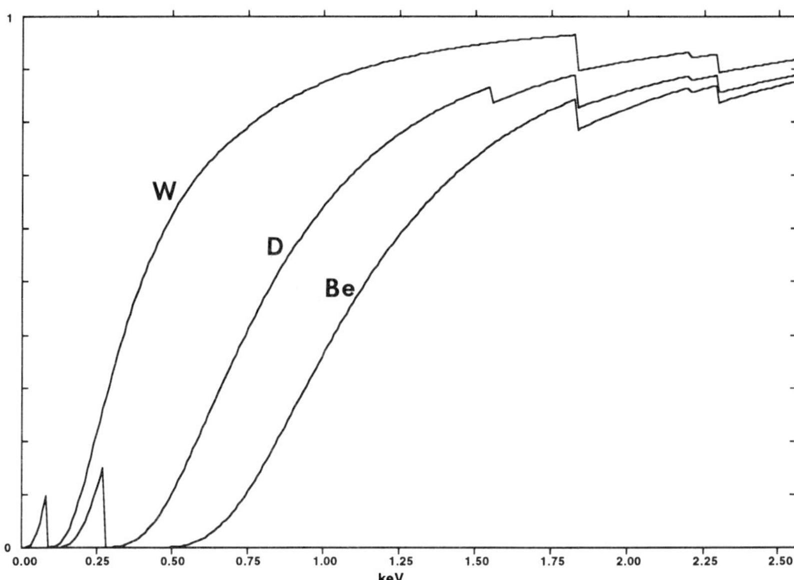


FIGURE 11.6.   Comparison of the efficiency of x-ray transmission through various types of windows. W = true windowless; D = 0.4-μm diamond; Be = 7.5-μm beryllium. All with 0.01-μm Au electrode and 0.1-μm silicon dead layer.

### 11.4.2. Surface Electrode

The front surface of the EDS detector, shown in Figure 11.4, is a gold electrode approximately 20 nm thick, which is used to apply bias across the detector. When x rays pass through this gold layer, absorption losses slightly attenuate the spectrum, an effect which is most notable near the gold $M$-edges, as shown in Figure 11.7 for the spectrum of K-411.

### 11.4.3. The X-Ray Detection Process

The physical process of x-ray photon detection is illustrated in Figure 11.8. The photon of energy $E_v$ is absorbed with the immediate ejection of an energetic photo-electron with energy $E_p = E_v - E_c$, where $E_c$ is the critical excitation (absorption) energy for the shell that is ionized. This energetic photoelectron propagates in the detector crystal with a range of nanometers to several micrometers depending on the initial value of $E_p$. As the photoelectron travels through the detector, it undergoes inelastic scattering, creating electron-hole pairs. The silicon atom which absorbed the photon is left in an ionized state, and it will subsequently lose its excess energy through electron transitions that result in the emission of an Auger electron (e.g., the $KLL$ transition for Si) or a characteristic x ray (e.g., Si $K\alpha$, $K\beta$). The energetic Si-$KLL$ Auger electron will also scatter inelastically as it propagates through the crystal, adding to the electron-hole production. The silicon $K$ x ray in most cases will be absorbed by a

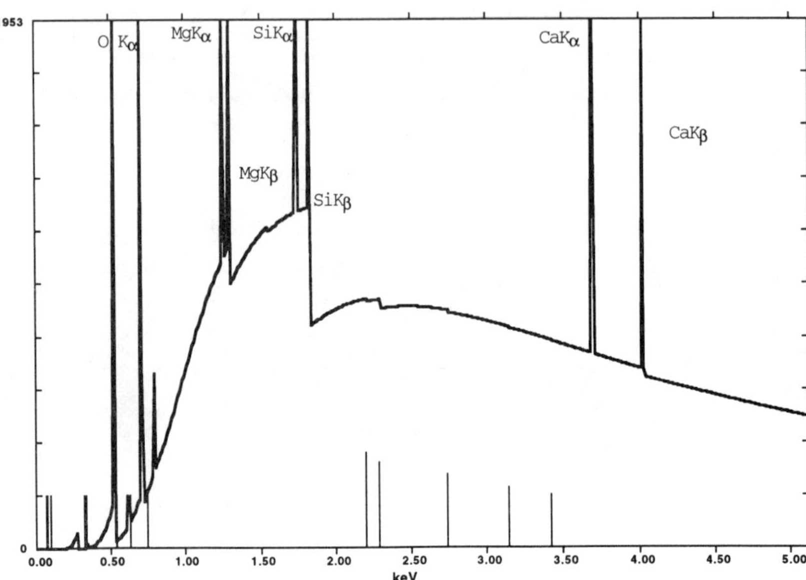

FIGURE 11.7.  Calculated spectrum of K-411 glass after passage through the gold surface electrode, showing discontinuities in the background corresponding to the gold *M*-shell absorption edges, which are marked as lines. The Si absorption edge arises from the detector and the specimen.

silicon *L*-shell electron, with ejection of another photoelectron and still more electron-hole pair production. The high voltage bias applied across the detector acts to separate the free electrons and holes before they can recombine, and this charge accumulates on the electrodes. Ideally, the charge from all these processes is proportional to the energy deposited by the photon in the detector. The ideal number of charge carriers created per incident photon with energy $E_v$ is

$$n = E_v/K \tag{11.1}$$

where *K* is 3.8 eV for silicon.

### 11.4.4. "Dead" Layer

A consequence of the surface of the silicon crystal is the existence of trapping sites for charge in an adjacent thin layer of the detector. These trapping sites permit some electron-hole pairs to self-annihilate, so that some of the charge is lost, thereby causing underestimation of the true value of the photon energy. This affected subsurface layer is often referred to as the "dead" layer, but it is more properly understood to be a partially active layer, whose efficiency is virtually zero at the Au-semiconductor interface and gradually rises to the efficiency of the bulk crystal over some thickness. The physics of the dead layer are described in much greater detail in Chapters 5 and 6 of this volume.[8,9] There are several consequences of the partially active silicon layer.

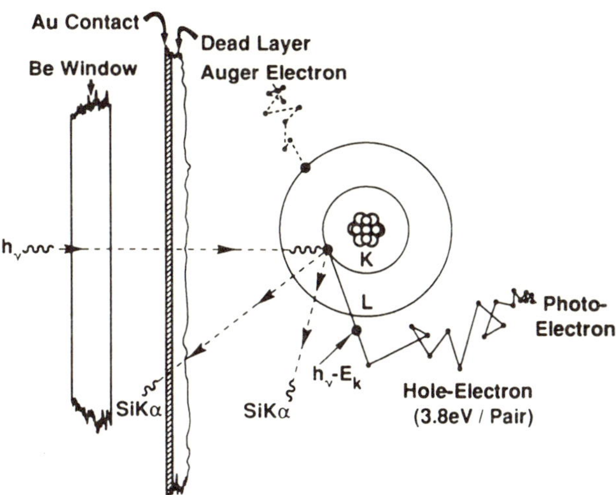


FIGURE 11.8. Schematic illustration of the photon detection process in the EDS, showing absorption of the x-ray photon and ejection and propagation of the photoelectron.

The layer acts like a thin window of silicon through which the x-ray spectrum must pass, leading to the introduction of absorption losses and spectral discontinuities at the Si $K$-absorption edge. The detection process serves to broaden the sharp edge, which has a true width less than 10 eV, into a broad structure that resembles a low intensity peak.

In addition, Si $K\alpha,\beta$–x-ray photons are emitted in the partially active layer as a result of photoelectric absorption of the spectrum photons, and half of the photons thus produced penetrate further into the active volume of the detector. Upon photoelectric capture within the active silicon, these Si $K\alpha,\beta$ photons appear indistinguishable from the true spectrum and thus contribute a "false" silicon component, the so-called "silicon internal fluorescence peak." This Si $K\alpha,\beta$ peak and the gold absorption edges can be seen in the experimentally measured spectrum of high purity boron shown in Figure 11.9. The relative magnitude of the silicon internal fluorescence peak will depend strongly on the thickness of the partially active silicon layer and on the relative amounts of characteristic and bremsstrahlung x rays with energies located just above the silicon $K$-edge (1.84 keV). Table 11.2 gives the apparent concentration of Si experimentally observed for a particular detector in our laboratory. To calculate the apparent Si concentration, DTSA was used to determine the intensity ratio $k = i_{sample}/i_{standard}$, with the standard intensity measured on a pure silicon standard, and then the NIST-ZAF matrix corrections embedded in DTSA were applied.

### 11.4.5. Peak Broadening

The photon energy is measured through the quantity of charge that is generated and collected. Electron-hole pairs are produced in discrete numbers. Because of

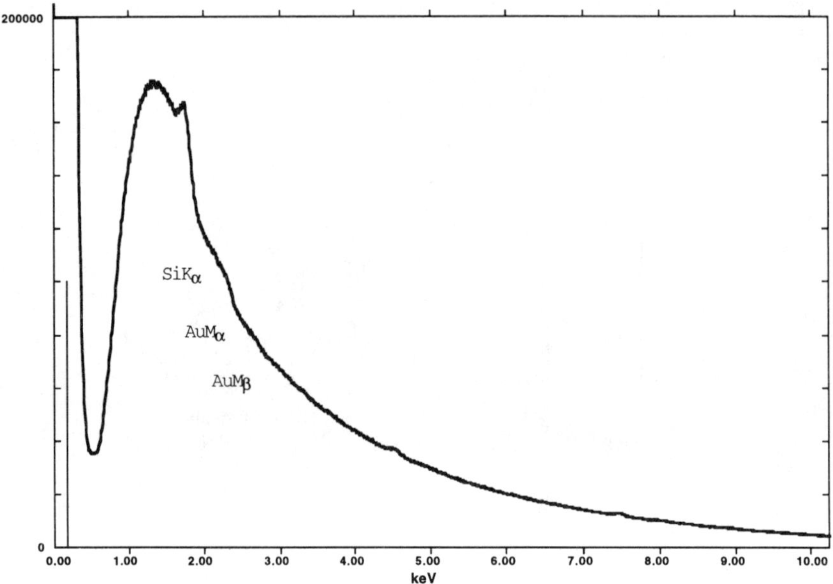



FIGURE 11.9.  Experimentally measured EDS spectrum of high purity boron, showing the Si $K$-shell absorption edge, the Si $K\alpha,\beta$ internal fluorescence peak, and the gold $M$-shell absorption edges. $E = 15$ keV.

random fluctuations, the number of carriers produced is not an absolute number for a given photon energy, but is subject to a statistical distribution. Furthermore, the measurement of the collected charge is subject to uncertainty due to the thermal noise of the amplification process. The distribution of the number of charge carriers for a single photon energy is reasonably well described by a Gaussian distribution, shown schematically in Figure 11.10:

$$y(E) = A_A \exp\left[-0.5(E - E_A / \sigma_A)^2\right] \tag{11.2}$$

where $\sigma_A$ is the standard deviation of the distribution, $A_A$ is the peak amplitude, $E_A$ is the peak energy, and the coordinate value y is the intensity at any value of the energy. $\sigma_A$ is given by

$$\sigma_A = \text{FWHM} / 2.355 \tag{11.3}$$

Fiori and Newbury[1] described the FWHM of the distribution for a silicon detector in terms of the two sources of noise by quadrature addition

$$\text{FWHM} \sim (C^2 E + N^2)^{0.5} \tag{11.4}$$

where $C$ is the measure of the uncertainty in the formation of the charge carriers from a photon of energy $E$, and $N$ is the electronic noise of the amplification process. If

$FWHM_R$ is the experimentally measured FWHM for a particular peak of energy $E_R$, then the FWHM for any other peak with energy $E_E$ can be calculated as:

$$FWHM_E^2 = 2.5(E_E - E_R) + FWHM_R^2 \qquad (11.5)$$

where FWHM and $E$ are in electron volts. Typically, the FWHM is specified for a detector at the energy of Mn $K\alpha$ (because of the availability of a convenient radioactive source, the isotope $^{55}$Fe), and (11.5) permits a calculation of the peak width for any other energy.

Peak broadening is perhaps the most serious unavoidable artifact of the EDS x-ray measurement process. The natural line width (FWHM) of an x-ray peak is of the order of 1 eV, depending on the peak energy (for Mn $K\alpha$, FWHM = 2.3 eV). For the typical Si(Li) EDS detector currently in service, the resolution measured at Mn $K\alpha$ is typically in the range 135–145 eV for the longest pulse shaping time (lowest maximum count rate and best resolution). The practical consequence of peak broadening is illustrated in Figure 11.11, where a Mn $K\alpha$ peak measured with an EDS with a FWHM of 140 eV and containing 1000 counts in the peak channel is actually derived from a peak only 2.3 eV wide and containing 14,898 total counts.

The effect of peak broadening on the spectrum of K-411 is illustrated in Figure 11.12(a), superimposed on the unbroadened spectrum. The spectrum is considerably transformed by broadening, most obviously in the appearance of the characteristic peaks. However, the action of broadening occurs at every channel, including those that are occupied by the background. A less obvious effect is the action of broadening to transform the sharp absorption edges into rounded structures that can appear to be peaks, as can be seen in the experimentally recorded spectrum of boron in Figure 11.9 and in the theoretical K-411 spectrum shown in Figure 11.12(b). Finally, the bremsstrahlung photons with energies at or just below the Duane-Hunt limit, which corresponds to the incident beam energy, are also broadened, as illustrated by Figure 11.12(c), so that the true Duane-Hunt limit is actually lower in energy than the apparent value given by the uppermost occupied channel.

Peak broadening imposes severe penalties upon practical microanalysis. Many situations are encountered in which peaks required for analysis are not resolved. Although peak deconvolution is possible, such procedures are inevitably limited by

TABLE 11.2.   Apparent Silicon Concentration Arising from Silicon Internal Fluorescence

| Target | Apparent Si concentration (mass fraction) | Target | Apparent Si concentration (mass fraction) |
|---|---|---|---|
| CaF$_2$ | 0.17 | Co | 0.27 |
| Sc | 0.20 | Ni | 0.43 |
| Ti | 0.19 | Cu | 0.63 |
| V | 0.19 | Zn | 0.36 |
| Cr | 0.26 | Ag | 0.37 |
| Mn | 0.34 | Sn | 0.29 |
| Fe | 0.26 | | |

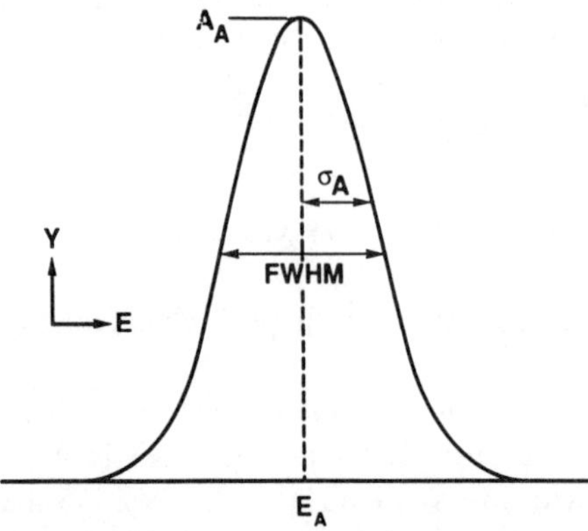



FIGURE 11.10. Theoretical Gaussian distribution used to describe an EDS peak. $\sigma_A$ is the standard deviation of the distribution, $A_A$ is the peak amplitude, $E_A$ is the peak energy, and the coordinate value $y$ is the intensity at any value of the energy.

statistical considerations if the actual concentration ratio of the components is large. For peaks separated by less than 50 eV, the preferred analytical approach is to use high resolution wavelength dispersive spectrometry, especially where the amplitude ratio of the component peaks is greater than 2:1. A second and less obvious consequence of peak broadening is the penalty suffered in the achievable limits of detection. Peak broadening causes the characteristic photons to be spread out across a band of the background. In any quantitative analysis procedure, the characteristic intensity must first be separated from the background. Various background subtraction algorithms are available, but eventually all such schemes are limited by the statistics of the background. For practical analytical conditions, the limit of detection with EDS is approximately 0.1% by weight, while for the same conditions the WDS limit of detection is approximately 0.01% because of the higher peak-to-background.

### 11.4.6. Peak Distortion

If a photon is absorbed in the partially active silicon just beneath the surface electrode, the loss of some of the electron-hole pairs due to recombination at defects results in an underestimate of the photon energy. These photons appear in bins on the histogram at lower apparent energies than their true value, creating a distortion from the ideal Gaussian shape on the low energy side of the peak, a phenomenon known as "incomplete charge collection." The effects of incomplete charge collection are most severe for low energy photons (energy < 3 keV). Silicon has a large mass absorption coefficient for low energy photons, resulting in shallow depth of absorption in the detector, and as a consequence, a significant fraction of the absorption occurs in the dead layer. An example of an experimental spectrum showing the incomplete charge effect for magnesium is shown in Figures 11.13(a) and 11.13(b).

Although the effects of incomplete charge collection are most obviously observable in the immediate vicinity of a peak in an electron-excited x-ray spectrum, the phenomenon actually extends over all energies below the peak, forming the so-called "background shelf." In addition to the incomplete charge collection occurring over the complete range of the charge deposited by the photon, the background shelf also arises

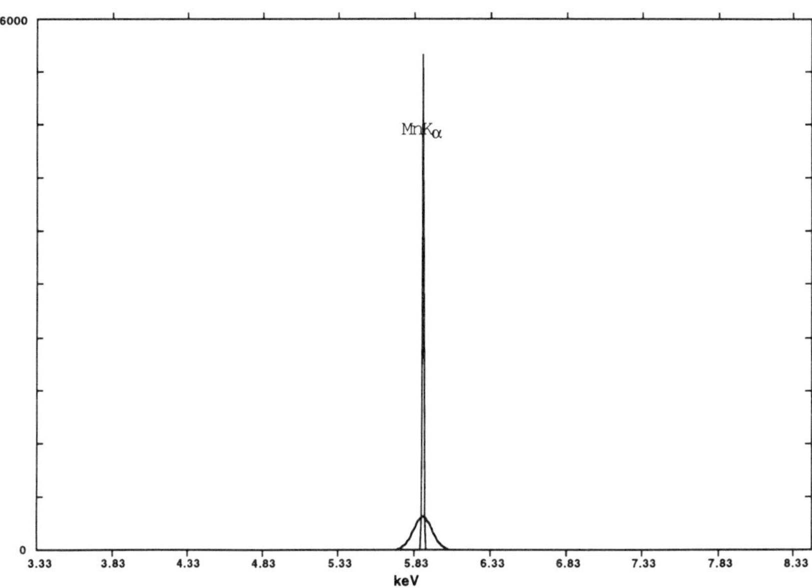


FIGURE 11.11.   Effect of peak broadening: a Mn $K\alpha$ peak with a FWHM of 140 eV and 1000 counts in the peak channel is actually derived from a generated peak less than 10 eV wide and containing 14,898 counts.

from the escape from the detector of bremsstrahlung generated by the photoelectron during its propagation and inelastic scattering in the detector. Because of the high bremsstrahlung background in electron-excited spectra, the background shelf is masked. The background shelf can be observed in spectra excited with radioactive sources, such as $^{55}$Fe.

### 11.4.7. Escape Peaks

The first step in the deposition of charge in the detector by photoelectric capture of the incident photon leaves the atom ionized in an inner shell. The ionized atom will subsequently undergo intershell electron transitions to return to the ground energy state, and the excess energy can be expressed as the emission of an Auger electron (for the Si $K$-shell, 96% of ionizations) or as a characteristic x ray (Si, 4%). The Auger electron is emitted with a substantial kinetic energy (for the Si $KLL$ Auger transition, $E_{kin} \approx 1.64$ keV). This energetic Auger electron scatters inelastically with a range of approximately 100 nm, so that it is highly likely to remain in the detector, creating electron-hole pairs just as does the photoelectron and contributing to the collection of charge proportional to the photon energy. The Si $K\alpha$ x rays have a much greater range in silicon, approximately 60 micrometers for 99% absorption, so there is a finite probability of escape of the Si x-ray from the detector. (Si $K\beta$ x rays are also created, but the relative yield is so low and the Si $K\beta$ energy is so similar to that for Si $K\alpha$ x rays that the effect is not significant.) If escape occurs, the charge collection for the

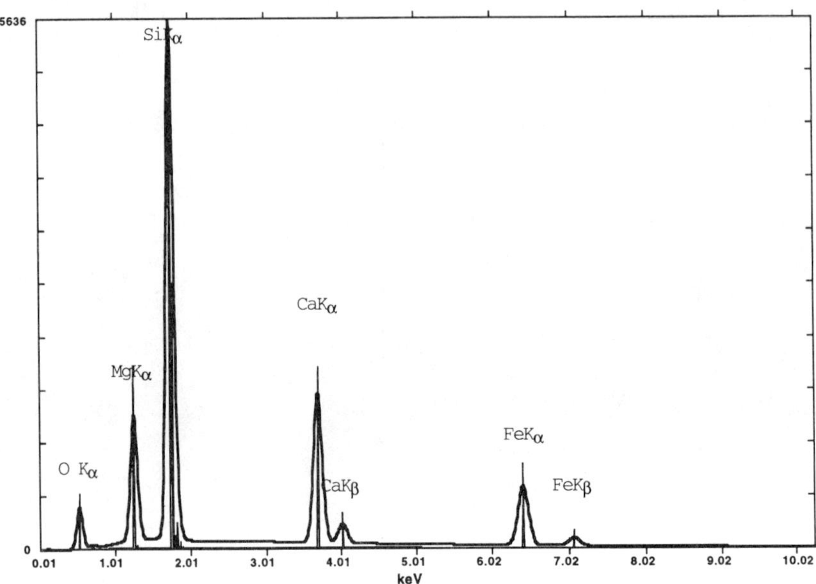

FIGURE 11.12.  (a) Spectrum of K-411 after the broadening action of the detection process, superimposed on the unbroadened spectrum (both autoscaled on the Si peak).

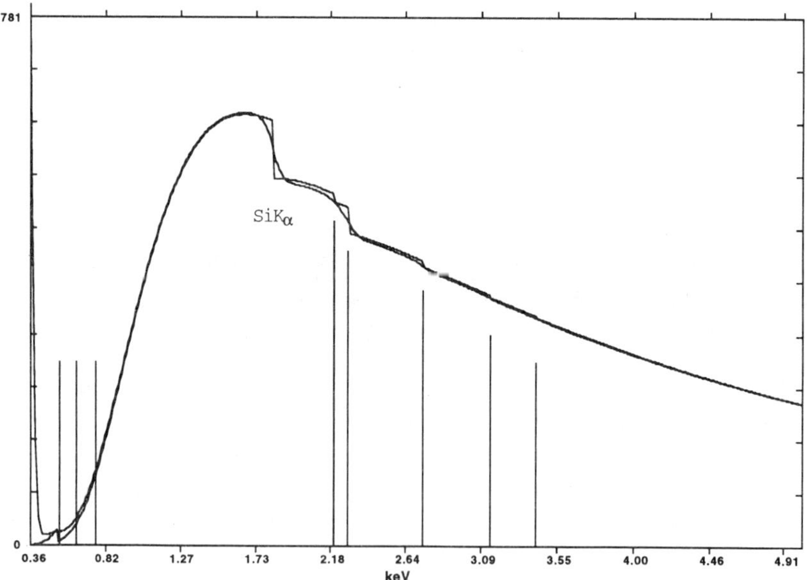

FIGURE 11.12.  (b) Appearance of the absorption edges arising from the gold surface electrode (carbon target).

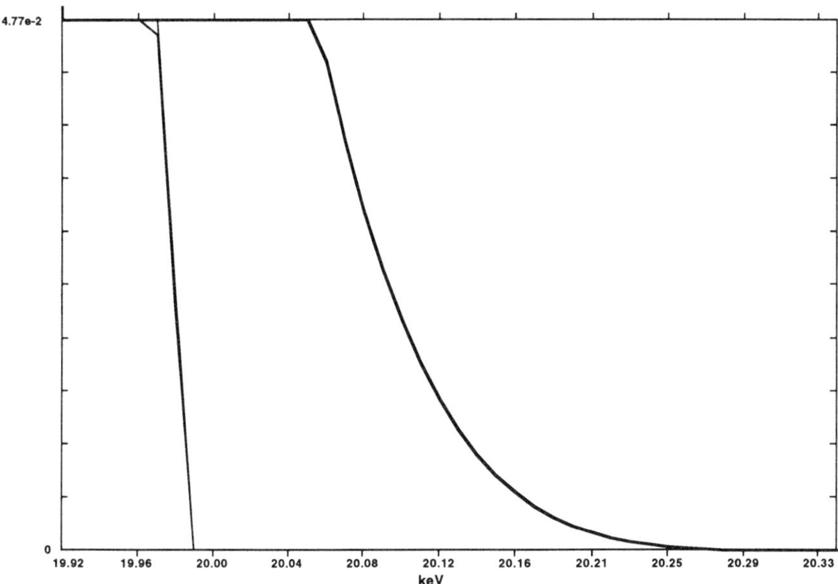


FIGURE 11.12.   (c) Appearance of the Duane-Hunt limit before (thin line) and after (thick line) broadening.

photon being measured is deficient by an amount equal to the energy of the Si $K\alpha$ x ray, 1.74 keV. Thus, a parasitic peak is formed below the parent peak at an energy

$$E_{esc} = E_{parent} - 1.74 \, \text{keV} \qquad\qquad (11.6)$$

An example of the single escape peak observed from scandium $K\alpha$ peak is shown in Figure 11.14(a). For a more complicated peak signature such as that from bismuth $M$, the escape peak is actually a composite of two peaks from Bi $M\alpha$ and Bi $M\beta$, as shown in Figure 11.14(b).

The relative magnitude of the escape peak depends on the energy of the parent peak. Photons from peaks with energies just above the absorption (critical excitation) energy of the detector material suffer maximum absorption and are most likely to be absorbed near the surface of the detector. Since the ionization events and the subsequent emission of Si $K\alpha$ photons occur near the front surface of the detector, the half of the Si $K\alpha$ x rays emitted isotropically toward the front surface have a higher probability of escape as compared to those emitted following absorption of a more energetic photon absorbed deeper in the detector. Examples of the ratio of the escape peak to the parent peak for several elements are given in Table 11.3.


## 11.5. ARTIFACTS THAT CAN BE REDUCED OR ELIMINATED

Once the analyst has become familiar with the physically unavoidable artifacts in energy dispersive x-ray spectrometry, over which there is no effective control, she/he is ready to consider those artifacts that arise from effects that can be eliminated, or at

                                                            

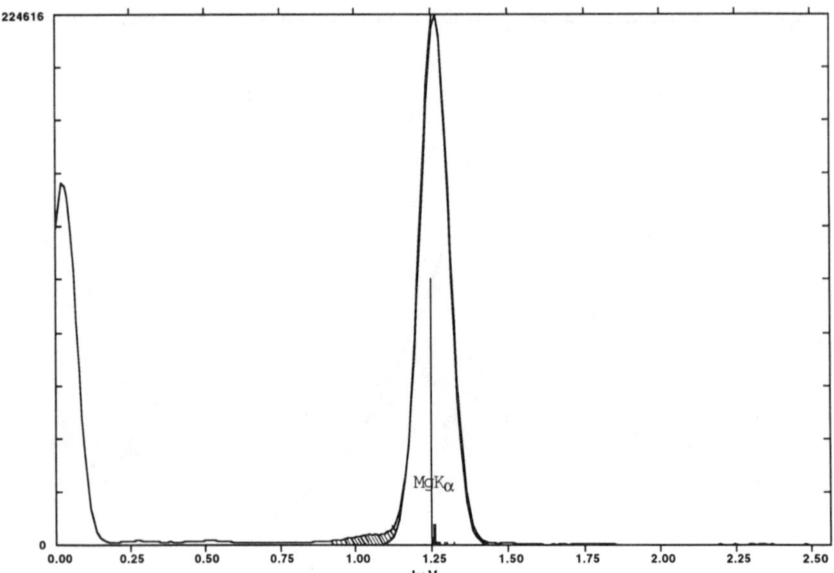


FIGURE 11.13.   Distortion of the low energy side of a Mg $K\alpha$ peak due to incomplete charge collection (hatched area). Experimentally measured spectrum (thick trace) compared to ideal Gaussian peak (thin trace). (a) Autoscaled to full peak height.

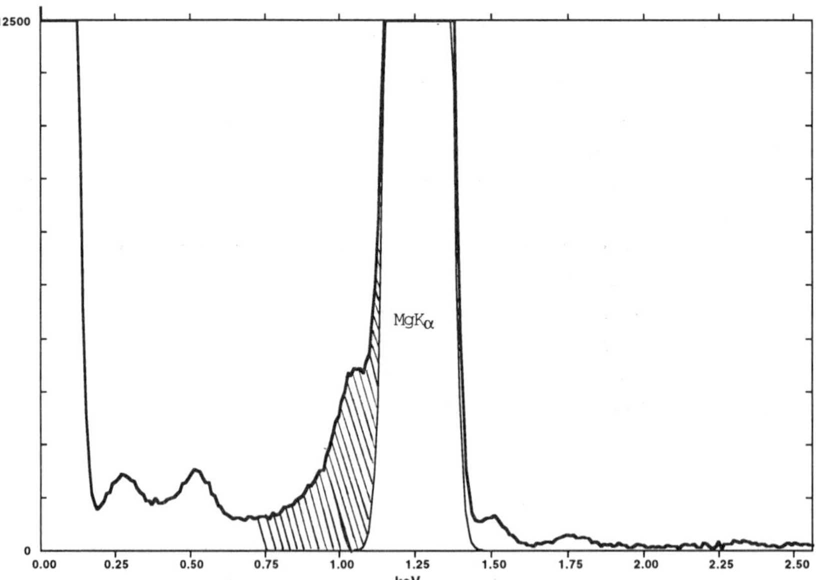


FIGURE 11.13.   (b) Vertical expansion.

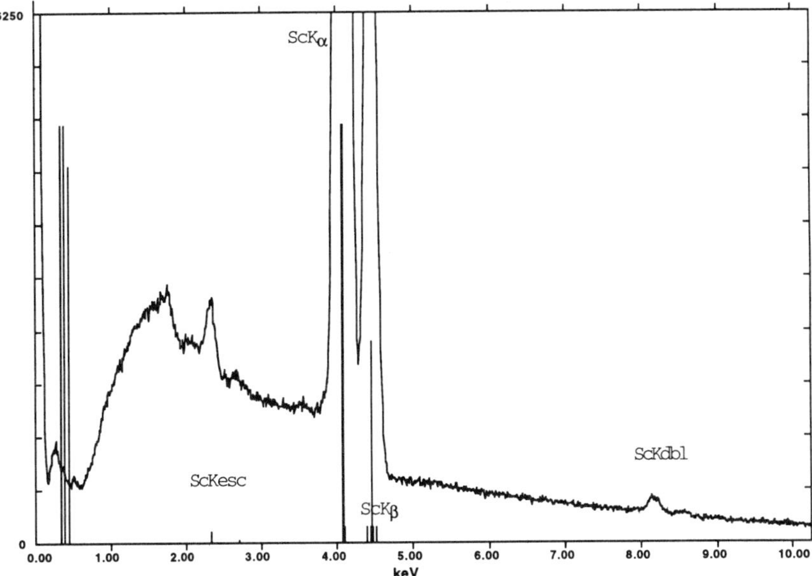

FIGURE 11.14.   Appearance of the Si escape peak: (a) Scandium.

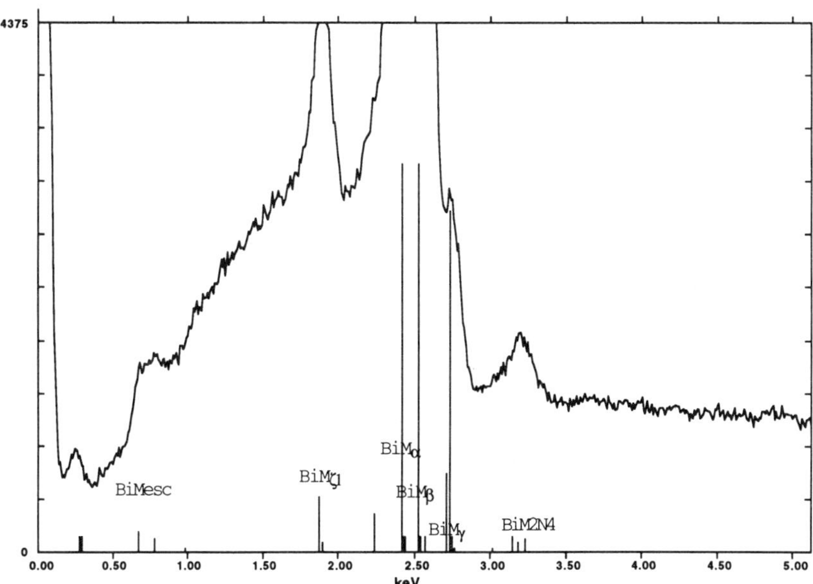

FIGURE 11.14.   (b) Bismuth.

least partially controlled, such as: stray radiation, direct entry of electrons into the detector, deadtime, pulse pileup, microphony/electromagnetic interference, and ground loops.

### 11.5.1. System Peaks

Ideally, the primary electron beam excites the specimen within a volume defined by the electron range and the range of secondary fluorescence by characteristic and bremsstrahlung x rays. In practice, the analyst must be aware of the possibility of excitation remote from the beam impact point from unexpected sources, such as electron scattering from the edge of the final aperture, x-ray transmission through the final aperture, electron backscattering from the specimen, etc. This phenomenon is most obvious when "system peaks" arise because the excitation actually extends to the microscope environment surrounding the specimen, including the specimen holder, stage, chamber walls, etc. (for more extensive discussions, see references 10 and 11). Remote excitation may also be a significant but less easily recognizable problem if the errant radiation is confined to a few millimeters from the beam but is still within the area of the specimen. In this case the x-ray spectrum may contain only peaks associated with the specimen composition itself, but local analysis will be compromised, for example, if the analytical requirement is to determine partitioning of a constituent between a local feature and the surrounding matrix.

Strictly speaking, x-ray excitation due to remote scattering is not an artifact of the EDS measurement system, but a failure of the microscope. The EDS is typically equipped with a collimator to restrict the area viewed by the spectrometer to a few square millimeters around the beam impact point. Nevertheless, good analytical procedure suggests that the analyst should evaluate the situation in a particular microscope with two tests. First, to evaluate a no-scattering condition, an "in-hole" spectrum count should be made with a Faraday cup for an SEM/EPMA, or with a suitable hole in a foil for the AEM, to assess stray radiation outside the incident beam. Second, to evaluate a scattering condition, a spectrum should be measured in the SEM/EPMA on a target consisting of a high atomic material embedded in a dissimilar matrix, e.g., high purity silver in aluminum, to assess the possible effects of remote excitation due to electron backscattering.

An example of such a measurement in an EPMA is shown in Figure 11.15, where the stray radiation due to remote excitation is seen to be negligible. In the AEM, such a test measurement can be carried out with a continuous material such as NIST

TABLE 11.3.   Silicon Escape Peak as a Fraction of the Parent Peak

| Element | Line | Energy | $I_{escape}/I_{parent}$ |
|---------|------|--------|-------------------------|
| P | $K\alpha$ | 2.015 | 0.0126 |
| Sc | $K\alpha$ | 4.088 | 0.0080 |
| Ti | $K\alpha$ | 4.508 | 0.0054 |
| Cr | $K\alpha$ | 5.409 | 0.0039 |
| Fe | $K\alpha$ | 6.400 | 0.0021 |
| Ni | $K\alpha$ | 7.470 | 0.0010 |

SRM-2063, which is a uniformly thin (approximately 76 nm) multielement glass film on a copper grid. Ideally, when the beam is placed on the film, there should be no excitation of the copper grid. The excitation of copper due to electron scattering and x-ray induced fluorescence can be measured with the beam placed at various distances (to a maximum of approximately 40 μm) from a grid bar. It must be recognized that such an experiment gives only a first estimate of the system peak situation in the AEM, since the electron scattering situation can vary strongly with the thickness and orientation of the specimen.

## 11.5.2. Electron Penetration

Since a large fraction of electrons backscattered from a specimen retain 50% of more of their incident energy, they may have sufficient energy to penetrate a 7.6- μm beryllium window and reach the detector, and they will certainly be capable of penetrating the ultra-thin materials of detectors capable of measuring low x rays. Such energetic electrons will scatter inelastically in the detector, producing electron-hole pairs and appearing as an apparent photon count in the spectrum. Direct electron contributions distort the background from the ideal shape that results from the bremsstrahlung generated during electron interaction in the specimen. To avoid this problem, manufacturers routinely use a magnetic field in front of the detector, obtained either from the magnetic field of the lens or from an accessory magnet/electromagnet and oriented to deflect electrons away from the detector window. The presence of a

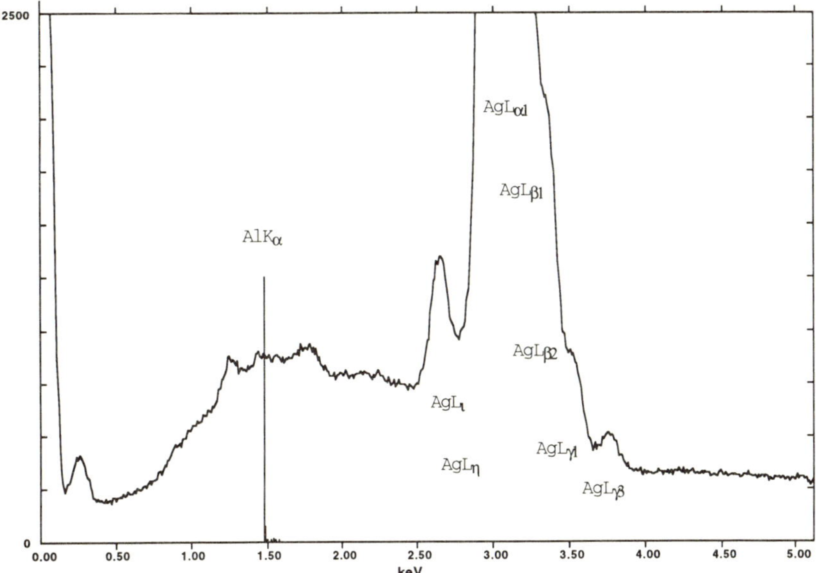


FIGURE 11.15. Spectrum recorded with the beam placed 25 μm from the edge of a 0.5-mm-diameter silver wire embedded in aluminum, expanded to show the absence of parasitic Al radiation from remote excitation (Ag *Lα* peak height = 73,500 counts). For these dose conditions, the Al standard intensity is 264,000 counts.

backscattered electron component can be assessed by comparing the measured background to the bremsstrahlung background calculated with DTSA. Figure 11.16 shows an example of a comparison between the theoretical prediction and the experimentally measured spectrum for an incident beam energy of 40 keV. The degree of correspondence suggests that high energy electrons contributed very little to the spectrum.

### 11.5.3. Dead Time

Although on the time scale of human awareness the x-ray spectrum appears to accumulate simultaneously at all energies, the pulses are in fact processed sequentially. The total time to process a pulse typically ranges from 10 µs to 50 µs, depending on the resolution that is required, with the trade-off of higher counting rates being purchased at the expense of poorer resolution. Sequential pulse processing leads to two major effects, "dead time" and pulse coincidence effects.

While the detector is "busy" processing the pulse produced by the absorption of a photon in the detector, it is effectively unavailable for detecting any other photons that arrive during the pulse processing time. The accumulated pulse processing time is referred to as "dead time." Even with constant incident beam current, the total spectrum count rate, and therefore the dead time, will vary with the efficiency with which characteristic peaks and bremsstrahlung x rays are excited, which in turn changes with the composition of the specimen. To place all measurements on the same dose base (electron flux × actual detector availability time), the dead time correction circuit of the signal processing chain acts to record the time intervals during which the

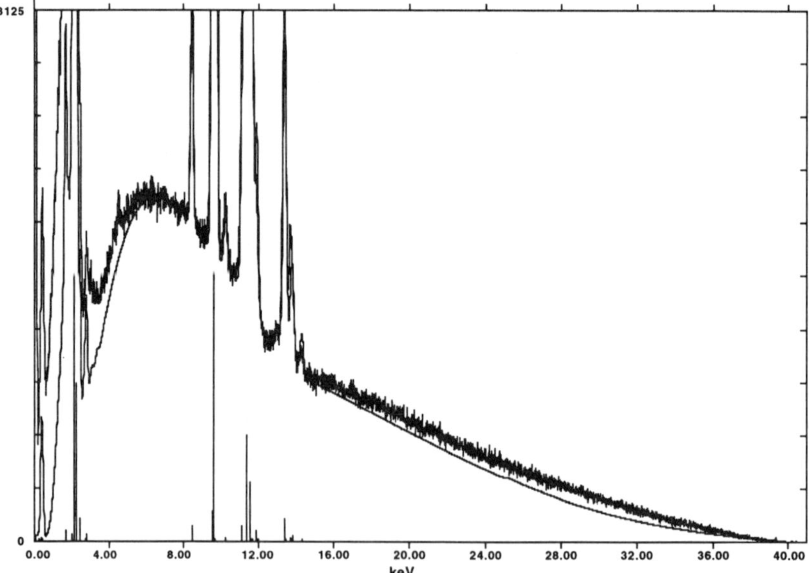

FIGURE 11.16.   Comparison of experimental and theoretical spectra for gold with an incident beam energy of 40 keV showing negligible contributions from direct electron entry.

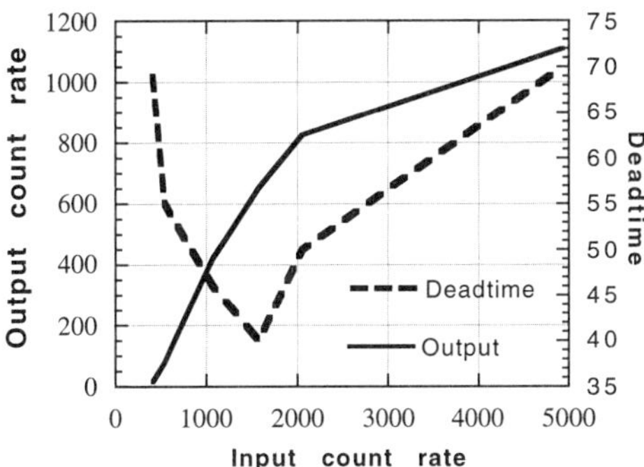


FIGURE 11.17.   Input-output characteristics from an improperly operating detector. (Data provided by John Small and Scott Wight, NIST.)

pulse processor is busy, and then adds additional time at the end of the true clock time to compensate for the loss. Actual spectrum accumulation time always exceeds true clock time or real time. While it is reasonable to expect that the automatic dead time correction performed by the computer-controlled x-ray analyzer is accurate, a careful operating procedure would include a periodic test of the output counts versus primary beam current. If an accurate and reproducible measure of the beam current is available, then a simple test can be performed to determine if the recorded output count (e.g., peak integral) scales with the dose (beam current × live time) or the input count rate as measured by the EDS system. For satisfactory operation, the output should be linear

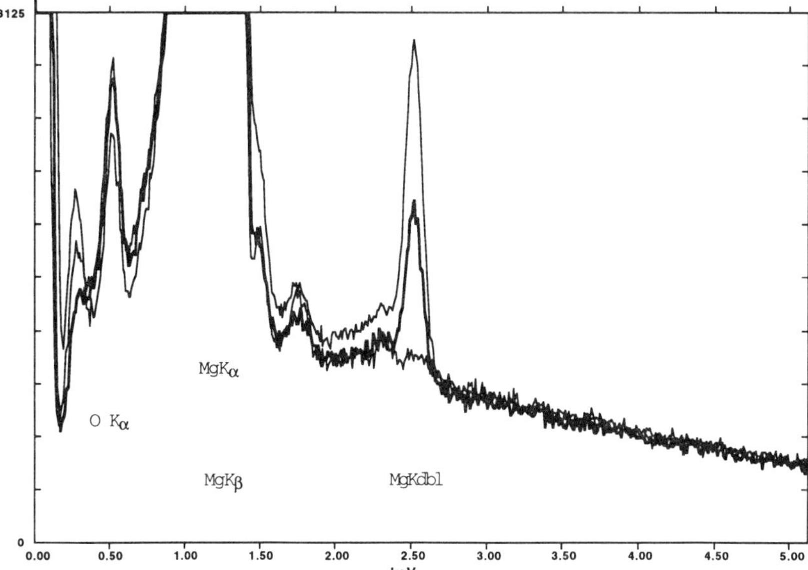


FIGURE 11.18.   Coincidence peak ("double energy") for magnesium: upper trace, 70% dead time; middle trace 47% dead time (thick line); lower trace, 14% dead time.

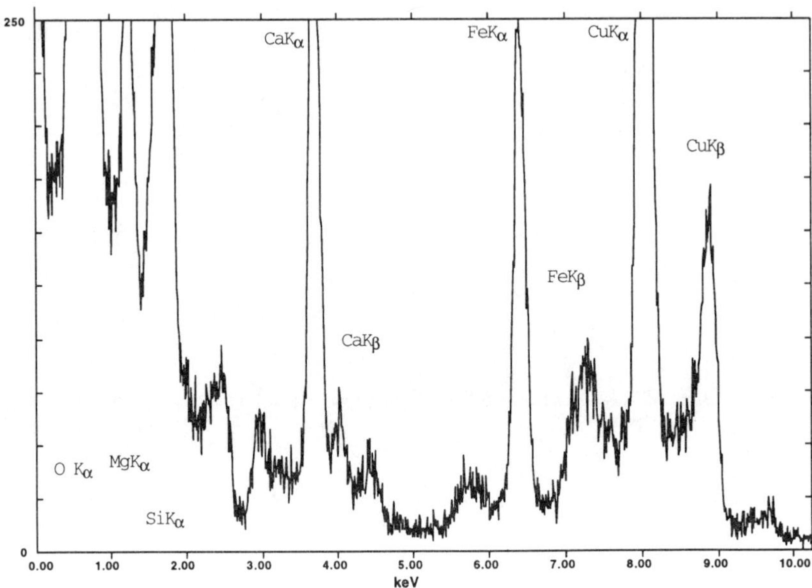


FIGURE 11.19. Spectral distortion due to "ground loop" interference from improper cable shielding. Specimen: NISTSRM-2063a thin film; beam energy: 300 keV; detector: intrinsic germanium in an analytical electron microscope. (Spectra provided by Eric Steel, NIST.) (a) Spectrum with background distortions and ghost peaks.

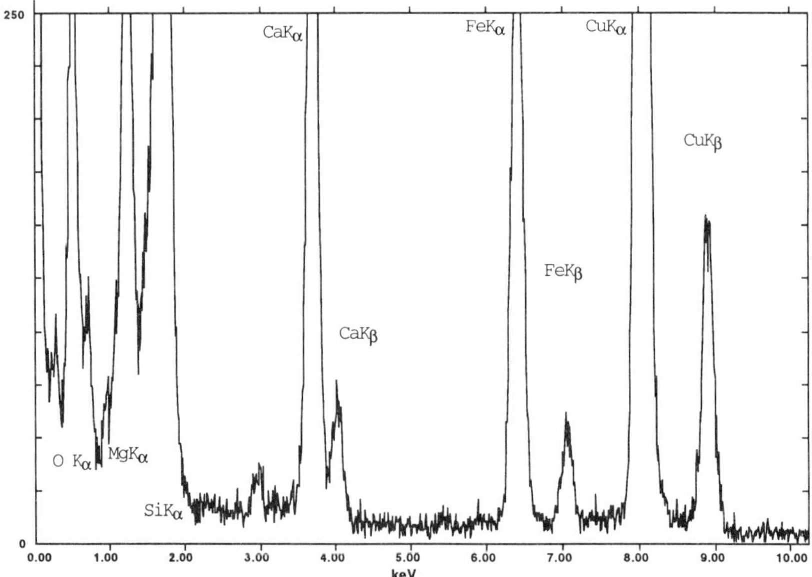


FIGURE 11.19. (b) Spectrum after proper cable shielding.

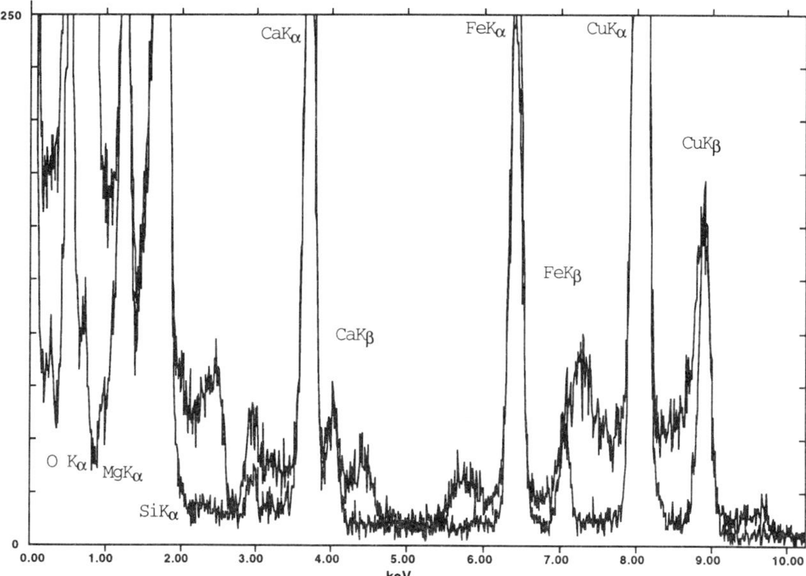


FIGURE 11.19.   (c) Overlay.

with dose for system dead times of at least 75%. An example of the response from a system with improperly functioning pulse processor is shown in Figure 11.17. The dead time behavior below an input count rate of 2,000 cps is extremely non-linear, and the deadtime reaches 100% (no output counts) for an input count rate of 500 cps. Moreover, the input-output response is non-linear over the range that is critical for quantitative measurements. The EDS system that produced these results was rejected and was subsequently corrected by the manufacturer.

Techniques for pulse processing are undergoing rapid changes with the introduction of digital processing methods (for example, see Chapter 9[12]). Digital processing offers the possibility of significantly reducing or eliminating several of the artifacts introduced by analog processing.

### 11.5.4. Pulse Pileup

To minimize problems that arise from pulse coincidence, the signal processing chain also constantly inspects the charge state of the detector to mark the arrival of a photon. When a second photon arrives during the measurement of a photon, the pulse inspection circuit determines if sufficient information has already been gained on the first pulse to measure it with acceptable accuracy. In this case, the measurement of the first photon is retained for the spectrum. If the inspection circuit determines that the detector charge has not returned to the baseline value in time for the measurement of the second photon, the measurement of the second photon is discarded. As the time separation between coincident photons is reduced, eventually the pulse pair resolution of the coincidence inspection circuit is exceeded, and the detector fails to distinguish between the two photons, so that an apparent photon with has the sum of the two photon pulses is added to the spectrum. Such coincidence events can obviously occur for photons with any two arbitrarily chosen energies within the excited range, but on the basis of probability, only high intensity peaks will create sufficient coincidence events to produce distinct artifact peaks in the spectrum. Moreover, the magnitude of coincident peaks will increase with the total spectrum count rate, since at high count rates the probability increases that two photons will enter the detector within the pulse pair resolution time. An artifact coincidence peak is illustrated for magnesium in Figure 11.18, and the effect of the count rate on increasing the relative height of the sum peak is also illustrated for spectra recorded at 14%, 47%, and 70% dead times. Pulse coincidence effects increase as the energy of the photon decreases, since the magnitude of the pulse from a low energy photon approaches the background noise and is less easily distinguished by the inspection circuit, becoming particularly significant below 1 keV. Modern pulse inspection circuitry effectively eliminates pulse coincidence peaks for parent peak energy above a few keV.

Clearly, the analyst has a measure of control over coincidence peaks. By operating at low system dead time, approximately 20% or less, sum peaks will only exist at very low levels for peaks with energies above approximately 1 keV. However, for most analytical situations, e.g., to reach low values of the limit of detection or to record peaks for subsequent deconvolution of components, the analyst generally wishes to operate at high counting rates to provide the best counting statistics within the time allowable for the measurement. Good analytical practice demands that whenever the identification of low intensity peaks is attempted, the possibility of sum peaks must always be considered. As an aid for rigorous qualitative analysis, the information on sum peaks should be listed by the computer-assisted analysis database, as illustrated in Figure 11.18, and should certainly be considered in automatic qualitative analysis.

The development of pulse detection/beam blanking has resulted in a significant improvement in the limiting count rate at which the dead time becomes significant. In such a system, upon detection of a photon, the electron beam is blanked to prevent a second photon from being generated. Although such a system permits limiting count rates that are greater by a factor of 5 or more compared to a conventional system, at high dead times coincidence peaks are still observed.

### 11.5.5. Ground Loops/Electromagnetic Interference/Microphony

The EDS detector typically processes $10^{-16}$–$10^{-15}$ C of charge to measure a photon in the energy range from 0.1 to 20 keV. Amplification circuits capable of measuring such small amounts of charge are sensitive to interference from a wide variety of sources, including acoustic (microphony) and electromagnetic. The general problem of "ground loops" in the electrical environment of the EDS, either in the microscope chassis or in the wiring of the EDS system, can lead to significant distortions in the recorded spectrum. An example is shown in Figure 11.19(a) of such a distortion which was found to be associated with the cable between the detector/pre-amplifier and the main amplifier assembly. The background was found to show large deviations from the correct physical behavior, and "ghost" peaks were found in association with each real peak. These distortions were eliminated by careful shielding of the cable, as shown in Figure 11.19(b). Background modeling with DTSA can be a very useful aid for recognition of non-physical effects in the measured spectrum.

## 11.5.6. Light Penetration

The Si(Li) detector and its FET amplifier are sensitive to long-wavelength electromagnetic radiation, including ultraviolet and visible light. With sufficient light illumination during x-ray measurement, coincidence events will lead to shifts in peak position and peak shape which can often be quite severe. Figure 11.20 shows an example of EDS spectra taken with a thin diamond window detector from ZnS, which is strongly cathodoluminescent with predominantly blue light emission. As the light emission is increased by increasing the scanned area (decreasing the magnification) at constant current, the peak becomes progressively shifted and distorted. To eliminate or at least minimize this effect, a reflective coating of aluminum is normally applied to the window. For this particular detector, the coating is either too thin or else has pinholes that permit light to penetrate. Thicker windows, such as 7.6-μm Be, are sufficiently opaque to light to eliminate the effect.

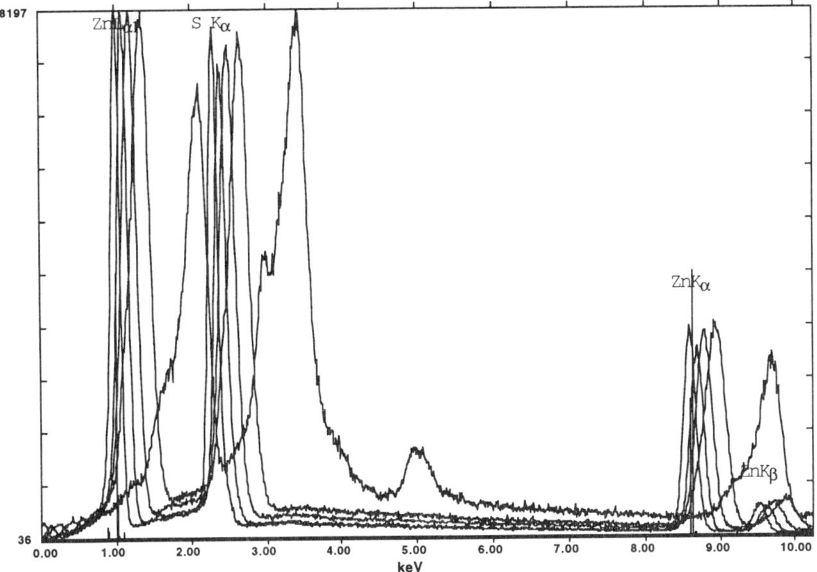


FIGURE 11.20.   Effect of light penetration into a Si(Li) detector. The specimen is ZnS, which is strongly cathodoluminescent in blue light. At constant incident beam current, the light emission increases as the area scanned increases (magnification decreases), causing the peak to shift in position and become more distorted.

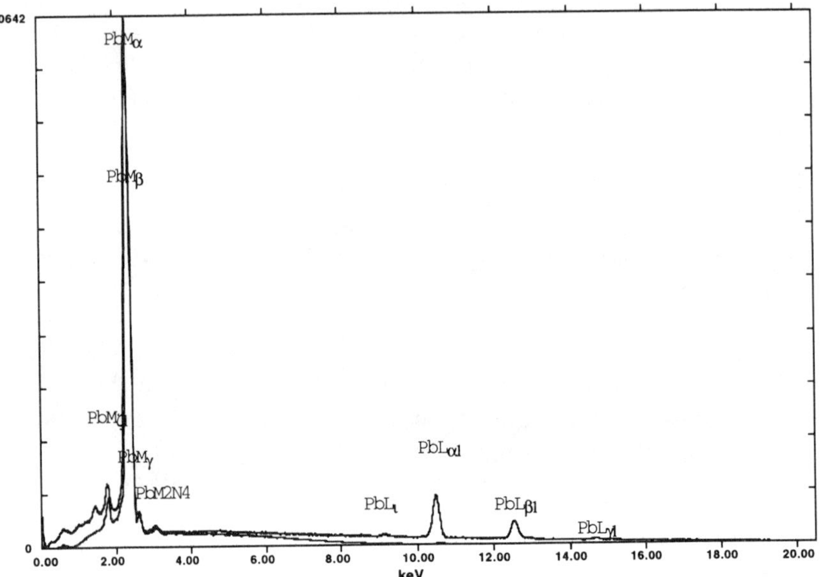


FIGURE 11.21.  Anomalous loss in spectrometer efficiency at intermediate and high x-ray energies. Specimen: lead excited with a beam energy of 20 keV. The spectrum from the defective detector, showing a severe decrease in efficiency above 5 keV, is compared to a spectrum from a detector with a normal response (efficiency ≈100% for 5–15 keV). (a) Scaled to Pb *M*.

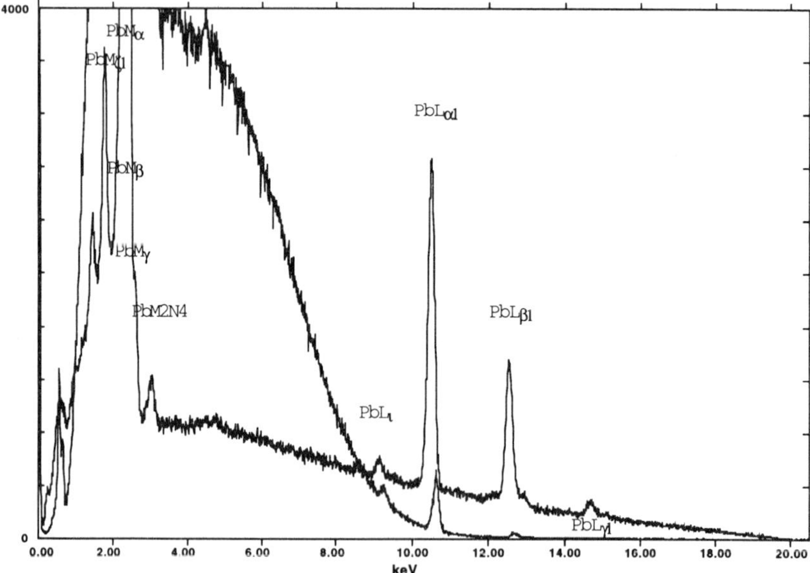


FIGURE 11.21.   (b) Expanded vertical scale to compare the background.

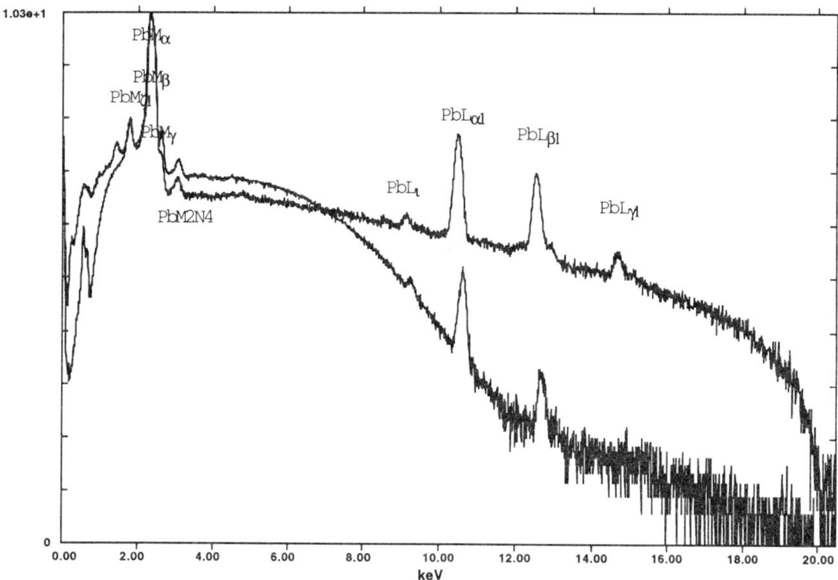


FIGURE 11.21.   (c) Logarithmic presentation of the intensity data.

### 11.5.7. High Energy Attenuation

A most unexpected artifact encountered in a Si(Li) detector delivered in December 1992 is illustrated in Figure 11.21. Normally, the efficiency of any Si(Li) detector is greater than 99% for photon energies from approximately 5 to 15 keV. High efficiency over this energy range is expected and would never be specified in a technical description for a procurement action. However, our experience suggests that nothing should be taken for granted. The response of the new detector was compared to that obtained from a "normal" detector for a lead metal target excited with a 20-keV electron beam, as shown in Figures 11.21(a)–11.21(c). The new detector had a large decrease in efficiency above 5 keV. There was such an attenuation in the measured x-ray intensities above 5 keV that, at the energies of the lead $L$-series, the Pb $L\gamma_1$ peak was lost even though it was easily visible in the "normal" spectrum! This problem was ascribed by the manufacturer to a defect in the pulse processor and was subsequently corrected. An alternative possibility would be that the effective detector thickness (i.e., the active volume for collecting charge) was severely reduced to approximately 100 μm as compared to the normal 3 mm. Such a situation might arise from improper fabrication or a failure in the biasing (see Chapter 5[8]). Obviously, the careful analyst must be prepared to test all aspects of the performance of a new EDS system.

## 11.6. LOW ENERGY SPECTROMETRY

The low energy portion of the x-ray spectrum below 1 keV deserves special consideration.[12,13] Access to this region by EDS has only become possible with the

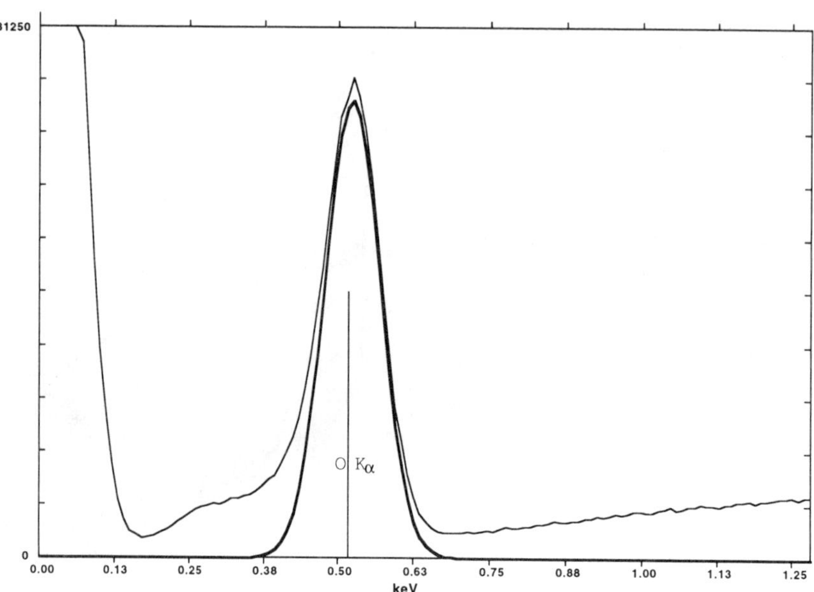


FIGURE 11.22.  Distortion of the low energy side of the oxygen $K$-peak due to incomplete charge collection. Specimen: $SiO_2$ with a beam energy of 15 keV. The spectrum has been background-subtracted, and a Gaussian profile (thick trace) has been fitted to the peak.

development of window materials that can transmit low energy x-ray photons with reasonable efficiency and the development of signal amplification circuits that can work with charge pulses that are very near the level of thermal noise. Artifacts are especially significant in the low energy regime, so much so that the accuracy of the measurement of both the x-ray energy and the x-ray intensity can be significantly affected. One reason for this is that, because of the shallow depth of absorption of low energy x rays in the EDS detector, incomplete charge collection significantly distorts the low energy side of the peak, as shown in Figure 11.22 for oxygen. Another reason is that, because the pileup inspection circuit must work with pulses that are just above the noise threshold, pileup rejection is badly compromised, leading to distortion of the high energy side of the peak. An extreme example is shown in Figure 11.23 for carbon, where a peak recorded at 45% dead time is compared to an ideal Gaussian calculated for the resolution predicted for carbon from the value at manganese, revealing extreme distortion of the high energy side of the peak. A further example of the difficulty in measuring low energy peaks is shown in Figure 24, where carbon peaks recorded at 13% dead time and 45% dead time are compared. A large peak shift as well as large distortions are noted for the high deadtime spectrum. At low energies, the calibration tends to become non-linear. For example, the carbon spectrum recorded at low deadtime in Figure 11.24 is not accurately calibrated despite the fact that nearby peaks such as oxygen showed excellent correspondence to the calibration. These phenomena are considered in more detail in Chapter 5 of this volume.[8]

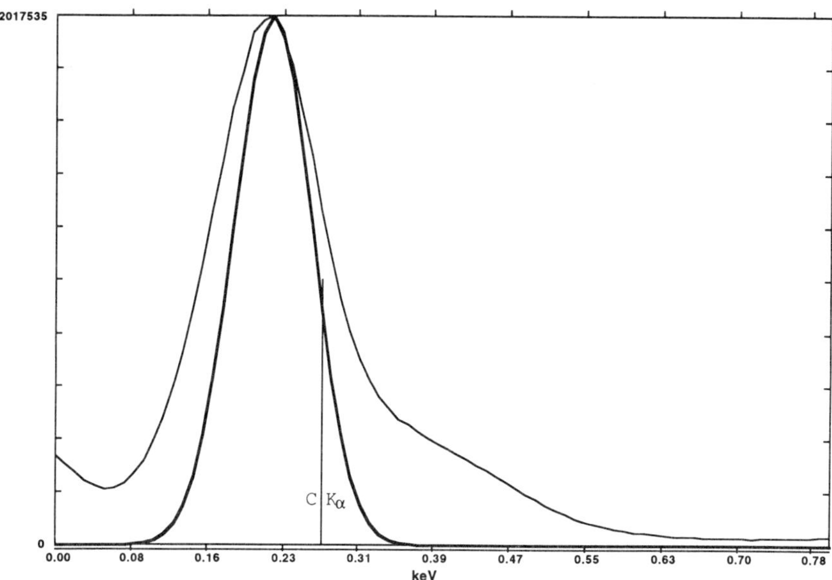



FIGURE 11.23. Distortion of the high energy side of a carbon $K$-peak recorded at 45% dead time. A Gaussian (thick trace) has been fitted to the carbon peak with a FWHM as appropriate to the value determined for Mn $K\alpha$. Note shift from calibrated C peak position.


## 11.7. CONCLUSIONS

Energy dispersive x-ray spectrometry has advanced considerably in the last 15 years, providing improved spectral resolution, count rate range, and access to low energy x rays. Indeed, spectral peaks for all elements in the periodic table starting with beryllium are now accessible with EDS. Despite the extensive advantages of EDS as a measurement tool, the practical situation as experienced in the microscopy laboratory still remains one of *caveat emptor*. As the experience in our laboratory has shown, EDS systems are still being delivered with serious performance deficiencies. Alternatively, the EDS as tested by the manufacturer may function well by itself on the laboratory test bench when excited with a radioactive source, but when that same EDS is placed on the electron column instrument, ground loops or electromagnetic interference may arise and degrade the performance. Moreover, the onus is placed upon the user to recognize deviations from acceptable behavior and to demand efforts from the manufacturer to correct the deficiencies. To recognize excursions from acceptable operation, the analyst must become familiar with the physical nature of the x-ray spectrum and the numerous deviations from the "true" spectrum that occur during measurement. First-principles spectrum simulation with computer-aided analysis, such as that available with Desk Top Spectrum Analyzer, can be an enormous help in understanding the nature of the spectrum in direct comparison to experimental measurements. When such an approach is followed, we find that the optimal final spectrum currently obtainable from EDS is in fact quite distorted from the true spectrum as a

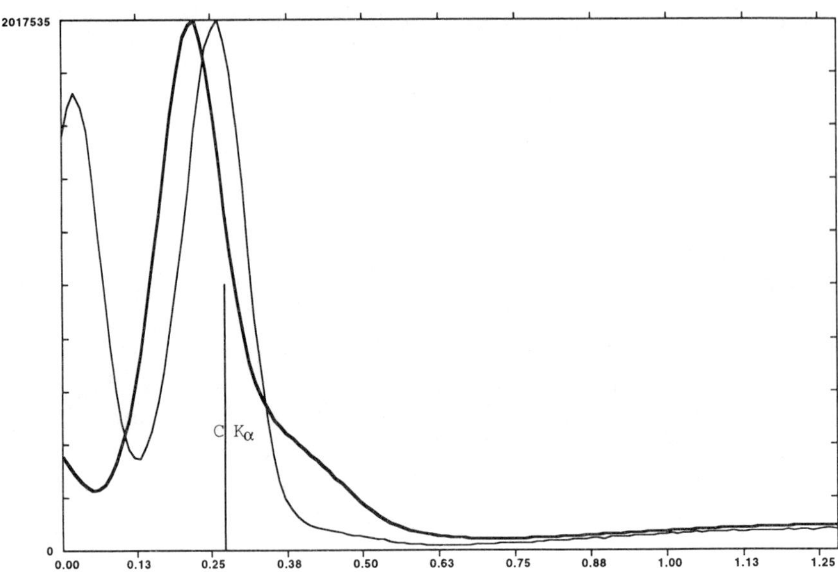



FIGURE 11.24.  Comparison of carbon peaks recorded at 13% dead time (thin trace) and 45% dead time (thick trace), showing extreme distortion at high dead time and peak position shift from calibration value.

result of the operation of various artifacts. The careful analyst must take into account the impact of the unavoidable artifacts of the EDS process upon the final qualitative and quantitative results. Equally important, the analyst must be on the lookout for the artifacts that are reducible or even avoidable with careful experimental practice. Analysis at the highest standard of performance requires knowledge of the complete x-ray generation and measurement process.

## REFERENCES


1. C. E. Fiori and D. E. Newbury, in: *Scanning Electron Microscopy/1978/I*, (O. Johari, ed.) SEM, Inc., Chicago, pp. 401–422 (1978).
2. R. Fitzgerald, K. Keil, and K. F. J. Heinrich, *Science* **159**, 528 (1968).
3. C. E. Fiori and C. R. Swyt, in: *Microbeam Analysis*—1989 (P. E. Russell, ed.) San Francisco Press, San Francisco, p. 236 (1989).
4. Desk Top Spectrum Analyzer, available from: National Institute of Standards and Technology, Gaithersburg, MD (1991).
5. M. Fabre de la Ripelle, *J. Phys.* (Paris) **10**, 319 (1949).
6. H. Bethe, *Ann. Phys.* **5**, 325 (1930).
7. J. A. Small, S. D. Leigh, D. E. Newbury, and R. L. Myklebust, *J. Appl. Phys.* **61**, 459 (1987).
8. D. C. Joy, in: *X-Ray Spectrometry in Electron Beam Instruments* (D. Williams, J. Goldstein, and D. Newbury, eds.) Plenum Press, New York, Chap. 5 (1995).

9. J. McCarthy, in: *X-Ray Spectrometry in Electron Beam Instruments* (D. Williams, J. Goldstein, and D. Newbury, Eds.) Plenum Press, New York, Chap. 6 (1995).

10. D. B. Williams, J. I. Goldstein, and C. E. Fiori, in: *Principles of Analytical Electron Microscopy* (D. C. Joy, A. D. Romig, Jr., and J. I. Goldstein, eds.) Plenum Press, New York, p. 123 (1986).

11. J. I. Goldstein, D. E. Newbury, P. Echlin, D. C. Joy, A. D. Romig, Jr., C. E., Lyman C. E. Fiori, and E. Lifshin, *Scanning Electron Microscopy and X-ray Microanalysis*, Second Ed., Plenum Press, New York, pp. 273–337 (1992).

12. R. W. Mott and J. J. Friel, in: *X-Ray Spectrometry in Electron Beam Instruments* (D. Williams, J. Goldstein, and D. Newbury, eds.) Plenum Press, New York, Chap. 9 (1995).

13. P. J. Statham and T. Nashashibi, *Microbeam Analysis—1988*, (D. E. Newbury, ed.) San Francisco Press, San Francisco, p. 50 (1988).

# 12

# Characterizing an Energy Dispersive Spectrometer on an Analytical Electron Microscope

*S. M. Zemyan and D. B. Williams*

## 12.1. INTRODUCTION

X-ray microanalysis in the analytical electron microscope is *invariably* performed with a Si(Li) or an intrinsic Ge (IG) energy dispersive spectrometer. Apart from the earliest electron microscope micro-analyzer instruments in the 1960s, which used a crystal or wavelength dispersive spectrometer, all commercial AEMs offer EDS as the only x-ray detector. The reason for this is the limited confines of the stage of a TEM to which the x-ray detector is interfaced when creating an AEM. The narrow polepiece gap of any TEM makes it impossible to get a detector closer to the specimen than about a centimeter. Consequently, the collection angle of the detector is small (typically 0.05–0.15 sr) thus limiting the x-ray counting statistics. Although small, the EDS collection angle is still very much larger than could be provided by a WDS. Since the x-ray count rate in AEMs is very low in the first place, because of the small probe currents and thin specimens, only an EDS can provide statistically meaningful x-ray data.

However, an EDS in an AEM is more likely to incur damage than in an SEM or EPMA—the other electron beam instruments that use EDS systems. There are several reasons for this damage. First, the electron beam energies (100–400 keV) in an AEM are much greater than in an SEM/EPMA, thus creating much brighter electron beams. Second, this intense beam of high energy electrons bombards a thin (electron-transparent) specimen that scatters the electrons (particularly in the forward direction). The

S. M. Zemyan and D. B. Williams • Department of Materials Science and Engineering, Lehigh University, Bethlehem, Pennsylvania 18015-3195

*X-Ray Spectrometry in Electron Beam Instruments*, edited by David Williams, Joseph Goldstein, and Dale Newbury. Plenum Press, New York, 1995.

scattered electrons can interact with regions of the specimen far from the incident beam. These remote areas, and any other part of the microscope that is hit by electrons, emit both characteristic x rays and bremsstrahlung x rays with energies up to that of the electron beam. Bremsstrahlung x rays of 100–400 keV energy can penetrate large amounts of material and fluoresce characteristic x rays from anywhere that they hit. Ideally the EDS should only "see" the x rays from the beam-specimen interaction volume, but it is not possible to prevent extraneous radiation from the AEM stage and other areas of the sample from entering the detector. Third, it is possible to generate an intense flux of high energy backscattered electrons (BSEs), particularly when the incident beam hits a bulk (non-transparent) portion of the specimen or a supporting grid bar. These BSEs can enter the EDS in varying amounts depending on the detector collection angle and the take-off angle relative to the BSE distribution. The resulting large electron-hole pair signal can overload the electronics, causing distortion of the x-ray spectrum, and the BSEs themselves may damage the Si(Li) detector by creating recombination centers such as point defects and by redistributing the Li composition profile.

The damage from these BSEs can be reduced if the detector is equipped with a well designed collimator and a protective shutter. Damaged detectors can often be revived by warming up with the bias voltage off *only after consultation with the manufacturer.* Emptying out the liquid $N_2$ dewar, filling it with hot water, and then drying with a hair dryer may solve the problem. However, after several such cycles, the detector performance may not return to acceptable levels (and the window seal may develop a leak due to the repeated thermal cycling). Then it is necessary to have the detector repaired.

Thus, for various reasons, the EDS performance is often compromised on an AEM and the lifetime of the detector is generally much shorter than in an SEM. Therefore, it is particularly important first to take the time to verify the performance of a new EDS system on an AEM and, second, to monitor the performance of the detector routinely. The operator must be able to detect when the detector performance falls below desirable levels and when it is time to replace or repair the detector. In order to be able to do this, however, there must be a baseline set of criteria for judging the performance of the EDS on an AEM, and such criteria are just beginning to become available. There are four requirements for the performance criteria:

1. They must be quick and easy to carry out.
2. They must be performed with specimens that are routinely available or easy to fabricate.
3. The test results should relate to the normal use of the AEM, i.e., the operator should know how the result of the test translates into the quality of measurement on a real specimen.
4. The tests should use standard protocols.

Currently, there is only one generally accepted standard test within the AEM community, and this is the definition of the energy resolution of the EDS (see next section). This paper summarizes some standard tests, that the SEM community has used for many years (see Lyman *et al.*[1]) but are not widely used in the AEM field, and also introduces some recently developed performance criteria aimed specifically at EDS systems on intermediate voltage (100–400 kV) AEMs.[2-4] Each test is described, including which specimen to use, what to measure, and how to interpret the results.

There are several fundamental parameters of both Si(Li) and IG detectors that can be specified, measured, and monitored to insure that a system is performing acceptably. In an SEM, which is relatively benign, Si(Li) detectors have been known to last ten years or more before requiring service or replacement. In contrast, as already noted, in an AEM the life of a detector often ends prematurely. So it is particularly important to monitor the detector performance on an AEM, in order that quantitative analyses made at very different times may be compared in a valid manner. These specifications, which will now be described, can be separated into detector variables, optimizing the detector-microscope interface, and signal processing variables.

## 12.2. DETECTOR VARIABLES

### 12.2.1. The Detector Resolution

The energy resolution $R$ of the detector is defined as the quadrature sum of several components:[5]

$$R^2 = P^2 + I^2 + X^2 \tag{12.1}$$

The term $P$ is a measure of the quality of the electronics in the pulse processing chain, defined as the FWHM of a randomized electronic pulser, $X$ is the FWHM-equivalent attributable to detector leakage current and incomplete charge collection within the detector, and $I$ is the intrinsic linewidth of the detector.[5] $I$ is controlled by fluctuations in the numbers of electron-hole pairs created by a given x ray and is defined as

$$I = 2.35(F\varepsilon E)^{1/2} \tag{12.2}$$

where $F$ is the Fano factor (which takes account of the deviation of the distribution of x-ray counts from Poisson statistics), $\varepsilon$ is the energy to create an electron-hole pair in the detector, and $E$ is the energy of the x-ray line. Because of the three factors, $P$, $I$, and $X$, the experimental resolution can only be defined under specific analysis conditions. The IEEE standard for $R$ is the FWHM of the Mn $K\alpha$ peak, generated (off the microscope) by an $^{55}$Fe source which produces $10^3$ cps with an 8- μs pulse-processor time constant.[6] If an $^{55}$Fe source is available, the detector resolution and peak-to-background ratio ($P/B$) of a new detector should be measured prior to interfacing to the column in order to confirm the manufacturer's stated values. However, it is more important to measure the detector resolution on the AEM column in the following manner.

Since Mn is not a particularly common thin-foil sample, we propose using a thin evaporated chromium film sample to check the resolution when the detector is on the column. An evaporated Cr film about 100 nm thick supported on a carbon film and a

Cu grid, as developed by Williams and Steel,[2] is ideal. Because Cr is next to Mn in the periodic table, the resolution of the Cr $K\alpha$ peak will be just a few eV less than that of Mn $K\alpha$ measured from the $^{55}$Fe source. Chromium is also stable with a resilient thin oxide film; it does not degrade in the electron beam and is very useful for other calibration checks and performance criteria.

Many EDS computer systems have an internal software routine which measures the resolution by calculating the FWHM of the chosen x-ray peak. Alternatively, as shown in Figure 12.1, first gather a Cr peak containing a large number of counts ($>10^5$), then select windows encompassing the peak between the channels on each side of the peak that contain half the maximum counts in the central channel. This defines the FWHM, which is taken as the energy resolution. However, it is also useful to gather the FWTM (full width at one tenth maximum), also shown in Figure 12.1. Typically, Si(Li) detectors have a FWHM resolution of ~140 eV at Mn $K\alpha$ and Figure 12.1 shows 136 eV. Because the value of $\varepsilon$ is lower for Ge (2.9 eV) than for Si (3.8 eV), IG detectors exhibit higher resolution than Si(Li) detectors, and Figure 12.1 shows an IG FWHM resolution of 118 eV at Cr $K\alpha$.

Resolution is defined primarily by the three terms in (12.1), but two other factors may affect the resolution, namely count rate and detector edge effects.[7] The detector resolution should be measured at each available pulse processing time constant ($\tau$). The resolution will degrade with decreasing $\tau$. Therefore, there is a compromise (at a fixed percentage dead time) between operating at a long $\tau$ in order to improve resolution and operating at short $\tau$ to improve the count rate. (There is also a more subtle counter effect,[7] which is that the resolution also degrades with increasing count rate at any $\tau$, and this effect is proportionately worse for longer $\tau$.)

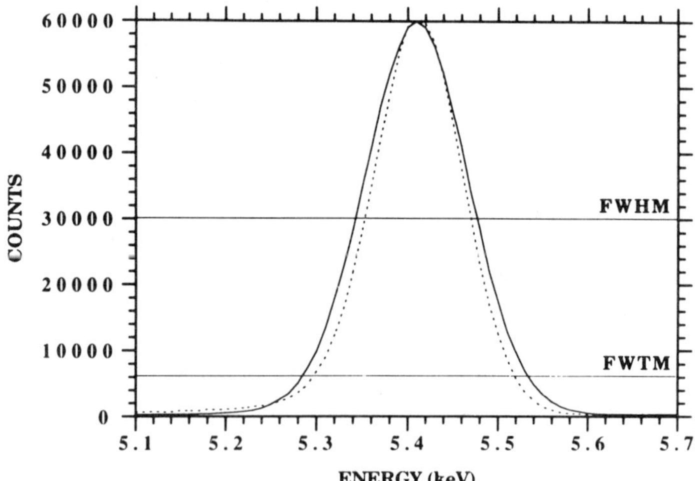


FIGURE 12.1.   Comparison of the resolution of Oxford Instruments Si(Li) detector (full line) and IG detector (dashed line) on the column of a Philips EM 430 AEM, using the Cr $K\alpha$ peak. The peaks were gathered at the same pulse processor settings. The FWHM values are 136 eV for the Si(Li) and 118 eV for the IG. The FWTM values are 252 eV for the Si(Li) and 223 eV for the IG. The ratios of FWTM/FWHM are 1.86 for Si(Li) and 1.89 for IG. (after Ref. 4, © Blackwells Scientific Publications).

Current EDS detectors are close to their theoretical resolution limit. If we assume that the electronics produce no noise, then $P = 0$ in (12.1), and if we ignore edge effects and incomplete charge collection (i.e., $X = 0$), then $R = I$. Assuming these approximations, then the theoretical resolution is given by (12.2). For Si, $F = 0.1$, $\varepsilon = 3.8$ eV, and the energy of the Mn $K\alpha$ line is 5.9 keV, which gives $R = 111$ eV as compared with 136 eV experimentally. So there is not much more room for improvement, and the resolution of EDS detectors will not approach that of crystal spectrometers which is 5–10 eV.

The resolution of the detector, as defined above in Figure 12.1, may degrade in service or as a result of a variety of reasons such as damage to the intrinsic region by high energy BSEs or bremsstrahlung x rays. Alternatively, bubbling of the liquid $N_2$ due to ice formation inside the dewar can degrade the resolution by generating acoustic noise. Ice formation can be minimized by filtering the liquid $N_2$ before putting it into the dewar and not using recycled liquid $N_2$. If the $N_2$ is bubbling, the detector can be warmed up with the bias voltage off and the dewar cleaned and dried, as already described.

### 12.2.2. Incomplete Charge Collection

The x-ray peak will not be a perfect Gaussian shape when displayed after processing and will usually have a low energy tail. The low energy tail arises because not all the electron-hole pairs created in the detector contribute to the charge pulse. This incomplete charge collection is caused by electron-hole pair recombination at the detector periphery or at various crystal defects such as the oxide-semiconductor interface on the front surface of the detector. Also, electron-hole pairs created in the dead layer may not be registered in the intrinsic region of the detector. The magnitude of the ICC can be determined from the ratio of the FWTM to the FWHM of the displayed peak, as defined in Figure 12.1 and shown in Figure 12.2. The Cr $K\alpha$ peak can be used and the theoretical value of the ratio for a true Gaussian is 1.823.

Using this criterion, both the Si(Li) and the IG detector spectra in Figure 12.1 display detectable ICC since the ratio is greater than 1.82 in both cases. As shown in Figure 12.2, the ratio is larger for x-ray peaks from the lighter elements. In Si(Li) detectors, the phosphorus $K\alpha$ peak shows the worst ICC effects because the P $K\alpha$ x ray fluoresces Si $K\alpha$ very efficiently and is thus most strongly absorbed near the front surface of the detector, in the dead layer. Similar problems in IG detectors arise for Mg and Al $K\alpha$ x rays, which efficiently fluoresce the Ge $L$ lines. Incomplete charge collection will also occur if a detector contains recombination sites such as point defects created by BSEs. Such defects may be annealed out by warming the detector, as described above. An IG detector should meet the same FWTM/FWHM ratio criterion as a Si(Li) detector. If the ratio is greater than 2 for the Cr $K\alpha$ peak, then there is something seriously wrong with the detector and it should be replaced.

### 12.2.3. Low Energy Efficiency

Over a period of time, ice and/or hydrocarbons will eventually build up on the cold detector surface, or on the window. If ice or hydrocarbon contamination does

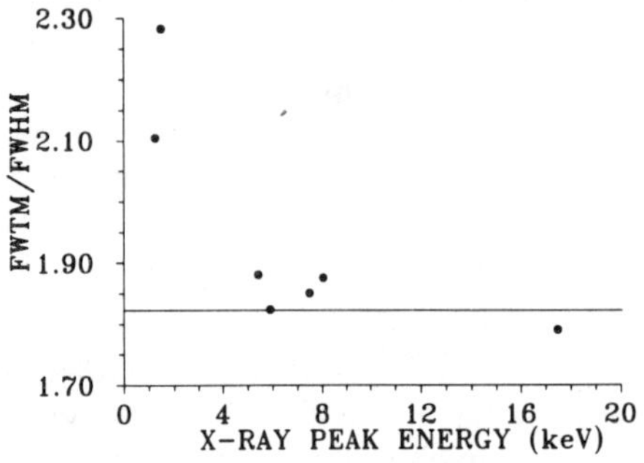



FIGURE 12.2.   Ratio of FWTM/FWHM for the $K\alpha$ lines of Al, Mg, Cr, Mn, Ni, Cu, and Mo. The theoretical ratio for a Gaussian peak is 1.823 and is represented by the horizontal line. The measurements were made with an Oxford Instruments IG detector on a Philips EM 430 AEM.

occur, it will reduce the detection efficiency for low energy x rays. While this is most likely to happen for a windowless detector, Be and ultra-thin window systems also suffer the same problem because of residual water vapor in the detector vacuum or because the window may be slightly porous. In all cases the problem is insidious because the effects may develop over many months and the degradation of the spectrum may not be noticed until differences in light element quantification are apparent from the same sample analyzed at different times. Therefore the low energy efficiency should be monitored regularly. The ratio of the Ni $K\alpha$ / Ni $L$ from a thin Ni or NiO film on a Be grid has been used as shown in Figure 12.3.[8] The ratio will rise with time if contamination or ice is building up on the detector and selectively absorbing the lower energy $L$-line. The ratio is most strongly affected by any ice layer, but will also differ for different detector dead layers, and different UTW materials, and different sample thicknesses, so it is not possible to quote an accepted figure-of-merit. However, as shown in Figure 12.3, for detectors with a Be window, a ratio between 10 and 20 is reasonable, while contaminated detectors have shown ratios greater than 400. This problem is discussed in detail by Michael[9] elsewhere in this text for the particular case of detectors in ultra-high vacuum environments.

It is best to measure the ratio immediately after installing a new (or repaired) detector and be aware that, as the ratio increases, any quantification involving similar low energy x-ray lines will become increasingly unreliable. At some arbitrary ratio, it becomes necessary to remove the ice or contamination. Automatic *in situ* heating devices that raise the detector temperature sufficiently to sublimate the ice make this process routine. If such a device is unavailable, it is necessary to warm up the detector as described above.

In summary the following detector performance criteria should be measured and continually monitored:

1.   The detector resolution on the column at the Cr $K\alpha$ line (typically 135–150 eV for Si(Li) and 115–130 eV for IG).
2.   The ICC defined by the FWTM/FWHM ratio of the Cr $K\alpha$ line (ideally 1.82).
3.   The ice buildup reflected in the Ni $K\alpha$ /Ni $L$ ratio.

If any of these figures of merit become significantly larger than the accepted values then the detector should be warmed up or, if this does not work, returned to the

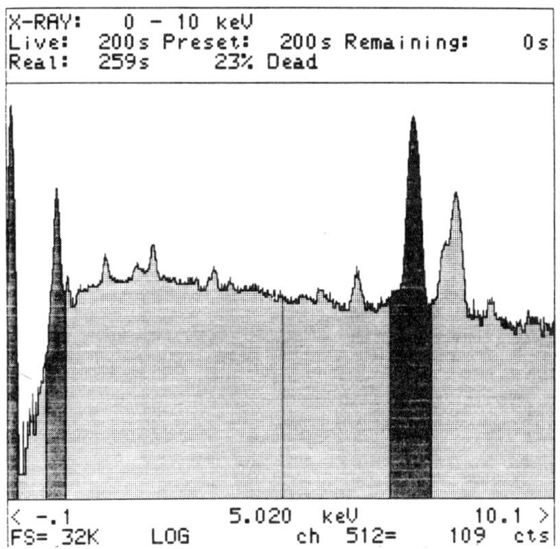


FIGURE 12.3.   Nickel spectra acquired from a thin Ni film on a Be grid in a Philips EM 400T AEM using an Oxford Instruments Si(Li) detector with a Be window. (a) The ratio of the intensities of the Ni $K\alpha$ to Ni $L$ is 12, indicating acceptable performance.

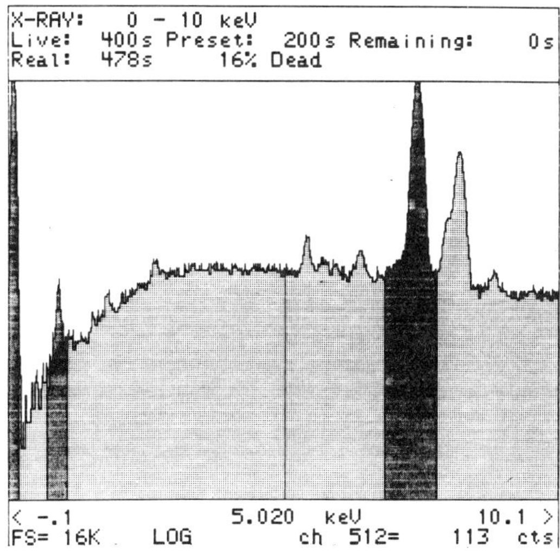


FIGURE 12.3.   (b) The ratio is 480, evidence of severe low energy attenuation due to ice and/or contamination on the window or the Si crystal. Acceptable performance was restored by conditioning the detector.

manufacturer for service. To minimize the dangers to the detector inherent in running an AEM:

1.   Do not generate high fluxes of x rays or backscattered electrons from thick regions of the specimen or grid bars (unless the detector is shuttered or retracted).
2.   Do not warm up the detector with the bias applied and without consulting the manufacturer.
3.   Do not use unfiltered or recycled liquid nitrogen.
4.   Do not cycle windowed detectors (either thermally or between atmosphere and vacuum) more than is necessary, since cycling will ultimately cause the window to leak.

## 12.3. OPTIMIZING THE DETECTOR-MICROSCOPE INTERFACE

When the EDS detector is attached to the AEM, certain performance parameters depend on a combination of the behavior of both instruments. These aspects, which are the most crucial to the performance of the AEM, are the most difficult to optimize, because it is often not clear which of the two instruments is at fault.

### 12.3.1. Detector Alignment

The detector must be aligned so that a line drawn from the center of the detector crystal, parallel to its axis, intersects a specimen at the eucentric height on the optic axis of the microscope. If the detector (or the collimator) is misaligned, then there will be a substantial drop in the intensity of generated x rays gathered by the detector. To insure the correct alignment, it is necessary to use the method proposed by Nicholson and Craven.[10] In this method, the instrument is operated in STEM mode at low magnification (<100X) and a scanning beam generates x rays over a large area of the specimen such as the Cr thin film already described (although, in this case, any uniform specimen is suitable). The resultant x-ray map is displayed on the STEM CRT or the EDS analysis system. If the EDS and collimator are correctly aligned, then the intensity of the x-ray signal will be constant as the beam scans off-axis. Conversely, if the detector or the collimator is not well aligned and some of the x rays are being shadowed from the detector, then the x-ray map will show large variations in intensity. If the intensity is asymmetric, it may be necessary to adjust the detector externally with shims to align it correctly or remove it and adjust the collimator. This procedure should always be carried out on initial installation by the manufacturer.

If it is not possible to create an x-ray map at low enough magnification, simply observe how the x-ray intensity varies from area to area on the sample. The best way to do this is to set the stage traverses to zero and select different specimen areas using the beam deflectors. The maximum intensity should be recorded in the middle grid square and for some distance around, or otherwise the detector is misaligned as shown in Figure 12.4. It is also instructive to do the same test with the sample moved up or down away from the eucentric plane, using the stage Z control. Again, the maximum intensity should be recorded at the eucentric plane.

### 12.3.2. Peak to Background Ratio (P/B) on the Column

There are many definitions of $P/B$,[2] but most methods exclude many counts in the peak, which is undesirable for AEM where count rate is often the limiting factor. The definition given by Fiori *et al.*[11] overcomes this limitation and gives a good approximation to the true, generated $P/B$, by integrating the full peak width over 700 eV (i.e., 70 channels, each 10 eV/channel), and taking the ratio of the total peak intensity to the average background intensity in a single 10 eV channel. This definition has been proposed as a standard $P/B$ ratio using the 100-nm film of $Cr$[2–4] as shown in detail in Figure 12.5. The background regions $B_1$ and $B_2$ are 700-eV windows extending between the 4.1- and 4.8-keV channels and 6.3 and 7.0-keV channels, respectively. The Cr peak window extends from 5.0 to 5.7 keV. (Note that depending on the way the windows are defined in a particular MCA system, a 700- eV window may extend over one less channel than defined here.) The Fiori $P/B$ ratio varies from instrument to instrument, and so it is necessary to propose some values as reasonable benchmark standards against which to compare other AEMs. Proposed values[4] are > 3500 at100 kV, > 4500 at 200 kV and >5500 at 300 kV, as shown in Figure 12.6. These values can vary substantially, e.g., if the specimen is tilted significantly more than 10° from the horizontal plane, or if the detector take-off angle is significantly less than 20°, or if the instrument is operated in TEM rather than STEM mode. The detector collection angle may also be a factor. Large collection angles catch some of the ICC at the detector periphery, so using an aperture on such detectors may improve the $P/B$. Ground loops may also degrade the $P/B$.

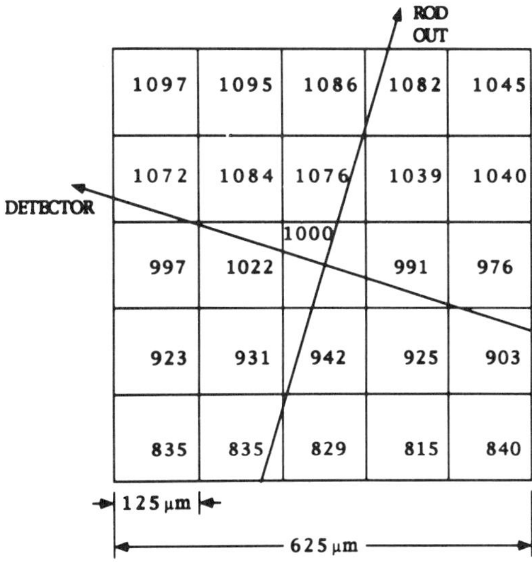


FIGURE 12.4.   Relative intensity of the Al $K\alpha$ (using an Al film on a Cu grid) over a 625 $\times$ 625 µm area. Measurements were made at low magnification (200X) by placing the beam in the center of each 200-mesh grid square and counting for a set live time. The position of the detector and specimen exchange rod are indicated. The measurements show a slight misalignment of the IG detector.

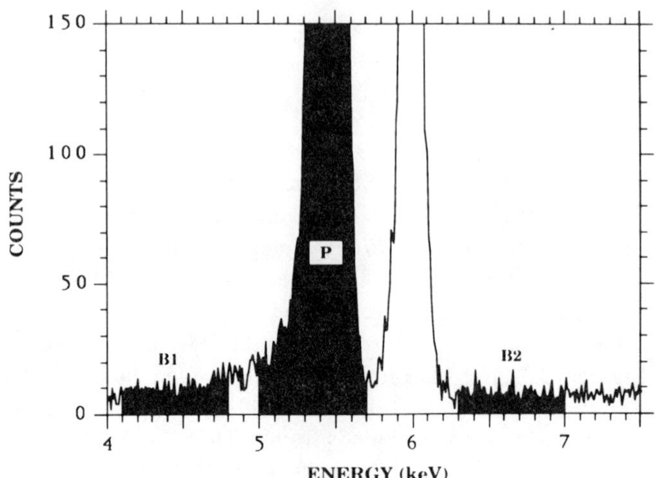


FIGURE 12.5. X-ray spectrum from the Cr standard showing the calculation of the Fiori $P/B$ ratio using the window method to define the background intensity. The peak intensity is integrated over 700 eV and the background intensity is obtained by dividing the average of $B_1$ and $B_2$ by 70 to get the intensity in a single 10-eV channel (after Ref. 4, © Blackwells Scientific Publications).

### 12.3.3. Relative Detector Efficiency

The relative detector efficiency is a measure of the number of x rays, measured in counts per second (cps), that are collected, detected, and processed by the EDS system. In a given time, this factor will depend on the specimen thickness, the beam current, and the solid angle of collection. Because of the dependence on thickness, it is necessary to use a specimen of standard thickness in order to compare results. Therefore, if the standard 100-nm Cr thin film is used and the beam current is measured, then the efficiency can be determined in cps/nA. If the detector collection angle is known, this efficiency can be normalized in cps/nA/sr. To determine this parameter, the beam current incident on the specimen must be measured, e.g., by inserting a Faraday cup into the beam after the final probe-limiting aperture or directly in the specimen plane. Values of the efficiency in cps/nA/sr, obtained in an intermediate voltage AEM, are 13,000 at 100 kV, > 9000 at 200 kV and > 8000 at 300 kV as shown in Figure 12.7.

## 12.4. PROCESSING VARIABLES

The performance of the EDS system is also governed by the processing of the detector output signal. There are three ways to insure that the pulse processing electronics are working properly. The first is the calibration of the energy range of the spectrum, although this will not change significantly from day to day unless the energy range or the time constant is changed. The second is the dead time correction circuitry,

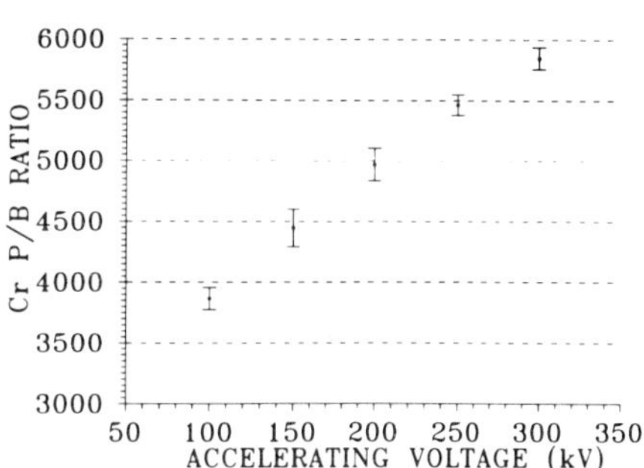

FIGURE 12.6. Cr *K*α *P/B* ratio versus accelerating voltage. Measurements were made in STEM mode in a Philips EM 430 AEM using an Oxford Instruments Si(Li) detector with a 20° take-off angle. The error bars are at the 95% confidence limits (after Ref. 4, © Blackwells Scientific Publications).

and the third is the maximum output count rate. Electronic circuit stability has improved to a level where these latter two steps need only be performed a few times a year, or if the detector has been repaired or a new detector installed.

### 12.4.1. Calibration of the Energy Display Range and Linearity

Obtain a spectrum from a thin foil of a sample that has a pair of $K$ x ray lines separated by about the width of the display range (e.g., Al-Cu for 0–10 keV). Alternatively, some systems use an internal electronic strobe to define zero, and in this case only one $K$-line at high energy is required. In either case, gather a spectrum and insure that the computer markers are correctly positioned at the peak centroid (e.g., Al $K$α at 1.49 keV and Cu $K$α at 8.04 keV). If the peak and marker are more than 1 channel (10 eV) apart, then the display needs to be recalibrated using the software routine supplied by the manufacturer. In typical software control systems, the energy calibration is trivial and rarely wanders by more than a channel during routine use.

In addition to the absolute energy calibration, the energy linearity should be checked. This requires plotting the peak centroid channel versus energy for several elemental peaks covering the desired energy range. The peaks chosen should be from easily fabricated elemental films or from discs of easily electropolished metals (e.g., $K$α lines from B, C, O, Mg for the low energy range and Ti, Cr, Mo, Ag, Au for the high energy range). Figure 12.8 shows such a plot[12] from a detector showing excellent

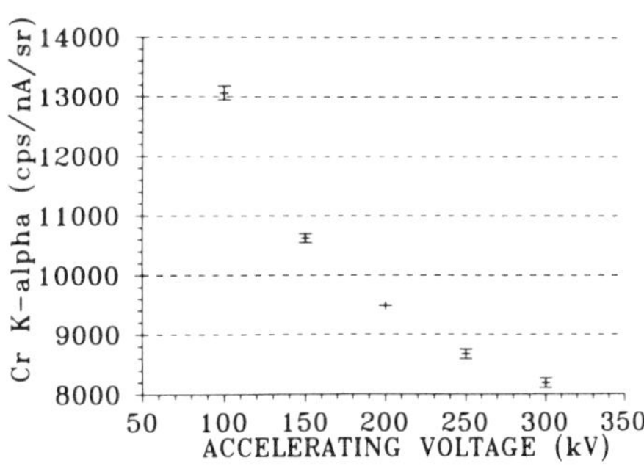

FIGURE 12.7. Cr *K*α count rate per unit beam current per unit solid angle of collection (in cps/nA/sr) versus accelerating voltage. This is a measure of the relative efficiency of the detector-AEM system. Measurements were made in STEM mode on a Philips EM 430 AEM using an Oxford Instruments Si(Li) detector with a 20° take-off angle. The error bars represent 95% confidence limits.

low energy linearity. In addition to the peak centroid variation, the $FWHM^2$ should also be linear with energy.

### 12.4.2. Checking the Dead Time Correction Circuit

If the dead time correction circuitry is working properly, the pulse processor will give a linear increase in output counts as the input counts increase for a fixed live time. First, select a pure-element uniform thin film that gives a strong $K\alpha$ peak, such as the Cr film used for resolution and P/B. If a disc specimen is used, all measurements must be made from the same point or from points of identical thickness (which is why an evaporated film is preferred). Second, select a live time (e.g., 50 s) and a beam current (measured by a Faraday cup) to give a dead time readout of about 10%. (If a Faraday cup is not available, use a calibrated exposure meter reading or the input count rate as a measure of the current.) Third, measure the total spectrum counts (minus the strobe if there is one) that accumulate in a preset *live time* (e.g., 50 s), which should give statistically significant counts. Fourth, repeat the experiment with higher input count rates. To increase the count rate, increase the beam current by choosing a larger diameter beam or larger C2 aperture. The dead time should increase as the input count rate goes up, but the live time will remain at the chosen value. If the full spectrum output counts are plotted against the beam current, then the plot should be linear as shown in Figure 12.9, at least up to 90% dead time. (Note that it will take increasingly longer clock times to attain the preset live time.) Such plots will probably become redundant as digital circuitry is employed in pulse processing circuits as described by Mott and Friel[13] elsewhere in this text.

### 12.4.3. Determination of the Maximum Output Count Rate

This test cannot be performed easily with the Cr thin film since not enough counts can be generated. Therefore, an electropolished disc specimen such as Mo should be used since the variable thickness can be used to change the total output count rate. First, gather a spectrum for a fixed *clock time* (e.g., 10 s), with a given dead time, say 10%. Record the input count rate (from the preamplifier to the (ADC), the output count rate (from the ADC to the pulse processor) and the percent dead time. Plot output count rate versus input count rate for each pulse processor setting (time constant $\tau$) as in Figure 12.10. A maximum output count rate of greater than 10 kcps should be achievable. While operating at a high dead time (>60%), it is advisable to record a spectrum to look for evidence of pulse pileup—a common artifact of EDS systems (a pileup of continuum intensity between 1 and 2 times the peak energy). Increase the dead time by acquiring from a thicker portion of the specimen and see how many counts accumulate in the full spectrum (again minus the strobe). The number of counts should rise to a maximum and then drop off as shown in Figure 12.10, because beyond a certain dead time, which depends on the system electronics, the detector will be closed more than it is open and so the counts in a given clock time will decrease. In Figure 12.10, this maximum is at about 60%, typical of modern systems, although in older EDS units[14] this peak can occur at as little as 30% dead time.[14] Repeat this experiment

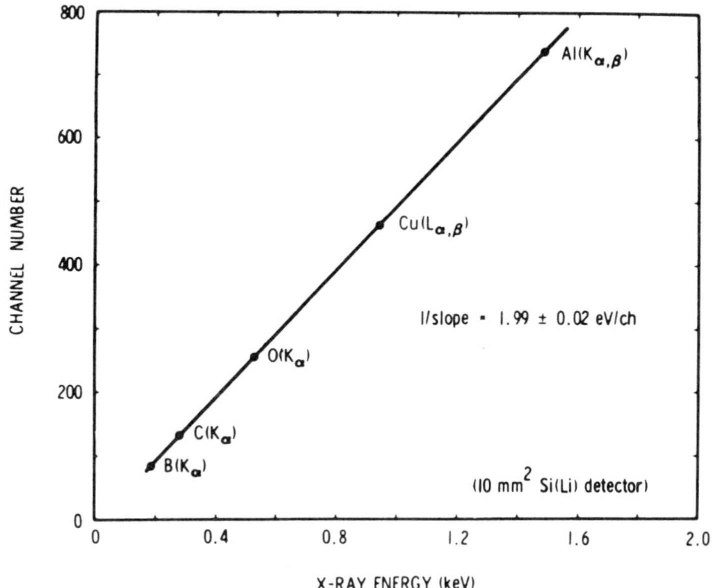


FIGURE 12.8. Experimentally determined linearity between pulse height and x-ray energy for a Si(Li) detector (after Ref. 12, © Elsevier North-Holland).

for different time constants, $\tau$, and the counts should increase as $\tau$ is lowered (at the expense of energy resolution) as also shown in Figure 12.10.

Clearly, operating the system at the maximum in such a curve results in the maximum possible x-ray counts from the specimen. Beware that, especially with older detectors, the resolution may degrade when the dead time is too high. On older detectors you may not want to operate above 10% dead time if you wish to avoid any degradation of resolution. With more modern detectors, a dead time of ~50% (i.e., near maximum throughput) should not severely degrade resolution. Generally, in the AEM, it is better to have more counts than to have the best energy resolution, so the shortest $\tau$ is recommended unless there is a peak overlap problem. Again, digital pulse processing should help improve such performance, as also described by Mott and Friel.[13] There is also the possibility of using megahertz beam blanking to increase the count rate at constant resolution, which enhances the overall system performance.[15] In summary, the processing performance parameters to monitor are:

1. The energy calibration and energy linearity of the MCA display.

2. The dead time circuitry, which controls the linearity of the output count rate as a function of beam current.

3. The maximum output count rate (governed by the choice of time constant) which should peak above ~60% dead time.

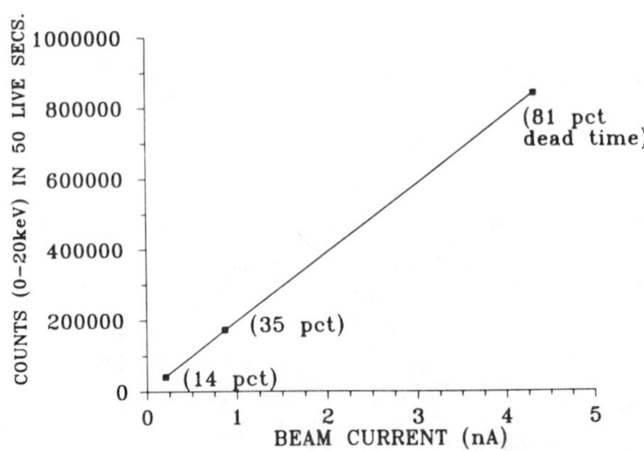


FIGURE 12.9. Total spectrum counts (not including the strobe) acquired from a Mo foil in 50 live seconds as a function of beam current. The results demonstrate the linearity of the dead time correction.

## 12.5. ARTIFACTS IN EDS SYSTEMS

A major drawback to EDS systems is the presence of artifacts in the spectrum introduced by the detector itself or by limitations in the pulse processing electronics. Also, there is the well known presence of systems peaks, introduced by stray radiation (both electrons and x rays within the microscope). These artifacts are well known in both the SEM and AEM communities and have been the topic of many discussions in textbooks and papers, so they will not be discussed in detail here. Detector artifacts comprise the escape peaks and internal fluorescence peaks (which will differ for Si(Li) and IG detectors) and the presence of incomplete charge collection. Pulse processing artifacts include the sum peak and the background shelf. For a full description of these artifacts, the classic paper by Fiori and Newbury[16] is recommended, as well as discussion in standard textbooks,[17,18] the article by Newbury in these proceedings[19] brings this topic up to date.

System peaks are typically discerned by placing the primary beam down a hole in the specimen and switching on the spectrometer to see what peaks can be detected when there should be no electron excitation.[18] A sterner test of the system involves placing a specimen in the beam, thus increasing scatter and increasing the possibility of remote excitation of x rays. For this test, use a specimen composed of light elements (e.g., C on a Be grid) or a sample of only one element that will produce detectable x rays (such as Cr or Ni film on a Be grid). Then the beam can be placed on the film to scatter into the post-specimen portion of the stage or on the grid to backscatter into the pre-specimen area of the stage.

Lyman and Ackland[20] have proposed using Cr film on a Au grid to test for stray radiation, and this procedure is recommended. In carrying out this test, different holders should be tested, different operational modes used (TEM versus STEM), and different electron optical conditions in the illumination system chosen (e.g., varying the probe size and the C2 aperture). Unless you are aware of the characteristics of your own AEM, then analysis is blind and the results cannot be trusted. Full qualitative analysis should be pursued before any quantitative analysis so that every peak in the

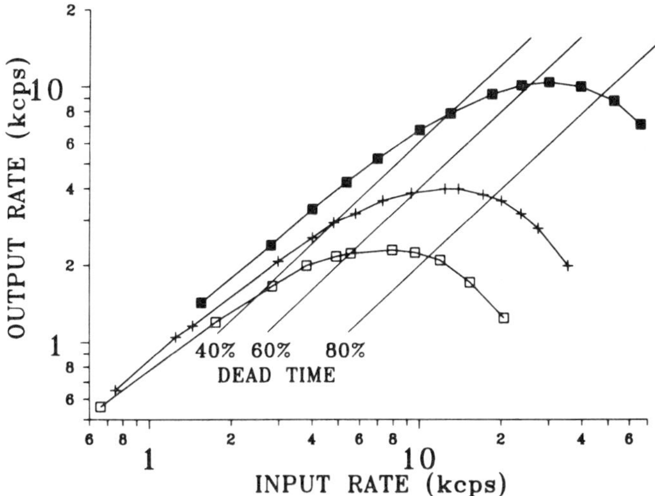



FIGURE 12.10.   Output count rate (from ADC to pulse processor) versus input count rate (from preamplifier to ADC) for three pulse processor settings. (■—long time constant, +—intermediate time constant, ◻—short time constant). The parallel lines through the curves correspond to constant dead time percentages and show that the maximum output occurs between 60% and 70% dead time. (Figure patterned after data first given in Ref. 14).

spectrum is identified and its source known (spurious or real) before proceeding with quantification.

## 12.6. SUMMARY

The EDS system interfaced to an AEM requires careful characterization prior to use for quantitative microanalysis. Standard tests are proposed here to determine if the system is optimized to produce the most counts in each analysis configuration. These tests should all be performed when a new AEM is installed, or when a new EDS system is interfaced to an existing AEM, in order to provide a baseline against which to monitor performance changes with time. Also, these tests may be used when trying to discriminate between the performance of different AEMs prior to purchasing. A few simple specimens are required:

1.  A thin (~100 nm) Cr film on a carbon film on a Cu grid for resolution, P/B ratio and efficiency tests.

2.  A Ni or NiO film for low energy efficiency.

3.  A Mo disc specimen for throughput and dead time testing.

4.  A Cr film on a Au grid for stray radiation and C or Cr on a Be grid for system peaks.

Not all of the tests need to be performed routinely, but on a regular basis, it is recommended that the resolution, P/B ratio, detector efficiency (particularly low energy efficiency for UTW or windowless detectors), and system peaks be measured and recorded. The electronic system performance, such as linearity, throughput and dead time checks, should not have to be performed regularly—perhaps every six months or after installation of a new system.

ACKNOWLEDGMENTS. The support of Sandia National Laboratories for the senior author, SMZ, is gratefully acknowledged, and DBW wishes to thank NASA for continued support through grant NAG 9-45.

## REFERENCES


1. C. E. Lyman, D. E. Newbury, J. I. Goldstein, D. B. Williams, A. D. Romig Jr., J. T. Armstrong, P. E. Echlin, C. E. Fiori, D. C. Joy, E. Lifshin, and K-R. Peters, *Scanning Electron Microscopy, X-Ray Microanalysis and Analytical Electron Microscopy: A Laboratory Workbook*, Plenum Press, New York (1990).
2. D. B. Williams and E. B. Steel in: *Analytical Electron Microscopy—1987* (D. C. Joy, ed.) San Francisco Press, San Francisco, p. 228 (1987).
3. S. M. Zemyan and D. B. Williams, in: *Proceedings of the 27th Annual MAS Meeting* (J. A. Small, ed.) San Francisco Press, San Francisco, p. 1236 (1992).
4. S. M. Zemyan and D. B. Williams, *J. Microsc.* **174**, 1 (1994).
5. G. F. Knoll, *Radiation Detection and Measurement*, J. Wiley and Sons, New York, p. 92 (1979).
6. ANSI/IEEE Standard 759, Standard Test Procedure for Semiconductor X-Ray Energy Spectrometer, Institute of Electrical and Electronic Engineers, New York (1984).
7. G. Bertolini and G. Restelli, in: *Atomic Inner-Shell Processes II: Experimental Approaches and Applications* (B. Crasemann, ed.) Academic Press, New York, p. 140 (1975).
8. B. G. Lowe, *Ultramicroscopy* **28**, 150 (1986).
9. J. R. Michael, in: *X-Ray Spectrometry in Electron Beam Instruments* (D. B. Williams, J. I. Goldstein, and D. E. Newbury, eds.) Plenum Press, New York, chap. 7 (1995).
10. W. A. P. Nicholson and A. J. Craven, *J. Microsc.* **168**, 289 (1993).
11. C. E. Fiori, C. R. Swyt, and J. R. Ellis, *Microbeam Analysis—1982* (K. F. J. Heinrich, ed.) San Francisco Press, San Francisco, p. 57 (1982).
12. R. G. Musket, *Nucl. Instrum. Methods* **117**, 385 (1974).
13. R. B. Mott and J. J. Friel, in: *X-Ray Spectrometry in Electron Beam Instruments* (D. B. Williams, J. I. Goldstein, and D. E. Newbury, eds.) Plenum Press, New York, chap. 9 (1995).
14. D. Vaughan, ed. *Energy-Dispersive X-ray Microanalysis: An Introduction*, Kevex Corp., Foster City, CA, p. 25 (1983).
15. C. E. Lyman, J. I. Goldstein, D. B. Williams, D. W. Ackland, S. von Harrach, A. W. Nicholls, and P. J. Statham, *Microbeam Anal.* **2** S234 (1993).
16. C. E. Fiori and D. E. Newbury, in: *Analytical Electron Microscopy—1981* (R. H. Geiss, ed.) San Francisco Press, San Francisco, p. 17 (1981).

17. J. I. Goldstein, D. E. Newbury, P. Echlin, D. C. Joy, C. E. Lyman, A. D. Romig, Jr., C. E. Fiori, and E. Lifshin, *Scanning Electron Microscopy and X-ray Microanalysis*, Second Ed., Plenum Press, New York (1992).
18. D. B. Williams, *Practical Analytical Electron Microscopy in Materials Science*, Philips Electron Optics, Mahwah, NJ, p. 57 (1984).
19. D. E. Newbury, in: *X-Ray Spectrometry in Electron Beam Instruments* (D. B. Williams, J. I. Goldstein, and D. E. Newbury, eds.) Plenum Press, New York, chap. 11 (1995).
20. C. E. Lyman and D. W. Ackland, in: *Microbeam Analysis—1991* (D. G. Howitt, ed.) San Francisco Press, San Francisco, p. 461 (1991).

# 13

# Wavelength Dispersive Spectrometry: A Review

*S. J. B. Reed*

## 13.1. INTRODUCTION

The electron microprobe prototype developed by Castaing incorporated a wavelength dispersive, or WD, type of x-ray spectrometer, making use of Bragg reflection by a curved crystal to select the required wavelength.[1] (The name was actually introduced later, to distinguish this from the energy dispersive or ED, type.) The curved crystal configuration enables reasonable intensity to be obtained with a point source since reflection occurs over a larger area than for a flat crystal.

Wavelength dispersive spectrometers have not changed in any major respect since their earliest application to electron microprobe analysis, but various improvements have been introduced. For example, some crystals used originally (e.g., mica and quartz) have been supplanted by others (e.g., TAP and PET) that give higher intensity. Also, the wavelength range has been extended to cover the $K$-lines of light elements (Be, B, C, N). Initially (in the 1960s), this was achieved by means of soap film "pseudocrystals" such as lead stearate. More recently (in the 1980s), multilayers consisting of alternating layers of light and heavy elements deposited by vacuum evaporation and offering substantially higher intensities have become available. The other main development that should be mentioned is the application of digital technology, which has transformed electron microprobe analysis in general and WDS in particular.

Since the early 1970s, ED spectrometers have competed with the WD type, which nevertheless still retains an important role: the much better WDS resolution not only enables complex spectra to be resolved but also allows smaller peaks to be detected above the continuum background. However, WD spectrometers suffer from two major disadvantages—low x-ray collection efficiency and serial operation (compared to the effectively

S. J. B. REED • Department of Earth Sciences, University of Cambridge, Cambridge CB2 3EQ, United Kingdom

*X-Ray Spectrometry in Electron Beam Instruments,* edited by David Williams, Joseph Goldstein, and Dale Newbury. Plenum Press, New York, 1995.

parallel mode of the ED type). The former necessitates the use of relatively high beam currents, especially for low elemental concentrations. The latter makes WDS generally slower than EDS, especially for recording large numbers of peaks.

In principle, the efficiency of WDS can be improved by using doubly curved crystals to obtain reflection over a larger area than with the usual cylindrically curved type.[2] If such crystals can be developed in practical form, the increased sensitivity will be especially beneficial for the analysis of materials that are damaged by high beam currents. It may also become feasible to use WDS for analyzing thin samples in the analytical electron microscope, where EDS is used exclusively at present. Work directed toward the practical realization of the benefits of doubly curved crystals is currently under way, as described below (Section 13.4.3). Also, several proposals have been put forward for parallel WDS, which could increase the efficiency of data collection, as discussed in section 13.8.1.

## 13.2. X-RAY DETECTION

In a WD spectrometer, x rays are detected by means of a proportional counter consisting of a gas-filled tube with a concentric anode wire. The most commonly used gas is argon with 10% methane. For long wavelength x rays, the gas is passed continuously through a "flow counter" with a thin (not completely impermeable) entrance window. For shorter wavelengths, a permanently sealed counter with a thicker window (usually made of beryllium) is used. This is normally filled with xenon, which has greater x-ray absorbing power than argon, but sometimes a flow counter filled with argon at a pressure of 2–3 atm is used instead. Although the replacement of these counters by solid-state detectors is a possibility, it does not seem imminent.

The mean output pulse height is proportional to the x-ray photon energy: this type of counter thus has energy-resolving capabilities. The energy resolution is much worse than that of the cooled solid-state detectors used in EDS but is useful for excluding high-order Bragg reflections by means of pulse height analysis, whereby only pulses with heights falling within a preset window are counted.

X-ray intensities are measured by counting pulses using standard digital circuitry. There is a certain dead time after the arrival of each pulse before another can be recorded. The dead-time of WD spectrometers is typically around 2 μs, which gives a 2% loss of counts at $10^4$ cps and 20% at $10^5$ cps. The latter should be regarded as an upper count-rate limit, especially for quantitative analysis. (Although a dead time correction can be applied, it is inadvisable to rely on it for large losses). Another reason for avoiding excessively high count rates is that the counter output pulses are subject to shrinkage owing to saturation effects, and this can give rise to inaccurate results when pulse height analysis is used.

## 13.3. BRAGG REFLECTION

The reflection of x rays by a crystal is illustrated in Figure 13.1. Constructive interference occurs between x rays diffracted by successive layers when the difference

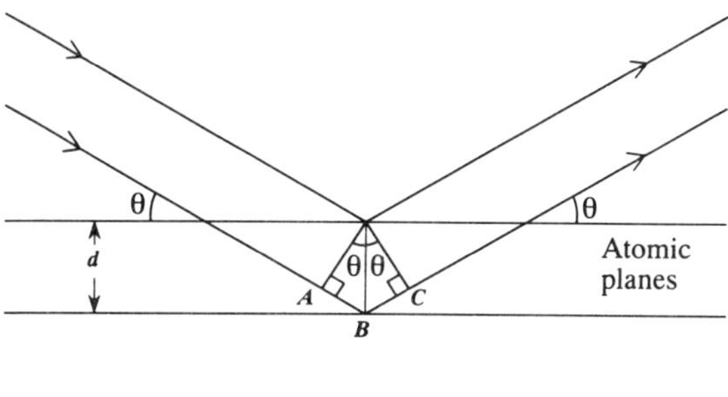


FIGURE 13.1. Bragg reflection: atomic planes of spacing $d$ in the crystal reflect x rays when difference in path length (ABC) is an integral number of wavelengths ($\theta$ = Bragg angle).

in path length is equal to an integral multiple of the wavelength $\lambda$, as expressed in Bragg's law:

$$n\lambda = 2d \sin \theta \qquad (13.1)$$

where $\theta$ (the "Bragg angle") is the glancing angle of incidence and reflection, $d$ is the interplanar spacing, and $n$ (= 1, 2, etc.) is the order of reflection. A small correction is required for the effect of refraction by the crystal[3] but this is almost negligible except in the case of multilayers used for long wavelengths.[4]

The reflection of x rays at angles close to the Bragg angle can be described by a curve of the form shown in Figure 13.2. The maximum efficiency may be quite high but drops off rapidly on each side. A useful approximation is to represent the reflection curve by a rectangle with an area equal to that under the curve (see Figure 13.2). The width of the rectangle, $\Delta\theta$, is typically in the region of $10^{-4}$ radian.

### 13.3.1. Choice of Crystal

In practical WD spectrometers, $\theta$ has a limited range (e.g., 12° to 65°), which defines the wavelength range for a given $d$ value. To cover all wavelengths of interest, crystals with different $d$ values are needed. Those currently used are lithium fluoride, or LiF ($d$ = 0.2013 nm), pentaerythritol, or PET ($d$ = 0.4371 nm) and thallium acid phthallate, or TAP ($d$ = 1.295 nm). Their wavelength ranges are shown in Figure 13.3. All elements down to fluorine ($Z$ = 9) can be detected. The $K$-line of oxygen can just be reached with TAP in some instruments, depending on the maximum $\theta$. For heavy elements, $L$- or $M$-lines are used instead of the $K$-lines, which are too short in wavelength.

True crystals are not available with larger $d$ spacings than TAP, so synthetic layered diffractors are used instead for very long wavelengths. Soap film pseudocrystals can be used for long wavelengths; the most common is lead stearate ($d$ = 5 nm),

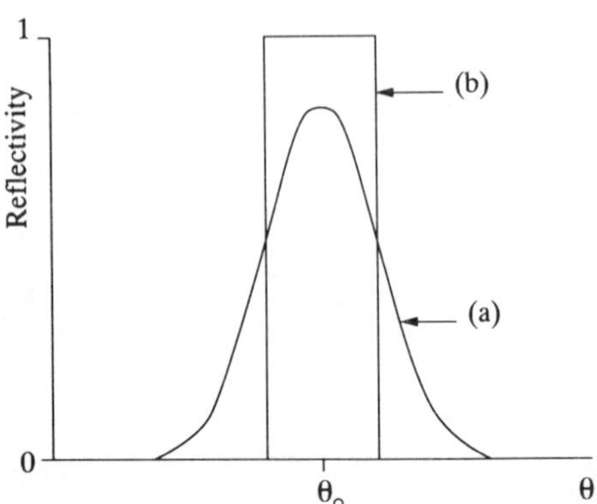

FIGURE 13.2.   Reflectivity of the crystal in the region of Bragg angle ($\theta_o$): (a) actual reflection curve, (b) idealized form.

which enables elements down to B ($Z = 5$) to be detected. Other varieties of soap film with larger $d$ values extend the range to Be ($Z = 4$). The more recently introduced evaporated multilayers[5] are available with various $d$ values and with different elements forming the diffracting layers. These are optimized for relatively narrow wavelength ranges, and it is desirable to use at least three different types to cover all of the light elements. They give substantially higher intensities than lead stearate, etc.,[6–9] but their wavelength resolution is rather poor and in some cases overlap by $L$- or $M$-lines of heavier elements is a problem. On the other hand, reflections of order higher than 2 are very weak, minimizing the likelihood of interference from high-order reflections of short-wavelength lines.

   An alternative approach to long-wavelength WDS is to use a diffraction grating, which requires a different mechanical arrangement.[10] Early work showed that good performance was attainable, as compared to stearate pseudocrystals, but this form of spectrometer has never achieved wide acceptance.

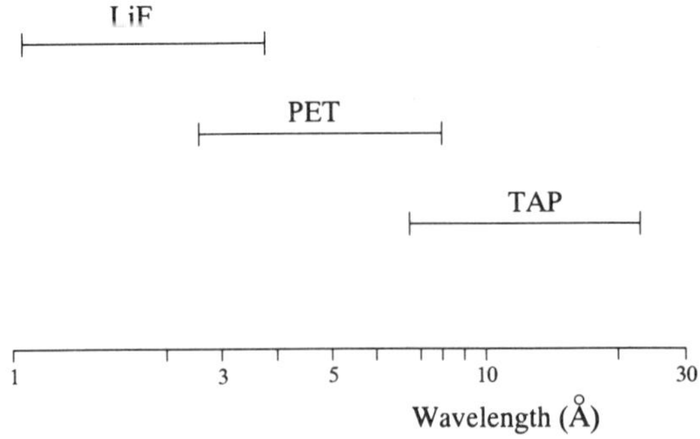


FIGURE 13.3.   Wavelength range of crystals used in WD spectrometers, for typical angular range (1 Å = 0.1 nm).

## 13.4. CURVED CRYSTAL GEOMETRY

The object of using a curved crystal is to obtain a constant Bragg angle for x rays originating from a point source. This can be achieved for all angles if source, crystal, and detector are disposed on a "Rowland circle" of radius $r$, and the crystal is curved to a radius $2r$ (Figure 13.4). The source-crystal distance is equal to $2r\sin\theta$ and is the same as the crystal-detector distance. It follows from Bragg's law (13.1) that this distance is equal to $r\lambda/2d$ (for $n = 1$). In most WD spectrometers the crystal moves along a straight line passing through the source, having the advantage of giving a constant x-ray take-off angle. The distance along the line is proportional to $\lambda$, making calibration simple. The Rowland circle has no physical existence, but a mechanism is provided which maintains the correct crystal angle and detector position as the crystal is moved up and down the linear track.

Although contemporary WD spectrometer designs conform to the above arrangement, other variants have been used in the past. For example, great mechanical simplification can be achieved by having a fixed Bragg angle, preset for a particular elemental line.[11] Being much cheaper and more compact, a larger number of such spectrometers can be fitted (e.g., up to 12), making analysis quicker but sacrificing flexibility in the choice of elements. (It is more practical to combine one or more fully variable spectrometers with a number of fixed ones.)

Another simplified form of WD spectrometer is the "semi-focusing" type,[12] in which the crystal rotates about a fixed axis to obtain different Bragg angles, with a 2:1 drive to the counter arm. For a crystal with a given curvature, focusing is correct only at one wavelength, although in practice it is adequate over a finite range. This shortcoming can be overcome if a flexible crystal of varying curvature is used.[13]

### 13.4.1. Johansson Geometry

The correct radius of curvature of the atomic planes in the Rowland circle configuration is $2r$, but the surface of the crystal should ideally lie on the Rowland

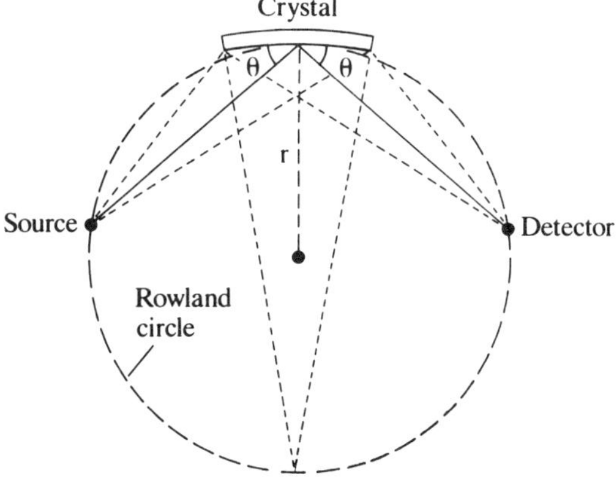


FIGURE 13.4.   Rowland circle geometry.

circle, with radius $r$. Achieving this "Johansson geometry" (Figure 13.5) involves grinding the surface of the crystal, which is relatively easy with LiF but is difficult for the softer PET and TAP crystals (though proprietary techniques exist). Often the simpler Johann configuration (see next section) is used to avoid this difficulty.

The effective angle of incidence for divergent rays is less than the "correct" angle by an amount $\delta\theta$, which is a function of the distance from the center-line of the crystal (Figure 13.6). Reflection occurs only where $\delta\theta$ falls within the width $\Delta\theta$ of the reflection curve of the crystal (Figure 13.2), which thus determines the reflecting area. This area can be calculated, given $\Delta\theta$. In practice, the width of the reflecting area is less than the width of the crystal, so not all of the surface reflects at once. The reflected intensity therefore does not benefit from increasing the width of the crystal. On the other hand, increasing the length (around the circumference of the Rowland circle) increases the intensity, but is limited by practical considerations.

As the spectrometer is scanned through a monochromatic x-ray line, reflection occurs first along the center line, and then the reflecting area splits into two bands that move outward to the edges of the crystal. The width of the recorded line profile is determined by the range of angles between the onset and cessation of reflection and is governed mainly by geometry rather than the width of the reflection curve (see Section 13.6.1).

### 13.4.2. Johann Geometry

In Johann geometry, the crystal is still curved to a radius of $2r$ but not ground (Figure 13.7). At points away from the center, the surface lies off the Rowland circle, leading to a geometrical aberration additional to that for divergent rays (discussed in the previous section). These errors are opposite in sign and the combined effect is that the locus of points at which the incidence angle is "correct" forms a cross. Both the reflecting area and the resolution can be calculated,[14,15] though real crystals may have geometrical imperfections that may considerably affect their behavior. The resolution is always worse than for the comparable Johansson case, owing to the additional aberration. Furthermore, increasing either the length or breadth of the crystal yields decreasing returns in intensity, while causing the resolution to deteriorate. Conversely,

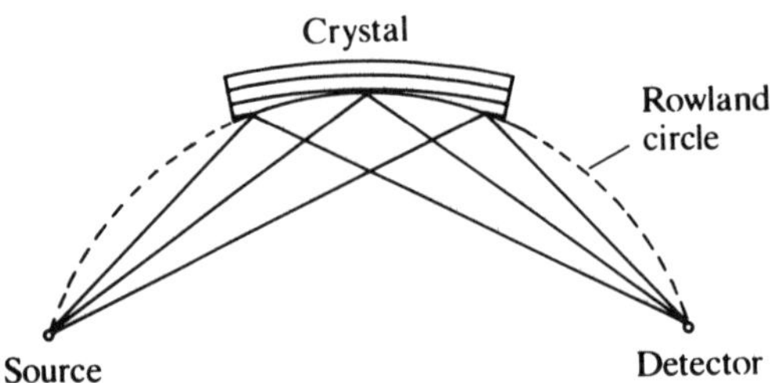


FIGURE 13.5.    Johansson geometry: crystal curved to radius $2r$ and surface ground to radius $r$ ($r$ = Rowland circle radius).

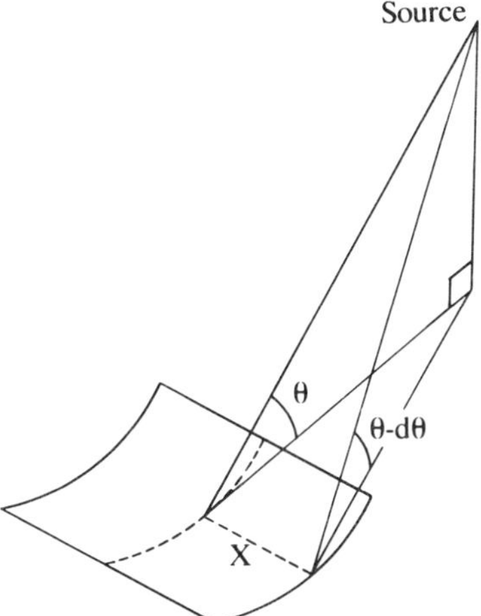


FIGURE 13.6.   Change in angle of incidence with divergence of x rays (cylindrical crystal).

resolution can be improved by limiting the effective area with an aperture, but only at the expense of intensity.

### 13.4.3. Doubly Curved Crystals

The aberration for rays oblique to the plane of the Rowland circle can be corrected in principle by making the crystal conform to a doubly curved (toroidal) form corresponding to the surface obtained by rotation about a line joining source and detector (Figure 13.8). Reflection then occurs over the whole surface, giving greatly increased intensity without sacrificing resolution.

Unfortunately, for any given value of the second radius, the geometry is perfect only for one value of $\theta$ (or wavelength) and the reflected intensity decreases rapidly on either side of this value. However, considerably enhanced intensity for all $\theta$ can be obtained with "Wittry geometry," in which the crystal surface is toroidal and the atomic

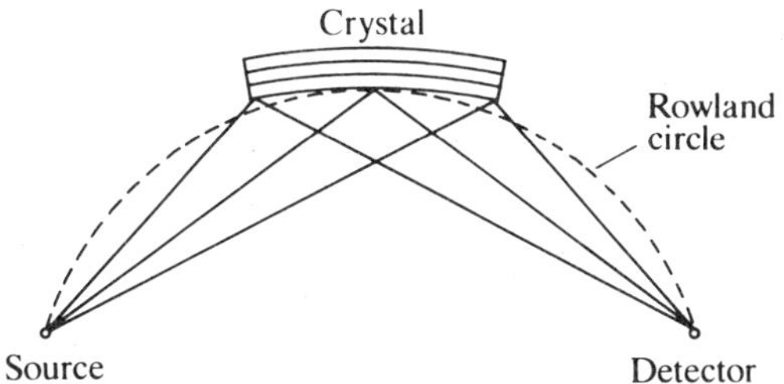


FIGURE 13.7.   Johann geometry: crystal curved to radius $2r$, but not ground as in Johansson geometry (Figure 13.5); at points away from centre of crystal, reflection at surface occurs off the Rowland circle.

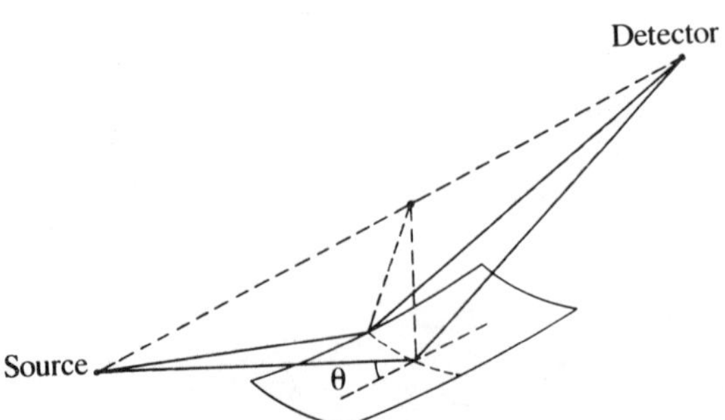


FIGURE 13.8.   Doubly curved crystal conforming to toroidal surface obtained by rotation about line joining source and detector, giving constant angle of incidence (cf. Figure 13.6).

planes are spherical.[2] It is difficult to make doubly curved crystals but one possibility is to use a segmented configuration.[16] An alternative approach is to use an oriented polycrystalline film deposited on a curved substrate.[17]

Wittry *et al.* calculated the properties of doubly-curved diffractors in the shape of the "logarithmic spiral" function, showing that reflection efficiency comparable to Wittry geometry is obtainable (over a limited range of Bragg angles).[18] Such diffractors should be easier to fabricate.

## 13.5. EFFICIENCY OF WD SPECTROMETERS

The intensity obtained with a WD spectrometer is dependent on the solid angle subtended by the crystal, the reflection efficiency, and the efficiency of the detector. The solid angle is given by $A/4r^2\sin\theta$, where $A$ is the area of the crystal and is plotted as a function of $\sin\theta$ for a typical spectrometer in Figure 13.9. The solid angle subtended by an ED detector is often smaller than that of the WD spectrometer: for example, for a 10 mm$^2$ detector 50 mm from the source, it is 0.004 steradian. However, higher intensities are obtained with the ED detector owing to the low reflection efficiency of the crystal in the WD spectrometer. Furthermore, the solid angle of the ED detector can be increased by moving it closer than the distance assumed above (though for most applications there is little practical advantage in this).

The crystal reflection efficiency is calculable in principle (see Section 13.4.1), but the width of the crystal reflection curve is not generally known, so it is more relevant to refer to empirical observations. The efficiency of a WD spectrometer can be determined by comparing line intensities in WD and ED spectra, using the latter for calibration, given that the solid angle of the ED detector is known and its efficiency can be assumed to be close to 100% for energies above 3 keV.[19,20]

Experimental intensity data obtained with a typical instrument are plotted in Figure 13.10 (note that these curves reflect variations in the generated intensity as well as the spectrometer efficiency). Comparison with ED intensities leads to the conclusion that the effective reflection efficiency in the WD spectrometer is of the order of 10%.

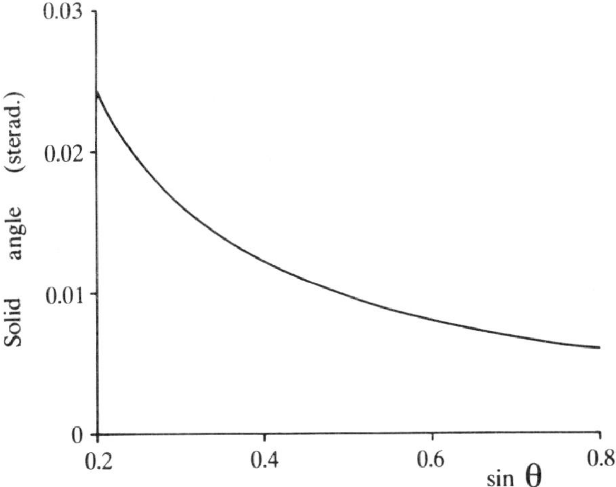


FIGURE 13.9.   Solid angle subtended at the x-ray source by the crystal in a typical WD spectrometer, as function of sin θ.

For a major improvement on this figure, a practical form of doubly curved crystal is required.

Figure 13.10 illustrates how, in the region where the wavelength ranges of two crystals overlap, the one with the larger $d$ value gives higher intensity because the Bragg angle is lower and hence the solid angle is larger. An example of this is Ca $K\alpha$, which is within the ranges of both PET and LiF, the intensity being higher for the former.

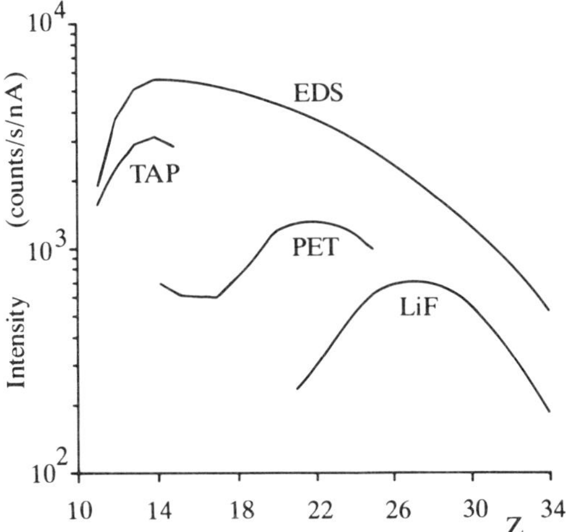


FIGURE 13.10.   Observed intensities of $K\alpha$ peaks obtained with a WD spectrometer using three different crystals, with EDS intensities plotted for comparison (for typical instrument).

## 13.6. LINE SHAPE

The widths of lines recorded by WD spectrometers are much greater than natural line widths;[21] observed line widths and shapes are thus determined predominantly by the properties of the curved crystal geometry. The line shape is typically more "pointed" than the Gaussian function, with more extended tails. A reasonable representation of this shape is given by the Voigt function, which is the convolution product of Gaussian and Lorentzian functions.[22] However, for practical calculations it is simpler to use the "pseudo-Voigt" function, in which Gaussian and Lorentzian functions are added.[22,23]

### 13.6.1. Resolution

It is convenient to express linewidths in energy units (eV), for comparison between ED and WD spectrometers (see Figure 13.11). The resolution of the latter is always better, but the efficiency is lower, as discussed in the previous section. Where the wavelength ranges of crystals overlap, that for which the Bragg angle is highest gives the better resolution (but lower intensity, as noted previously). For a practical example of this, see Figures 13.12 and 13.13.

As noted in Section 13.4, resolution can be improved in principle by limiting the reflecting area, which can be achieved by means of a slit in front of the counter. However, in practice, the loss of intensity for a worthwhile improvement in resolution is large. For most applications, there is in any case little need for better resolution, which has several disadvantages, including requiring more accurate setting of the Bragg angle and increased sensitivity to several factors, including the effect of

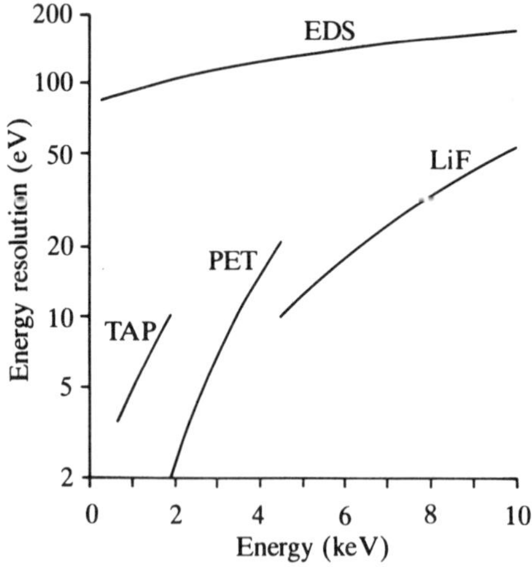


FIGURE 13.11.   Resolution (in energy units) of WD spectrometers using three different crystals, with EDS resolution plotted for comparison.

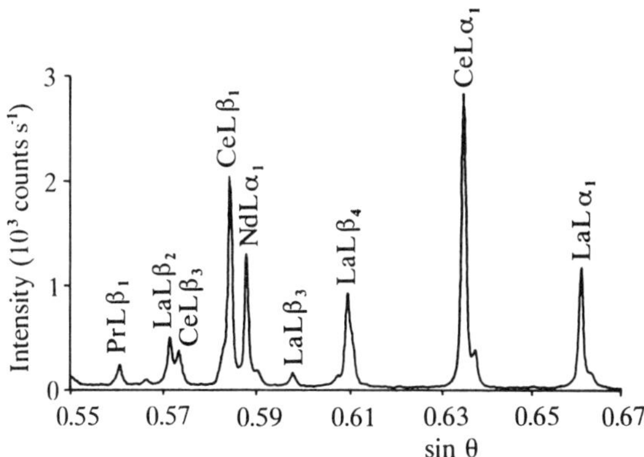

FIGURE 13.12.   Wavelength dispersive spectrum of rare earth elements in monazite, recorded with LiF crystal.

temperature on crystal $d$ values, changes in source position (see next section), and wavelength shifts related to chemical bonding.

## 13.7. WAVELENGTH DISPERSIVE SPECTROMETERS FOR ELECTRON MICROPROBE ANALYSIS

Electron microprobes have been fitted with WD spectrometers since the invention of the technique. A fully equipped present day instrument has both WD and ED spectrometers, the latter being useful for rapid qualitative analysis and for analyzing

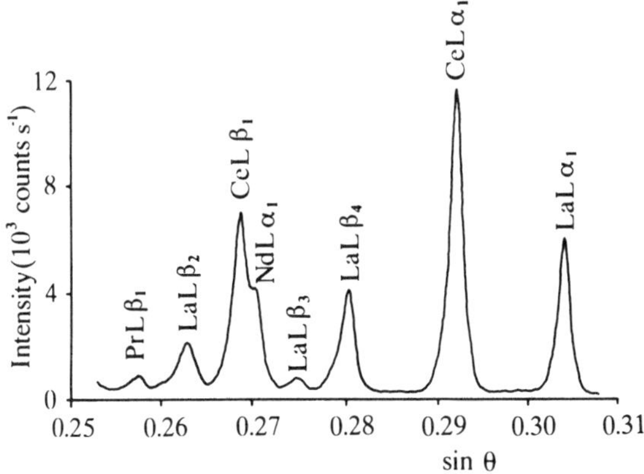

FIGURE 13.13.   Same spectrum as in Figure 13.12, recorded with PET crystal (note poorer resolution but higher intensity).

materials that cannot tolerate a high beam current. (Scanning electron microscopes mostly have ED spectrometers, but one of the WD type is also sometimes fitted.)

If a WD spectrometer is mounted vertically (Figure 13.14), differences in specimen height of only a few micrometers can cause changes in Bragg angle large enough to affect measured peak intensities. The problem can be solved by adjusting the "$z$" movement of the specimen stage to focus the image of the surface in the optical microscope. In the case of an instrument lacking such a microscope (e.g., an SEM), or for samples with very irregular surfaces, it is preferable to orient the spectrometer such that movement of the source in the $z$ direction does not change the Bragg angle (see Figure 13.14). The plane of the Rowland circle must be inclined to the horizontal to give a reasonable x-ray take-off angle (e.g., 40°). In either orientation, the Bragg angle is sensitive to lateral movement in one direction, which must be taken into account when locating the beam for quantitative analysis and when recording scanning x-ray images. In the latter case, the difficulty can be overcome by changing the Bragg angle in synchronism with the scan[24] or by moving the specimen instead of the beam.

In a typical electron microprobe, up to five vertical WD spectrometers can be fitted, compared with only two in the inclined orientation. The advantage of multiple spectrometers is that the need to change crystals in the course of multielement analyses is avoided; also time can be saved by measuring several peaks simultaneously.

Wavelength dispersive spectrometers are usually operated under vacuum to avoid attenuation of soft x rays by absorption in air. However, high vacuum is unnecessary—the pressure obtained with a rotary pump alone is adequate. The spectrometer chamber can be isolated from the high vacuum of the column by a thin window.

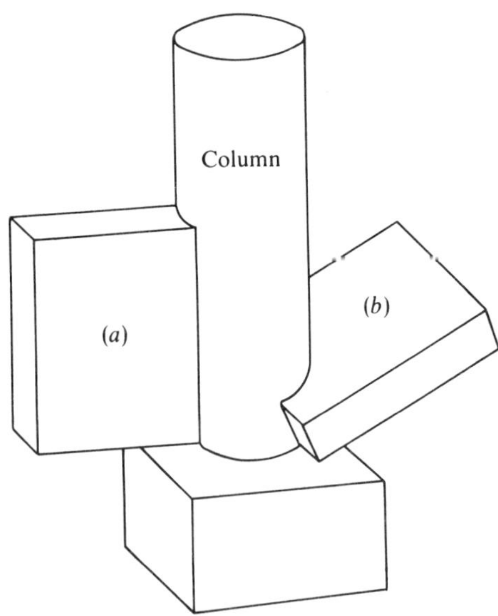


FIGURE 13.14.　Different WD spectrometer orientations (a) Vertical. (b) Inclined.

The spectrometer mechanism may be driven by a stepping motor controlled by a computer. Higher drive speeds can be achieved, however, by using a nonstepping type of motor, with a digital encoder reading the spectrometer position. The minimum step size should not be greater than $10^{-5}$, expressed in terms of sin $\theta$.

It is usual to provide more than one crystal for each spectrometer, in order to give operational flexibility. Typically either two or four crystals are mounted on a turret. In some designs the crystal carriage must be driven to a specific position for changing crystals, but in others this operation can be carried out in any position.

## 13.8. WAVELENGTH DISPERSIVE SPECTROMETERS FOR ANALYTICAL ELECTRON MICROSCOPY

In the analytical electron microscope, a thin specimen prepared as for TEM is used, and elemental analysis is conducted in similar fashion to conventional electron microprobe analysis. The spatial resolution is higher because the incident electrons are scattered much less in the specimen. The x-ray intensities obtained, however, are lower because of the relatively short path of the electrons within the specimen. Though the prototype AEM developed by Duncumb[25] used WD spectrometers, later versions have invariably utilized ED spectrometers because of their greater x-ray collection efficiency (solid angles exceed 0.1 steradian in current models).

The advent of field emission electron guns has enabled the current in a beam of a given diameter to be increased markedly. This has inspired a revival of interest in WDS for the AEM, since the higher current compensates to some extent for the low spectrometer efficiency. However, for this approach to be fruitful, the efficiency must

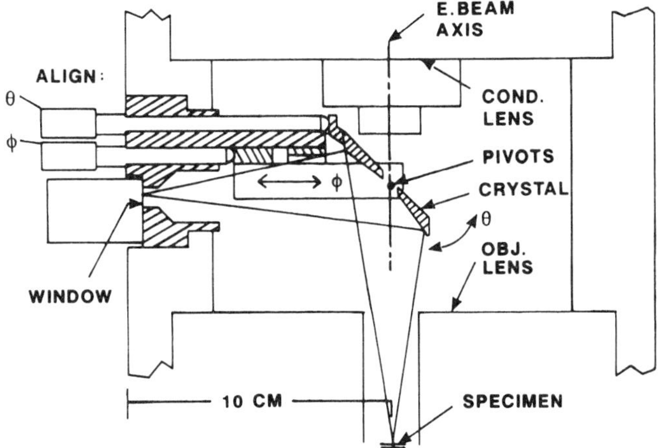


FIGURE 13.15. Wavelength dispersive spectrometer designed for analytical electron microscope[27]: the doubly curved LiF crystal subtends a large solid angle to give high intensity; the hole in the crystal allows electron beam to pass; the Bragg angle is fixed at an appropriate value for Ca $K\alpha$ x rays (with fine adjustment).

still be increased substantially over that presently available. Also, spectrometer design requirements for AEM may differ in some respects from those of the conventional electron microprobe (for example, the materials used must be compatible with the ultra-high vacuum used in field emission instruments).

Various design studies for WD spectrometers adapted to AEM have been suggested. For example, Goldstein *et al.* have discussed an arrangement whereby the crystal is housed inside the column, giving a high solid angle with a crystal of moderate size.[26] However, as noted in Section 8.4, a large solid angle alone does not give high intensity; double curvature of the crystal is necessary to achieve this. In the spectrometer constructed by Sun *et al.*, a doubly curved crystal is mounted directly above the specimen, with a hole in the crystal for the beam, giving a solid angle of 0.075 steradian[27] (Figure 13.15). The LiF crystal is at a fixed angle (with a fine tuning adjustment) and is intended for the detection of the Ca $K\alpha$ line only, this being subject to interference from the K $K\beta$ peak in ED spectra (notably for biological samples with high K-contents.)

### 13.8.1. Parallel Collection

The conventional WD spectrometer is actually a "monochromator" that selects one wavelength at a time. By using other configurations, a range of wavelengths can be detected in parallel. In the arrangement shown in Figure 13.16, the x rays are detected by means of a position-sensitive detector after dispersion by the crystal.[28] An alternative arrangement is to use a flat crystal, also with a position-sensitive detector.[29] In this case, x rays of a given wavelength form an arc in the plane of the detector, which can be recorded by a two-dimensional charge-coupled device (CCD) array.[30] Advantages of such spectrometers include the absence of moving parts and the simultaneous recording of several lines. However, there is not a major gain in x-ray collection efficiency, since the size of the reflecting area for any one wavelength is

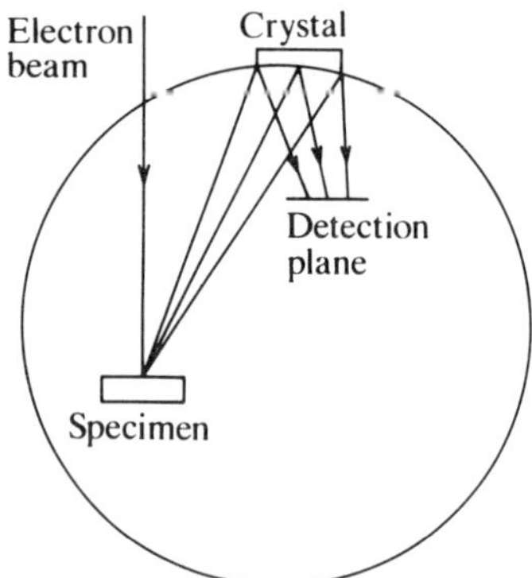


FIGURE 13.16.   Proposed configuration of parallel collection WD spectrometer (after Ref. 28, © 1991 San Francisco Press).

limited. (The properties of several different parallel collection configurations have been calculated by Wittry and Li.[31])


## 13.9. WAVELENGTH DISPERSIVE ANALYSIS

A brief account will now be given of the way in which WD spectrometers are used for analysis. Assuming that it is known which elements are to be included in the analysis, a decision must be made as to the allocation of elements to different spectrometers (in the case of an instrument with multiple spectrometers). For some elements there is only one choice of crystal for the principal characteristic x-ray line, whereas others fall in wavelength regions where two crystals overlap. In the latter case, the choice depends on whether intensity or resolution is most important, the former being highest at low Bragg angles and the latter at high angles (see Sections 13.5 and 13.6). Another factor is that the analysis time can be minimized by distributing the elements evenly amongst the available spectrometers.

Certain considerations regarding beam current should be mentioned. A high current gives high x-ray intensities, enabling the analysis time to be minimized or higher precision to be obtained in a given time. However, possible limitations imposed by the susceptibility of the specimen to beam damage must be taken into account. Also, count rates exceeding around $10^5$ cps are undesirable (see Section 13.2). This commonly restricts the beam current to approximately 100 nA, as far as major elements are concerned. However, a higher current (if the specimen allows) is advantageous for minor and trace elements, though it will usually be necessary to measure the standards at a lower current to avoid excessive count rates.


### 13.9.1. Peak Intensity Measurements

For quantitative analysis, x-ray intensities are measured with the spectrometers set on the relevant peaks. Although peak positions are derived in the first instance from tables, it is necessary to find the precise position empirically. This is an ideal application of computer control: the spectrometer can be stepped across the region containing the peak and the position of maximum intensity determined automatically. This is usually done in two stages, the approximate peak position being obtained from the first scan and more accurately from the second, which covers a narrower range, with longer counting times. "Peak-seeking" is done on the standards at the start of a session. The positions are stored and used for the analysis of the samples (for which peak-seeking is liable to be difficult because of the small size of the peaks in many cases). Sometimes it may be desirable to use slightly different settings for specimens and standards to allow for wavelength shifts related to chemical bonding. Instead of relying on the position being perfectly constant, small shifts can be taken into account by making measurements at a series of closely spaced wavelengths across the peak and calculating the maximum intensity.[32] This allows not only for chemical shifts, but also for thermal drift and variations in the source location.

Both the position of the intensity maximum and the shape of the $K$-lines of light elements ($Z < 10$) are subject to chemical bonding effects. For quantitative analysis, peak areas are required, so it is necessary to convert measured peak intensities to areas using empirical "profile factors"[33] or "area/peak factors,"[34] which may differ significantly between specimen and standard.

Since it is often necessary to use one spectrometer for more than one element, the reproducibility of the spectrometer angle should be of a high order to avoid the necessity for peak-seeking prior to each intensity measurement. Furthermore, changing crystals in the course of an analysis should be avoided if possible since this may result in peak positions changing.

Intensities are measured by counting pulses from the proportional counter. The statistical precision is given by: $\sigma = n^{0.5}$, where $\sigma$ is the standard deviation and $n$ is the number of counts. For $n = 10^5$, the relative precision defined as $\pm 2\sigma$ is thus $\pm 0.6\%$. This is adequate for most purposes, given that the overall accuracy as governed by other factors is rarely better than $\pm 1\%$. For major elements, a counting time of a few seconds may therefore be sufficient. For minor and trace elements, much longer times are needed, depending on the required detection limit and the time available per analysis.

### 13.9.2. Background Measurement

Characteristic x-ray peaks are superimposed on a background caused by the continuous x-ray spectrum (also known as the "continuum" or "bremsstrahlung"). The background level is a function of the continuum intensity (which is approximately proportional to mean atomic number) and the spectrometer efficiency. For major elements, the peak-to-background ratio may be several hundred, in which case the effect of background is trivial. However, for lower concentrations, the contribution of the background must be subtracted from the measured peak intensity.

Background is usually measured by offsetting the spectrometer on each side of the peak and calculating the intensity at the peak position by linear interpolation. This allows for slope in the background, but sometimes (especially at low Bragg angles) there is also significant curvature. A factor correcting the background obtained by linear interpolation for curvature can be derived from measurements on a sample in which the element concerned is absent, but which is compositionally similar to the analyzed material. The effect of curvature can be minimized by using the smallest possible offset. (Note that it is not necessary to be completely clear of the tails of the peak, since their contribution has no effect on the specimen/standard peak intensity ratio provided the peak shape is constant.)[14]

### 13.9.3. Interferences

Despite the high resolution of the WD spectrometer, either peak or background intensities (or both) are sometimes affected by a neighboring peak. If the apparent background intensity is enhanced, the background-corrected intensity as obtained by linear interpolation is reduced, so the concentration obtained will be too low. This error can be eliminated by means of an "overlap correction," whereby the effect is estimated from the measured intensity of the overlapping peak. The same principle can be applied to the positive error obtained when the overlap affects the measured peak intensity of the element concerned more than it does the background. Usually provision for corrections of this sort is limited or absent in commercial software, in which case they must be applied retrospectively by the user.

Geological samples which contain all the rare earth elements (REE) simultaneously provide a good example of interference effects. Some interferences between REE can be avoided by using the $L\beta$ line rather than the $L\alpha$ line for analysis, but others are unavoidable. One approach to this problem is to determine empirical "overlap factors" from measurements on pure oxides.[35] These factors give the intensity from the interfering element appearing at the peak wavelength of the interfered-with element as a fraction of the intensity of the measured peak of the latter, and are used to correct the intensities measured on the analyzed samples. In a variant of this approach, overlap factors are derived from mathematical peak shape functions.[36] Modeling peak shapes is not simple, however, especially when satellite lines are taken into account.[22] (For a more general discussion of "spectrum stripping," see Ref. 37).

Another alternative is to treat interferences in terms of concentrations. The interference factor, defined as the ratio of the apparent concentration of the interfered-with element caused by interference to the concentration of the interfering element, can be determined empirically from a pure element or other suitable standard. In this case, "ZAF" factors must be included in the calculation to allow for interelemental differences in the relationship between intensity and concentration.[38]

## REFERENCES


1. R. Castaing, *Doctoral Thesis*, University of Paris (1951).
2. D. B. Wittry and S. Sun, *J. Appl. Phys.* **67**, 1633 (1990).
3. B. K. Agarwal, *X-Ray Spectroscopy*, Second Ed., Springer-Verlag, Berlin (1991).
4. S. Luck, D. S. Urch, and D. H. Zheng, *X-Ray Spectrom.* **21**, 77 (1992).
5. J. A. Nicolosi, J. P. Groven, D. Merlo, and R. Jenkins, *Opt. Eng.* **25**, 964 (1986).
6. P. J. Potts and A. G. Tindle, *Mineral. Mag.* **53**, 357 (1989).
7. K. Kawabe, S. Takagi, M. Saito, and S. Tagata, in: *Microbeam Analysis—1988* (D. E. Newbury, ed.) San Francisco Press, San Francisco, p. 341 (1988).
8. J. J. McGee, J. F. Slack, and C. R. Herrington, *Am. Mineral.* **76**, 681 (1991).
9. L. Contardi, S. S. Chao, J. Keem, and J. Tyler, *Scanning Electron Microsc.* **1984/II**, 577 (1984).
10. J. B. Nicholson and M. F. Hasler, *Adv. X-Ray Anal.* **9**, 420 (1966).
11. J. B. Nicholson, H. Neuhaus, and M. F. Hasler, in: *The 5th International Congress on X-Ray Optics and Microanalysis* (G. Mollenstedt and G. H. Gaukler, eds.) Springer-Verlag, Berlin, p. 269 (1969).
12. J. V. P. Long, and V. E. Cosslett, in: *X-Ray Microscopy and Microradiography* (V. E. Cosslett, A. Engstrom, and H. H. Pattee, eds.) Academic Press, New York, p. 435 (1957).
13. H. A. Elion and R. E. Ogilvie, *Rev. Sci. Instrum.* **33**, 753 (1962).

14. S. J. B. Reed, *Electron Microprobe Analysis*, Second Ed., Cambridge University Press, Cambridge p. 75 (1993).

15. W. Z. Chang and D. B. Wittry, *J. Appl. Phys.* **74**, 2999 (1993).

16. D. B. Wittry and S. Sun, *J. Appl. Phys.* **69**, 3887 (1991).

17. D. B. Wittry, W. Z. Chang, and T. W. Barbee, Proceedings of the 50th Annual Meeting EMAS (G. W. Bailey, J. Bentley, and J. A. Small, eds.) San Francisco Press, San Francisco, p. 1732 (1991).

18. D. B. Wittry, W. Z. Chang, and R. Y. Li, *J. Appl. Phys.* **74**, 3534 (1993).

19. R. B. Marinenko, K. F. J. Heinrich, R. L. Myklebust, and C. E. Fiori, in: *Microbeam Analysis—1980* (D. B. Wittry, ed.) San Francisco Press, San Francisco, p. 56 (1980).

20. R. B. Bolon, M. D. McConnell, and M. E. Gill, in: *Microbeam Analysis—1979* (D. E. Newbury, ed.) San Francisco Press, San Francisco, p. 204 (1979).

21. C. Brogren, *Ark. Fys.* **23**, 219 (1963).

22. T. C. Huang and G. Lim, *Adv. X-Ray Anal.* **29**, 461 (1986).

23. G. Remond, P. Coutures, C. Gilles, and D. Massiot, *Scanning Microsc.* **3**, 1059 (1989).

24. C. R. Swyt and C. E. Fiori, in: *Microbeam Analysis—1986* (A. D. Romig, Jr. and W. F. Chambers, eds.) San Francisco Press, San Francisco, p. 482 (1986).

25. P. Duncumb, in: *The Electron Microprobe* (T. D. McKinley, K. F. J. Heinrich, and D. B. Wittry, eds.) Wiley, New York, p. 490 (1966).

26. J. I. Goldstein, C. E. Lyman, and D. B. Williams, *Ultramicroscopy* **28**, 162 (1989).

27. S. Sun, J. M. Tormey, and D. B. Wittry, in: *Microbeam Analysis—1990* (J. R. Michael and P. Ingram, eds.) San Francisco Press, San Francisco, p. 415 (1990).

28. C. E. Fiori, S. A. Wight, and A. D. Romig, in: *Microbeam Analysis—1991* (D. G. Howitt, ed.) San Francisco Press, San Francisco, p. 327 (1991).

29. H. Ebel, M. Mantler, N. Gurker, and J. Wernisch, *X-Ray Spectrom.* **12**, 47 (1983).

30. N. J. Zaluzec and M. G. Strauss, *Ultramicroscopy* **28**, 131 (1989).

31. D. B. Wittry and R. Y. Li, *Rev. Sci. Instrum.* **64** 2195 (1993).

32. A. K. Ferguson and D. K. B. Sewell, *X-Ray Spectrom.* **9**, 48 (1980).

33. W. Weisweiler, *Mikrochim. Acta* **1975/1**, 611 (1975).

34. G. F. Bastin and H. J. M. Heijligers, *X-Ray Spectrom.* **15**, 135 (1986).

35. R. Amli and W. L. Griffin, *Am. Mineral.* **60**, 599 (1975).

36. P. L. Roeder, *Can. Mineral.* **23**, 263 (1985).

37. G. Remond, J. L. Campbell, R. H. Packwood, and M. Fialin, (1994) *Proc. 11th Pfefferkorn Symposium, Scanning Microsc. Suppl.* **7**, in press.

38. J. I. Donovan, D. A. Snyder, and M. L. Rivers, *Microbeam Anal.* **2**, 23 (1993).

# 14

# Synthetic Multilayer Crystals for EPMA of Ultra-light Elements

*G. F. Bastin and H. J. M. Heijligers*


## 14.1. INTRODUCTION

Quantitative wavelength dispersive EPMA of the ultra-light elements boron, carbon, nitrogen, and oxygen with a conventional lead stearate crystal ($2d$ spacing approximately 10 nm) is severely hampered by a number of problems. Among these the most troublesome are the problems associated with low count rates, sensitivity to peak shape alterations, and the efficient transmission of higher-order reflections of alloying elements. Acceptable count rates can be produced for elements such as boron or carbon on elemental standards, but only at the expense of increasing the beam current to rather excessive values of 100–300 nA. Even under such extreme conditions, however, the count rates for nitrogen are usually unacceptably low. This is caused by the extreme absorption of N $K\alpha$ x rays in both the stearate crystal and the detector window; both contain appreciable amounts of carbon, and the wavelength of N $K\alpha$ falls right into the C $K$-absorption edge. The situation for O $K\alpha$ x rays is much more favorable in this respect. In addition, the O $K\alpha$ peak is rather close to the lower mechanical limit (short wavelength limit) of most types of spectrometers making it generally impossible to have proper access to the background on that particular side of the peak. This makes a correct background determination impossible.

The problems with the analysis of N $K\alpha$ are not restricted to low count rates alone. In addition to the extremely low peak-to-background ratios, frequently approaching unity or even lower in some binary nitrides (ZrN, $Nb_2N$), there can also be appreciable problems in the correct determination of the background. These may be caused in part by a strong curvature in the background. On top of this, a number of

G. F. BASTIN AND H. J. M. HEIJLIGERS • Laboratory for Solid State Chemistry and Materials Science, University of Technology, NL-5600 MB Eindhoven, The Netherlands

*X-Ray Spectrometry in Electron Beam Instruments*, edited by David Williams, Joseph Goldstein, and Dale Newbury. Plenum Press, New York, 1995.

higher-order reflections from an alloying element may show up and interfere with the N $K\alpha$ peak itself even if severe discriminator settings in the pulse height analyzer are being used. Notable examples in this respect are combinations of nitrogen with the elements Zr, Nb, and Mo. In nitride compounds containing these elements, it is almost a hopeless task to subtract the proper background from the N $K$ spectrum.

We have pointed out on many occasions[1–5] that the major problem in the analysis of ultra-light elements is not so much the variation in peak position that can be observed when moving from standard to specimen, but rather the extreme variation in peak shapes which can occur as a result of changes in chemical bonds. The worst elements in this respect are boron and carbon, with the effects tending to decrease with increasing atomic number of the light element. An additional problem in the analysis of boron can be the simultaneous dependence of peak position and peak shape on the crystallographic orientation of the specimen with respect to the spectrometer. This can cause the emitted B $K\alpha$ peak to move back and forth upon rotation of the specimen under the electron beam as well as changing the shape of the peak in a cyclic manner. This is due to the presence of polarized components in the B $K\alpha$ peak which can be partially filtered out by the analyzer crystal.[6] Filtering is most efficient when the angle of incidence of the x rays on the analyzer crystal is 45°. Unfortunately, with its $2d$ spacing of 10 nm, the stearate crystal produces its B$K\alpha$ peak for an angle of incidence dangerously close to this optimum value, making it very sensitive to shape alterations.

The combination of effects mentioned so far can make accurate quantitative EPMA of the ultra-light elements with a conventional lead stearate crystal cumbersome and very time consuming. Fortunately, new synthetic multilayer crystals have become commercially available recently and our experiences with three of these crystals will be the subject of this paper. First, we shall discuss the basic requirements for a good analyzer crystal in wavelength dispersive analysis. Next, we shall elaborate on the impact these requirements have on quantitative EPMA of the ultra-light elements. Finally, we shall come to a straightforward comparison of performance between the conventional stearate crystal and its modern synthetic multilayer replacements. We will compare analysis of the various light elements starting with the easiest element (oxygen) and gradually increasing the level of complexity of analysis up to the element boron.

## 14.2. BASIC REQUIREMENTS FOR A GOOD WDS ANALYZER CRYSTAL

The basic requirements for a good WDS analyzer crystal for EPMA can be summarized as follows:

1.  The crystal should produce the highest possible peak count rates per unit of beam current (cps/nA).

2.  The background produced should be as low as possible.

The combination of items 1 and 2 will lead to the best possible peak-to- background ratio, and this in turn is related to the detectability limit of the element in question.

3. The best possible energy (wavelength) resolution is required. Good energy resolution is in general beneficial because it allows separation of peaks of neighboring wavelengths.

The question is now whether the same requirements apply for WDS of ultra-light elements. As far as high peak count rates are concerned, they are also highly desirable for ultra-light element radiation. However, we have to realize that the count rates in the latter case will always be substantially lower than for medium-to-high-$Z$ elements because of a number of fundamental physical limitations (e.g., low x-ray fluorescence yields for low-$Z$ element radiation).

The importance of having low continuum backgrounds is even more pronounced for ultra-light element radiation since the peak count rates are bound to be lower. As a consequence, the importance of a correct background determination and subtraction becomes all the more important. As a matter of fact, it is frequently not so much the absolute magnitude of the background, but rather the ease with which the background can be subtracted from the emission spectrum that should be the basis for the judgment of the performance of a specific crystal. Keeping in mind that ultra-light element radiation produces rather wide peaks covering a fairly large range in wavelengths, it is useful to introduce the concept of a "clean background," meaning that the analyzer crystal should preferably produce

No higher-order reflections of alloying elements.

No spectral artifacts.

No strong curvatures or kinks in the background.

The importance of these demands is illustrated in Figure 14.1, which shows the N $K\alpha$ spectrum recorded from ZrN with a conventional stearate crystal and the corresponding background on pure Zr. In addition to the fact that the peak-to-background ratio is exceedingly low to begin with (of the order of unity), we must face the problem of subtracting a complicated background characterized by a multitude of interfering reflections in addition to a curved continuous background. In such compounds it is almost a hopeless task to determine the net intensities in a correct manner.

The ideal background should be such that the background intensity varies in a linear manner with wavelength and no artifacts or spectral interferences occur. The background under the peak can then be determined easily by the routine process of linear interpolation between values at equidistant positions on both sides of the peak. As far as energy resolution is concerned, we can take one of two approaches:

1. If we are interested in studying the fine structure of emission bands of ultra-light element radiations in order to obtain information on the chemical bond, then we need the best possible resolution.
2. If we are interested in measuring the emitted intensities for the purpose of quantitative analysis, then it can be an absolute nuisance to have an analyzer crystal with excellent resolution. This is related to the main problem in EPMA of ultra-light elements: peak shape alterations. The magnitude with which the latter effects can be observed is closely related to the spectral resolution of the analyzer crystal, because good resolution means that more fine structure in the peak can be resolved, which in turn means that more components can be seen outside the immediate maximum of the peak.

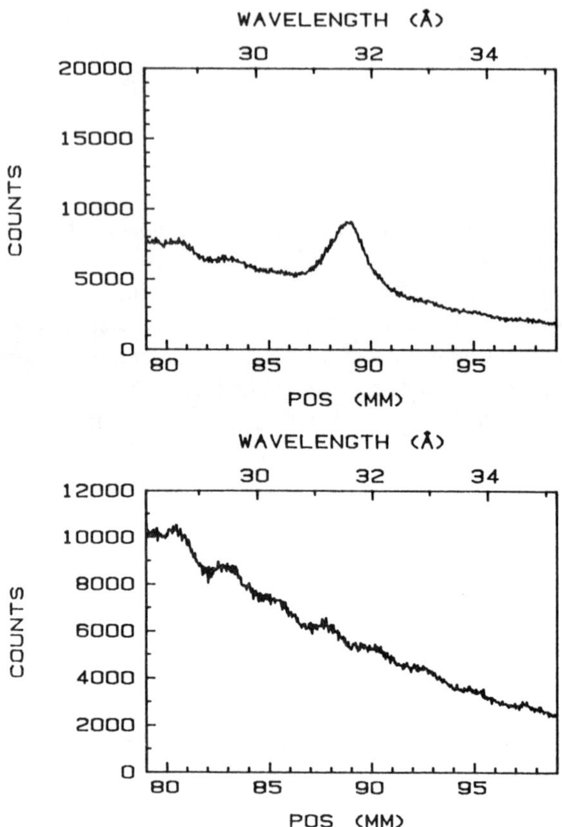


FIGURE 14.1.   N $K\alpha$ peak from ZrN (top) and corresponding background from pure Zr (bottom), recorded with the lead stearate crystal at 10 kV and 300 nA beam current.

Peak shape alterations usually result in the development of strongly asymmetrical peaks and, what is worse, they also lead to the fact that the *peak* intensity is usually no longer a correct measure of the *integral* emitted intensity. As a consequence, it is imperative to carry out *integral* intensity recordings over the full wavelength region of ultra-light element radiation. Consequently, for quantitative purposes it is necessary to determine *integral* (*area*) k-ratios, rather than *peak* k-ratios. Fortunately, there is a fixed ratio between the area k-ratio and the peak k-ratio for a specific compound relative to a specific standard, and we have called this ratio the area/peak factor (APF). As it turns out, the APF is a very convenient parameter to measure peak shape alterations between specimen and standard. An APF value smaller than 1 simply means that a narrowing of the peak is observed when moving from standard to specimen, while the reverse applies for APF values larger than 1. It is necessary to point out that the concept of FWHM is useless for the purpose of measuring peak shape alterations,

TABLE 14.1.  Overview of Synthetic Multilayer Crystals

| Conventional crystal: Stearate (STE, $2d$ 10.0 nm) | | | | | |
|---|---|---|---|---|---|
| F $K\alpha$ | O $K\alpha$ | N $K\alpha$ | C $K\alpha$ | B $K\alpha$ | Be $K\alpha$ |
| Wavelength (nm)  1.832 | 2.362 | 3.160 | 4.470 | 6.760 | 11.40 |
| Replacements: | W/Si (LDE) $2d$ 5.98 nm | | | Mo/B$_4$C (OVH) $2d$ 14.7 nm | |
| | Ni/C ($2d$ 10.0 nm) | | | | |

because in the vast majority of cases, shape alterations are concentrated at or near the foot of the peak by the development of extra components.

From the discussion so far it follows that, apart from the net peak intensities and the qualities of the background, the peak shape alterations (expressed in APFs) should be one of the crucial items on the basis of which the performance of a synthetic multilayer crystal should be compared to that of the conventional stearate crystal it is supposed to replace.

## 14.3. COMPARISON OF PERFORMANCE BETWEEN STEARATE CRYSTAL AND SYNTHETIC MULTILAYER REPLACEMENTS

Table 14.1 gives an overview of the three multilayer crystals that were tested and the light elements and wavelength ranges they cover. The dashed line for the LDE crystal indicates that this crystal is not specifically recommended by the manufacturer for the analysis of carbon, although its $2d$ spacing would make it suitable for this purpose. All multilayer crystals were supplied by Ovonic Synthetic Materials Company, Troy, Michigan. Our first multilayer was the LDE crystal, later followed by the OVH crystal. Both were originally mounted in an additional dedicated ultra-light element spectrometer on our JEOL 733 Superprobe, which was also equipped with a stearate crystal. After the installation of a new microprobe (JEOL 8600), the stearate crystal was replaced by the Ni/C crystal.

### 14.3.1. Analysis of Oxygen

For several years we had the advantage of having the W/Si (LDE) and the Mo/B$_4$C (OVH) multilayers next to our conventional stearate (STE) crystal on our JEOL 733 microprobe. This enabled us in many cases to make straightforward comparisons of the performances of crystals under identical circumstances. The results for this comparison on the O $K$ peak emitted from Fe$_2$O$_3$, a compound that we favor as oxygen standard, are shown in Figure 14.2.

It is obvious that the peak intensity obtained with LDE is at least three times higher than with STE. On the other hand, the development of the tail on the left side of the STE peak which is not present for the LDE crystal, indicates that the spectral resolution of LDE must be somewhat poorer than that of STE. It is interesting to point out how efficient the LDE crystal is for oxygen detection. While the presence of a native oxide skin on polished

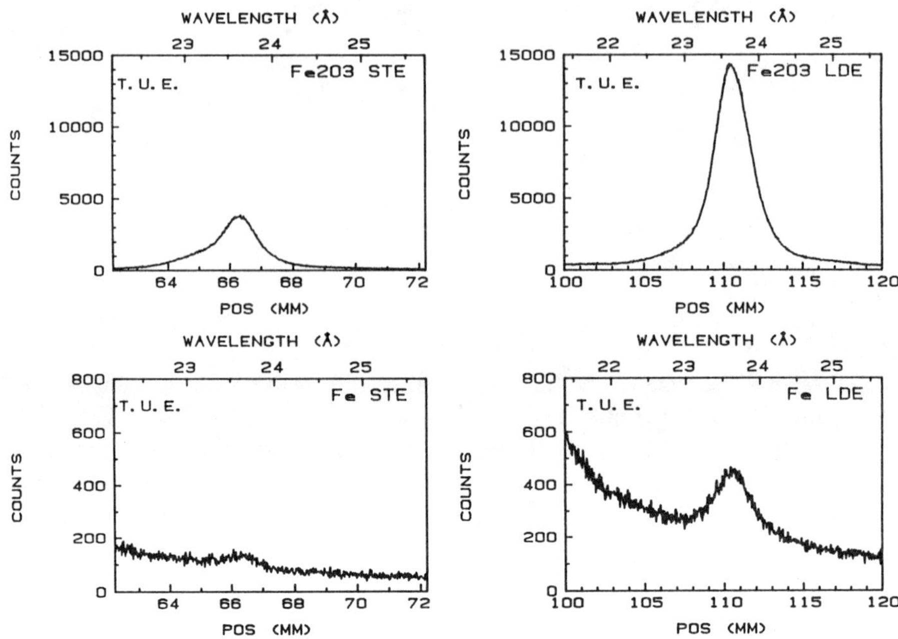


FIGURE 14.2. Oxygen spectra (top) and backgrounds (bottom) recorded from $Fe_2O_3$ with the stearate crystal (STE, left) and W/Si multilayer (LDE, right) at 10 kV and 50 nA.

iron (several nanometers thick) is clearly visible with LDE, it is hardly noticeable with STE. As our experience shows it is not difficult to detect the presence of oxygen even on freshly polished noble metals such as gold or platinum.

As we mentioned in Section 14.1, there can be problems with a correct background determination for O $K\alpha$ on STE due to the proximity of the lower limit switch of the spectrometer. Because of its lower $2d$ spacing, the LDE crystal presents no problems in this respect since the position of the peak is shifted toward the middle of the range and the background is easily accessible on both sides of the peak.

As far as peak shape alterations are concerned, the effects were found to be relatively small, certainly on LDE. There is every indication they are more pronounced with STE. However, as a result of the problems with the background determination on STE, the APF values could not be measured with sufficient accuracy. Table 14.2 gives a selection of APF values, relative to $Fe_2O_3$, for LDE, which shows more or less the maximum variation observable.

The interference of higher-order reflections of alloying elements with the light element peak was found to be only a minor problem in the case of O $K\alpha$. Some cases where problems might occur with STE are $Al_2O_3$ (third order Al $K\alpha$) and ZnO (second order Zn $L\alpha$). We found the LDE crystal very efficient in suppressing third order reflections and fairly efficient in reducing second order reflections in the wavelength range of O $K$. For the rest of the multiorder reflections, the backgrounds obtained with both crystals were in general smooth.

TABLE 14.2.   Variation in APF Relative to $Fe_2O_3$ for Some
Binary Oxides Observed with the LDE Crystal

| Compound | APF |
|----------|-----|
| $B_6O$ | 1.063 |
| $SiO_2$ | 1.044 |
| $SnO_2$ | 0.974 |
| PbO | 0.965 |

As Table 14.1 indicates, the Ni/C multilayer should also be a possible candidate to replace the STE crystal for the analysis of oxygen. We did some preliminary tests in which we compared the performance of Ni/C with that of LDE (W/Si) on $Fe_2O_3$. Typical results are shown in Figure 14.3. Although the Ni/C crystal provided approximately 3.5 times higher peak count rates than LDE, its *P/B* ratio as well as its spectral resolution were so much worse than that of LDE that we really do not consider the Ni/C crystal as a suitable replacement for either STE or LDE.

Based on this analysis of oxygen we can conclude that the LDE (W/Si) multilayer is a significant improvement over STE because it provides 3–4 times higher peak count rates and it is fairly efficient in suppressing higher orders of reflection. Furthermore, the detectability limit for oxygen is significantly improved, since oxygen can be detected on virtually any polished specimen with LDE. We consider the W/Si multilayer, with a *2d* spacing of 5.98 nm, as the most useful analyzer crystal for O *K*α x rays.

### 14.3.2. Analysis of Nitrogen

For nitrogen, which is the most difficult of the ultra-light elements to analyze, the situation is particularly bad with a conventional STE crystal. As mentioned in Section 14.1, the wavelength of N *K*α is such that it falls right in the C *K*-absorption edge, and carbon is abundantly present in both the STE crystal and the detector window. As an

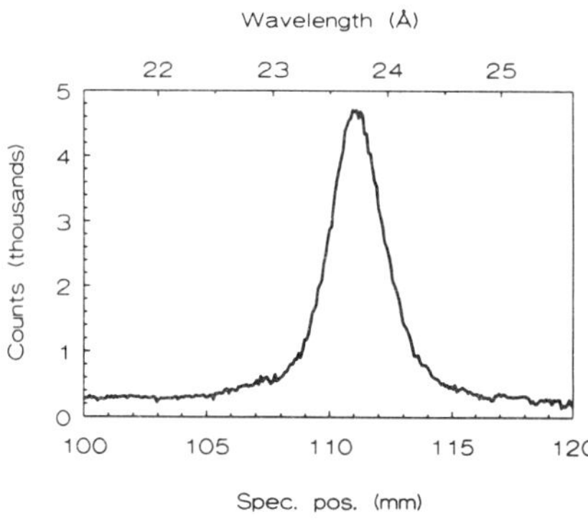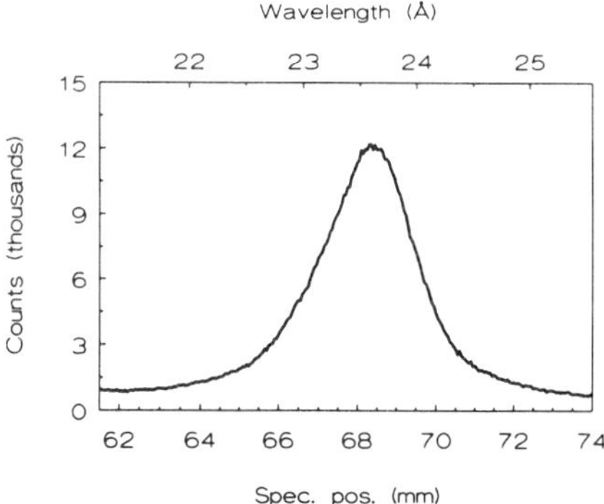

FIGURE 14.3.   Oxygen spectra recorded from $Fe_2O_3$ with the W/Si crystal (LDE) (left) and the Ni/C crystal (right) at 10 kV and 50 nA.

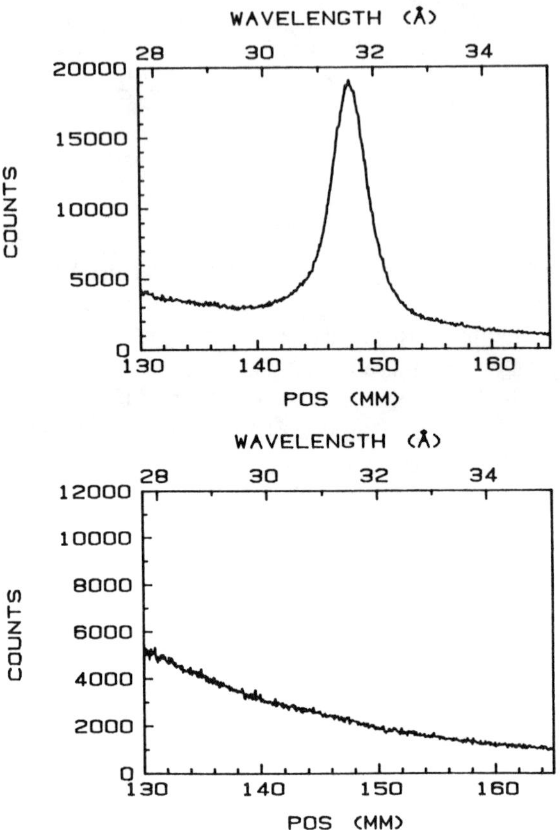

FIGURE 14.4.  N $K\alpha$ peak from ZrN (top) and corresponding background from pure Zr (bottom), recorded with the W/Si multilayer crystal (LDE) at 10 kV and 300 nA beam current.

immediate result, the peak count rates that can be obtained on STE are exceedingly low. In addition, the background under the peak can be very high and complex. The complexity can be caused either by strong curvature or by the presence of multiple higher-order reflections (see Figure 14.1). Therefore, extremely low peak-to-background ratios are frequently observed with STE. Notoriously difficult examples are the nitrides of the elements Zr, Nb, and Mo, in which $P/B$ ratios below 1 are no exception. According to our experience, it is almost impossible to do accurate quantitative EPMA on such compounds with a STE crystal, and if there is ever improvement to be gained by trying new multilayer crystals this is certainly the case for nitrogen.

Comparable to the situation for O $K\alpha$ the LDE crystal also provides a significant increase in peak count rates for N $K\alpha$ (Figure 14.4) — typically, by a factor of 2.7. Such an increase is important because it improves the detectability limit for nitrogen substantially. Equally important, however, is the fact that the background produced on LDE is so much cleaner.

TABLE 14.3.   Variation in APF Relative to $Cr_2N$ for Some
Binary Nitrides Observed with the STE Crystal in Comparison
to the LDE (W/Si) Crystal

| Compound | APF | |
|---|---|---|
| | STE | LDE |
| BN(hex) | 1.052 | 1.101 |
| $Si_3N_4$ | 1.071 | 1.084 |
| CrN | 1.018 | 1.018 |
| ZrN | 0.964 | 0.978 |
| HfN | 0.973 | 0.984 |

Obviously, the LDE crystal is most efficient in suppressing higher-order reflections in the wavelength range of N $K\alpha$. The combination of both effects leads to a significant improvement in $P/B$ ratio, which in turn makes accurate quantitative EPMA of nitrogen possible. The effects of peak shape alterations are not very pronounced on either STE or LDE (see Table 14.3).

As far as the Ni/C crystal is concerned, it is supposed to be a suitable replacement for STE on the basis of its $2d$ spacing (Table 14.1). In our preliminary experience with this crystal, similar remarks apply as made earlier for the analysis of oxygen. Although the peak count rates on Ni/C are 6–7 times higher than on LDE, the background and spectral resolution are so much worse than on LDE that this crystal cannot be recommended to replace the LDE crystal and presumably not even the STE crystal.

Based on this analysis of nitrogen, we can conclude that: the W/Si crystal offers a most significant improvement over the conventional stearate crystal in terms of improved peak count rates and suppression of higher-order reflections. As a matter of fact, we consider the W/Si crystal a must for quantitative EPMA of nitrogen, which can hardly be performed with a stearate crystal. In our opinion it is questionable whether the Ni/C crystal is a suitable replacement for stearate.

### 14.3.3. Analysis of Carbon

The shape alterations in carbon emission peaks can be very pronounced indeed with a STE crystal, as we have shown on various occasions. [1–4] Individual APF values, relative to an $Fe_3C$ standard, can be as low as 0.715 (ZrC) and as high as 1.379 (vitreous carbon) (Table 14.4). In addition, there appears to be a strong variation in APF with the atomic number of the alloying element. Notoriously strong carbide formers such as Ti and Zr produce the narrowest C $K\alpha$ peaks (lowest APFs), while weaker carbide formers such as B and W produce much broader peaks in their C $K\alpha$ emission spectrum, compared to $Fe_3C$.

Although the W/Si (LDE) crystal is not specifically recommended for the analysis of carbon, we did compare its performance to that of STE on a number of binary carbides. As was to be expected, we found the peak intensities on LDE to be only 60–70% of those on STE. Furthermore, we observed that the spectral resolution on LDE was poorer than that on STE, in that less detail could be seen in the peaks. In addition, the emission peaks were much more symmetrical on LDE. The latter obser-

TABLE 14.4.  Variation in APF Relative to Fe3C for
Some Binary Carbides Observed with the STE Crystal
in Comparison to the LDE (W/Si) Crystal

|  | APF | |
| --- | --- | --- |
| Compound | STE | LDE |
| Vitreous carbon | 1.379 | — |
| $B_4C$ | 1.048 | 1.010 |
| $\alpha$-SiC | 0.861 | 0.933 |
| TiC | 0.723 | 0.868 |
| $V_2C$ | 0.775 | 0.873 |
| VC | 0.773 | 0.873 |
| ZrC | 0.715 | 0.880 |

vations led us to the expectation that perhaps the APFs on LDE would be much closer
to unity, which would be very important from the point of view of quantitative EPMA.
This expectation was confirmed by our measurements (Table 14.4).

Therefore, in spite of the lower peak count rates, the LDE crystal can be an
attractive option for the analysis of carbon. We found an additional advantage in some
cases, notably the lighter carbides, that the backgrounds produced with LDE were
simpler than those with STE. Whereas the STE crystal gives a kinked ($B_4C$) or strongly
curved (SiC) background, the background on LDE is smooth and flat. A somewhat
disappointing feature of LDE for the analysis of C $K\alpha$ is that, contrary to what we have
seen for O $K\alpha$ and certainly for N $K\alpha$, suppression of higher-order reflections does not
occur.

Based on this analysis of carbon we can conclude that the count rates on LDE
are approximately 30% lower than on STE and higher-order reflections are no longer
suppressed. On the other hand, LDE is much less sensitive to peak shape alterations
than STE, which makes it an attractive option for quantitative EPMA of carbon.

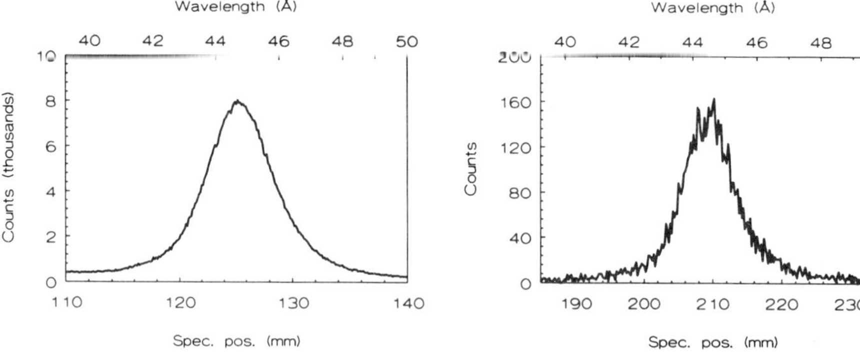


FIGURE 14.5.  Carbon spectra recorded from vitreous carbon with the Ni/C crystal (left) and the W/Si crystal
(LDE, right) at 10 kV and 7.6 nA.

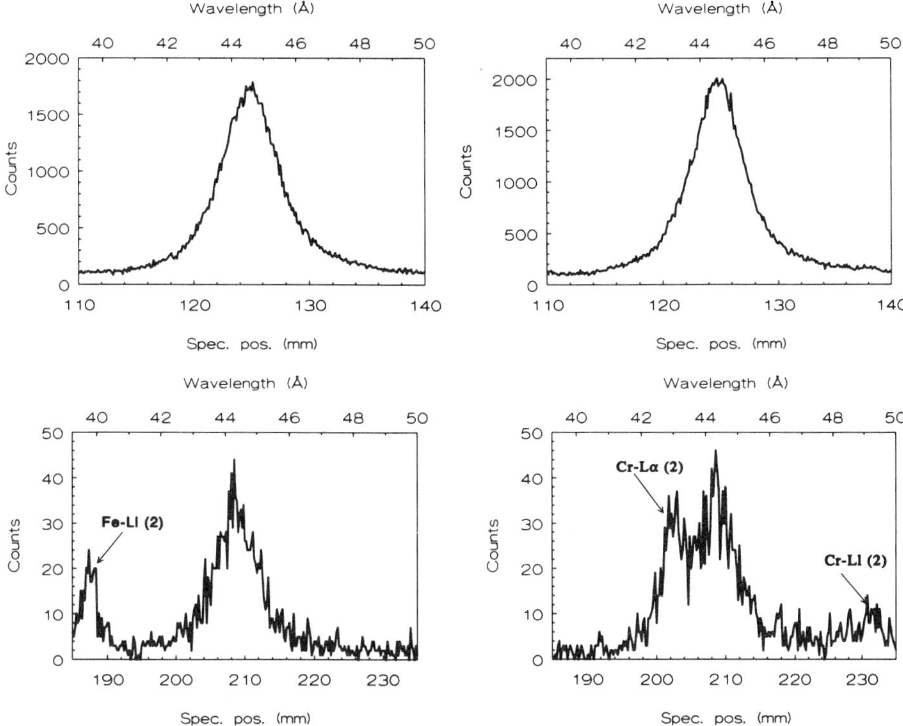



FIGURE 14.6.   Carbon spectra recorded from $Fe_3C$ (left) and $Cr_{23}C_6$ (right) with the Ni/C crystal (top) and the W/Si crystal (LDE, bottom) at 10 kV and 40 nA. Note the efficient suppression of the second order Fe-Ll line and the second orders of the Cr $L\alpha$ and Cr $Ll$ lines by the Ni/C crystal and the huge increase in peak count rates on the Ni/C crystal.

According to Table 14.1, the Ni/C multilayer should be a suitable crystal for the analysis of carbon. Our preliminary experience with this crystal indicates this is indeed the case. As Figure 14.5 shows, there is a huge increase (40 times) in peak count rate compared to LDE. At the same time, judging from the increase in peak width, the spectral resolution of Ni/C is poorer than that of LDE. The LDE resolution is already poorer than that of STE. It might be expected, therefore, that the effects of peak shape alterations on Ni/C would be smaller than on either LDE or STE. Unfortunately, no numerical APF values are available. We also found indications that higher-order reflections are efficiently suppressed by Ni/C, which makes it even more suitable for the analysis of carbon. These effects are illustrated in Figure 14.6. We conclude that the Ni/C crystal offers tremendous improvements for the analysis of carbon.

### 14.3.4. Analysis of Boron

Peak shape alterations in the B $K\alpha$ emission peak can be very extreme indeed with a conventional STE crystal, as Figure 14.7 shows. In a number of cases, at least two different peaks are visible and sometimes even three. The most interesting compound is

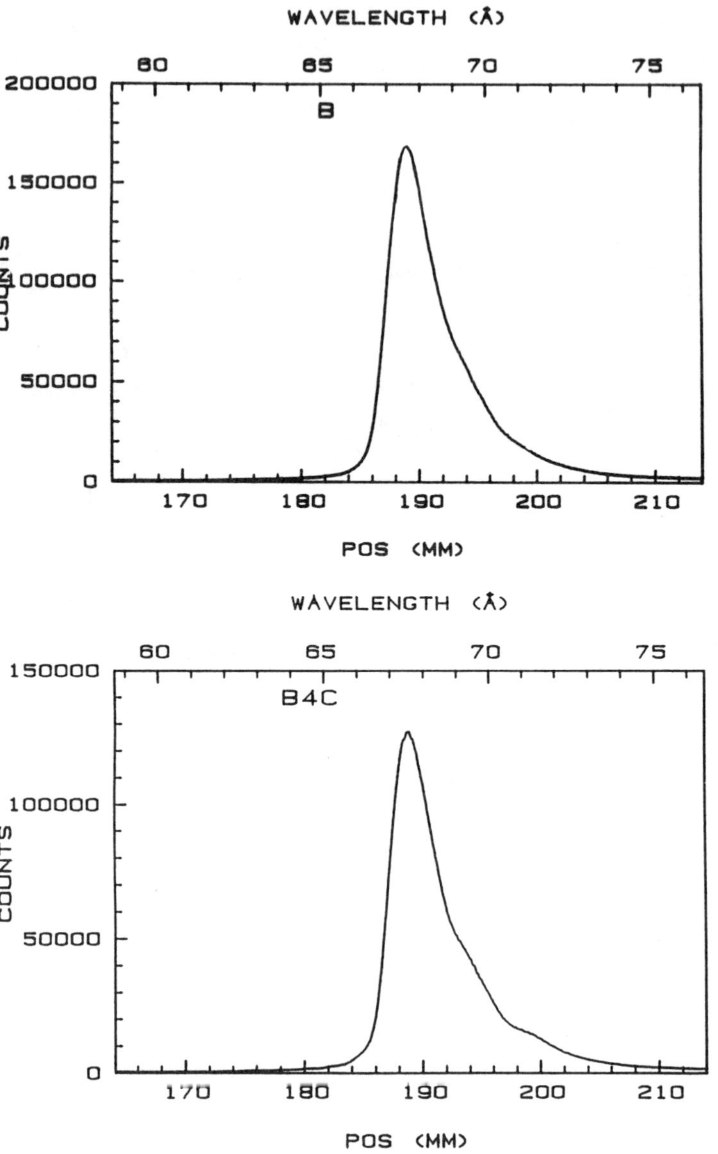


FIGURE 14.7.   Variations in the shape of B $K\alpha$ emission peaks between elemental B, $B_4C$, BN (hexagonal), and $LaB_6$. Spectra recorded with a conventional stearate crystal at 10 kV and 300 nA.

hexagonal BN, where two additional peaks are visible that are well separated from the main peak. It is important to realize that these additional peaks really belong to the B $K\alpha$ emission band and have to be included in intensity measurements.

Although these extreme peak shape alterations in themselves form a tremendous obstacle for accurate quantitative EPMA, the situation can be complicated further by the possible dependence of the peak shape on the crystallographic orientation of the

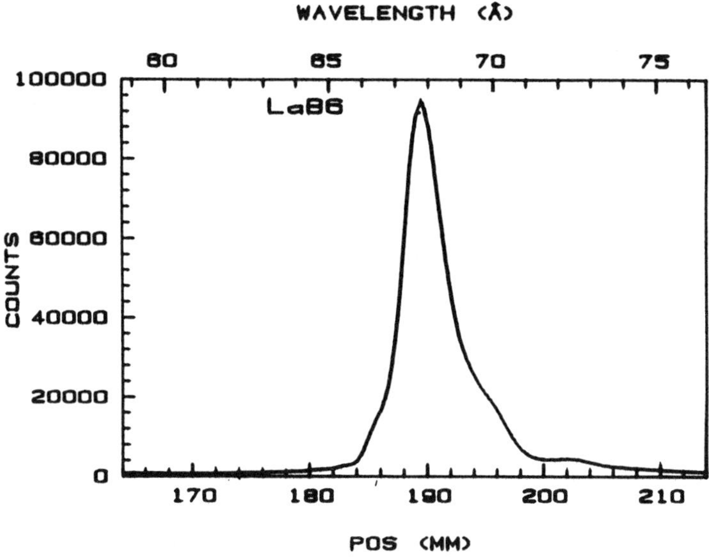


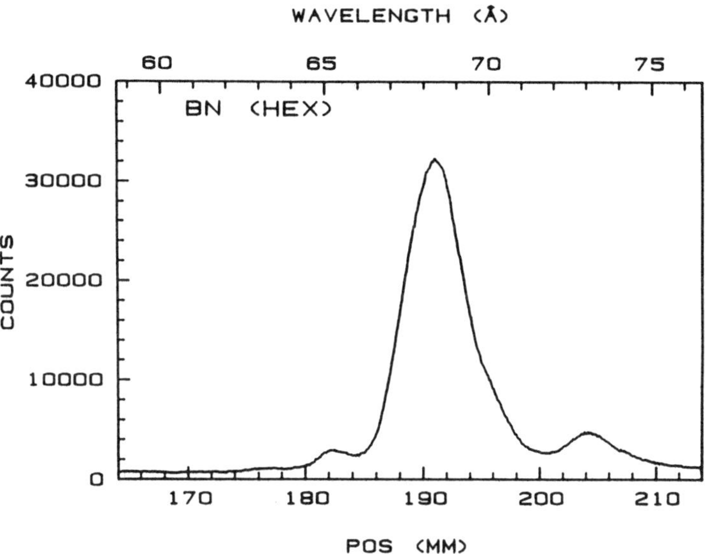


FIGURE 14.7. *(Continued)*.

specimen with respect to the analyzer crystal. This effect is illustrated in Figure 14.8 for $ZrB_2$.

Rotation of the specimen in its own plane under the electron beam can make the B $K\alpha$ peak shift back and forth and at the same time the peak can become narrower or wider. This peculiar phenomenon is extremely dangerous for quantitative EPMA because it means that compounds exhibiting these effects cannot even be used as

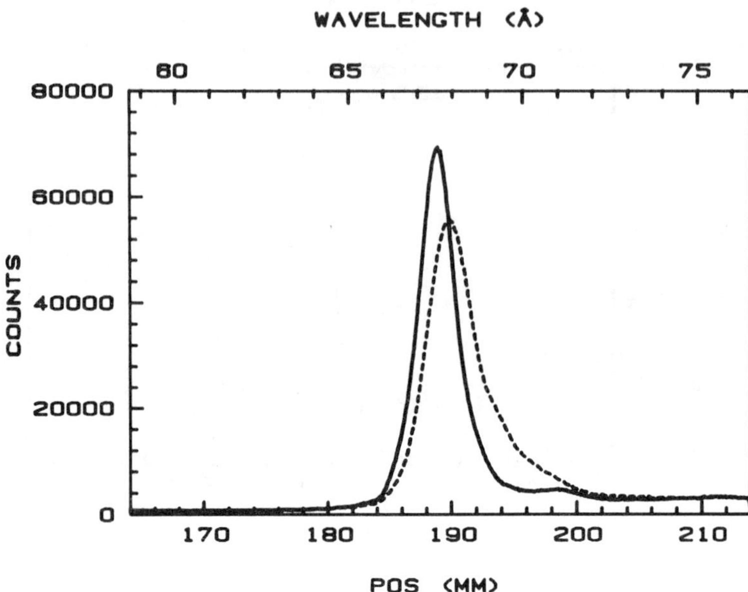


FIGURE 14.8.  Variation in the shape of the B $K\alpha$ peak emitted from $ZrB_2$ as a function of the crystallographic orientation of the specimen with respect to the analyzer crystal. Stearate crystal, 10 kV and 300 nA.

standards in the same systems. The origin of these phenomena[6] can be found in the presence of polarized components in the B $K\alpha$ emission band and the filtering action the analyzer crystal can exercise on specific components, depending on the orientation of the specimen relative to the analyzer crystal. The filtering action is most pronounced when the angle of incidence of the x rays on the analyzer crystal is 45°. As it happens, for a stearate crystal with $2d$ spacing of 10 nm, this angle is 42.5°. As a result, there is not only a tremendous variation in APF, relative to boron, from one compound to another [with values[3] ranging from 0.665 ($ZrB_2$) up to 1.31 (NiB)], but on top of that the APF for a specific compound can vary significantly, depending on crystallographic orientation. The latter effect can, in principle, be expected in all borides with a crystal symmetry lower than cubic. Needless to say, it is extremely difficult to work with a stearate crystal under these conditions.

In efforts to improve the situation by devising new crystals due attention should should be paid to the phenomena observed with the stearate crystal. It seems that the following considerations play a crucial role:

1.  The angle of incidence of the B $K\alpha$ x-rays on the analyzer crystal should be changed away from the value of 45°. This can be done most suitably by increasing the $2d$ spacing. It is to be expected that the dependence of APF on the crystallographic orientation of the specimen will then be reduced.

2.  The effects of peak shape alterations should be reduced altogether. The only way to achieve this seems to be by lowering the spectral resolution.

3.  Any increase in peak count rate would be welcome.

Judging the two candidate multilayer replacements OVH (Mo/$B_4C$) and Ni/C (see Table 14.1) in the light of these requirements, it is expected that the Ni/C crystal will not meet the crucial first demand because its $2d$ spacing is exactly the same as that of the STE crystal. Therefore, the angle of incidence of the B $K\alpha$ x rays on the crystal will have the same rather critical value of 42.5° as the STE crystal. So far, however, we have not made any detailed comparisons, as far as this aspect is concerned.

The OVH crystal, on the other hand, with its $2d$ spacing of 14.7 nm will shift the B $K\alpha$ peak to a position where the angle of incidence of the x rays on the crystal is 27.4°. Merely on the basis of this fact it is to be expected that the dependence of the peak shape alterations on the crystallographic orientation of the specimen will be reduced. Our experimental findings fully confirm these expectations, as Table 14.5 shows.

It is clear from Table 14.5 that the OVH crystal indeed reduces the effects of crystallographic orientation on the peak shape alterations because the observed ranges in APF are significantly smaller than with STE. Besides, in the vast majority of cases, the numerical APF values are much closer to unity. This must be attributed to the poorer spectral resolution of the OVH crystal, which must be considered an advantage in this case. Apart from this, the OVH crystal gives a huge increase in peak count rate for B $K\alpha$. Both phenomena are clearly demonstrated in Figure 14.9, where the stearate crystal is seen to produce a strongly asymmetrical peak due to the presence of several distinct components, thus indicating a better spectral resolution. These effects go unnoticed in the much broader and almost symmetrical peak produced on OVH.

As stated before, the lower spectral resolution of OVH is generally an advantage because it leads to less pronounced peak shape alteration effects. Besides, in the wavelength range of B $K$, it is rather rare to come across cases of interference with higher-order reflections, which would require the best possible spectral resolution. In addition, we found evidence for suppression of higher-order reflections by OVH.

There are cases, however, where the lower resolution of OVH turns into a disadvantage, notably in combinations of boron with Nb and Mo where the $M\zeta$ lines

TABLE 14.5.  Variation in APF Relative to Elemental Boron for Some Binary Borides Observed with the OVH (Mo/B4C) Crystal in Comparison to the Stearate Crystal. Where a range in APF is given this refers to the maximum variation that could be observed depending on the crystallographic orientation of the specimen

| | APF | |
| Compound | STE | OVH |
| --- | --- | --- |
| $B_4C$ | 1.014 | 1.061 |
| $AlB_2$ | 1.152–1.095 | 1.054 |
| TiB | 0.690–0.835 | 0.855–0.915 |
| $TiB_2$ | 0.799–0.945 | 0.922–0.965 |
| $Fe_2B$ | 1.242 | 1.085 |
| $ZrB_2$ | 0.665–0.915 | 0.904–0.948 |
| $LaB_6$ | 0.898 | 0.951 |

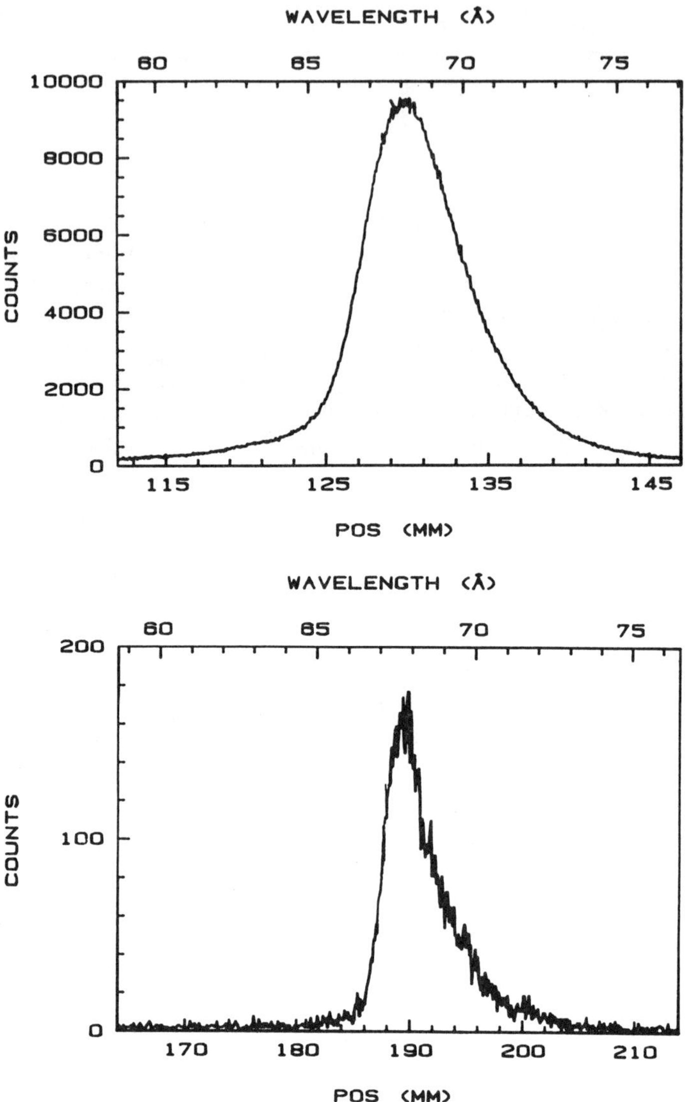


FIGURE 14.9. Direct comparison between the B $K\alpha$ emission peaks recorded with the OVH (Mo/B$_4$C) multilayer crystal (top) and the stearate crystal (bottom) from elemental boron. Accelerating voltage was 15 kV, and beam current 10 nA. Note the huge increase in count rate and the poorer spectral resolution on OVH as compared to stearate.

of the metals (first orders) seriously interfere with the B $K\alpha$ peak. Especially in the case of MoB, where the $M\zeta$ line could easily be distinguished from B $K\alpha$ with stearate, the OVH crystal is not able to resolve both components and only one (combined) peak is produced (Figure 14.10). In such cases, which are rare, it would be most difficult to interpret the asymmetry of the peak on OVH in terms of the possible presence of an interfering line.

Although the OVH crystal is a very attractive option for the analysis of boron, because of the significant reduction in peak shape alteration effects it offers, the remaining effects are still of such a magnitude (see Table 14.5) that they have to be taken into account for quantitative EPMA.

Based on this analysis of boron we can conclude that the OVH ($Mo/B_4C$) multilayer offers a significant improvement over STE because it provides more than 30 times higher peak count rates and it is fairly efficient in suppressing higher order reflections. A beneficial side effect of the improved count rates is the possibility of simultaneous measurement of ultra-light and heavier elements. The excessive beam currents (300 nA) normally required for the analysis of boron with stearate usually make it impossible to measure heavier elements in the same run, because the huge count rates produced by these heavier elements would saturate the detector.

Perhaps the most important aspect of OVH, however, is its ability to reduce the effects of peak shape alterations and their dependence on crystallographic orientation. Such performance cannot be expected from the Ni/C multilayer on the basis of its lower $2d$ spacing. Besides, our preliminary observations show that the peak count rates for B $K\alpha$ on Ni/C are significantly lower (approximately 2.5 times) than on OVH. We conclude therefore, that the OVH multilayer with a $2d$ spacing of 14.7 nm is the most useful analyzer crystal for B $K\alpha$ x rays. The OVH crystal can also be used successfully for the analysis of beryllium, where it produces very much improved count rates as compared to the organic crystals presently used.


## 14.4. GENERAL CONCLUSIONS

From the evidence presented here it can be concluded that synthetic multilayer crystals can offer genuine improvements over the conventional stearate crystal in terms of increased peak count rates, reduced sensitivity to peak shape alteration effects, and suppression of higher order reflections. The extent to which each of these beneficial effects can be observed depends very strongly on the specific multilayer considered as well as on the wavelength range to be measured. It is, therefore, not realistic to expect that one synthetic multilayer crystal can provide optimum performance for all of the wavelengths ranging from F $K\alpha$ up to Be $K\alpha$. According to our experience, the W/Si multilayer (with $2d$ spacing of 5.98 nm) is the best choice for the analysis of nitrogen and oxygen (and presumably fluorine too), offering significant improvements over stearate on all major items. For carbon, the Ni/C multilayer gives the best performance, while for boron (and beryllium), the $Mo/B_4C$ crystal ($2d$ spacing of 14.7 nm) is highly recommended.

One property that all multilayer crystals have in common is their poorer spectral resolution as compared to stearate. Apart from a few exceptions, however, this is considered an advantage. Further advantages of synthetic multilayers are that they are sturdy, mechanically stable, and long-lived compared to the delicate stearate crystal. In addition, their $2d$ spacings can, in principle, be tailored to meet specific demands.

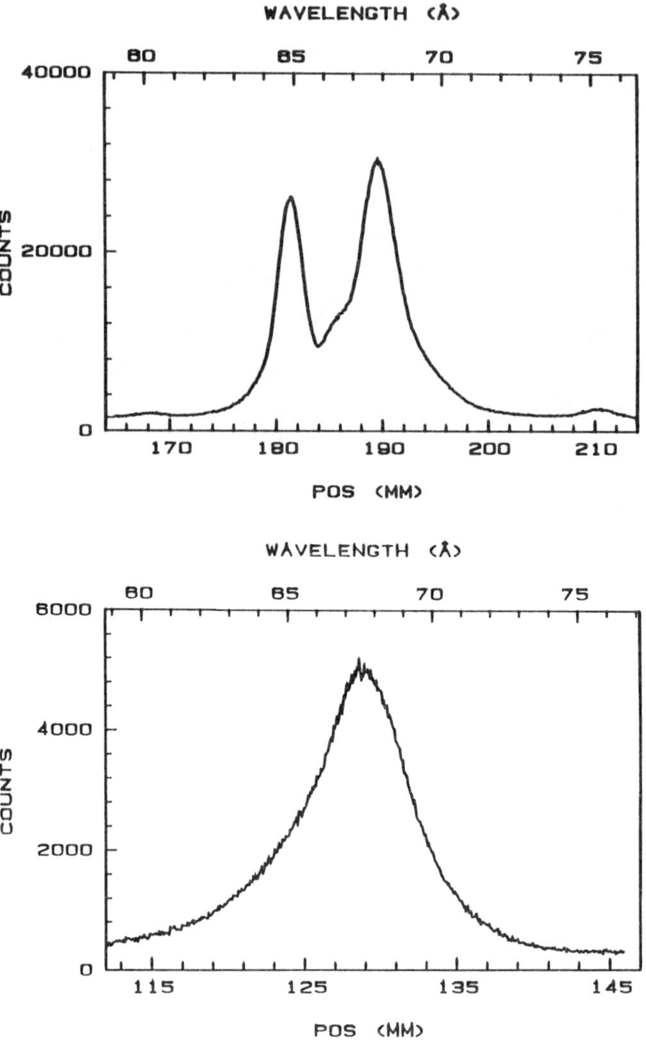

FIGURE 14.10.  B *K*α (second peak) and Mo *M*ζ (first peak) spectra recorded from MoB with the stearate crystal (top) and OVH crystal (bottom). Accelerating voltage 10 kV. Beam current for STE 300 nA, for OVH 10 nA. Note the lower spectral resolution of OVH which is not able to separate the Mo-Mζ peak from the B *K*α peak.

## REFERENCES


1.  G. F. Bastin and H. J. M. Heijligers, *X-Ray Spectrom.* **15**, 135 (1986).
2.  G. F. Bastin and H. J. M. Heijligers, *J. Microsc. Spectrosc. Electron.* **11**, 215 (1986).
3.  G. F. Bastin and H. J. M. Heijligers, *Scanning* **12**, 225 (1990).

4.  G. F. Bastin and H. J. M. Heijligers, in: *Quantitative Electron Probe Microanalysis of Ultra-Light Elements, Electron Probe Quantitation (K. F. J. Heinrich and D. E. Newbury, eds.) Plenum Press, New York, p. 145 (1991).*

5.  G. F. Bastin and H. J. M. Heijligers, *Mikrochim. Acta* **12** [Suppl.] 19 (1992).

6.  G. Wiech in: *X-Ray Emission Spectroscopy* (P. Day, ed.) Emission and Scattering Techniques, NATO Adv. Study Inst., Reidel Publishing Company, Ser. C, 103 (1981).

# 15

# A von Hamos-Type Parallel Collection Wavelength Dispersive Spectrometer for Microbeam Analysis

*A. M. Panin and M. W. Lund*

## 15.1. INTRODUCTION

Wavelength dispersive spectrometers (WDS) of the Johann and Johannson types are used in microbeam analysis of samples in the (TEM).[1,2] A proportional counter (PC) is usually used as a collector of x-ray quanta in WDS. Such spectrometers have a reasonably good geometrical aperture of the order of $\Omega/4\pi \simeq 10^{-3}$, where $\Omega$ is the solid angle of effectively reflecting area of a crystal. They have good spectral resolution ($\Delta\lambda/\lambda \simeq 10^{-3}$), but they are not parallel detectors, i.e., they cannot simultaneously measure the intensities of different spectral lines. Taking into account scanning across the spectra, the real aperture of a Johann spectrometer is approximately $10^2$–$10^3$ times less. A quantitative interpretation of spectra for thermally or radiation-changeable compounds (or alloys of low and high vapor pressure elements) is difficult for such spectrometers.

We propose using a von Hamos-type[3] WDS with a solid-state, position-sensitive x-ray detector (PSXD), such as a one-dimensional CCD or a photodiode (PD) array, placed on the axis of a cylindrical spectrometer (Figure 15.1). The axis of the spectrometer should coincide with the source position to minimize aberrations.

Von Hamos-type x-ray spectrometers have been successfully applied for the diagnostics of point plasma sources, such as laser plasma[4,5] and micropinch plasma[6].

A. M. PANIN • Brigham Young University, Provo, Utah 84602.   M. W. LUND • MOXTEK, Inc., Orem, Utah, 84057.

*X-Ray Spectrometry in Electron Beam Instruments*, edited by David Williams, Joseph Goldstein, and Dale Newbury. Plenum Press, New York, 1995.

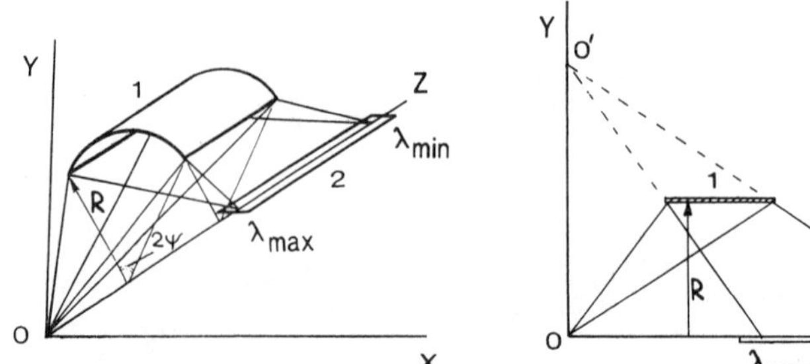


FIGURE 15.1.   Von Hamos spectrometer geometry with the axial focusing: 0—point source; 1—crystal bent over circular cylinder with axis $z$; 2—detector plane.

A von Hamos-type cylindrical spectrometer is a so-called vertical focusing spectrometer (it focuses x rays onto an axis which is parallel to the dispersion axis). It has good collection efficiency; the numerical value of the aperture is of the order of $10^{-4}$ for axis focusing geometry (at a given wavelength). By aperture, we mean the part emitted by the source in $4\pi$ flux collected by the spectrometer for a given waveband.

The dispersion equation for a von Hamos-type spectrometer with axial detection is similar to that for the plane crystal spectrometer scheme:

$$\lambda(z) = \frac{2d}{n\sqrt{(z/2R)^2 + 1}} \; ; z(\lambda) = 2R[(2d/n\lambda)^2 - 1]^{1/2} \qquad (15.1)$$

Here $d$ is the crystal spacing, $n$ is the order of reflection, $R$ is the crystal radius, and $z$ is the distance along the spectrometer axis between the source and the point on the detector plane at which $\lambda$ is focused (see Figure 15.1).

So, if we use a 2-inch (50 mm) linear charge-coupled device or photodiode array, and choose $R$ of 25 mm or smaller, we may cover Bragg angles from 20° to 60°, and a waveband from 0.37 to 0.87 fraction of $2d$ spacing in the first-order and from 0.19 to 0.43 in the second-order reflection.

For a very common crystal LiF ($2d = 4.028$ Å) the waveband in the first order would be from 1.5 to 3.5 Å, for Ge crystal ($2d = 6.532$ Å) from 2.42 to 5.68 Å, for mica crystal ($2d = 19.9$ Å) from 7.4 to 17.7 Å, and so on. The 25-μm pixel size of most available detector arrays allows us to keep high spectral resolution ($\lambda / \delta\lambda \simeq 10^3$) for a crystal radius as small as 25 mm. Crystal radius should be small enough to fit microscope port size to take the advantage of covering wide Bragg angle range. Usually port diameter is in the 35–100-mm range, so the spectrometer crystal should have a radius in the 15–50-mm range if the crystal is coaxial with the port. Using molecular crystals or multilayers in combination with a phosphor-coated detector array allows us to extend the upper wavelength limit to 70–80 Å (for light element analysis). A relative collected intensity $P$ of von Hamos geometry may be estimated as follows. Suppose we want to compare $P$ for $R = 25$ mm von Hamos spectrometer with $\psi = 90°$

(see Figure 1a) and $P$ for $R = 400$ mm Johannson spectrometer with a $20 \times 40$ mm$^2$ crystal. Let both have the same type crystal, and both cover the same spectral band (say, $\theta_{Bragg}$ from $20°$ to $60°$). Both are given the same time to collect a signal (i.e., to scan across the spectral range in the case of Johann-type spectrometer). Solid angle for Johannson spectrometer is $5 \times 10^{-4}$ versus $5 \times 10^{-2}$ for von Hamos one (with a 2 inch detector). Because the crystal is the same, the reflectivity is the same as well. Assuming that both spectrometers have detectors with similar quantum efficiencies, we can come to the conclusion that it is the ratio of crystal solid angles which gives the ratio of collected quanta in the covered spectral range. In our example, the von Hamos-type spectrometer has a hundred-fold larger solid angle than the Johannson spectromer. X-ray scattering flux from a crystal will be essentially stronger for a von Hamos spectrometer due to the proximity of a crystal to a source, and the larger crystal solid angle viewed by a detector. However, due to smaller detector area for a von Hamos spectrometer ($1.3 \times 50$ mm$^2$) compared to the $20 \times 50$ mm$^2$ proportional counter aperture for a Johannson type, the $P/B$ ratio for a von Hamos-type spectrometer may be close to that for Johannson type. Scattered radiation is often weaker than a continuum spectra of the specimen itself, and in this case high $P$ rather than $P/B$ is the advantage.

Thus, a von Hamos type parallel detection spectrometer is a very compact WDS and could be integrated with a microscope or microanalyzer. It would cover a wide enough waveband (to further expand it, several crystals may be used) and would still possess a good aperture and good spectral resolution for analytical microbeam analysis.

One shortcoming of a von Hamos-type spectrometer (especially for spectroscopy of extended sources) is that it produces strong geometrical aberrations (if radius of the spectrometer is to be small). Let us consider some of these aberrations and their influence on the spectral and spatial resolution of the small-radius von Hamos spectrometer.

In our model, the shape of a real crystal may be approximated by different smooth surfaces. They may be close to a circular cylinder to model the von Hamos-type spectrometer or may be of another shape to model another type of x-ray spectrometer. The model has an exact solution that enables us to calculate the form of the focal curve for arbitrary crystal shape and for arbitrary arrangement of the x-ray source, crystal, and detector plane. The only limitation is that the cross section of the crystal should be of constant shape along some axis (i.e., have general cylindrical symmetry). In most cases the model gives a good approximation of the real bend of a crystal (we may call it one-dimensional bent), because the radii of curvature of a real bent crystal usually differ from one other along and across the dispersion axis (such cases may be termed quasi-one-dimensional bent).

## 15.2. THEORY

Consider a ray trace analysis of a crystal x-ray spectrometer. Let the reflecting planes with spacing $d$ be oriented parallel to the crystal surface, which has the form

$$f(x, y, z) = 0 \tag{15.2}$$

Let the detector plane be $xz$ ($y = 0$) (see Figure 15.2). The point source is placed at the origin. The left-handed system of coordinates was chosen only for figure clarity.

The smoothness of the crystal enables us to assume that function $f(x, y, z)$ is analytical and differentiable in the region and can be expressed in the form:

$$y = y(x, z) \tag{15.3}$$

Using the coordinate $x = x_1$ as a variable, let us find the following for a given wavelength $\lambda$ and a given order of reflection $n$:

1. Find the point $(x_1, y_1, z_1)$ on the crystal surface where the radial vector $r_1$ intersects it at the Bragg angle, $\theta_{Bragg}$,

$$\theta = \text{arc cos}\,\frac{n\lambda}{2d} = \frac{\pi}{2} - \theta_{Bragg} \tag{15.4}$$

2. Find the normal, $n_0$, to the crystal surface at the point $(x_1, y_1, z_1)$, and then find the intersection of this normal, $n_0$, with the detector plane $y = 0$.

3. Find the intersection of the ray reflected at $(x_1, y_1, z_1)$ at the Bragg angle with the detector plane $y = 0$.

If the reflecting surface is assumed to be cylindrical (but not necessarily a circular cylinder surface) this can be expressed as follows:

$$f(x, y, z) = f(x, y), \quad y(x, z) = y(x) \tag{15.5}$$

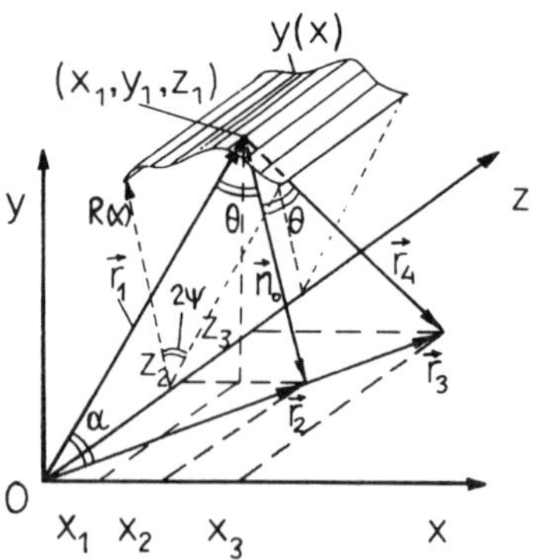


FIGURE 15.2.   Aberration calculation geometry (cylindrical bent, i.e., crystal shape $y(x)$ does not depend on $z$ coordinate). Here, O = point source, $r_1$ = incident ray, $n_0$ = normal to the crystal surface, $r_4$ = reflected ray, $xz$ = detector plane, $r_3$ = image point.

This condition considerably simplifies the solution. From Figure 15.2 we find that

$$y_1 = y(x_1), \quad x_2 = x_1 + y_1 \left. \frac{dx}{dy} \right|_{x_1} \tag{15.6}$$

and a normal vector, $n_0$, in our case is equal to

$$n_0^2 = y_1^2 + (x_2 - x_1)^2 = y_1^2 \left( 1 + \left( \left. \frac{dx}{dy} \right|_{x_1} \right)^2 \right) \tag{15.7}$$

According to the cosine theorem

$$r_2^2 = r_1^2 + n_0^2 - 2r_1 n_0 \cos\theta \tag{15.8}$$

or, in terms of rectangular coordinates,

$$x_2^2 + z_2^2 = x_1^2 + y_1^2 + z_1^2 + n_0^2 - 2n_0 \cdot \cos\theta \cdot \sqrt{x_1^2 + y_1^2 + z_1^2} \tag{15.9}$$

Rewriting (15.8) as follows ($z_2 = z_1$) we could find the value of the incident ray vector, $r_1$:

$$r_1 = \sqrt{x_1^2 + y_1^2 + z_1^2} = \frac{(x_1^2 + y_1^2 + n_0^2 - x_2^2)}{2n_0 \cos\theta} \tag{15.10}$$

and then find $z_1$ the coordinate of intersection of this vector with the crystal surface, and $r_2$, the sum of vector $r_1$ and normal $n_0$:

$$z_1 = \pm\sqrt{r_1^2 - x_1^2 - y_1^2}, \quad r_2 = \sqrt{z_1^2 + x_2^2} \tag{15.11}$$

Then, introducing $\alpha$ as the angle between $r_1$ and $r_2$ (see Figure 15.2),

$$\alpha = \text{arc cos} \left( \frac{r_1^2 + r_2^2 - n_0^2}{r_1 \cdot r_2} \right) \tag{15.12}$$

Using the sine theorem, we may find the length of vector $r_3$, which is the vector connecting the origin with the image point in the detector plane $xz$:

$$r_3 = \sqrt{x_3^2 + z_3^2} = r_1 \sin(2\theta) / \sin(2\theta - \alpha) \tag{15.13}$$

Image point coordinates — our goal — are then easily obtained from

$$x_3 = r_3 x_2 (x_2^2 + z_2^2)^{-1/2}, \quad y_3 = 0, \quad z_3 = z_2 x_3 / x_2 = r_3 z_2 (x_2^2 + z_2^2)^{-1/2} \qquad (15.14)$$

Since the problem has a mirror symmetry, both $z_2$ and $z_3$ may be positive or negative.

Substituting a shift of the origin

$$x'_1 = x_1 + \Delta x, \quad y'_1 = y(x_1 + \Delta x) \qquad (15.15)$$

In equation (15.5), we get the coordinates $(x'_3, z'_3)$ of the image for a source, $S'$ which is placed at an arbitrary point $(\Delta x, \Delta y, \Delta z)$ rather than at the origin:

$$x'_3 = x_3(x'_1) \qquad (15.16)$$

$$z'_3 = z_3(x'_1) + \Delta z \qquad (15.17)$$

Thus, for an arbitrary reflecting surface, $y(x)$ the locus of reflected rays intercecting the detector plane, $y = 0$, is the set of points $(x'_3, 0, z'_3)$ for a given wavelength $\lambda = 2d \cos\theta/n$. As $x_1$ varies within the crystal aperture, this set of image points $(x'_3, 0, x'_3)$ generates a trace curve on the detector plane. This curve is by definition the monochromatic image of a point source $S'$ on the detector plane. In other words, this curve, or spectral line in the detector plane, is the locus of all rays with given wavelength reflected by the crystal and crossing the detector plane.

If, for example, a flat crystal is placed at a distance of $y = 1$ from a source, then the ray trace for a Bragg angle of, say, $\theta_{Bragg} = \pi/4$ is a circle on the detector plane:

$$x^2 + z^2 = 4 \qquad (15.18)$$

and for other wavelengths it is a family of concentric circles:

$$x^2 + z^2 = (2\cot(\theta_{Bragg}))^2 \qquad (15.19)$$

Obviously, for another crystal shape the spectral line may be a curve of complex shape with loops, self-crossings, return points, gaps, etc.

## 15.3. SOME GEOMETRICAL ABERRATIONS OF THE VON HAMOS SPECTROMETER

Now we may apply the above trigonometry to the von Hamos spectrometer in axial detection geometry to find out that aberrations we may expect and what their influence will be on the spectral resolution. Let both point source 0 and detector plane $y = 0$ coincide with the spectrometer axis $z$ (Figure 15.1). For an ideal von Hamos spectrometer, the reflecting surface $y(x)$ is a right circular cylinder of radius $R$:

$$y = R\sqrt{1 - (x/R)^2} \tag{15.20}$$

In this case, all rays of a given wavelength from a point source are focused at one image point $(0, 0, 2R \cot \theta_{Bragg})$. A shift $\delta x$ of the source position from the spectrometer axis in the $x$ direction (i.e., perpendicular to the cylinder axis) causes an opposite shift $-\delta x$ of the image onto the detector plane. Aberration converts an image of the point source into a small parabola-like curve, the length of which is limited by the aperture of the spectrometer crystal (see Figure 15.3, 15.4). The coordinates of the blur curve center (i.e., image point $(x_i, y_i, z_i)$ for reflection from crystal center $x_1 = 0$) are given by the (15.16) and (15.17):

$$(x_i, y_i, z_i) = (-\delta x, 0, \sqrt{(2R \cot \theta_{Bragg})^2 - (2\delta x)^2}) \tag{15.21}$$

Thus, the image center shift is

$$(\delta x, \delta y, \delta z) = (-\delta x, 0, \sqrt{(2R \cot \theta_{Bragg})^2 - (2\delta x)^2} - 2R \cot(\theta_{Bragg}))$$

$$= (-\delta x, 0, -(\delta x)^2 / R \cot(\theta_{Bragg})) \tag{15.22}$$

Thus, from (15.22) it is seen that a von Hamos spectrometer gives an inverse image of an extended source, with aberrations of order $\Delta z$. This feature of a von Hamos spectrometer is used to make an image spectrogram of a small object. It was also used in laser plasma spectroscopy[4,5] and in micropinch discharge plasma investigations[4] to obtain plasma spectra having spectral and spatial resolution at the same time.

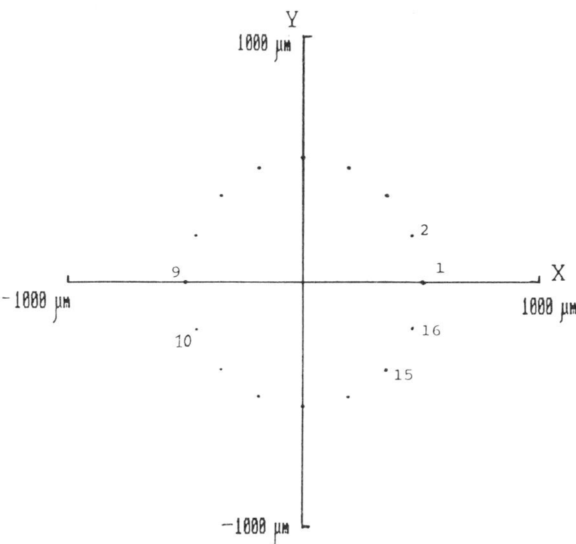


FIGURE 15.3. Sixteen source positions in the $xy$ plane to calculate aberrations of von Hamos spectrometer (Figure 15.1) in the detector plane $xz$. Distance from sources to the spectrometer axis is 0.5 mm.

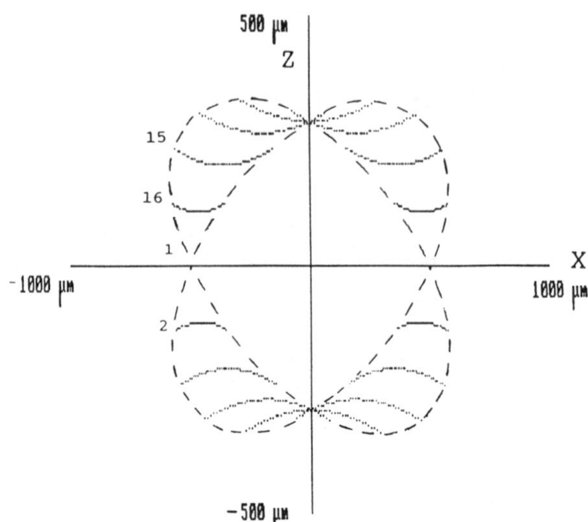


FIGURE 15.4.   Single wavelength images of the sources from Figure 15.3. for an ideal circular cylinder von Hamos spectrometer. Crystal radius $R$ = 100 mm, diffraction angle $\theta_{Bragg}$ = 60°, crystal aperture $\psi$ = 60°, cylinder axis is $z$ (see Figure 15.1). Coordinate $z$ is shifted on $2R \cot(\theta_{Bragg})$, dashed ellipses are images of family ellipses.

When a point source is not far away from the axis, the aberrations are small because, as is seen from (15.21) and (15.22), the blur curve size is squared proportionally to the square of shift $\delta x$.

Numerical calculations using the above formulas were performed for the source positions as shown in Figure 15.3 and for two spectrometer radii: $R$ = 100 mm and $R$ = 25 mm. These values are arbitrary to some extent, although we believe that a von Hamos-type compact spectrometer for an analytical electron microscope should have a crystal radius somewhere in between these two. For $R$ = 100 mm, Bragg angle of reflections $\theta_{Bragg}$ = 60°, and a crystal aperture (i.e., cylinder sector angle) $2\psi$ = 60°, the results of these calculations are shown in Figures 15.4 and 15.5. The shift of the source from the spectrometer axis was chosen to be 0.5 mm. This figure is again chosen arbitrarily, but we believe that this is a realistic estimation of a spectrometer/source alignment accuracy. The calculated blur curves are shown in Fig. 15.4. The parabola-like aberration curve remains small ($5 \times 2$ µm), while the source remains in the detector plane (Fig. 15.5). Since the dispersion axis of the spectrometer is the $z$ axis, the spectral resolution along this axis in this case remains very high: $\lambda / \Delta\lambda \simeq R / \delta z > 10^4$. A 5-µm blur in the $x$ direction is small, and spatial resolution in this case may also be considered relatively high.

If a source leaves the detector plane, the image curve immediately expands, in proportion to the first order of vertical shift. The size of the image is limited by the aperture of the crystal. For our example, the $\delta y$ = 0.5 mm shift of a source out of the

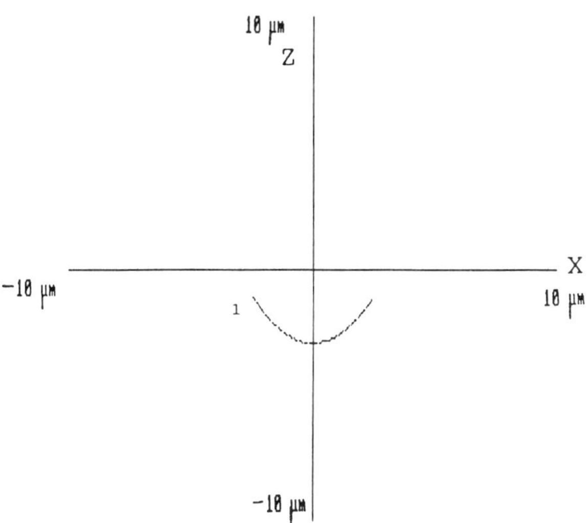


FIGURE 15.5.    Magnified image of the #1 source, which is in the detector plane $xz$ (source position is $x, y$ = 0.5, 0 mm).

detector plane causes the image curve to expand to a $520 \times 50$ µm size (see curve 5 or 13 in Figure 15.4). The spectral resolution decreases to $2 \times 10^3$, because neighboring wavelengths overlap along the $z$ axis. Nevertheless, this value is still high enough for an analytical WDS instrument. A spectrometer with such spectral resolution will be capable of resolving, for example, the $K\alpha$ doublet for elements above sulfur.

For detector plane shifted in the $y$ direction, aberrations for the same spectrometer are shown in Figure 15.6.

For a smaller radius spectrometer, as well as for a more glancing Bragg angle, the geometrical aberrations along the spectrometer axis $z$ become considerably greater [see (15.22)], whereas in the perpendicular direction $x$, the blur is about the same (with the exception of the case when the source is in the $y = 0$ detector plane), as shown in Figures 15.7 and 15.8.

For the von Hamos spectrometer with small radius, we must use thin crystals to avoid high stress and cracking while the crystal is being bent. The necessary thickness may be estimated using a simple mechanical model of a cylindrically bent crystal with thickness $s$ over radius $R$. Atoms on the interior of the crystal cylinder undergo compression along their surfaces; thus interatomic distance $a$ in this direction becomes $a(R - s/2)/R$ after bending, whereas on the outer surface, distance between atoms becomes $a(R + s/2)/R$. Thus the internal stress on either surface is linear proportional to $s/R$ ratio, and reducing $R$ requires a directly proportional reduction in the thickness of the crystal, $s$. If a 1-mm-thick crystal may be bent elastically over a 500-mm radius, then for a 25-mm radius spectrometer, we should use a 0.05-mm thick crystal. Usually the thickness-to-radius ratio, $s/R$, is close to $10^{-3}$ for safe bending of most crystals. For

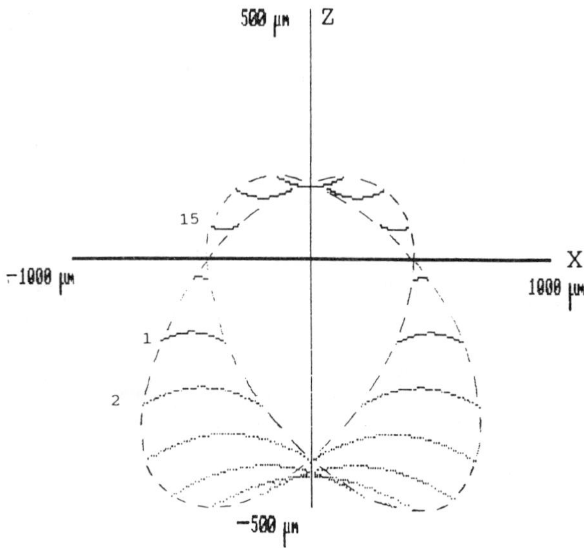

FIGURE 15.6.  Same aberrations as in Figure 15.4, but for the detector plane $xz$ shifted 0.25 mm in the $y$ direction.

our $R = 25$–$100$ mm von Hamos spectrometer, we must use 0.02–0.1-mm-thick crystals, which are more like films. These are not easy to manufacture with a uniform thickness across a large (centimeters) area. The uniformity is important, because any variation in thickness may cause variation of the radius along the surface of the bent

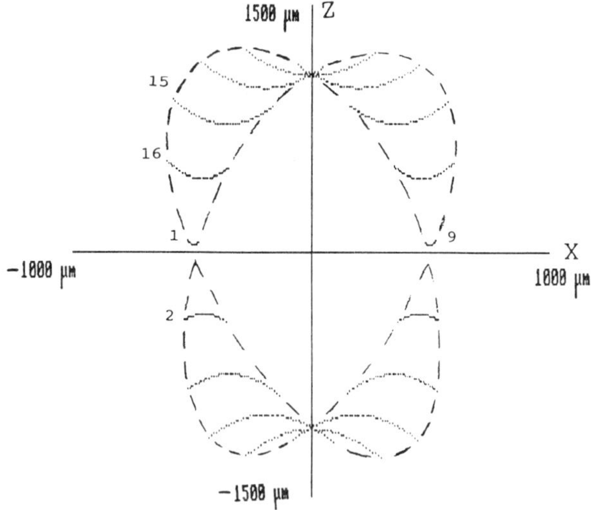

FIGURE 15.7.  Same aberrations as in Figure 15.4, but for a smaller, spectrometer radius of $R = 15$ mm and a more glancing Bragg angle, $\theta_{Bragg} = 25^\circ$.

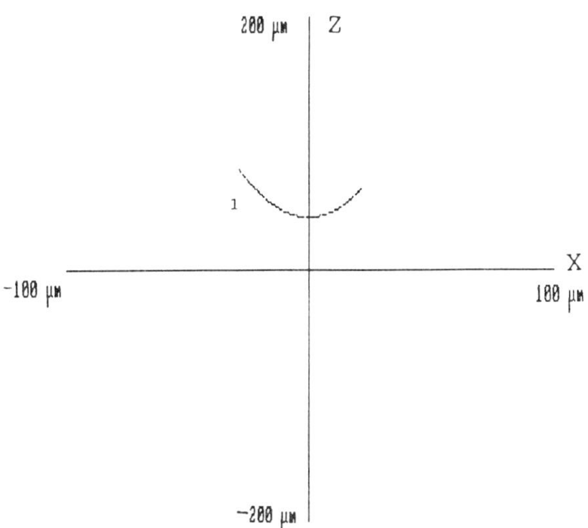


FIGURE 15.8.   Same aberrations of a source in the detector plane as in Figure 15.5, with $R = 15$ mm, $\theta_{Bragg} = 25°$.

crystal. This variation will cause strong geometrical aberrations of the image in the spectrometer focus. In addition, such thin crystals can easily be deformed beyond their elasticity limit while being handled, also causing residual stress, radius variation, and image aberrations.

Because of the importance of crystal shape to proper focusing, let us consider the aberrations of a von Hamos spectrometer due to small deformations of the reflecting crystal. Thin crystals are often slightly wavy because of residual stress; this is especially true for soft crystals. When they are bent into a circular cylinder, these defects cause geometrical aberrations of the image spot in addition to those aberrations considered above. Because the spectrometer focuses (or collects) reflected radiation in the $x$ direction, waviness (or deviations from the circular cylinder) of the crystal along the $x$ direction causes much higher image aberration onto the detector plane than does waviness along the $z$ direction.

Waviness can be modeled by adding an x-sine wave to the circular cylinder surface in our model:

$$y = R\sqrt{1 - x^2/R^2} + B \sin (2\pi Mx / R) + A \qquad (15.23)$$

Here $R$ is the crystal radius, $B$ is the amplitude of the waviness of the crystal surface, $M$ is the density (or frequency) of these waves, and $A$ is a possible shift of the spectrometer axis with the detector plane $xz$. The shape of the aberrant curves or spots are shown in Figure 15.9 for the same source positions and spectrometer ($R = 100$ mm) as in Figure 15.3. The amplitudes of the waviness are 3, 10, and 30 μm for Figures

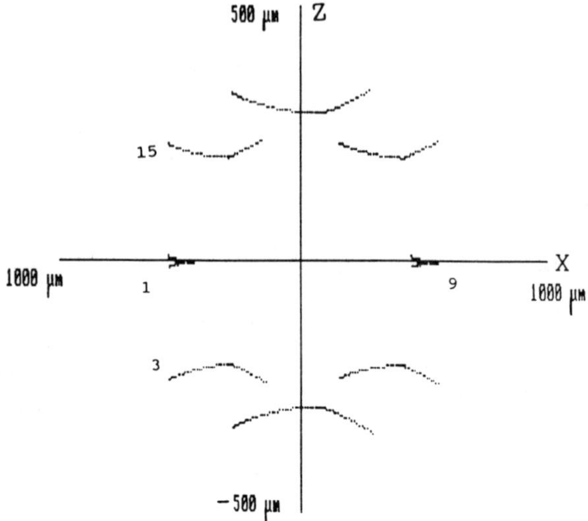

FIGURE 15.9  Geometrical aberrations of a slightly wavy crystal. Same spectrometer and source positions as in Figures 15.3 and 15.4. (a) 3 μm waviness of the crystal surface.

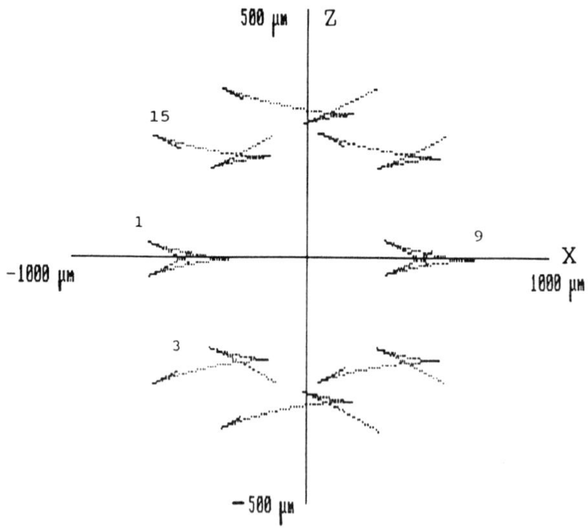

FIGURE 15.9.  (b) 10 μm waviness of the crystal surface.

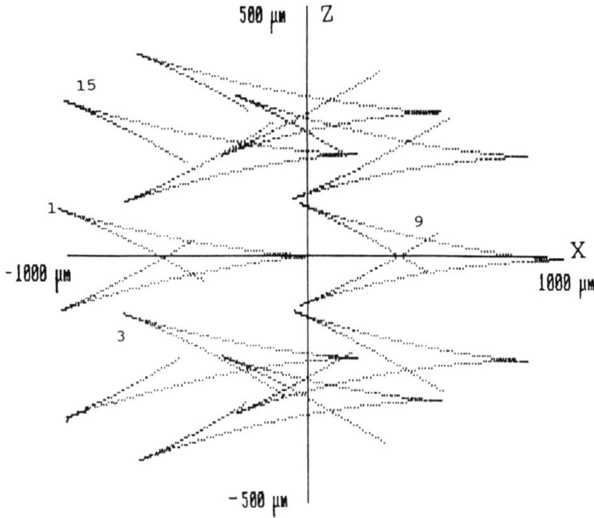



FIGURE 15.9.   (c) 30 μm waviness.

9(a), 9(b), and 9(c), respectively, and frequency of waviness $M$ is 1.5 periods for a crystal aperture of $\psi = 60°$. These figures are also arbitrary to some extent; the real crystal may have smaller or larger (and non-periodic) waviness. Nevertheless, calculated aberrations are quite general and such modeling is good for demonstration purposes. The image curve in this case begins to look like a star or swallow-tail shaped curve with a number of return points. This number of return points in our model is $2M = 3$.

If the source is polychromatic, then the stars of different wavelength overlap each other in the dispersion axis direction $z$ and the spectral resolution depends on the size of the star along the dispersion axis $z$. This size is proportional to the amplitude $B$ and frequency $M$ of the crystal waviness. Spectral resolution $P$ for our particular spectrometer remains higher than $P = 10^3$ until the amplitude $B$ is smaller than the value $R/2PM$, which, in our case, is 15 μm. Spectral resolution of the real spectrometer will depend, of course, on deviations of the real crystal from an ideal cylinder.

In general, any crystal conventionally used in x-ray wavelength dispersive spectroscopy may be used for a small-radius von Hamos-type spectrometer. Mica is a suitable crystal to bend over small radii, and thin pieces of mica can be bent safely over radius $R = 15$ mm or even less. The disadvantage of mica is that it has high reflectivity over higher orders of diffraction (up to the 11th) compared to its reflectivity in the first and second orders.

For detection of light elements (Be–F) either molecular crystals or multilayers may be used. For multilayer reflectors the spectral resolution $P$ is essentially less than for crystals and is approximately equal to the number of effectively reflecting layer pairs, $N$. Because of absorbtion, $N$ is typically in the 20–100 range, depending on the

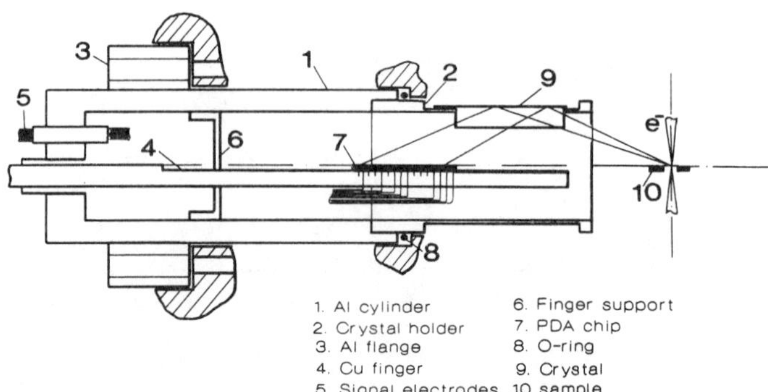


FIGURE 15.10.    Design of the von Hamos spectrometer for analytical electron microscope or microanalyzer.

wavelength and on the multilayer design. Nevertheless, multilayers offer a good reflectivity (10–50%) in the soft x-ray band.

## 15.4. SPECTROMETER DESIGN

A von Hamos spectrometer design for an electron microscope or a microanalyzer is shown in Figure 15.10. A spectrometer consists of a cylindrical case (Item1) with a flange (Item 3), a changeable crystal (or multilayer) holder (Item 2) with mounted crystal (Item 9). A one-dimensional CCD or PDA array (Item 7) seems to be a suitable detector for a small-radius ($R = 15$–50 mm) von Hamos WDS. Unfortunately, the intensity of x-ray flux from a sample in a microscope or microanalyzer is 6–8 orders of magnitude lower than the intensity of conventiomal x-ray tubes (due to lower current and thinner target). Thus, the detector should have very low noise, i.e., should be able to work in a single-quanta detection mode (as proportional counters do). To achieve this, the existing silicon-based CCD or PDA detector should be cooled down to $-70°C$ or lower. Because such temperatures are difficult to achieve by thermoelectric coolers, liquid nitrogen cooling should be used. In our design, the PDA chip is mounted on a cold finger (Item 4). Although it is desirable to have a preamplifier in the same detector chip, we could not find an affordable PDA with a built-in preamplifier. A 1024-pixel, one-dimentional PDA with $25 \times 2500$ μm pixel size will be used for a prototype spectrometer. The integration time of the spectrometer will depend on the intensity of x-ray source and may be an hour or more. A 486 PC will be used to collect and process data.

Because of the imaging property of von Hamos spectrometers, the diameter of the focused microbeam or the area scanned by the beam during collection time should be smaller than the detector pixel size (25 μm for most CCDs or PDAs) to avoid degradation in spectral resolution. For a spectrometer with a multilayer this requirement is not as strict because of a wide rocking curve: for a multilayer spectrometer with $R = 15$ mm, the source size could be as large as 200–300 µm.

## REFERENCES


1. J. I. Goldstein, D. E. Newburg, P. Echlin, D. C. Joy, C. E. Fiori, and E. Lifshin, *Scanning Electron Microscopy and X-ray Microanalysis*, Plenum Press, New York (1981).
2. G. F. Bastin and H. J. M. Heijligers, *X-ray Spectroscopy* **15,** 135 (1986).
3. L. von Hamos, *Naturwissenschaften* **20,** 750 (1933).
4. A. P. Shevelko, *Quantum Electronics* (Sov.) **4,** 2013 (1977).
5. B. Yaakobi, R. E. Turner, H. W. Schnopper, and P. O. Taylor, *Rev. Sci. Instrum.* **50,** 1609 (1979).
6. A. M. Panin, in: *Proceedings of the 10th International Colloquium on UV and X-Ray Spectroscopy of Astrophysical and Laboratory Plasmas* (E. Silver and S. Kahn, eds.) Cambridge University, Cambridge, UK, p. 575 (1993).

# 16

# Fitting Wavelength Dispersive Spectra with the NIST/NIH DTSA Program

*R. L. Myklebust*

## 16.1. INTRODUCTION

Labar has shown that relative x-ray line intensities within an x-ray family are important for both qualitative and quantitative microprobe analysis.[1] However, the measurement of many of these relative intensities is difficult due to peak overlaps. X rays generated in solid materials by the electron beam in an electron probe microanalyzer may be measured either by an energy dispersive spectrometer or by a wavelength dispersive spectrometer. Although the EDS spectrum can contain all of the x-ray energies from zero to the energy of the electron beam, only a relatively small number of channels in the spectrum may be from the particular characteristic x-ray line of interest. The detector electronics broaden the line into a Gaussian- shaped peak whose width is so broad that in many cases it will overlap other x-ray peaks from the same element or other elements in the specimen. A WDS uses a diffracting crystal to separate the x rays by wavelength according to Bragg's Law:

$$n\lambda = 2d \sin \theta \tag{16.1}$$

where $n$ is the order of the diffraction, $\lambda$ is the x-ray wavelength, $d$ is the crystal spacing and $\theta$ is the diffraction angle. The WDS, therefore, measures only one wavelength at a time (actually, because of crystal imperfections, a narrow band of wavelengths, approximately 10 eV in width for a LiF crystal) and must be mechanically scanned as a function of $\theta$ to obtain a spectrum of the x-ray line of interest. Since all of the WDS

R. L. MYKLEBUST • National Institute of Standards and Technology, Gaithersburg, Maryland 20899.

*X-Ray Spectrometry in Electron Beam Instruments*, edited by David Williams, Joseph Goldstein, and Dale Newbury. Plenum Press, New York, 1995.

measurements are from the peak of interest, no time is wasted measuring x rays outside of the peak, and a significantly larger number of x rays are counted than with the EDS. In addition, because the pulse width for WDS is less than 1 μs, as compared to 10–50 μs for EDS, the electron beam current may be increased to obtain an even higher count with the WDS. At high beam currents, the EDS system dead time becomes so high that the EDS may shut off entirely. There are, therefore, two principal benefits of using a WDS for measuring x-ray peaks: first, the superior peak resolution for separating overlapped peaks, and second, the much higher number of x rays counted in the peak of interest. This double benefit should significantly increase the statistical accuracy of the measurements as was demonstrated by Fiori.[2]

## 16.2. PEAK FITTING

Whatever spectrometer is used for x-ray measurements, the area of the peak of interest must be extracted from the data. The Desk Top Spectrum Analyzer (DTSA) program, developed by C. E. Fiori, C. R. Swyt, and R. L. Myklebust, was originally designed for displaying energy dispersive x-ray spectra, performing mathematical analyses on these spectra, and generating simulated spectra from basic physics. The

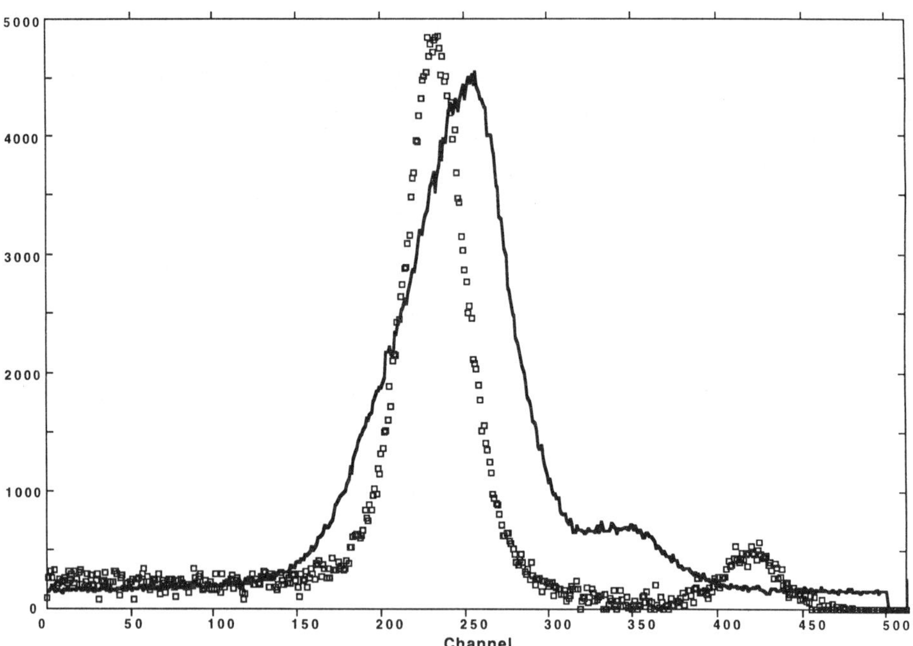

FIGURE 16.1.   Solid curve is a WDS spectrum of first order Au $L\alpha_1$ and $L\alpha_2$ measured with a LiF crystal on a Cameca microprobe with a beam energy of 25 keV. Dotted curve is a WDS spectrum of second order Au $L\alpha_1$ and $L\alpha_2$ measured under the same conditions. There are 500 data points in each spectrum measured by G. Remond.

program operates as a multichannel analyzer with EDS spectra displayed on an energy scale (usually 10 eV per channel). The horizontal axis is therefore linear in energy (usually with channel zero equal to 0 eV), so that the characteristic x-ray peaks will be displayed at the correct energy positions in the spectrum. Two peak fitting procedures are available in DTSA: a linear least squares procedure that finds the peak area by scaling a reference spectrum of the peak of interest to the measured peak, and a non-linear sequential simplex procedure that finds the peak area by iteratively comparing a mathematical peak model to the measured peak. If there are overlapping peaks from the same element, the reference spectrum will contain all of them and the area found will be the sum of all the peaks in the reference for that element. Since the areas of individual peaks were required for this study, the sequential simplex procedure was used because it produces separate areas for each member of a family of lines.

### 16.2.1. Wavelength-to-Energy Scale Conversions

While x-ray spectra collected with a WDS have the same physical characteristics as x rays collected with an EDS, the measurements are not linear with energy and the peak shapes are different. In addition, the spectrum energy scale will be reversed if the spectrometer was scanned from low to high wavelength, the usual procedure. Thus, the first step is to convert the WDS spectrum to a linear energy scale similar to that used for EDS spectra, so that we can use all of the EDS spectrum handling, energy markers, and

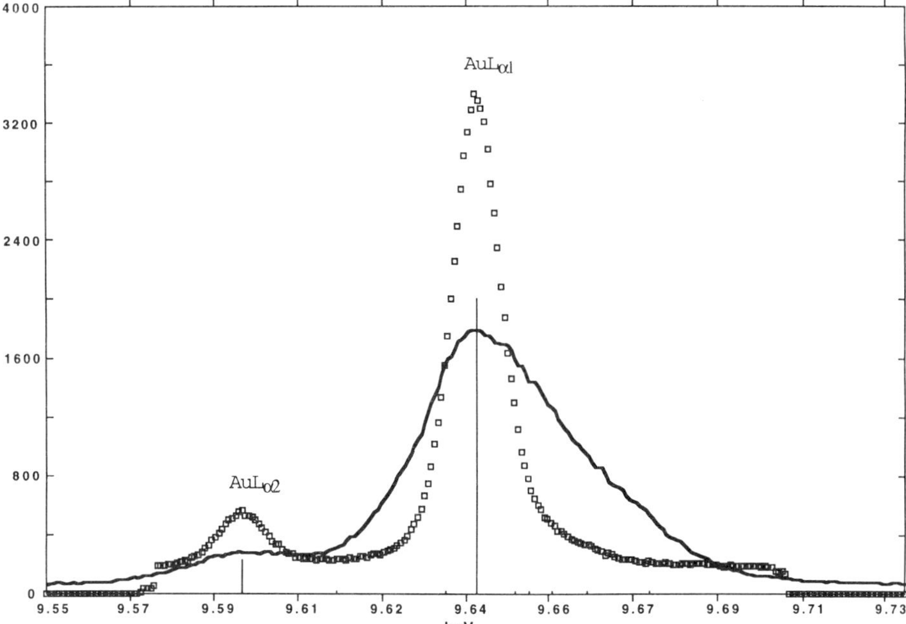

FIGURE 16.2. The two spectra in Figure 16.1 are shown after conversion to the energy scale. The dotted curve is second order and the solid curve is first order.

fitting routines with only slight modifications. Reversing the spectrum is an easy matter; however, converting the non-linear energy scale obtained by this operation is not trivial. The WDS spectrum is linear in $\theta$ as shown in Figure 16.1; therefore, the number of counts in any channel in the spectrum must be properly distributed into the appropriate channels in an $I_E$ versus $E$ spectrum that is linear in energy. Ideally, the non-linear $I_\theta$ versus $\theta$ spectrum should be divided by taking equal energy integrals of the $I_\theta$ versus $\theta$ spectrum. We perform an approximation to this by assuming the counts in a non-linear ($\theta$) channel are evenly distributed across the energy range of the channel and then dividing the counts into the linear spectrum channels. While this procedure is not perfectly accurate it is a reasonable first approximation. The conversion is done with Bragg's Law (16.1) to convert the $\theta$ values to wavelength, $\lambda$, for each of the data channels. The wavelengths are then converted to energy with the following equation:

$$E\,(\mathrm{keV}) = 12.398\,/\,\lambda \tag{16.2}$$

The data at each of the energies are then proportionally divided between equally spaced energy channels to produce spectra that are linear in energy, as shown in Figure 16.2. All of the functions that are used to extract information from an EDS spectrum can now be applied to the WDS spectrum that has been transformed to a spectrum that is linear in energy.

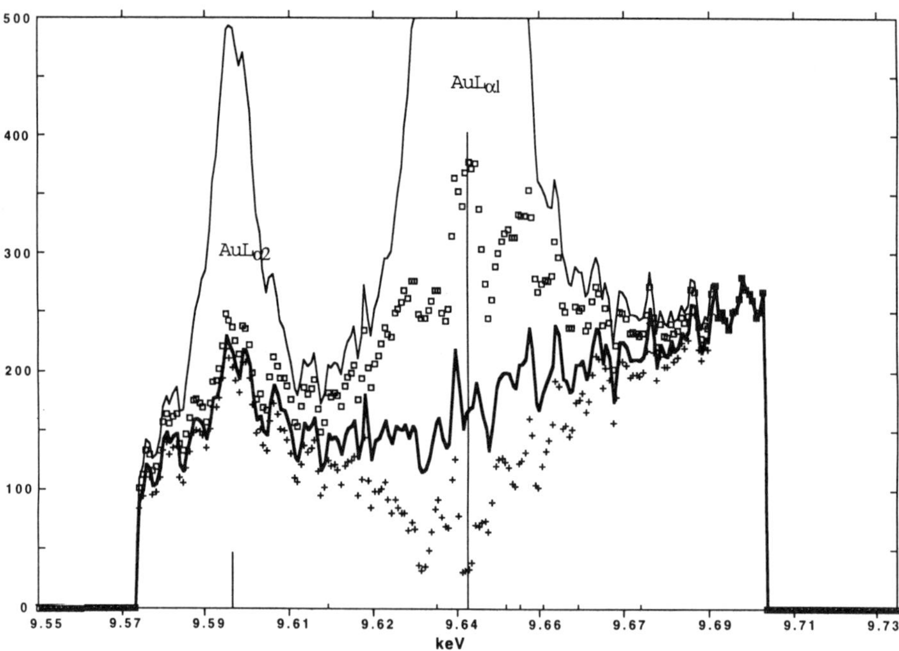


FIGURE 16.3.   The second order Au $L\alpha_1$ and $L\alpha_2$ peaks at 12.5 keV fitted with different hypermet numbers. ++ line has hypermet = 0.4, solid line has hypermet = 0.55 and squares line has hypermet = 0.8. The peak resolution is 19 eV.

Figures 16.1 and 16.2 show the gold $L\alpha_1$ and $L\alpha_2$ x-ray lines measured by first order diffraction and second order diffraction. The separation of the $L\alpha_1$ and $L\alpha_2$ lines can easily be seen in the first order spectrum (Figure 16.1) and the two lines are completely separated in the second order spectrum (Figure 16.2). In this case the same diffracting crystal was used for both first ($n = 1$) and second ($n = 2$) order. Therefore, the $\sin\theta$ for the second order must be twice that of the first order, resulting in a wider separation of the two x-ray peaks. The reason for this can be demonstrated by differentiating 16.1:

$$n\Delta\lambda = 2d \cos \theta \, \Delta\theta \tag{16.3}$$

and substituting for $2d/n$ from (16.1):

$$\Delta\theta = \Delta\lambda \sin \theta / (\lambda \cos \theta) = \tan \theta \, \Delta\lambda / \lambda \tag{16.4}$$

If $\lambda$ is the average wavelength of two peaks, and $\Delta\lambda$ is the difference in wavelength of the two peaks, then $\Delta\theta$ is the difference in the spectrometer angle of the two peaks. Since $\Delta\lambda / \lambda$ will be constant for any two given peaks, $\Delta\theta$ varies as $\tan\theta$. Therefore, the best spectrometer resolution is obtained at high diffraction angles. Second order diffraction is one method to increase the spectrometer resolution and, if possible, another method is to use a crystal with a smaller d-spacing. Unfortunately,

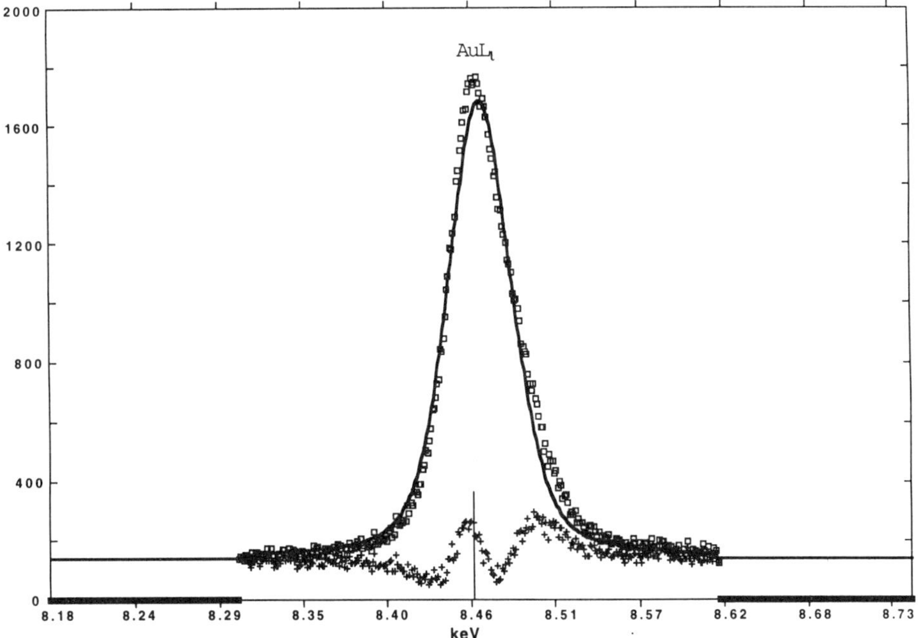


FIGURE 16.4.   The Au $Ll$ peak at 25 keV fitted with hypermet = 0.55 and resolution = 0.6. The squares are the data, the ++ are the residuals, and the solid curve is the resulting fitted peak.

because of the smaller solid angle of the spectrometer, higher order diffraction is less efficient, which reduces the number of x rays measured and makes longer counting times or higher beam currents necessary to obtain the same statistics as first order diffraction.

### 16.2.2. Peak Shape

The sequential simplex peak fitting procedure in the DTSA program was originally designed to fit Gaussian peak shapes to the x-ray peaks in an EDS spectrum, as described by Fiori.[3] Since WDS peaks are not simply Gaussian in shape, a Lorentzian and a Gaussian must be combined to obtain a peak shape that approximates a WDS x-ray peak, as demonstrated by Remond.[4]

$$P_i = xG_i + (1 - x)L_i \qquad (16.5)$$

In (16.5), $P_i$ is the intensity of the combined peak at channel $i$, $G_i$ is the intensity of a pure Gaussian at channel $i$, $L_i$ is the intensity of a pure Lorentzian at channel $i$, and $x$ is the "hypermet number," or the fraction of the combined peak that is Gaussian. The hypermet number must be determined by the user by fitting peaks that have no overlaps with any other peak. Different hypermet numbers are tested until the best fit is obtained, as determined by the residuals of the fit. The data shown in Figure 16.3

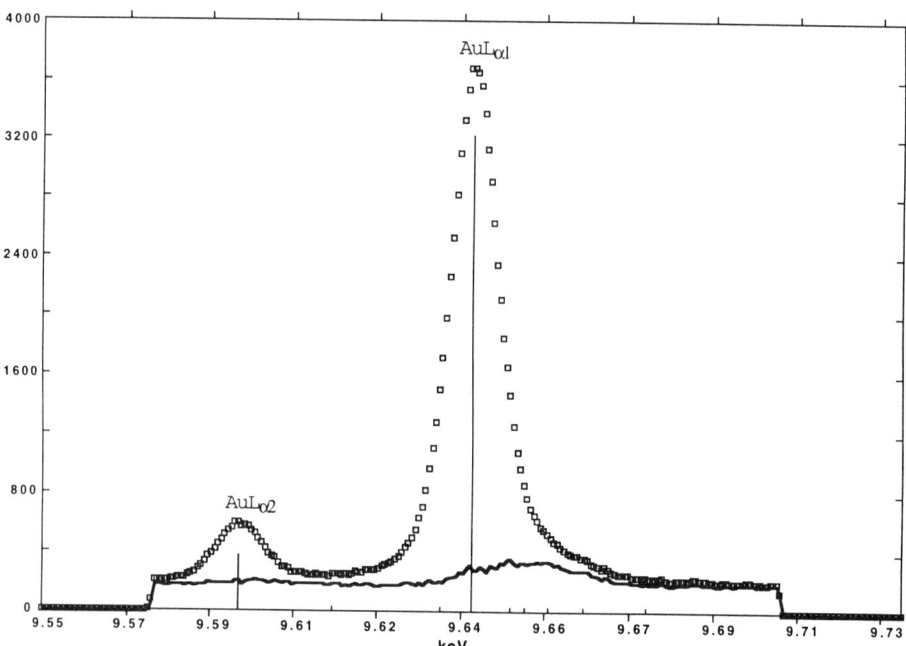


FIGURE 16.5.   (a) The second order Au $L\alpha_1$ and $L\alpha_2$ peaks at 25 keV fitted without satellite lines. The solid curve shows the residuals of the fit. The peak resolution is 19 eV.

are the residuals for three different fits to the gold $L\alpha_1$ and $L\alpha_2$ peaks showing the effect of varying $x$ in (16.5). It may also be necessary to vary the starting peak resolution since a resolution that is too far off may produce a false result. To minimize a tedious process, however, both of these parameters may be varied within the simplex setup of the DTSA program. The entire procedure must be repeated for each crystal on each spectrometer at each wavelength since both the peak resolution and the hypermet number may vary for each setup. The hypermet number seems to remain reasonably constant for a particular crystal; however, the resolution varies with $\tan\theta$, as shown by (16. 4). Figure 16.4 is a first order gold $L1$ peak at 25 keV, together with the residuals and the fitted peak. There are no minor x-ray peaks or satellite peaks near this peak. The variations in the residuals are due to crystal imperfections in the spectrometer that become more predominant as the spectrometer is moved to lower $\theta$ values.

### 16.2.3. Minor Family Members

Because of the high resolution and sensitivity of WDS, the minor x-ray peaks and the satellite peaks must be included in the simplex fit if the beam energy is sufficiently high to allow them to be excited. The WDS spectrum shown in Figure 16.5(a) is Au $L\alpha_1$ and $L\alpha_2$ measured at 25 keV, together with the residuals of a fit done without including satellite peaks and with $x = 0.6$. Figure 16.5(b) shows the residuals

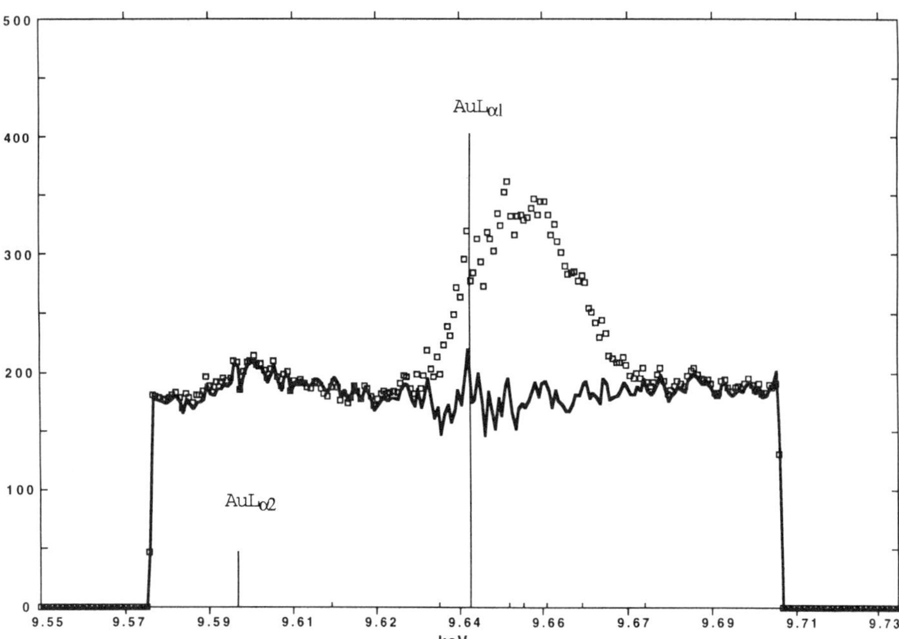

FIGURE 16.5.   (b)The second order Au $L\alpha_1$ and $L\alpha_2$ peaks at 25 keV were fitted with and without including satellite lines.The solid curve shows the residuals of the fit with satellite lines and the □□ curve shows the residuals without satellites.

TABLE 16.1.  Fit Results at 25 keV

| X-ray line | Integrated counts (with satellites) | Integrated counts (without satellites) | Energy (eV) |
|---|---|---|---|
| Au $L\alpha_1$ | 60007 | 59327 | 9711.8 |
| Au $L\alpha_2$ | 6705 | 6751 | 9625.9 |
| Au $L\alpha_3{}^{\wedge}{}_z$ | 414 | — | 9700.5 |
| Au $L\alpha\alpha$ | 95 | — | 9774.9 |
| Au $L\alpha'$ | 385 | — | 9731.0 |
| Au $L\alpha_s$ | 0 | — | 9660.4 |
| Au$L\alpha^{\wedge}{}_{ix}$ | 1117 | — | 9739.5 |
| Au $L\alpha^{\wedge}{}_x$ | 668 | — | 9752.5 |
| Au $L\alpha^{\wedge}{}_y$ | 1016 | — | 9725.7 |
| Sum | 70407 | 66078 | |
| $L\alpha_2 / \Sigma L\alpha_1 + L\alpha_2$ | 0.100 | 0.102 | |

for the same spectrum fitted with and without including the satellite peaks. The hump in the residuals of Figures 16.5(a) and 16.5(b) occurs in the energy range of most of the Au $L\alpha$ satellite lines. Table 16.1 contains the integrated counts for each of the peaks for the two fits. In this case, the Au $L\alpha_1$ and $L\alpha_2$ counts for the two fits are very similar but the total of all peaks differs due to the presence of the satellite peaks.

Satellite peaks arise from multiple inner shell ionizations (mostly double ionizations). There are two principal processes for doubly ionizing an atom.[5,6]. In the first process, the primary beam electron knocks out two electrons from the inner shells. For the $L\alpha_1$ and $L\alpha_2$ lines, the electrons would be removed from the $L_{III}$ and $M_{IV}, M_V$ shells if the beam energy is greater than the sum of the energies of the $L_{III}$ and $M_{IV}, M_V$ shells. In the second process, the primary electron knocks out one electron from the $L_I$ shell followed by a Coster-Kronig transition of an $L_{III}$ shell electron with the emission of an Auger electron from the $M_{IV}$ or $M_V$ shell. In this case, the difference in energy between the $L_I$ and $L_{III}$ shells must exceed the sum of the energies of the $L_{III}$ and $M_{IV}, M_V$ shells. The doubly ionized atom then emits a satellite x ray in the same way as the first method. According to Richtmyer and Ramberg,[7] there are 29 satellites in the gold $L\alpha$ spectrum. We have made no attempt to include all of these, but have tried to select the ones reported to be predominant.

Since Coster-Kronig transitions are more likely to occur for gold than double ionization by primary beam electrons, reducing the beam energy to a level that will not excite the $L_I$ or $L_{II}$ shells will eliminate all Coster-Kronig transitions and greatly reduce the intensities of the satellite peaks.[8] Figure 16.6(a) shows a Au $L\alpha_1$ and $L\alpha_2$ spectrum measured at 12.5 keV, together with the residuals of a fit done without including satellite peaks and with $x = 0.6$. Figure 16.6(b) shows the same spectrum with the residuals for a fit that does include the satellite peaks (even though satellites should not appear in this case as the beam energy is too low), together with the residuals for a fit that excludes the satellite peaks. The $L_I$ shell for gold is 14,353 eV and the sum of $L_{III}$ and $M_{IV}, M_V$ shells is 14,125 eV. Notice that the residuals from these two fits are almost the same with no evident hump in Figure 16.6(a); however, the fit

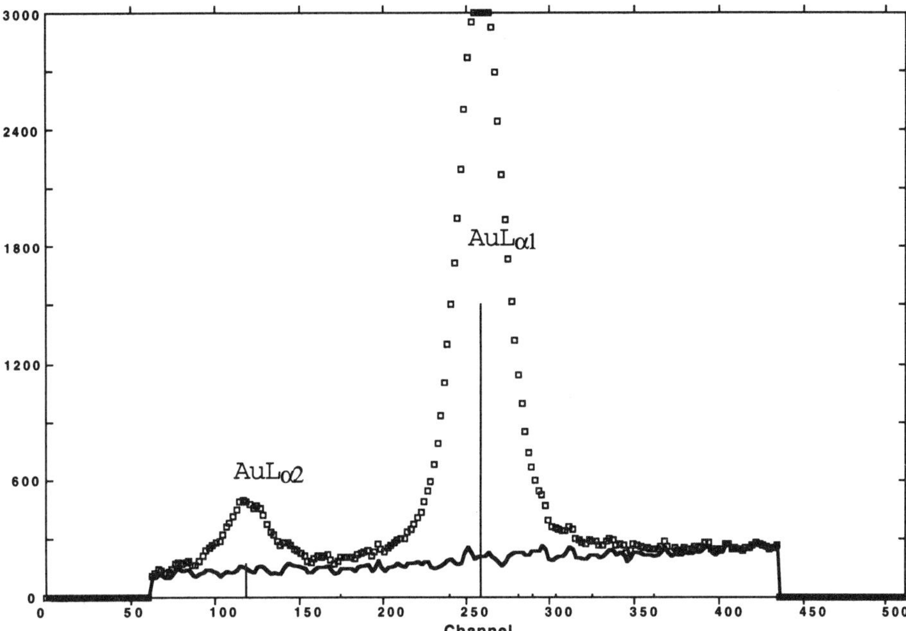


FIGURE 16.6. (a) The second order Au $L\alpha_1$ and $L\alpha_2$ peaks at 12.5 keV fitted without satellite lines. The solid curve shows the residuals of the fit. The peak resolution is 19 eV.

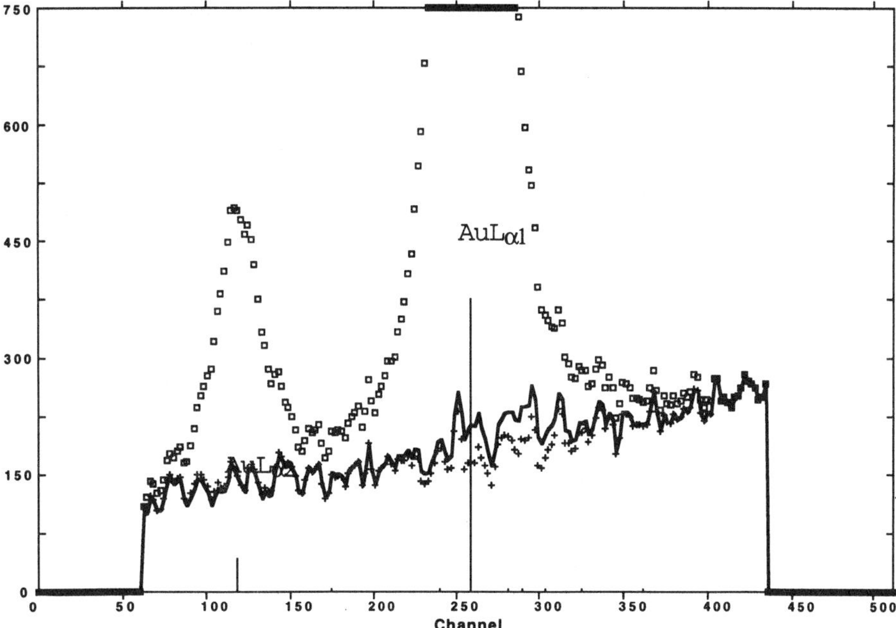


FIGURE 16.6. (b) The second order Au $L\alpha_1$ and $L\alpha_2$ peaks at 12.5 keV fitted with satellite lines. The solid curve shows the residuals of the fit without satellite lines and the □□ curve shows the residuals with satellites.

TABLE 16.2.  Fit Results at 12.5 keV

| X-ray line | Integrated counts (with satellites) | Integrated counts (without satellites) | Energy (eV) |
|---|---|---|---|
| Au $L\alpha_1$ | 51765 | 51990 | 9711.8 |
| Au $L\alpha_2$ | 5770 | 5769 | 9625.9 |
| Au $L\alpha_3{}^{\wedge}{}_z$ | 59 | — | 9700.5 |
| Au $L\alpha\alpha$ | 70 | — | 9774.9 |
| Au $L\alpha'$ | 856 | — | 9731.0 |
| Au $L\alpha_s$ | 3 | — | 9660.4 |
| Au $L\alpha^{\wedge}{}_{ix}$ | 459 | — | 9739.5 |
| Au $L\alpha^{\wedge}{}_x$ | 349 | — | 9752.5 |
| Au $L\alpha^{\wedge}{}_y$ | 980 | — | 9725.7 |
| Sum | 60311 | 57759 | |
| $L\alpha_2 / \Sigma L\alpha_1 + L\alpha_2$ | 0.100 | 0.100 | |

including satellites, as shown in Figure 16.6(b), has actually caused a dip in the residuals. Table 16.2 contains the integrated counts for each of the peaks for the two fits. In this case, the Au $L\alpha_1$ and $L\alpha_2$ counts for the two fits are still similar, and the total of all peaks is again different due to the inclusion of the satellite peaks in the fit. The values obtained for satellite peaks in this fit are due to artifacts in the fitting procedure. This agrees with the theory for the generation of satellites; however, no claim is made for the intensities found since the resolution of the spectrometer used is not high enough to allow the satellite peaks to be completely resolved. It may be possible to obtain some meaningful data by this method, but many more measurements will be required. The values of $L\alpha_2 / \Sigma L\alpha_1 + L\alpha_2$ listed in Tables 16.1 and 16.2 are all similar; therefore, results obtained for the diagram x-ray lines may be satisfactory.

## 16.3. CONCLUSIONS

The procedures involved here are not trivial, and fitting the WDS spectra to obtain the best results requires great care both in data collection and in actual fitting. The DTSA program has allowed us to process this data in a manner that while not simple, is at least manageable. The results of this work will enable us to better determine the relative x-ray line weights for $L$- and $M$-series lines and possibly even some of the satellite lines. There are also important lessons for the analyst who is using WDS as an analytical tool. At low $\theta$ angles, the intensity will be greater than at high $\theta$ angles because of the larger spectrometer solid angle; however, the peak resolution will be much better at the higher $\theta$. In addition, the peak profiles may be more irregular at the low $\theta$ angles, because crystal imperfections become more evident as both the solid angle and the resolution increase. These effects become important for analysis of minor elements when the analysis peak is overlapped by a peak from a major element.

## REFERENCES


1. J. L. Lábár, C. E. Fiori, and R. L. Myklebust, *Microbeam Anal.*, **2**, 169 (1993).
2. C. E. Fiori and C. R. Swyt, in: *Proceedings of the 50th Annual Meeting of the Electron Microscopy Society of America* (G. W. Bailey, J. Bentley, and J. A. Small, eds.) p. 1636 (1992).
3. C. E. Fiori, R. L. Myklebust, and K. Gorlen, NBS Special Publication 604, National Bureau of Standards, Washington, D.C., p. 233 (1981).
4. G. Remond, Ph. Coutures, G. Gilles, and D. Massiot, *Scanning Microsc.* **3**, 1059 (1989).
5. N. A. Dyson, *X-Rays in Atomic and Nuclear Physics*, Cambridge University Press, Cambridge, UK (1990).
6. E. H. Burhop, *The Auger Effect and Other Radiationless Transitions*, Cambridge University Press, Cambridge, UK (1952).
7. F. K. Richtmyer and E. G. Ramber, *Phys. Rev.* **51**, 925 (1937).
8. R. D. Deslattes, *Aust. J. Phys.* **39**, 845 (1986).

# 17

# Application of Layered Synthetic Microstructure Crystals to WDX Microanalysis of Ultra-light Elements

*R. Rybka and R. C. Wolf*

## 17.1. INTRODUCTION

In the last several years, the use of layered synthetic microstructure (LSM) crystals has resulted in considerable improvement in WDX analysis of light elements. The advantages of LSM crystals compared to conventional lead stearate or lead octodecanoate crystals have been discussed and presented by other investigators, e.g., Bastin and Heijligers.[2–6] Layered synthetic microstructure crystals are fabricated by vapor deposition of alternate layers of heavy and light elements onto a highly polished substrate. Some of the more commonly used types include: W/Si (LSM-060; $2d \approx 60$ Å), Ni/C (LSM-080; $2d \approx 80$ Å), and Mo/B$_4$C (LSM-200; $2d \approx 200$ Å). Of these, the LSM-080 is most versatile, since it provides a wide wavelength range ($\approx 22$–$72$ Å) with good performance for O, C, and B, as well as better performance for N than is possible with stearate-type crystals. Another advantage of this crystal is its ability to suppress higher orders of high energy x-ray lines. The LSM-060 has a narrower wavelength range ($\approx 17$–$55$ Å), but provides better spectral resolution for x-ray peaks of O, N, and C as well as better sensitivity for O and N. However, it does not suppress higher order x-ray lines as well as LSM-080. The LSM-200 crystal enables the analysis of Be and provides the highest sensitivity for B.

It is the objective of this work to illustrate the use of LSM crystals for several important applications requiring microanalysis of ultra-light elements and to report performance results from several new LSM crystals.

R. RYBKA AND R. C. WOLF • Microspec Corporation, Fremont California 94539

*X-Ray Spectrometry in Electron Beam Instruments*, edited by David Williams, Joseph Goldstein, and Dale Newbury. Plenum Press, New York, 1995.

## 17.2. EXPERIMENTAL RESULTS

Data reported in this paper were measured using a Microspec WDX-3PC wavelength dispersive spectrometer mounted to a Leica Stereoscan S200 SEM at a 35° angle of inclination.

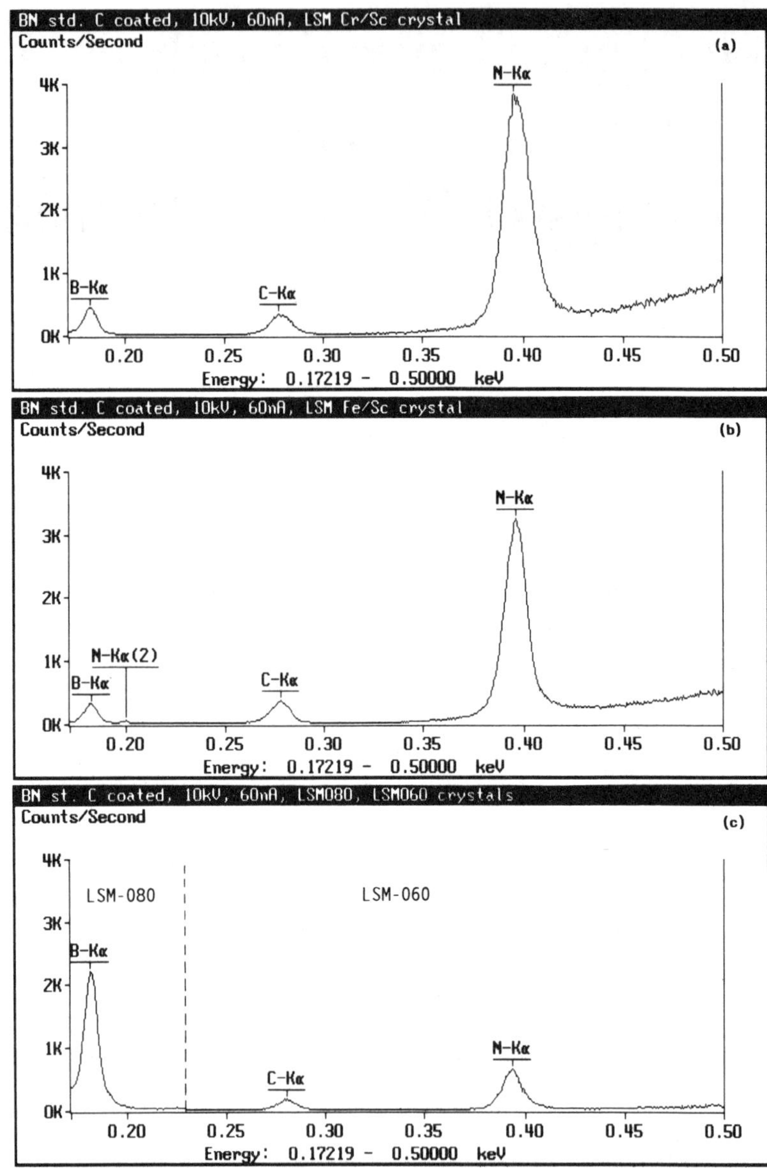


FIGURE 17.1.   Spectra recorded on BN using (a) LSM Cr/Sc, (b) LSM Fe/Sc, (c) LSM-080 and LSM-060 crystals. SEM electron beam voltage: 10 kV; sample current: 60 nA.

TABLE 17.1.   Comparison of Performance for N $K\alpha$ Measured from Various LSM Crystals

| Crystal type | Peak intensity (CPS/$\mu$A) | $(P - B)/B$ Ratio | Resolution (FWHM, eV) | Sensitivity (ppm) |
|---|---|---|---|---|
| LSM Cr/Sc | $6.2 \times 10^4$ | 19.5 | 16.5 | 170 |
| LSM Fe/Sc | $5.3 \times 10^4$ | 17.0 | 12.7 | 200 |
| LSM-060 | $9.8 \times 10^3$ | 27.0 | 11.8 | 370 |
| LSM-080 | $1.9 \times 10^4$ | 5.5 | 13.9 | 580 |

### 17.2.1. Analysis of Nitrogen

One of the most important, yet most difficult, ultra-light elements for x-ray microanalysis is nitrogen. In many analytical situations, the identification of nitrides is required. Recent studies using new prototype LSM crystals fabricated from Cr/Sc and Fe/Sc materials with $2d$ spacings of 78–80 Å have shown improved results for analysis of nitrogen.

A comparison of counting efficiency for the N $K\alpha$ line using LSM Cr/Sc, LSM Fe/Sc, and LSM W/Si (LSM-060) is shown in Figure 17.1(a–c), respectively. These spectra were recorded under identical SEM conditions on a BN sample which had been carbon coated. The measured values of peak intensity, $(P - B)/B$ ratio, resolution, and sensitivity for N $K\alpha$ spectra shown in Figure 17.1 are given in Table 17.1. The sensitivity was calculated using the following equation:

$C_{DL} \geq 3.29/(t \times P \times P/B)^{1/2}$, where $t$ is total counting time (1000 s), $P$ is the count rate (100 nA specimen current) and $P/B$ is the peak-to-background ratio.[7]

These data indicate that the LSM-060 crystal achieves the highest resolution and $(P - B)/B$ ratio for N $K\alpha$, although its peak intensity is considerably lower. Because of much higher intensity, the LSM Cr/Sc crystal achieves highest sensitivity for use in the analysis of nitrogen. However, the Fe/Sc LSM crystal has only slightly lower intensity and $(P - B)/B$ ratio for N$K\alpha$ than does Cr/Sc, but has better spectral resolution for all light element spectra than does Cr/Sc. In comparison to Ni/C (LSM-080), the Fe/Sc and Cr/Sc LSM crystals exhibit superior performance for nitrogen, but poorer performance for oxygen, carbon, and boron.

### 17.2.2. Analysis of Nitride Inclusion

Figure 17.2(a) is an SEM micrograph of an area analyzed on the surface of a fatigue fracture sample of 17-7PH stainless steel. This micrograph shows several inclusions of prismatic shapes located in a surface depression.

Figure 17.2(b) shows an x-ray distribution (dot map) of this area for N $K\alpha$ measured with the Cr/Sc LSM crystal using 10 kV SEM electron beam voltage and 5 nA absorbed sample current. This fracture sample was analyzed qualitatively and quantitatively. A portion of the qualitative spectrum recorded using an LSM-060 crystal is given in Figure 17.3(a,b). The prismatic inclusion was identified as a complex (Ti, Fe, Cr, Ni) nitride containing trace amounts of Mn, Si, Al, O, and C. Quantitative analysis of the inclusion was difficult since its surface was a rough, uneven fracture

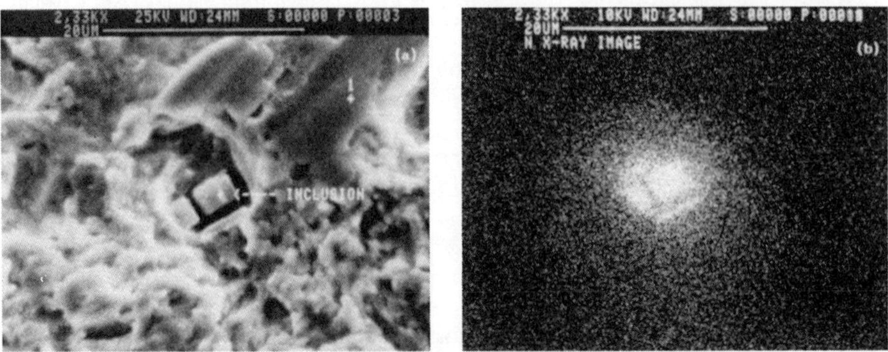

FIGURE 17.2.   Nitride inclusions in a 17-7PH stainless steel fatigue fracture sample. (a) Secondary electron image. (b) X-ray image of N $K\alpha$ using the LSM Cr/Sc crystal. SEM electron beam voltage: 10 kV; sample current: 5 nA.

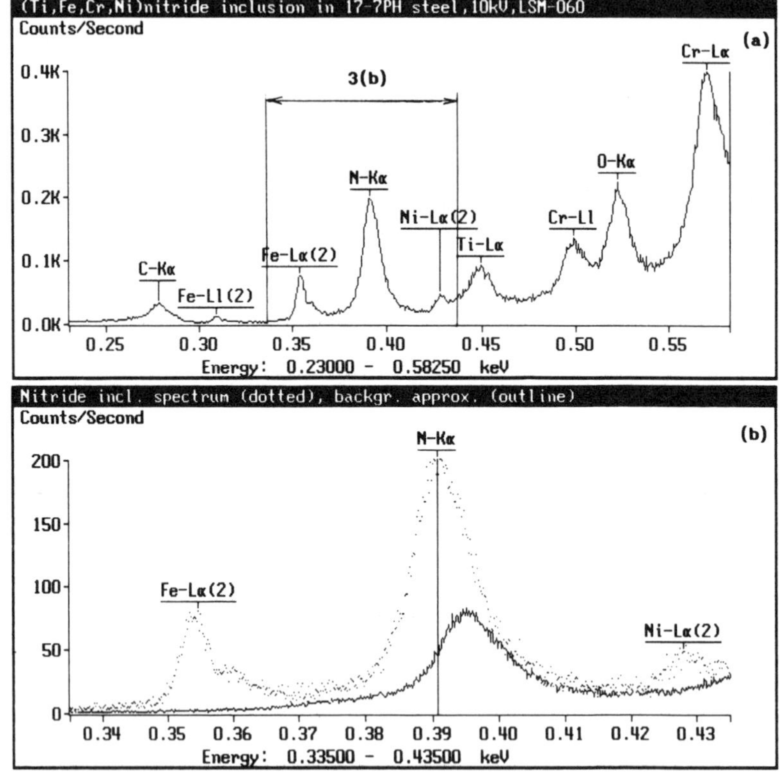


FIGURE 17.3.   (a) Qualitative spectrum of the nitride inclusion recorded using the LSM-060 crystal. (b) Expanded portion of spectrum (dotted) and approximated background of N $K\alpha$ contributed by Ti $Ll$ (solid line).

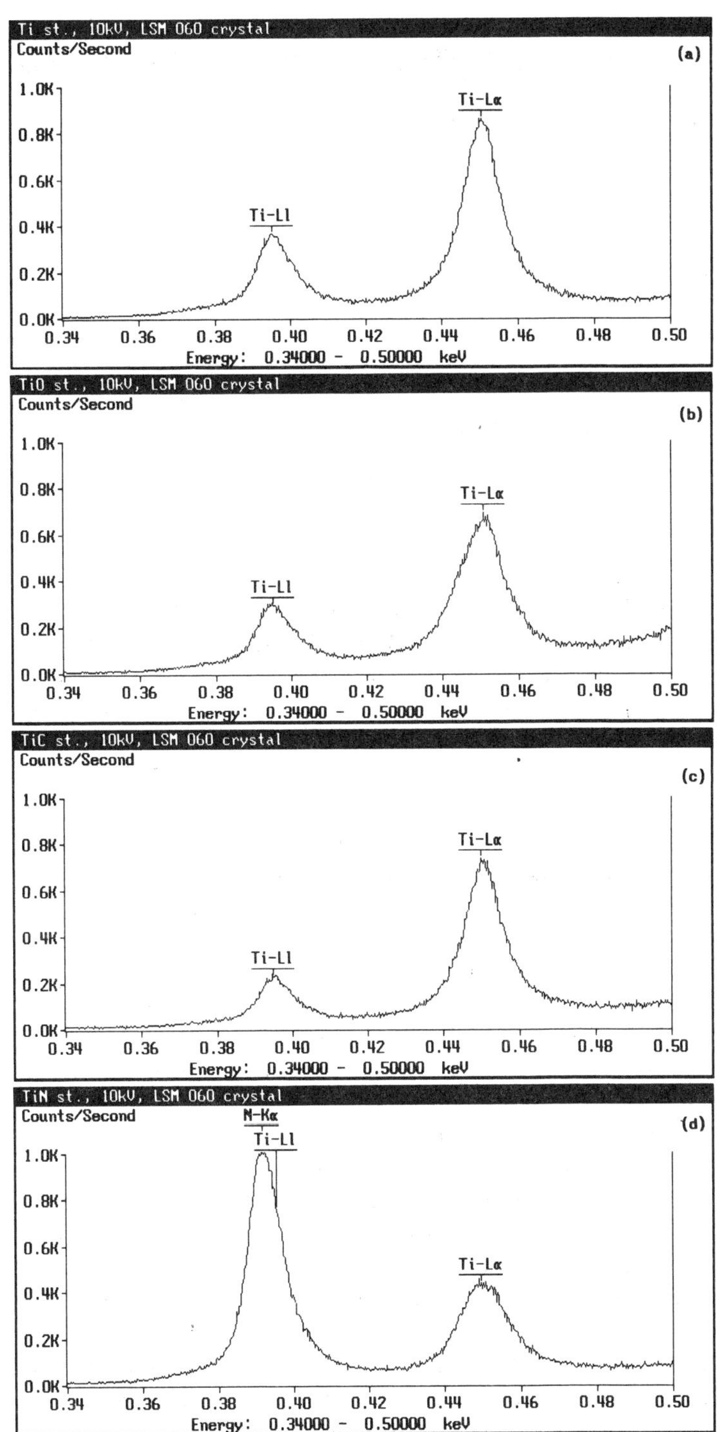


FIGURE 17.4.   Spectra of Ti $L\alpha$ and Ti $Ll$ lines. Lines recorded on (a) pure Ti, (b) TiO, (c) TiC, and (d) TiN standards using LSM-060 crystal. SEM electron beam voltage: 10 kV; sample current: 100 nA.

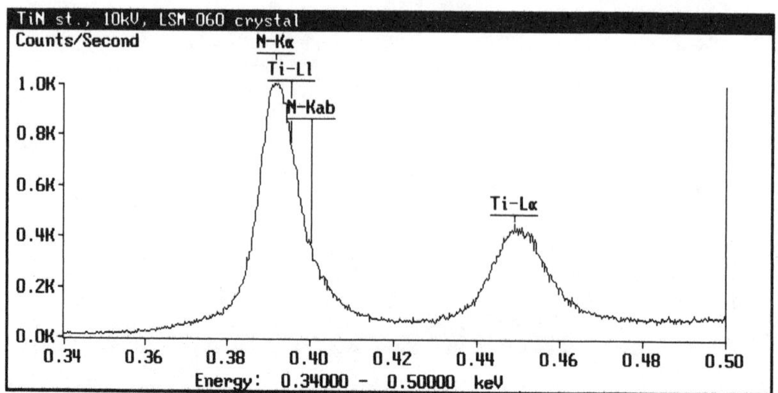



FIGURE 17.5.   TiN spectrum showing spectral positions of N $K\alpha$, Ti $L\alpha$ and Ti $Ll$ lines with respect to the N $K$-absorption edge.

and the x-ray take-off angle could not be determined accurately. Therefore, the results of the analysis given below should be considered semi-quantitative.

In order to optimize spatial resolution and emission on the inclusion, an electron beam voltage of 10 kV and beam current of 20 nA were used. In addition, to reduce attenuation of emitted x rays by surrounding surface unevenness and effects of secondary fluorescence, the sample was tilted 20° toward the spectrometer to provide a higher effective take-off angle.

A major difficulty in this analysis involved determining the correct background for nitrogen since the Ti $Ll$ peak interferes with the N $K\alpha$ peak.[8,9] In the qualitative spectrum of the nitride inclusion, Figure 17.3(a,b), the Ti $L\alpha$ peak is clearly distinguishable, but the Ti $Ll$ peak is combined with the N $K\alpha$ peak. In order to determine a background for nitrogen it is necessary to approximate the portion of N $K\alpha$ background contributed by the Ti $Ll$ line. This requires evaluating the shape, spectral position (wavelength), and amplitude (intensity) of the Ti $Ll$ peak shown in Figure 17.3(b). If the shape or spectral position of the Ti $Ll$ varies with chemical bonding, it would not be possible to approximate a background for the N $K\alpha$ peak. However, spectra recorded from pure Ti, TiO, and TiC standards (Figures 17.4(a,c), respectively) show that there is negligible variation of peak shape or spectral position (peak shift) for the Ti $Ll$ due to chemical bonding. Therefore, it is reasonable to assume that the peak shape and spectral position of Ti $Ll$ from TiN also exhibits negligible shape or position variations. This conclusion should be equally applicable for the case of a complex nitride inclusion in which Ti is the principal nitride-forming element. The background determination for N $K\alpha$ from the inclusion is then reduced to approximating the intensity of the interfering Ti $Ll$ peak.

It is possible to use the Ti $L\alpha$/Ti $Ll$ intensity ratio measured from a pure Ti standard to approximate the Ti $Ll$ intensity from the nitride inclusion. However, in this case it is necessary to account for large absorption differences which occur. The Ti $L\alpha$ undergoes much higher absorption in nitrogen than does Ti $Ll$ due to the presence of the N $K$-absorption edge, which lies between these two peaks as shown in Figure 17.5. An absorption factor for

TABLE 17.2.    Results of Analysis of Nitride Inclusion in 17–7PH Steel

| Element | Line | Weight percent | Atomic percent | Count error[10] (sigma wt%) |
|---------|------|----------------|----------------|------------------------------|
| C | $K\alpha$ | 0.15 | 0.49 | 0.04 |
| N | $K\alpha$ | 13.3 | 36.5 | 0.52 |
| O | $K\alpha$ | 1.30 | 3.10 | 0.03 |
| Ni | $L\alpha$ | 2.32 | 1.50 | 0.04 |
| Al | $K\alpha$ | 0.34 | 0.48 | 0.005 |
| Si | $K\alpha$ | 0.17 | 0.23 | 0.009 |
| Ti | $K\alpha$ | 36.5 | 29.4 | 0.32 |
| Cr | $K\alpha$ | 8.70 | 6.50 | 0.11 |
| Mn | $K\alpha$ | 0.58 | 0.41 | 0.05 |
| Fe | $K\alpha$ | 30.7 | 21.2 | 0.27 |
| | | Total: 94.06 | | |

SEM electron beam voltage: 10 kV; effective take-off angle: 60.7° (ave.)
SEM electron beam current: 20 nA; correction: $\phi(\rho z)$[12,13]

Ti $L\alpha$ in the nitride inclusion was calculated using the $\phi(\rho z)$ model of the quantitative x-ray microanalysis correction procedure. An absorption factor ($A$ factor) of 0.603 was determined after three iterations (Ti $Ll$ background contribution was ignored during the first iteration). This indicates that the Ti $L\alpha$/Ti $Ll$ ratio in the nitride is only about 60% of that from pure Ti due to the high absorption of Ti $L\alpha$ in nitrogen. In the ZAF corrections for Ti $L\alpha$ in this nitride, fluorescence effects ($F$ factor) and atomic number effects ($Z$ factor) were considered negligible. It is important to note that the absorption of Ti $Ll$ in pure Ti and in the nitride are very nearly the same since the mass absorption coefficient for Ti $Ll$ in pure Ti does not vary greatly from the total mass absorption coefficient for Ti $Ll$ in the nitride.

Using the calculated absorption factor ($A = 0.603$), it was possible to approximate the Ti $L\alpha$/Ti $Ll$ intensity ratio for the nitride and thus approximate the portion of intensity attributable to Ti $Ll$ measured with the N $K\alpha$ peak from the inclusion. An outline (solid line) of this approximated peak is shown in Figure 17.3(b) superimposed on the measured spectrum from the inclusion for N $K\alpha$. A background correction for N $K\alpha$ could then be made to allow for the analysis of N. The results, given in Table 17.2, show that the inclusion is a complex nitride with major constituents of Ti, Fe, Cr, and Ni, in addition to nitrogen. Due to the above mentioned problems involving take-off angle uncertainty and nitrogen background determination, the total concentration differs somewhat from 100%. Although semi-quantitative, this analysis is useful in further defining the chemical composition of the nitride inclusion.

### 17.2.3. Analysis of Borophosphosilicate Films

Another important application of LSM crystals is the analysis of borophosphosilicate (BPSG) films. The use of BPSG glass films in integrated circuit technology has gained wide acceptance in the semiconductor industry as a viable passivation layer on silicon substrates. The concentrations of boron and phosphorus in BPSG films affect the physical properties and characteristics of these films, and therefore accurate chemical analysis of these two elements in relation to the BPSG fabrication process is very important.

Because the thickness of BPSG films used as passivation layers is usually only 0.2-2.0 μm, analysis by other techniques (e.g., x-ray fluorescence) is difficult due to depth of penetration effects. The electron probe (SEM/WDX) microanalysis technique has been applied as a suitable non-destructive method for determination of chemical composition of these films.[11]

A BPSG film 0.8 μm thick deposited on a silicon substrate (wafer) was analyzed using the WDX technique. To provide conductivity, the sample and standards were coated simultaneously with a carbon layer approximately 150 Å thick. Qualitative WDX analysis indicated the presence of B, P, Si, and O, and a trace amount of nitrogen in the sample. The nitrogen is suspected to have originated from the $N_2$ atmosphere used during the BPSG fabrication process. The phosphorus and silicon were analyzed using a PET crystal, while the other elements were analyzed using LSM crystals. Oxygen and nitrogen were analyzed using an LSM-060 crystal. For purposes of comparison, boron was analyzed qualitatively using both the LSM-200 and LSM-080

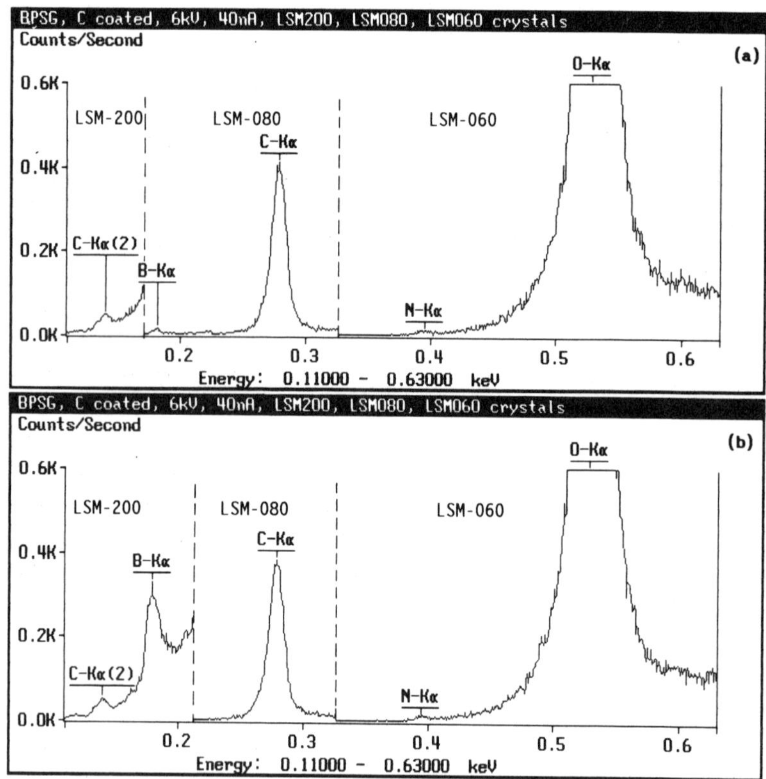



FIGURE 17.6.   Spectra from BPSG film sample recorded using LSM-200, LSM-080, and LSM-060 crystals (concentration of B: 2.35 wt%). (a) B $K\alpha$ line recorded by LSM-080 crystal. (b) B $K\alpha$ line recorded by LSM-200 crystal. SEM electron beam voltage: 6 kV. Sample current: 30 nA.

TABLE 17.3. Calculated Values of Maximum Depths of X-Ray Generation (Rxg) and X-ray Emission (Rxe) in 0.8 μm-Thick BPSG Film[12,13]

| X-ray line | Rxg (μm)/Rxe (μm) | | | | |
|---|---|---|---|---|---|
| $V_o$ (kV) | B $K\alpha$ | P $K\alpha$ | Si $K\alpha$ | O $K\alpha$ | N $K\alpha$ |
| 3 | 0.20/0.13 | 0.13/0.13 | 0.13/0.13 | 0.18/0.16 | 0.19/0.13 |
| 5 | 0.45/0.21 | 0.36/0.36 | 0.36/0.36 | 0.42/0.40 | 0.43/0.35 |
| 6 | 0.60/0.24 | 0.53/0.50 | 0.53/0.50 | 0.60/0.50 | 0.60/0.44 |
| 8 | 0.97/0.28 | 0.90/0.82 | 0.90/0.85 | 0.95/0.78 | 0.97/0.62 |
| 10 | 1.40/0.30 | 1.25/1.15 | 1.25/1.20 | 1.38/1.05 | 1.40/0.76 |
| 12 | 1.90/0.30 | 1.60/1.50 | 1.60/1.55 | 1.80/1.35 | 1.85/0.80 |
| 15 | 2.80/0.30 | 2.50/2.25 | 2.50/2.40 | 2.60/1.50 | 2.70/0.80 |

crystals. Results are shown in Figure 17.6(a,b). All spectra were recorded under the same SEM conditions. The C $K\alpha$ x rays are emitted from the conductive carbon coating. As the spectra show, the B $K\alpha$ intensity measured from the LSM-200 is about 15 times greater than the B $K\alpha$ intensity measured from the LSM-080 crystal although the $(P - B)/B$ ratio from the LSM-080 crystal is 2.5 times greater. This results in more than twice the sensitivity for boron using an LSM-200 crystal (i.e., 55 ppm for LSM-200 versus 130 ppm for LSM-080).

The BPSG sample was also analyzed quantitatively. Quantitative analysis of surface layers with thicknesses from submicron to several micrometers requires careful selection of electron beam voltage to ensure that the maximum depth of x-ray generation is less than the film thickness so that only x rays generated in the film are analyzed. Values of maximum depth of x-ray generation (Rxg) and maximum depth of x-ray emission (Rxe) for electron beam voltages of 3, 5, 6, 8, 10, 12, and 15 kV were calculated for P $K\alpha$, Si $K\alpha$, O $K\alpha$ and N $K\alpha$ lines, using the Bastin *et al.*[12] and Rehbach and Karduck[13] modified Gaussian $\phi(\rho z)$ model of the x-ray depth distribution function. The results are given in Table 17.3.

The values for Rxe are smaller than the values for Rxg because of absorption of generated x-ray radiation occurring inside the sample mass. This is particularly true for B $K\alpha$, which is very highly absorbed by most materials, including the BPSG film. As a result, the Rxe value for boron in BPSG is approximately 0.3 μm and does not vary significantly with SEM electron beam voltage, although the Rxg value increases with increasing SEM beam voltage. Data from Table 17.3 indicate that for analysis of a 0.8-μm-thick BPSG film, the SEM electron beam voltage should not exceed 7 kV. X-ray intensities for B, P, Si, and O $K\alpha$ lines were measured using electron beam voltages given in Table 17.3 and are plotted in Figure 17.7. An electron beam current of 25 nA was used for these measurements.

The plots show different effects for each of the x-ray lines. A maximum intensity for B $K\alpha$ is measured at voltages between 2.5 and 3.0 kV, because most of the B $K\alpha$ radiation is generated at depths close to the surface (i.e., 0.3 μm) and undergoes minimal absorption. As beam voltage is increased, more of the B $K\alpha$ x rays are generated at greater depths (>0.3 μm) and undergo greater absorption before exiting the sample. The intensities for Si $K\alpha$ and P $K\alpha$ increase with increasing accelerating voltage. For voltages higher

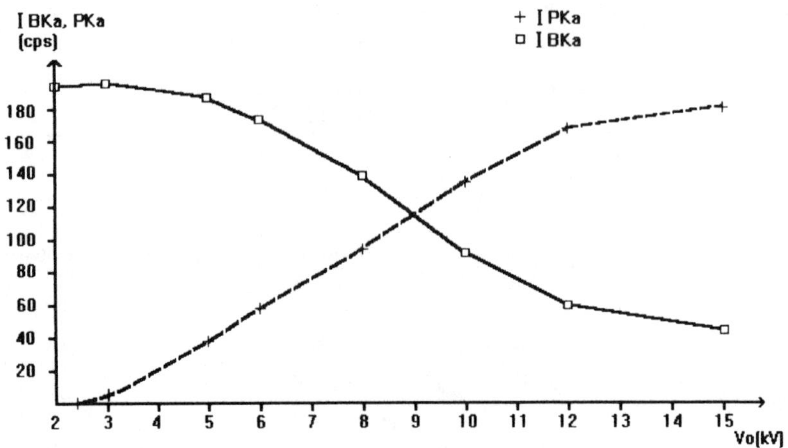


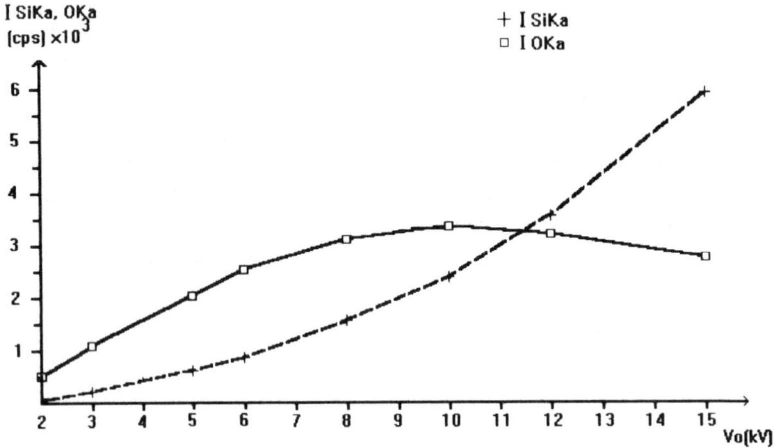


FIGURE 17.7.   Intensities of B $K\alpha$, P $K\alpha$, Si $K\alpha$ and O $K\alpha$ lines measured on a BPSG film sample as a function of SEM electron beam voltage; electron beam current: 25 nA.

than 10 kV, a considerable portion of electrons penetrates through the BPSG layer to the silicon substrate, causing a more rapid increase of Si $K\alpha$ intensity. Conversely, the rate of increase of P $K\alpha$ becomes less since there is no phosphorus in the substrate, only in the BPSG layer. In addition, the O $K\alpha$ intensity reaches a maximum at 10 kV and decreases for higher voltages, again showing the effects of penetration through the BPSG layer. For voltages lower than 5 kV, intensities for P and Si are so low that, because of long counting times required, quantitative analysis becomes impractical.

Based on the above considerations, a quantitative analysis of the BPSG film was made for B, P, Si, O, and N using an electron beam voltage of 6 kV and electron beam

TABLE 17.4. Results of Quantitative Analysis of 0.8 μm-Thick BPSG Film

| Element | Line | Weight percent | Atomic percent | Count error[10] (sigma wt%) | Count precision (relative %) |
|---------|------|----------------|----------------|------------------------------|-------------------------------|
| B | $K\alpha$ | 2.35 | 4.22 | 0.04 | 3.5 |
| P | $K\alpha$ | 3.77 | 2.36 | 0.06 | 3.3 |
| Si | $K\alpha$ | 37.6 | 25.98 | 0.38 | 2.0 |
| O | $K\alpha$ | 54.6 | 66.23 | 0.36 | 1.3 |
| N | $K\alpha$ | 0.87 | 1.21 | 0.19 | 43 |
| | | Total: 99.19 | | | |

SEM electron beam voltage: 6kV; effective take-off angle: 35.0 deg;
SEM electron beam current: 25 nA; correction: $\varnothing(\rho z)^{12,13}$

current of 25 nA. Results for this analysis are given in Table 17.4. For analysis of Si and O, a high purity $SiO_2$ standard was used; for N, pure BN was used; for P, pure InP was used; and for B, a glass standard containing 12.8 wt.% $B_2O_3$ was used. For analysis of boron, the LSM-200 was used rather than LSM-080 since this crystal gives higher sensitivity for boron. Mass absorption coefficients used in the $\varnothing(\rho z)$ data correction model[12,13] for Si and P $K\alpha$ lines are those reported by Heinrich[1] and for B, O, and N $K\alpha$ lines are those reported by Bastin and Heijligers.[2–4]

An important consideration in quantitative analysis is the level of precision required. For most analyses, a relative precision of less than 3% for major elements and less than 20% for minor elements is satisfactory, although it will depend on the specific situation. Counting precision for concentrations of analyzed elements can be expressed by the equation $(2\sigma/C) \times 100\,(\%)$, where $C$ is the measured concentration of the analyzed element, $\sigma$ is the counting error (sigma wt%),[10] and the confidence level is 95%. The counting error is based on total peak and background counts collected from the sample and standard. This is dependent upon the count rates attainable as well as count times selected.

In relation to the BPSG fabrication process, a high degree of analytical precision for boron and phosphorus is required, even though these two elements are minor constituents in the BPSG film. The quantitative results given in Table 17.4 show 2.35 wt% boron and 3.77 wt% phosphorus present in the film. Counting precision of better than 4% relative were obtained for all elements analyzed with the exception of nitrogen which was detected in trace amount only. As Table 17.4 shows, the counting precision for boron was 3.5%. Count times of 100 s on peak, background, standard, and sample were used to attain this result. To obtain greater precision, longer count times would be necessary (e.g., counting precision of 1% for boron would require approximate count times of 1200 s instead of 100 s), assuming beam conditions were the same. Since count rates for phosphorus using the PET crystal were higher than the count rates for boron, shorter count times (e.g., BPSG sample: 50 s peak, 20 s background; InP standard: 20 s peak, 5 s background) were satisfactory to obtain a counting precision of 3.3%.

To prevent damage to the analyzed area of the BPSG film, an electron beam current of 25 nA was used for analysis. As the above analysis illustrates, the ability of the LSM-200 crystal to give high count rates at low-to-moderate beam currents (e.g., 10-50 nA) make its use important for analysis of boron. For example, use of an

LSM-080 crystal for the above BPSG analysis would have required a beam current of 125 nA and a count time of about 100 s to achieve the same level of precision. Analysis of BPSG using lead stearate or lead octodecanoate crystals would require beam currents of several hundred nanoamps to achieve equivalent precision while maintaining reasonable count times. At higher values of beam current (i.e., > 50 nA), permanent damage to the analyzed area may become a serious problem. In addition, at higher beam currents, the electron beam diameter becomes larger and spatial resolution is poorer. However, it should be noted that loss of spatial resolution may not be a problem, depending upon homogeneity of the BPSG film and other sample characteristics.

### 17.2.4. Analysis of Beryllium

The LSM-200 crystal, with $2d \approx 200$ Å, can also be used for analysis of beryllium (Be $K\alpha = 114$ Å). Results of analysis of a Be-Cu eutectic alloy, which show distribution of Be present in the eutectic phase, are given in Figure 17.8. The x-ray spectrum was recorded

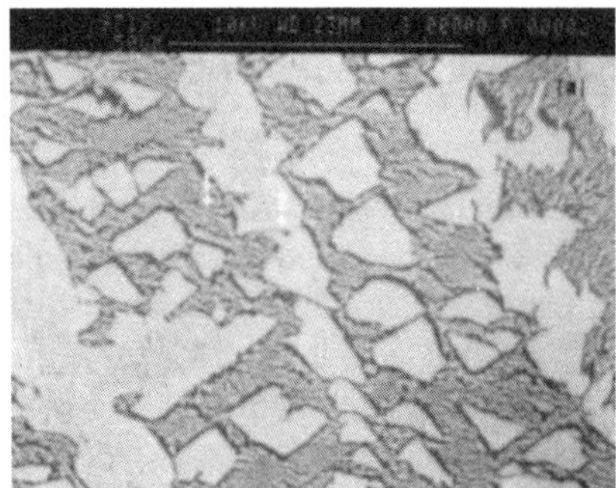
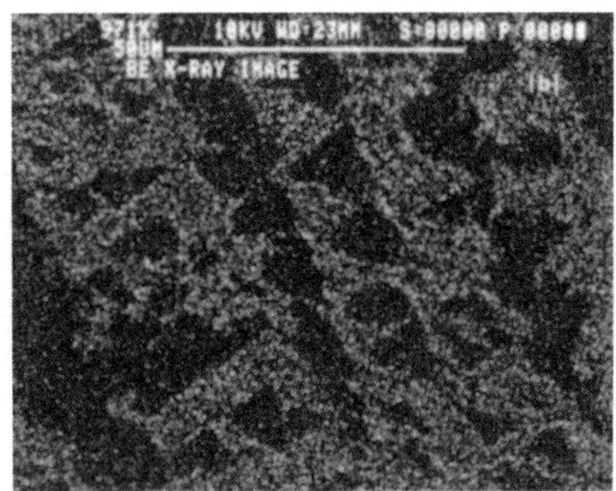
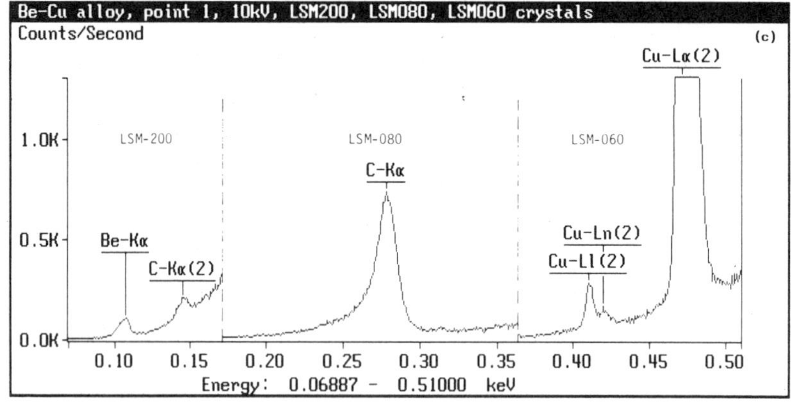


FIGURE 17.8.   Analyzed area from the surface of polished and chemically etched sample of Be-Cu alloy. (a) Secondary electron image, sample current: 5 nA. (b) X-ray image of Be $K\alpha$, sample current: 5 nA. (c) Qualitative spectrum recorded using LSM-200, LSM-080, and LSM-060 crystals. SEM electron beam voltage: 10 kV; sample current: 50 nA.

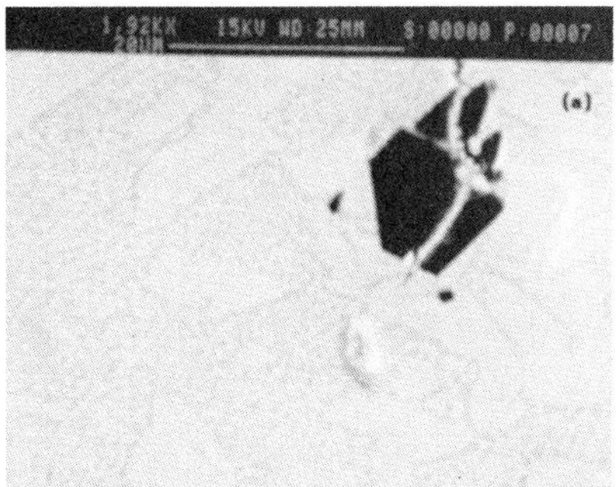
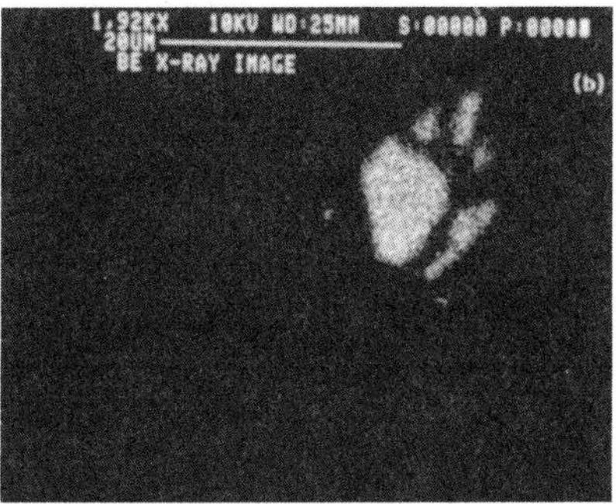
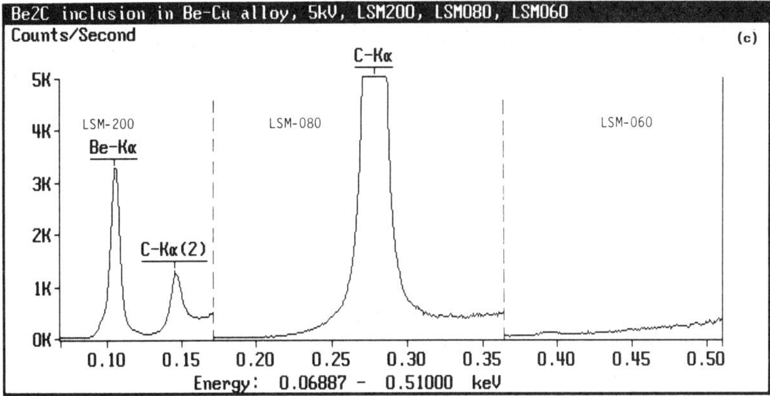

FIGURE 17.9.   Analysis of $Be_2C$ inclusion in Be-Cu alloy. (a) Secondary electron image, sample current: 5 nA. (b) X-ray image of Be $K\alpha$, sample current: 5 nA. (c) Qualitative spectrum of $Be_2C$ inclusion recorded using LSM-200, LSM-080, and LSM-060 crystals. SEM electron beam voltage: 5 kV; sample current: 100 nA.

using the LSM-060, LSM-080, and LSM-200 crystals with the electron beam at position "1" in the eutectic phase [Figure 17.8(a)]. This phase contains approximately 5.5 wt% Be, 2.5 wt% C and > 90 wt% Cu. The solid solution phase [bright areas in the secondary image, Figure 17.8(a)] contain only about 1.5 wt% Be with the balance Cu. The Be $K\alpha$ x-ray dot map in Figure 17.8(b) clearly shows the difference between 1.5 wt % and 5.5 wt % Be.

The Be-Cu alloy also contains Be carbide inclusions as shown by the darker phase in Figure 17.9(a) and identified by the Be $K\alpha$ dot map and x-ray spectrum in Figure 17.9(b,c) respectively. In order that the volume irradiated by the electron beam remains completely within the carbide inclusion, an electron beam voltage of 5 kV was used. Quantitative analysis of Be and C in this carbide requires consideration of two other important phenomena often encountered when analyzing ultra-light elements:

1.  Wavelength shifts of Be $K\alpha$ and C $K\alpha$ peaks measured on sample and pure element standards (Be and vitreous C), Figures 17.10(a,b).

2.  Variation in peak shape (widths) for C $K\alpha$ peaks measured on the sample and standard, Figure 17.10(b).

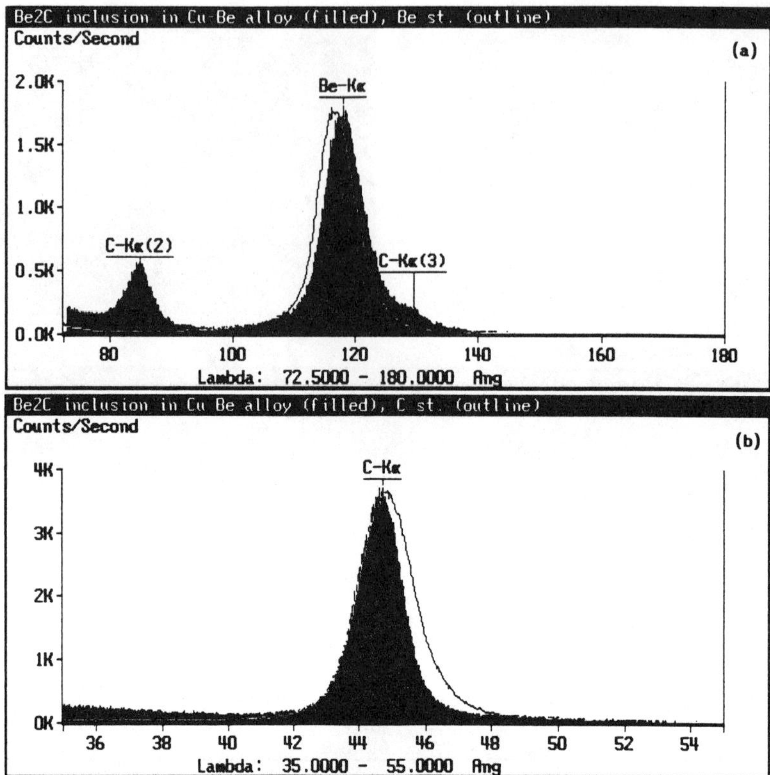


FIGURE 17.10. (a) Wavelength shift of the Be $K\alpha$ line recorded on $Be_2C$ inclusion (filled plot) and on pure Be standard (outline plot). (b) Wavelength shift and peak width change of C $K\alpha$ line measured on $Be_2C$ inclusion (filled plot) and vitreous carbon standard (outline plot).

To account for the wavelength shift effect, individual peak (wavelength) settings were determined for both sample and standard before intensities were measured. To compensate for peak shape variations in the C $K\alpha$ spectra, the integral intensities were determined for the sample and standard. From these integral intensities, the area peak factor[5] was calculated and used to adjust measured peak intensity for quantitative analysis. Effects of carbon contamination were minimized by using a turbomolecular pumped vacuum system with efficient foreline trap and by careful sample preparation.

TABLE 17.5. Results of Quantitative Analysis of Be2C Inclusion in Be-Cu Alloy. SEM electron beam voltage: 5 kV; effective take-off angle: 35.0°; SEM electron beam current: 20 nA; correction: $\phi(\rho z)$[12,13]

| Element | Line | Weight percent | Atomic percent | 100 nA for Be (sigma wt%) | Count error[10] Count precision (relative %) |
|---|---|---|---|---|---|
| Be | $K\alpha$ | 59.1 | 66.43 | 0.32 | 1.1 |
| C | $K\alpha$ | 39.8 | 33.57 | 0.28 | 1.4 |
| | | Total: 98.9 | | | |

Although effects of carbon contamination can be significant when analyzing trace amounts of carbon, these effects are less when analyzing higher concentrations of carbon (e.g., as in carbides). Since carbon is present as a major constituent in $Be_2C$, in the present analysis, effects of carbon contamination did not increase carbon results by more than 1% relative (i.e., $\approx 0.4$ wt%). Results of quantitative analysis of the Be carbide inclusion are given in Table 17.5. The atomic percentages for Be and C show clearly the correct stoichiometry for $Be_2C$. Mass absorption coefficients used in the $\phi(\rho z)$ data correction model[12,13] are those reported by Henke *et al.*[14] for the Be $K\alpha$ line and by Bastin and Heijligers[5] for the C $K\alpha$ line.

### 17.2.5. Improved X-Ray Imaging

The high intensities provided by LSM crystals enable x-ray dot maps to be performed on ultra-light elements at low beam current levels similar to those used for heavier elements. As an example, Figure 17.11 shows scanning images from a cross section of a Nb-Al-Ti-C composite which has been reinforced with SiC fibers. An LSM-080 crystal with high counting efficiency for C $K\alpha$ was used for the carbon dot

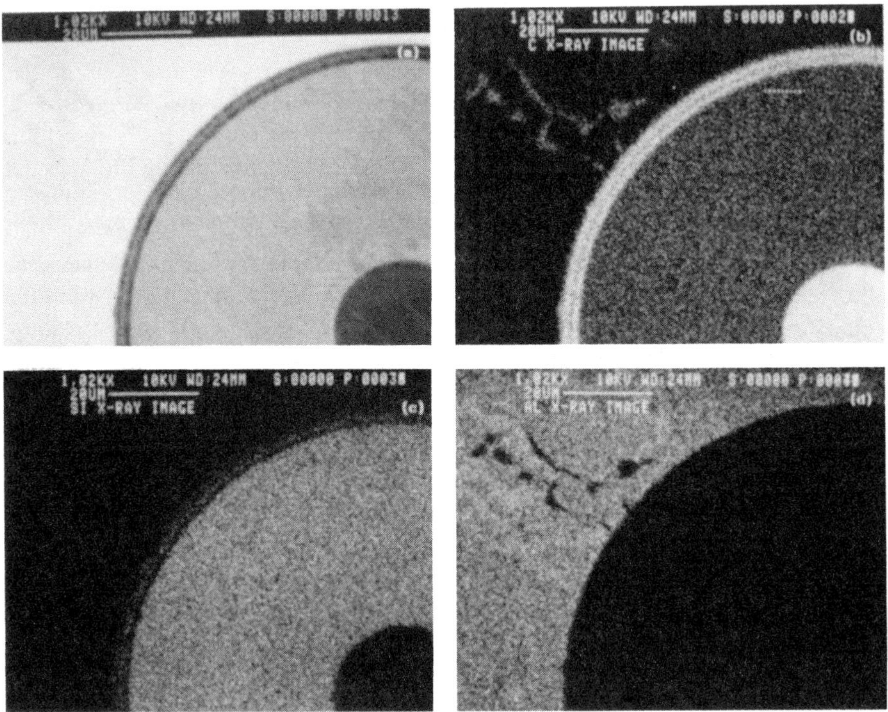

FIGURE 17.11.   Analysis of an area from the surface of cross-sectioned and polished composite of Nb, Al, Ti, and C reinforced by SiC fibers. (a) Secondary electron image. (b) C $K\alpha$ x-ray image, 20 min acquisition time. (c) Si $K\alpha$ x-ray image, 20 min of acquisition time. (d) Al $K\alpha$ x-ray image, 8 min acquisition time. SEM electron beam voltage: 10 kV, sample current: 5 nA.

map so that only 5 nA beam current at 10 kV accelerating voltage was required. The x-ray images for Al and Si were made using the same beam conditions. As Figure 17.11(b) shows, the good spatial resolution ($\approx$ 1 $\mu$m) reveals the presence of a double carbon layer at the circumference of the fibers.

## 17.3. CONCLUSIONS

The application of layered synthetic microstructure crystals to the analysis of ultra-light elements (e.g., O, N, C, B, and Be) in several different samples has been presented. These studies represent several years of analytical experience using LSM crystals and demonstrate greater sensitivity than is possible with conventional lead stearate or lead octodecanoate pseudocrystals. The higher diffracting intensities of LSM crystals make it possible to perform analysis of ultra-light elements at low electron beam currents, similar to those required for heavier elements (i.e., atomic number >10). The use of lower electron beam current and careful selection of accelerating voltage make it possible to improve spatial resolution (to ~1 $\mu$m) for analysis of ultra-light elements. In addition, the use of lower beam currents reduces the possibility of electron beam damage to the analyzed area of delicate samples (e.g., BPSG films).

The processes by which LSM crystals are fabricated allow the $2d$ spacing to be customized and diffracting efficiency to be optimized, as shown by results from special Cr/Sc and Fe/Sc LSM crystals optimized for nitrogen analysis. These processes make it possible to develop LSM crystals which provide optimum performance specifically for one or two ultra-light elements in a small wavelength range. Newly developed multicrystal spectrometers make it possible to mount up to three different LSM crystals together with conventional crystals such as LiF, PET, and TAP to enable the analyst to optimize performance for the full range of ultra-light elements.

It is well documented that LSM crystals are useful in suppressing higher order diffractions in certain analytical situations.[5,6,15] Layered synthetic microstructure crystals are also much more durable and less likely to be damaged than stearate or decanoate type crystals. They are not damaged by moisture or oil condensates and, with reasonable care, can be cleaned with organic solvents. Although spectral resolution (i.e., FWHM) using LSM crystals is not as good as stearate or decanoate type crystals (approximately 50% less), in most analytical situations, other advantages greatly outweigh this disadvantage.

Research and development of new LSM materials with improved performance in terms of analytical sensitivity and resolution, as well as efforts to extend the spectral range, especially to shorter wavelengths, are continuing.


## REFERENCES

1. K. F. J. Heinrich, in: *Proceedings of the 11th ICXOM* (Brown and R. H. Packwood, eds.) Univ. Western Ontario, p. 67 (1986).
2. G. F. Bastin and H. J. M. Heijligers, *Microanalysis of Boron in Binary Borides*, Internal Report, Eindhoven University of Technology (1986).

3. G. F. Bastin, and H. J. M. Heijligers, *Quantitative Electron Probe Microanalysis of Oxygen*, Internal Report, Eindhoven University of Technology (1989).

4. G. F. Bastin and H. J. M. Heijligers, *Quantitative Electron Probe Microanalysis of Nitrogen*, Internal Report, Eindhoven University of Technology (1988).

5. G. F. Bastin, and H. J. M. Heijligers, *Quantatitative Electron Probe Microanalysis of Carbon in Binary Carbides*, Internal Report, Eindhoven University of Technology (1990).

6. G. F. Bastin and H. J. M. Heijligers, in: *Electron Probe Quantitation* (K. F. J. Heinrich and D. E. Newbury, eds.) Plenum Press, New York, p. 145 (1991).

7. T. O. Ziebold, *Anal. Chem.* **39**, 858 (1967).

8. P. Duncumb and D. A. Melford, in: *X-Ray Optics and Microanalysis*, (R. Castaing, P. Deschamps, and J. Philibert, eds.) Hermann, Paris, p. 240 (1966).

9. G. F. Bastin and H. J. M. Heijligers, *Scanning* **13**, 325 (1991).

10. Z. Kotrba, *X-Ray Spectrom.* **7**, 195 (1978).

11. R. B. Wolf, The Construction of Calibration Curves for Boron and Phosphorus in Borophosphosilicate Glass using Wavelength Dispersive X-Ray Spectroscopy, Internal Report, San Jose State University (1990).

12. G. F. Bastin, H. J. M. Heijligers, and F. J. J. van Loo, *Scanning* **8**, 45 (1986).

13. W. Rehbach and P. Karduck, in: *Microbeam Analysis* (D. E. Newbury, ed.) San Francisco Press, San Francisco, pp. 285-287 (1988).

14. B. L. Henke, P. Lee, T. J. Tanaka, R. L. Shimabukuro, and B. K. Fujikawa, *Atomic Data* **27**, 1 (1982).

15. C. E. Lyman, D. E. Newbury, J. I. Goldstein, D. B. Lifshin, and K. R. Peters, *Scanning Electron Microscopy, X-Ray Microanalysis and Analytical Electron Microscopy, Plenum Press, New York (1990).*

# 18

# An Evaluation of Quantitative Electron Probe Methods

*K. F. J. Heinrich*

ABSTRACT

This publication describes a comparative test of several procedures of electron probe microanalysis commonly used at present, on the basis of results obtained from specimens of presumably known composition. The strategy used is to separate as far as possible the effects of the atomic number, absorption, and fluorescence on the results of subsets from a set of measurements containing 1826 entries, selected as described later in detail.

## 18.1. ELECTRON PROBE MICROANALYSIS

### 18.1.1. The Specimen

Electron probe microanalysis is based on the observation and evaluation of characteristic x rays emitted by a specimen that is irradiated by accelerated (typically 3–30 keV) and focused electrons. Typical instruments use an electron beam normal to the specimen surface, and analyses are performed on specimens thicker than the range in depth of x-ray emergence. It is usually assumed that the composition within the measured microvolume is homogeneous and that the specimen conducts electricity to the extent that no problems arise from electrostatic charges.

There is great practical interest in procedures which differ from the classical specimen-beam arrangement, first proposed by the inventor of EPMA, R. Castaing.[1] Normal electron beam incidence is not necessary in principle; microanalysis can be

K. F. J. HEINRICH • National Institute of Standards and Technology (ret.). Present address: Rockville, Maryland 20850

*X-Ray Spectrometry in Electron Beam Instruments*, edited by David Williams, Joseph Goldstein, and Dale Newbury. Plenum Press, New York, 1995.

"

performed with electron beams at an oblique angle. The analytical accuracy is not as easy to evaluate in such cases, because we do not have available the quantity of analytical data on known specimens comparable to that available in cases with a normal beam incidence. The beam inclination cannot be taken into account by mere geometrical considerations, since the reemission of high energy electrons (electron backscatter) increases at oblique electron beam impact. Hence, the calculation of the x-ray generation losses due to backscatter must be modified, and efficient tests of the accuracy cannot be performed at present.

Another possibility is the investigation of specimens the surface of which has not been altered and is not flat. Although procedures using the measurement of continuous radiation as an intensity standard have been proposed,[2,3] such procedures are outside the scope of the present communication. Other configurations of interest which are not treated here are thin or layered specimens, particles, and fibers. Again, we do not have a set of measurement data comparable with that the "classical configuration." For the problems arising from low electrical conductivity, see Bastin.[4]

### 18.1.2. The Standard

Castaing proposed the use of elementary standards and the measurement of intensity ratios of the x-ray lines of interest between specimen and standard. In this way, uncertainties in parameters common to the two targets, such as the fluorescence yield, cancel or are minimized.[1] This mode of operation is generally accepted, although in some areas, including the analysis of minerals, standards of matching composition are frequently used.[5] There is a trade-off between the possible errors in the evaluation procedure for an intensity ratio from a specimen and an elementary standard (hence of different composition), and the possible errors in postulating the composition of a microscopic volume in a multielement standard (a stoichiometric compound or a homogeneous material analyzed on a macroscopic scale, of composition close to that of the specimen). With the energy dispersive solid-state detectors (such as lithium-drifted silicon detectors) the efficiency of the detector system can be estimated much more accurately than with a curved-crystal spectrometer with gas detectors. Hence, in principle, it is possible to dispense with the analytical standard. In practice, the detector system used in standardless analysis should be calibrated from time to time since deposition of ice and other impurities, or changes in the detector itself, may cause variations in the detector efficiency as a function of photon energy. Again, this subject is outside the boundaries of this communication.

### 18.1.3. Choice of Expressions for Concentrations and Errors

When the possibility of quantitative microprobe analysis was first studied, it was postulated that the x rays produced within the specimen were approximately proportional to the mass fraction of the emitting element (first approximation of Castaing). The mass fractions thus became the traditional expression for the concentration of elements in the specimen. They are usually called "concentrations," although this definition differs from that used in chemistry, which denotes moles or weight percent per unit of volume in a liquid solution.

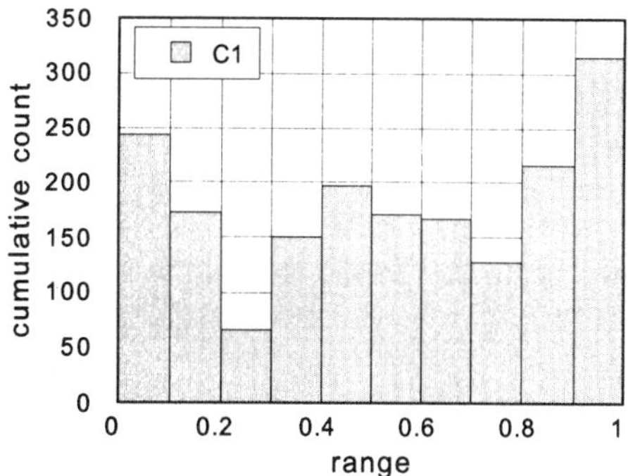


FIGURE 18.1.   Distribution of weight fractions of the analyzed element in the set.

Errors can be expressed either on an absolute scale or relative to the amount measured:

$$\varepsilon_{abs} = C_{calc} - C_{true}; \ \varepsilon_{rel} = \varepsilon_{abs} / C_{true} \tag{18.1}$$

In a population of data such as that used in the present work, where the concentrations vary from 0.01 or less to almost 1 (Figure 18.1), the choice matters. The error distribution with respect to concentration depends on the causes of error. The fluorescence correction, and the relative error related to it, are largest for large concentrations of the exciting element and therefore, in a binary specimen, for small analyte concentrations. Absolute errors due to instrumental instability increase with the concentration of the analyte, while matrix-related errors decrease as the composition of the specimen approaches that of the standard.

In this communication, absolute errors are reported. For the characterization of error distributions, we use the costumary statistics: mean and standard distribution. In Tables 18.2, 18.4, and 18.7, we also report the root mean square,[6] an expression of error distribution around the zero error ("true") value rather than the mean, thus permitting an appreciation of accuracy with a single statistic.

## 18.2. ERRORS

### 18.2.1. Classification of Errors

The following possible sources of error may affect the estimate of the accuracy of the method:

Specimen-related errors
1.   The estimate of the specimen composition.
2.   The estimate of the standard composition.

3.  Inhomogeneity of the specimen and/or the standard.
4.  Surface contamination and irregularities.
5.  Electrostatic effects in specimens of poor electrical conductivity.

Instrumental errors

6.  Deviations from the assumed geometry.
7.  Statistical variations of the measured x-ray intensity.
8.  Instability of the apparatus, particularly of the electron beam excitation.
9.  Effects of coincidence losses in detectors and associated circuits.

Measurement errors

10. Inclusion of x-ray emissions other than the characteristic line of interest:
    a. Background.
    b. Line interferences.
    c. Satellite lines.

Errors in the correction procedure

11. Effects of parameters involved in the calculation, e.g., fluorescent yields and x-ray mass absorption coefficients.
12. Effects of the models themselves.

The error statistics depend on

1.  The elements and lines observed.
2.  The nature of the specimens considered.
3.  The concentration distribution of the elements measured.

Certain errors (such as beam drift and coincidence losses) can be minimized by careful operation and revealed by the observation of measurements on a single target; some (e.g. line overlap, errors in the fluorescence correction) occur only with certain combinations of elements; some, such as those due to lack of electrical conductivity, depend on the physical properties of the specimen. Others vary with operating conditions (e.g., those related to absorption of x rays in the specimen, which increases with the depth of penetration). The effects of surface artifacts increase with decreasing electron energy; this is probably the cause, or one of the causes, of increased analytical error at low overvoltages (see Table 18.1).

Special problems may arise with the specimen preparation, including polishing, etching, coating of the specimen surface, and deposition of impurities, especially in the preparation of soft biological tissue.


### 18.2.2. The Data Set

The data set used in the present work is the sum of several sets of measurements of relative x-ray intensity from binary specimens provided by various authors.[7] In all cases, pure elements were used as standards. Included are data available from the literature, as well as unpublished data obtained through the kindness of various authors, such as those produced at the National Bureau of Standards in 1986 by R. Marinenko and H. Konuma. Values were excluded according to one or several of the following criteria:

1.  Most data obtained on instruments having an x-ray emergence angle below 30°. Such instruments, and measurements obtained with them, are dated and

therefore suspect, not only with respect to absorption losses (which increase with decreasing angle), but also in other aspects of data collection. A few values were incorporated because of special important situations they cover.

2. Single values not corroborated by measurements of other compositions within the binary system under identical conditions, or of the same specimen at different operating voltages.
3. Sets of measurements that show lack of internal consistency.
4. Measurements on specimens of low electrical conductivity.
5. Measurements on specimens other than those which are of uniform composition within the range of x-ray excitation, and limited by a flat surface normal to the electron beam (e.g., layered or coated specimens, particles, thin films, and those measured at inclined beam incidence).

The sources of the data are listed at the end of Appendix 18.1. The total set contains 1826 measurements, sequentially ordered according to the following priorities:

1. Atomic number of element measured.
2. Atomic number of second element.
3. Weight fraction of first element.
4. Operating voltage.
5. X-ray emergence ("take-off") angle.
6. X-ray line, in the following order: $K\alpha$, $K\beta$, $L\alpha_1$, $L\beta_1$, $M\alpha_1$, $M\beta_1$, which are coded 1, 2, 3, 4, 8, and 9 (the lines coded 5–7 were not used in the set).

The set may not be representative of the type of specimens analyzed in a given laboratory. For instance, the error statistics in silicate minerals may differ from those in alloys; however, there is no large body of data on the analysis of minerals with the use of element standards.


### 18.2.3. Data Reduction Procedures

Various competing procedures for electron probe data evaluation are available. All must take into account several aspects of the electron-target interaction, mainly

1. The deceleration of electrons in the target.
2. The loss of energetic electrons due to their backscattering.
3. Inner-shell ionization and subsequent emission of x rays.
4. Absorption of these primary x-rays within the specimen.
5. Additional generation of characteristic x rays due to x-ray fluorescence and other secondary processes; absorption of these x rays within the specimen.

In the time-honored ZAF procedure, the data evaluation ("correction") is done in three steps which are performed for specimen and standard and recalculated in subsequent iterations:

The *atomic number correction*, or $Z$ correction, which takes into account aspects 1, 2, and 3 above.

The *absorption correction*, or $A$ correction, which addresses itself to aspect 4 above.

The *fluorescence correction*, or *F* correction, which considers the most important kind of secondary x-ray production, that due to x-ray fluorescence excited by characteristic lines (aspect 5 above).

In the description of the procedures, the underlying physics are usually discussed, but a rating of their accuracy requires testing them on results of measurements on specimens of known composition. It is difficult to estimate the accuracy of a procedure unless the components of the correction related to these physical aspects can be evaluated independently. A useful evaluation can be obtained by testing subsets of a large data base in which one correction dominates the accuracy of the determinations. However, it is impossible to separate completely their individual effects. It is difficult to distinguish, for instance, the effects of faulty x-ray mass absorption coefficients from those of inaccurate absorption correction models. In fact, the causes of error may be misidentified, as can be seen below.

The combination of the $Z$, $A$, and $F$ corrections in an analytical procedure was formulated by Philibert and Tixier under the name of ZAF procedure, and is still widely used.[8] It is customary to formulate these three corrections as factors ($F_Z$, $F_A$, $F_F$) which, when multiplied by the weight fraction of the element being measured, $C_a$, provide an estimate of the relative intensity, $k$:

$$I_{\text{sample}} / I_{\text{element}} \equiv k = C_a \cdot F_Z \cdot F_A \cdot F_F \tag{18.2}$$

J. Henoc described an additional correction for fluorescence due to the x-ray continuum. This effect is usually less important then the others, and is omitted in most correction schemes. Henoc's correction procedure is imbedded in a FORTRAN program called COR, which, while incorporating the continuum fluorescence, follows basically the lines of ZAF.[9]

When the factor for fluorescence by the continuum ($C$) is included, (18.2) becomes

$$I_{\text{sample}} / I_{\text{element}} \equiv k = C_a \cdot F_Z \cdot F_A \cdot (F_F + F_C - 1) \tag{18.3}$$

It is better to abandon at this point the concept of multiplicative factors, and to write

$$I_{\text{sample}} = I_P f_P + I_F' + I_C' \quad \text{and} \quad I_{\text{standard}} = I_P^\circ f_P^\circ + I_C^{\circ'} \tag{18.4}$$

where $I_P$ and $I_P^\circ$ are the generated primary intensities for specimen and standard, $f_P$ and $f_P^\circ$ the primary absorption factors for specimen and standard, $I_F'$ is the emitted characteristic fluorescence for the specimen, and $I_C'$ and $I_C^{\circ'}$ are the emitted fluorescence intensities for specimen and standard.

Some recent procedures, such as the PAP and PPS procedures by Pouchou and Pichoir,[10] and the phi-rho-$z$ method, advocated by J. Brown and colleagues,[11,12] are assumed by their authors to conceptually differ from the ZAF procedure. However, the three parts on which the classic ZAF procedure is based can also be separated in these approaches, although a differentiation of the stopping power effect (aspects 1 and 3 above) and the backscatter effect (aspect 2 above) is not explicit in the phi-rho-z method. For these reasons we will address this subject after having treated the variations of ZAF, including the method of Scott and Love[13] and the PAP and PAPS methods.

To separate as far as possible the effects of the diverse mechanisms, we selected subsets of the total set in which one of the ZAF factors was predominant and tested the procedures, and combinations thereof, to determine the conditions which resulted in the smallest distribution of errors.

### 18.2.4. The Test Procedure

The test procedure, T-CONVERT, that was used here is written in TurboPascal for the Apple Macintosh SE computer. It offers menus for the selection of procedures and parameters, reads a file of data, and calculates the intensity ratios predicted according to the selections made. It writes another file with the new results, and on request, prints a table providing the characteristics of each entry, the calculated $k$ value, the error in the weight fraction of the measured element, and the factors $F_S$, $F_R$, $F_A$, $F_F$, and $F_C$, which correspond, respectively, to the electron stopping power, backscatter, x-ray absorption, characteristic fluorescence, and continuum fluorescence calculations. The source is noted, and a final column shows flags for problems that may have arisen.

The calculation produces relative intensities ($k$ values), but the analyst is interested in the error in the estimated weight fractions, $\varepsilon_C$. To obtain this error, the parabolic approximation is used which assumes that in a binary specimen, the following relationship is valid over the entire range:

$$C_{true}/(1 - C_{true}) = \alpha \cdot k_{exp}/(1 - k_{exp}); \tag{18.5}$$

where $C_{true}$, the true concentration, and $k_{exp}$, the ratio of experimental intensities from specimen and standard, are parameters known for the binary of question. Once the factor $\alpha$ has been established, the following equation can be postulated for an estimate of the absolute error in concentration, $C_{est} - C_{true}$:

$$C_{est} - C_{true} = [1/1 + (1/\alpha)(1 - k_{calc})/k_{calc}] - C_{true}. \tag{18.6}$$

$C_{est}$ is the estimate of concentration derived from the calculated intensity ratio, $k_{calc}$. Since $C_{est}$ is fairly close to $C_{true}$ the the parabolic approximation will not introduce a significant error in $C_{est}$.

T-CONVERT offers the following choices (the abbreviations in parentheses are those used in Tables 18.2 and 18.3);

For the calculation of the stopping power term ($S$):
1.   Average of the energy between $Eo$ and $Eq$ (ave).
2.   The logarithmic-integral procedure of Philibert and Tixier[8] (li).
3.   The PAP procedure[10] (pap).
4.   The Scott procedure[13] (sct).

5. Numerical integration of the Bethe equation in 10 steps between $Eo$ and $Eq$ (ni).
6. A method derived by Duncumb from the results of Monte Carlo calculations.[14]

For the backscatter correction term ($R$):
1. Duncumb's formula[15] (dun).
2. Yakowitz' formula[16] (yak).
3. Formula of Scott et al.[13] (sct).
4. PAP formula[10] (pap).
5. A formula by Myklebust[17] (my).

For the absorption model ($Fa$):
1. Philibert-Duncumb-Heinrich formula[18] (pdh).
2. Scott *et al.*[13] (sct).
3. PAP[10] (pap).
4. PPS[10] (pps).
5. Quadratic formula[19] (q).
6. Duplex formula[20] (dpl).

For characteristic fluorescence (flu):
1. Complete calculation[21] (c).
2. Reed's formula[22] (r).
3. Omit the correction (-).

For continuum fluorescence (con):
1. Heinrich's formula[23] (kh).
2. Variation for Small's continuum data[3] (not used).
3. Omit (-).

For the choice of the J-factor (Figure 18.2) (J):
1. Wilson ($J = 11.5\ Z$)[24] (wil).
2. Bloch ($J = 13.5\ Z$)[25] (blo).
3. Duncumb[18] (dun).

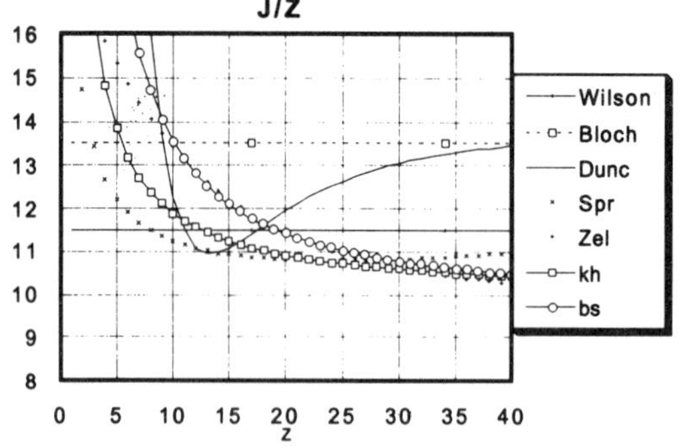


FIGURE 18.2.    Models for the *J*-factor.

4. Berger and Seltzer[26] (bs).
5. Heinrich (unpublished: $J = 9.94\ Z + 19.52$) (kh).
6. Springer[27] (spr).
7. Zeller[28] (zel).

An option of varying $J$ with electron energy is provided.[29]

For the inner-shell ionization cross section (ISI; Green-Cosslett's model was used in all cases except for numerical integration)

1. Green-Cosslett[30] (co).
2. Gryzinski[31] (gr).

For the fluorescent yield ($w$, of significance only in the complete fluorescence and continuum fluorescence procedures)

1. Burhop[32] (bu).
2. Heinrich (unpublished) (he).
3. Howarth[33] (ho).
4. Laberrigue[33] (not used).
5. Wapstra[33] (not used).

For x-ray mass absorption coefficients ($\mu$)

1. The values used in program FRAME (modified by Myklebust from Heinrich[34]) (fm).
2. Heinrich[35] (kh).
3. PAPMAC[10] (not used in the present work).
4. On-line numerical input (not used).

### 18.2.5. The Atomic Number Subset

Because the atomic number effect affects all measurements, its evaluation must precede those for absorption and fluorescence, which are not always significant. The following cases were purged from the data set in order to form the atomic number subset:

1. When the absorption factor for the specimen was below 0.85.
2. When the difference between the two atomic numbers was below four (this excludes cases of strong $K$-$K$ fluorescence and of negligible atomic number effect).
3. When the overvoltage (Eo / Eq) was below 1.5 (The inclusion of such cases would significantly deteriorate the accuracy of the overall result; see last entry in Table 18.1).

This subset consists of 498 cases. The element pairs represented in it are reasonably distributed, as shown in Figure 18.3. The statistics of all combinations of procedures which were tested are reported in Table 18.1.

Figure 18.4 shows the limits of errors of 50% and 90% of the test data. The errors there reported include, of course, the errors in the assumed standard compositions and measurement errors, as well as those caused by the correction procedures.

The options used to calculate the values shown in Table 18.1 are arranged

TABLE 18.1.　Results with the Atomic Number Subset

| Name | $S$ | $R$ | $F_a$ | flu | con | $J$ | Joy | $\mu$ | w | ISI | Mean | S.D. |
|---|---|---|---|---|---|---|---|---|---|---|---|---|
| (498cases) | | | | | | | | | | | | |
| zaf1 | ave | dun | pdh | - | — | wil | — | fm | | | −.0097 | .0150 |
| zaf2 | ave | dun | pdh | r | — | wil | — | fm | | | −.0081 | .0146 |
| zaf3a | ave | dun | pdh | r | — | blo | — | fm | | | −.0010 | .0161 |
| zaf4 | ave | dun | pdh | r | — | dun | — | fm | | | +.0016 | .0207 |
| zaf4b | li | dun | pdh | r | — | dun | — | kh | | | +.0016 | .0203 |
| zaf4a | li | dun | pdh | r | — | bs | — | kh | | | +.0005 | .0133 |
| zaf5 | li | dun | pdh | r | — | bs | — | fm | | | +.0009 | .0133 |
| zaf6 | li | dun | q | r | — | bs | — | fm | | | +.0016 | .0136 |
| zaf7 | li | dun | dpl | r | — | bs | — | kh | | | +.0007 | .0140 |
| zaf7a | ni | dun | pdh | r | — | bs | — | kh | | co | +.0005 | .0133 |
| zaf7b | ni | dun | phd | r | — | bs | — | kh | | gr | +.0005 | .0132 |
| zaf7c | li | yak | phd | r | — | bs | — | kh | | | −.0022 | .0140 |
| zaf7d | li | dun | pdh | c | — | bs | — | kh | bu | | −.0001 | .0139 |
| zaf7e | li | dun | pdh | c | — | bs | — | kh | he | | −.0001 | .0139 |
| zaf7f | li | dun | pdh | c | — | bs | — | kh | ho | | −.0001 | .0139 |
| zaf7g | li | dun | pdh | c | kh | bs | — | kh | bu | | +.0001 | .0152 |
| zaf7g1 | ni | dun | dpl | c | kh | bs | — | kh | bu | gr | +.0003 | .0154 |
| zaf7g2 | li | dun | dpl | r | — | bs | — | kh | | | +.0008 | .0140 |
| zaf8 (Scott) | sct | sct | sct | r | — | wil | — | kh | | | −.0037 | .0117 |
| zaf8a | sct | dun | dpl | r | — | wil | — | kh | | | −.0081 | .0119 |
| zaf8b | sct | dun | dpl | r | — | bs | — | kh | | | +.0010 | .0122 |
| zaf8c | sct | pap | sct | r | — | wil | — | kh | | | −.0078 | .0120 |
| zaf8d | sct | dun | pdh | r | — | bs | — | kh | | | +.0009 | .0114 |
| zaf9 (PAP) | pap | pap | pap | r | — | zel | — | kh | | | −.0023 | .0115 |
| zaf9a | pap | pap | pap | r | — | bs | — | kh | | | +.0001 | .0110 |
| zafa | pap | pap | pap | r | — | spr | — | kh | | | −.0052 | .0145 |
| zafb | pap | pap | pap | r | — | kh | — | kh | | | −.0040 | .0120 |
| zaf10 (PPS) | pap | pap | pps | r | — | zel | — | kh | | | −.0020 | .0146 |
| zaf11 | pap | pap | dpl | r | — | bs | — | kh | | | −.0003 | .0113 |
| zaf11a | pap | dun | pap | r | — | zel | — | kh | | | −.0019 | .0119 |
| zaf12 | pap | dun | dpl | r | — | zel | — | kh | | | −.0024 | .0119 |
| zaf13 | pap | dun | dpl | r | — | kh | — | kh | | | −.0000 | .0114 |
| zaf14 | pap | dun | pdh | r | — | bs | | kh | | | −.0002 | .0112 |
| zaf15 | pap | dun | sct | r | — | bs | — | kh | | | +.0005 | .0112 |
| zaf17 | pap | my | dpl | r | — | bs | — | kh | | | +.0022 | .0124 |
| Uo | pap | dun | dpl | r | — | bs | — | kh | | | +.0008 | .0145 |
| (89 cases) | | | | | | | | | | | | |

approximately in the historical sequence. Each line involves the calculation of the intensity ratios from the postulated composition of the 498 entries, the calculation of the errors in concentration, and the statistics (mean and standard deviation).

Figure 18.4 shows the medians and ranges of 50% and 90% of the errors for selected lines of Table 18.1.

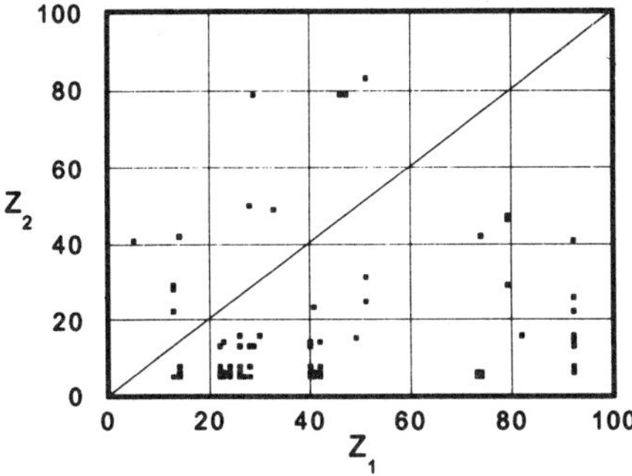

FIGURE 18.3.   Atomic numbers of sets of element pairs in the subset for the $Z$ correction.

*Commentaries to the Test Results in Table 18.1.*

*zaf1–4.* In this early approach, the stopping power is calculated at a single electron energy halfway between the operating beam energy and the critical excitation potential of the line in question. The omission of fluorescence in the first set, and the choices of $J$ (except for that of Duncumb) do not strongly affect the standard variation. The formula for $J$ proposed by Duncumb produces a significant increase in the standard deviation.

*zaf5–17.* The series using the logarithmic-integral technique or numerical integration yields better results than the the group zaf1–3. The complete calculation of characteristic fluorescence[21] does not improve the results ($K$-$K$ fluorescence is excluded from the data set). The choice of fluorescence yield is not significant, and the introduction of fluorescence due to the continuum[23] actually worsens the statistics. (This correction would be important in the determination of small amounts of an element in a low-atomic weight matrix, using high-energy lines, such as, for instance, determining 1% Cu in Be using the Cu $K\alpha$ line. Such situations are not present in the

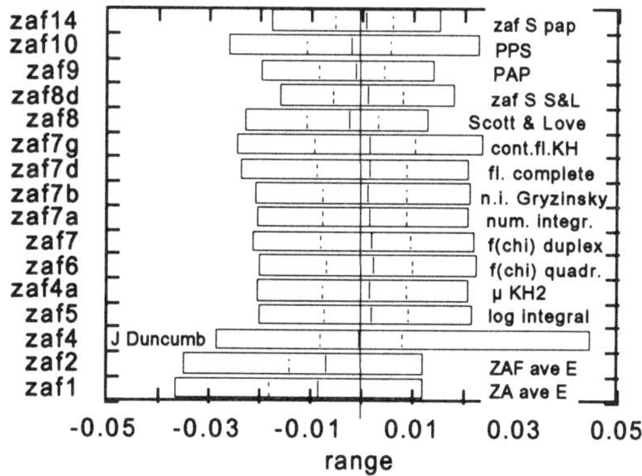

FIGURE 18.4.   Median (solid line) and range distributions for 50% (dashed line) and 90% (box limits) of errors in selected data sets from Table 18.1. For a key of methods see Table 18.1.

test set.) The choices of the absorption correction ($f_A$) and of the x-ray absorption coefficient ($\mu$) do not greatly affect the results, as can be expected in a set in which absorption factors below 0.85 were excluded. This fact shows that the approach of partial data sets is useful.

The choice of backscatter model is not very significant either (compare zaf11 with zaf6, and those from zaf8 on). This was also suspected at the onset of the exercise.

Significant reductions of the standard deviation are achieved in the method of Scott (zaf8) and the PAP model of Pouchou and Pichoir. The improvement is due to the fact that these authors have replaced the Bethe equation of energy loss by empirical approaches which are more realistic at low electron energies, while converging towards the Bethe equation at higher energies. That this is the reason for the improvement was shown in sets not reported here, in which factors other than the $S$ factor were replaced by those of other authors, without deterioration of the statistics (or even marginal improvement). The rise in standard deviation observed when the PAP equation for $S$ was replaced by the PPS equation of the same authors was surprising since this equation only affects the absorption term.

A separate test of cases with overvoltage $Eo/Eq$ below 1.5 shows that the standard deviation is higher in these cases, but not dramatically so.

*Conclusions.* An atomic number correction based on either Scott's or the PAP formula, with the conventional Reed fluorescence correction and with omission of the continuum fluorescence correction, is appropriate for the next part of the study, which deals with the absorption correction. The $J$ factors proposed by Berger and Seltzer and those of Zeller give similarly good results. The choices of backscatter correction, fluorescence yield, and inner-level ionization cross section are, at this point, irrelevant.

The phi-rho-z approach of J. Brown and that of Packwood[12] were incorporated in T-CONVERT at this point, so that their atomic number and absorption corrections could be compared with those of the ZAF methods.


### 18.2.6. The Absorption Subset of the Data Set

To study the effects of the absorption of primary x rays in the specimen a second subset (absorption correction subset) was selected according the following criteria:

1. The absorption factor for the specimen had to be equal to or lower than 0.85 (calculated by the duplex procedure)[20] in order to exclude cases in which the absorption uncertainty would not be expected to be significant.

2. Cases in which the difference between the two atomic numbers was below 4 were eliminated (this excludes the strong $K$-$K$ fluorescence).

3. Cases involving primary emission from elements of atomic number below 10 were eliminated. While the absorption correction is particularly important for these cases, complications arise from uncertainties in mass absorption coefficients,(4,10) from the emission in bands that are wider than the resolution of spectrometers and vary in position and shape with the nature of the chemical bonds, from potential effects of low specimen conductivity, and from the fact that essentially only one source of data (Bastin) is available. Inclusion of the emissions

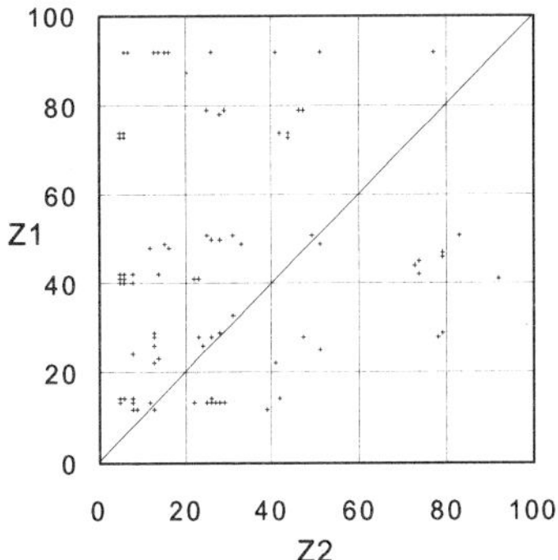

FIGURE 18.5.   Atomic numbers of sets of element pairs in the subset for the absorption correction.

of elements of atomic number below 10 would therefore cloud the problem to be treated here; they should be handled separately from the present study.

The absorption subset consists of 812 cases. The distribution of the element pairs represented in it is shown in Figure 18.5. The subset is affected by the errors of measurement, the errors in the estimated composition and in the atomic number correction, and the errors due to the absorption. It should be noted that the atomic-number correction only refers to the variation of primary x-ray generation (per unit of concentration) as a function of matrix composition. The absorption correction also contains an effect related to the matrix composition, since the distribution of depth within the specimen of x-ray emission (phi-rho-z curve) varies with the matrix; hence the distribution of length of x-ray trajectories in the specimen is also a function of matrix composition.

Experience indicates that errors in the absorption correction by any method correlate better with the *ratio* of the absorption factors of binary specimen and elementary standard ($F_A = f_{spec}/f_{std}$) than with the absorption factor of the specimen itself. Figure 18.6 shows a strong correlation between this ratio and the ratio of atomic

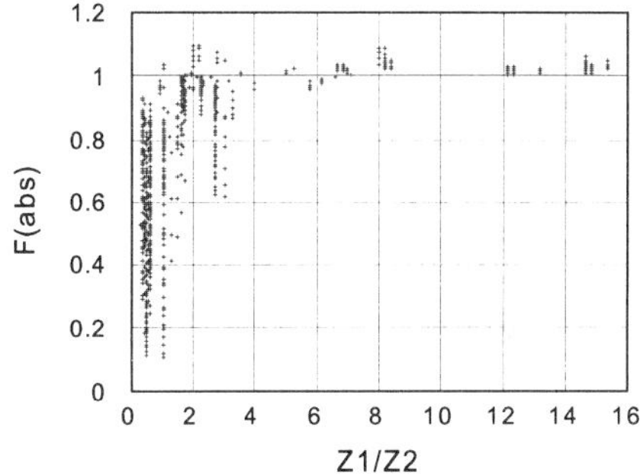

FIGURE 18.6.  The absorption correction factor $F_A$ as a function of the ratio of atomic numbers.

TABLE 18.2.    Results with the Absorption Correction Subset

| Name | $S$ | $R$ | $F_a$ | fl | $J$ | $J_v$ | μ | w | II | Mean | S.D. | RMS |
|---|---|---|---|---|---|---|---|---|---|---|---|---|
| $f^* < 0.90$, | | | | | | | | | | | | |
| 1060 cases | | | | | | | | | | | | |
| azf1—ex | pap | dun | pdh | r | zel | — | kh | | | −.0020 | .0406 | |
| $f^* < 0.85$, $Z_1 > 10$ | | | | | | | | | | | | |
| 812 cases | | | | | | | | | | | | |
| ZAF: | | | | | | | | | | | | |
| azf22 3.5 s | ave | dun | pdh | r | bs | — | kh | he | co | +.0041 | .0169 | .0174 |
| azf13 | li | dun | pdh | r | bs | — | kh | he | co | +.0041 | .0171 | .0176 |
| azf14 4 s | ni | dun | q | r | bs | — | kh | he | co | −.0056 | .0184 | .0193 |
| azf15 | ni | dun | dpl | r | bs | — | kh | he | co | −.0074 | .0161 | .0177 |
| azf16 | ni | dun | sct | r | bs | — | kh | he | co | −.0046 | .0145 | .0152 |
| azf17 | ni | dun | sct | r | wil | — | kh | he | co | −.0068 | .0143 | .0159 |
| Duncumb: | | | | | | | | | | | | |
| azf18a | ni | dun | dun | r | d2 | + | kh | he | co | +.0006 | .0138 | .0138 |
| PAP: | | | | | | | | | | | | |
| azf1a | pap | pap | pap | r | zel | — | kh | he | co | −.0016 | .0134 | .0135 |
| azf10 | pap | pap | pap | r | bs | — | kh | he | co | −.0002 | .0186 | .0186 |
| PPS: | | | | | | | | | | | | |
| azf24 | pap | pap | pps | r | zel | — | kh | he | co | +.0005 | .0150 | .0150 |
| azf7 | pap | pap | dun | r | zel | — | kh | he | co | +.0036 | .0145 | .0150 |
| azf21b | pap | pap | br | r | wil | — | kh | he | co | −.0007 | .0147 | .0165 |
| azf8 | pap | pap | sct | r | wil | — | kh | he | co | −.0008 | .0150 | .0168 |
| azf30 | pap | pap | pw | r | zel | — | kh | he | co | −.0002 | .0186 | .0186 |
| azf3 | pap | dun | sct | r | zel | — | kh | he | co | −.0036 | .0134 | .0139 |
| azf4 | pap | dun | pps | r | zel | — | kh | he | co | +.0024 | .0152 | .0154 |
| azf5 | pap | dun | dpl | r | zel | — | kh | he | co | −.0064 | .0149 | .0162 |
| azf6 | pap | dun | q | r | zel | — | kh | he | co | −.0046 | .0174 | .0180 |
| azf1b | pap | dun | pdh | r | bs | — | kh | he | co | +.0058 | .0173 | .0191 |
| azf27 | pap | dun | dun | r | d2 | — | kh | he | co | +.0064 | .0192 | .0202 |
| Scott: | | | | | | | | | | | | |
| azf2b | sct | sct | sct | s | wil | — | kh | he | co | −.0029 | .0131 | .0134 |
| azf2a | sct | sct | sct | s | blo | — | kh | he | co | +.0054 | .0167 | .0184 |
| azf11 | sct | sct | sct | s | bs | — | kh | he | co | −.0009 | .0151 | .0151 |
| Packwood: | | | | | | | | | | | | |
| azf3 | pw | la | pw | s | wil | — | kh | he | co | −.0014 | .0206 | .0207 |
| Brown: | | | | | | | | | | | | |
| azf21a | br | — | br | s | bs | — | kh | he | co | −.0130 | .0190 | .0230 |
| azf12 | br | — | br | s | wil | — | kh | he | co | −.0130 | .0190 | .0230 |

The abbreviations br and pw refer to Brown (ref. 11) and Packwood (ref. 12).
For Packwood's backscatter correction, his first variant is taken.

numbers. This correlation renders difficult the evaluation of the absorption correction model as a function of atomic number, since regardless of the model large absorption occurs more frequently when the ratio of atomic numbers is close to one.

In some cases, the use of the formulas of the authors for the absorption term, $f(\chi)$, resulted in failure of the computer program. It is improbable that these failures are caused by computer roundup error. A very simple numerical integration in 10–20 steps of the term $\varphi(\rho z) \exp(-\chi \cdot \rho z)$ eliminated the problem.

### 18.2.7. Discussion of the Absorption Correction

As expected, most errors reported in the absorption table (Table 18.2) exceed those in the atomic number table (Table 18.1). The root mean squares (RMS) tend to converge toward a value of 0.013, which is probably the sum of experimental errors and errors in the estimate of composition of the binaries; this value compares well with 0.011, the value toward to which the data of Table 18.1 converge.

The lowest values of RMS were obtained with the method of Love and Scott (azf2b),[12] the PAP procedure of Pouchou and Pichoir (azf1a),[7] and the new method of Duncumb (azf18a).[14] What distinguishes these three models from the classical ZAF procedures is a greater attention to the shape of the phi-rho-z curve, particularly at shallow depth, and the modifications of the traditional Bethe equation for stopping power.

A probability curve of the error distribution of any of the above methods shows non-linearity in the extremes, suggesting outliers, particularly on the high side (Figure 18.7). It would be tempting to exclude some of the suspect values. If this were done, the RMS would be reduced almost to the level of the RMS of Table 18.1, as shown in Table 18.4 for one method. This result suggests that the residual error in the better absorption methods is very small indeed. However, truncating the set of binary data would not change the conclusions that can be drawn from the original set.

This exercise also gives us a clearer picture of the nature and role of the factor $J$, the mean energy loss per ionization. In the three methods which give the smallest error, changes were made either in the equation for stopping power or in the factor $J$ that figures in it. Bethe's equation, which is commonly used in ZAF procedures, was formulated and experimentally confirmed for energies considerably higher than those found in the target interacting with the electron beam. At low electron energies, this classical model fails since the energy losses in an interaction cannot be larger than the kinetic energy of the electron. Joy has proposed a formula for the factor $J$ in Bethe's equation in which the effect of diminishing electron energy is taken into account.[29] Duncumb conserves the form of the Bethe equation while, following Joy, he modifies the value of $J$ as a monotonic function of electron energy. The methods of Love and Scott and of Pouchou and Pichoir are based on an empirical modification of the stopping power formula; each of them had selected previously an

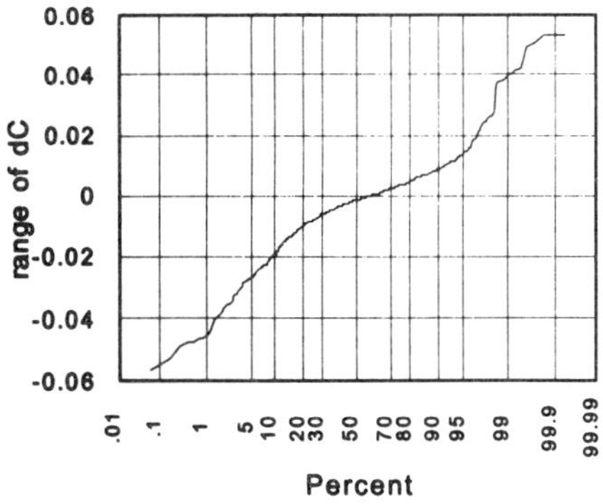


FIGURE 18.7. Probability plot of the error distribution in the set azf2b.

expression for the mean excitation energy $J$. As can be observed in the Tables, the choice of expression for J can strongly affect the accuracy of a given method. This parameter "is a well-defined parameter of the theory, which ... depends only upon the ground- and excited-state wave functions of a stopping material, independent of the energy and other characteristics of a stopping material."[26] In the context of the making of a stopping power model for electron probe analysis, however, this fundamental parameter is converted into a part of the empirical accommodation, since the energy loss equation is adjusted after a model for $J$ has been chosen. Each of the three most successful models uses a different expression for $J$, and in each case changing the expression for $J$ causes the error distribution to broaden. The expressions for $J$ and for the stopping-power model are therefore interdependent and should not be separated.

Love and Scott's and Duncumb's methods follow the traditional path of the ZAF procedure, in which the different parts for atomic number, absorption, and fluorescence calculation are clearly differentiated. The PAP method is formally derived in a somewhat different way, and its authors do not explicitly and separately calculate the absorption and atomic number expressions. However, PAP can be easily transcribed into the classical ZAF form, and doing so would render the method easier to understand. An explicit statement of the absorption factors is useful in the judgement of the experimental measurement conditions: many measurements reported in the data set are of interest for the method evaluation, but they are not good examples for choosing adequate conditions of measurement. In particular, one should minimize the absorption correction if this were possible without producing other problems. However, this is difficult to do if the absorption factors are not explicitly calculated.

*The phi-rho-z Methods.* As can be deduced from Tables 18.1 and 18.2, the performance of the phi-rho-z methods for the set of data used in this study shows higher error distributions than the ZAF-type calculations. The averaging of properties of elements for the multielement targets is a possible source of error. Since tracer experiments for multielement (or binary) matrices were never performed, no experimental proof of the averaging methods can be derived. It should be recalled that cases of very soft radiation, of energy below 1 kV, are excluded from the data set.

*Speed of Calculation.* While I have not systematically investigated the time invested in each procedure, this is a factor of possible interest when a large number of measurements is converted, as is the case in quantitative area scans. T-CONVERT has not been optimized for maximum speed but I believe that the fragmentary results observed at this point indicate significant differences in the speed at which these calculations are processed. It is not clear, however, how important the speed of calculation is, even for area scans. Usually the time required for data collection exceeds that for calculation.

## 18.3. EVALUATION OF THE CHARACTERISTIC FLUORESCENCE CORRECTION

### 18.3.1. The Procedure

The correction for characteristic line fluorescence is only necessary when an emitted line from an element other than that being measured is strong and close to the

TABLE 18.3.   Cases of Significant Fluorescence ($Ff > 0.5\%$)

| Element pair | Emitter | Excited by | Cases | Max. $F$ |
|---|---|---|---|---|
| 12–13 | 12 $K\alpha$ | 13 $K$-lines | 14 | 1.08 |
| 24–26 | 24 $K\alpha$ | 26 $K$-lines | 56 | 1.36 |
| 28–29 | 28 $K\beta$ | 29 $K\beta$ only | 07 | 1.03 |
| 29–30 | 29 $K\alpha$ | 30 $K\beta$ only | 01 | 1.006 |
| 14–42 | 14 $K\alpha$ | 42 $L$-lines | 07 | 1.009 |
| 28–78 | 28 $K\alpha$ | 78 $L$-lines | 04 | 1.03 |
| 29–79 | 29 $K\alpha$ | 79 $L$-lines | 63 | 1.07 |
| 29–13 | 29 $L\alpha_1$ | 13 $K$-lines | 04 | 1.03 |
| 41–23 | 41 $L\alpha_1$ | 23 $K$-lines | 14 | 1.015 |
| 50–26 | 50 $L\alpha_1$ | 26 $K$-lines | 01 | 1.005 |
| 51–25 | 51 $L\alpha_1$ | 25 $K$-lines | 51 | 1.012 |
| 79–47 | 79 $L\alpha_1$ | 47 $K$-lines | 06 | 1.018 |
| 79–47 | 79 $M\alpha$ | 47 $L$-lines | 08 | 1.016 |
| 49–51 | 49 $L\alpha 1$ | 51 $L$ except $L\alpha_1$ | 03 | 1.010 |
| Total: 193 | | | | |

edge that emits the measured line. In that case, the exciting line produces ionizations at the measured level, in addition to those produced by the primary electrons. Table 18.3 shows the cases of fluorescence in our binary data collection which affect the result to more than 0.5% (193 cases). The full calculation of the fluorescence contribution proceedes by the following steps:

1. Determine which elements produce lines that cause appreciable fluorescence.
2. Calculate the absolute intensities generated by these lines.
3. Determine the emergent fluorescent contribution due to each line.
4. Add all contributions and place them in ratio to the emergent primary intensity of the line which is measured.

The ratio determined in step 4 is called the fluorescence correction factor. The procedure described here can be applied only when the ratio of primary radiation of the two lines involved can be correctly estimated. If so, the largest uncertainty in the calculation is in the selection of the factors determining the emission of primary x rays of both the emitting and the indirectly excited element.

The most important case is that of the excitation of $K$-lines by one or more $K$-lines of other elements. This type of excitation may cause very strong fluorescence (up to 30% of the primary emission). However, the uncertainty in the primary emissions is insignificant since strong excitation is only obtained when the atomic numbers of the exciting and the excited elements are close, so that errors in calculating the primary emission tend to cancel. The same is true of the excitation of $L$-lines by $L$-lines, and of $M$-lines by $M$-lines. There is little difficulty in calculating these interactions with the required accuracy.

When the exciting lines are produced by a level other than that of the fluorescent emission (e.g., $K$-lines excited by $L$-lines or *vice versa*), the errors in the calculation of the ratios of primary intensities are larger. However, such cases are less frequently significant, since the fluorescent intensities are lower.

Occasionally, not all lines of the exciting level are capable of ionizing the excited element. For instance, in the case of Ni-Cu binaries, only the Cu $K\beta$ radiation can excite the nickel $K$-level. However, such cases are exceptional, and the intensities of the fluorescent emission are usually so low that they can be ignored without significant penalty.

### 18.3.2. The Test Procedure

We compare three approaches to the correction of fluorescence due to characteristic radiation. The first follows exactly the steps pointed out above, while the second approach uses a simplified procedure proposed by Castaing[1] and modified by Reed.[22] The full calculation of primary intensities is avoided and simplifications are applied to an approach by Green and Cosslett to calculating primary intensities.[30] The complete equation is as follows:

$$\frac{I_f'(a)}{I_p'(a)} = 0.5C_b \frac{\mu(a)}{\mu(ab)} \frac{r(a)-1}{r(a)} \, \omega(b) \frac{A(a)}{A(b)} \left(\frac{U_b-1}{U_a-1}\right)^{1.67} \tag{18.7}$$

where the first term is the ratio between emergent fluorescent and primary radiation of element a excited by element $b$, $C_b$ is the concentration of the exciting element, $\omega(b)$ is the fluorescent yield for element $b$ and the shell of interest, $r(a)$ is the appropriate weight of of line (relative line intensity; e.g., for the $K$-lines: $K\alpha/(K\alpha + K\beta)$, $A(a)$ and $A(b)$ are the atomic weights of elements $a$ and $b$, and the $U$ terms are their respective overvoltages, or ratios between operating and minimum exciting energy. $\mu(a)$ and $\mu(ab)$ are the mass absorption coefficients for the measured radiation of $a$ and for the elements $a$ and $b$, respectively.

TABLE 18.4.   Comparison of Errors in the Three Subsets

| Subset | Entries | Mean | S.D. | RMS |
|---|---|---|---|---|
| absorption | 1060 | −0.0005 | 0.0133 | 0.0133 |
| atomic number | 498 | + 0.0056 | 0.0171 | 0.0176 |
| fluorescence1 | 193 | −0.0016 | 0.0143 | 0.0143 |
| fluorescence2 | 186 | −0.0037 | 0.0083 | 0.0090 |
| fluorescence3 | 81 | + 0.0002 | 0.0079 | 0.0079 |
| fluorescence4 | 80 | −0.0003 | 0.0066 | 0.0066 |
| fluorescence5 | 106 | + 0.0118 | 0.0146 | 0.0157 |

where:
fluorescence1: all cases of the set included.
fluorescence2: the cases of Z 51–25 excluded.
fluorescence3: cases of $K$-$K$ and $L$-$L$ excitation only.
fluorescence4: same, with case 215 excluded.
fluorescence5: cases of $K$-$L$, $L$-$K$, and $M$-$L$ excitation.

TABLE 18.5.   Calculated Fluorescence Effects

| El.1 | El.2 | Conc.El.1 | Method | Fluor.yield | Fluor. factor |
|------|------|-----------|--------|-------------|---------------|
| Fe | Ni | 0.1 | complete | Burhop | 1.2825 |
|    |    |     |          | Heinrich | 1.2672 |
|    |    |     |          | Reed | 1.2939 |
|    |    |     | Reed (1968) | Burhop | 1.2682 |
|    |    |     |          | Heinrich | 1.2537 |
|    |    |     |          | Reed | 1.2791 |
|    |    |     | Reed (1990) | Burhop | 1.2621 |
|    |    |     |          | Heinrich | 1.2479 |
|    |    |     |          | Reed | 1.2727 |
| Mg | Al | 0.1 | complete | Burhop | 1.0417 |
|    |    |     |          | Heinrich | 1.0974 |
|    |    |     |          | Reed | 1.0639 |
|    |    |     | Reed (1968) | Burhop | 1.0373 |
|    |    |     |          | Heinrich | 1.0870 |
|    |    |     |          | Reed | 1.0571 |
|    |    |     | Reed (1990) | Burhop | 1.0414 |
|    |    |     |          | Heinrich | 1.0966 |

This approach considers fluorescence of $K$- and $L$-lines due to primary emission of $K$- or $L$-lines of appropriate energy. It does not include the cases in which only part of the $K$- or $L$-emission acts on the measured emission, and it neglects the effects from or on $M$-lines. These omissions are usually inconsequential since the effects of partial excitation, and the interactions involving $M$-lines, are smaller than the typical $K$-$K$, $L$-$K$ and $K$-$L$ interactions.

In 1990, Reed presented a thorough analysis of fluorescence correction and proposed a slight modification of the procedure for the important $K$-$K$ interaction.[36] This modified Reed procedure is the third alternative used.

Before discussing the details, we compare the errors observed in the fluorescence subset with those observed in the absorption and atomic number subsets (Table 18.4). This table indicates that the $K$-$K$ and $L$-$L$ fluorescence subset has a substantially smaller error distribution than the first two subsets. This is surprising since absorption and atomic number errors are certainly present in the fluorescence subset as well. However, the fluorescence factor is significant only when the primary absorption is moderate. Hence, the errors attributable to the absorption correction are not large in this subset. Furthermore, in the 81 cases of $K$-$K$ and $L$-$L$ excitation, the errors attributable to the atomic number correction cannot be large since significant fluorescence of this kind only occurs when the difference in atomic number is not large. Furthermore, most of the available dataset for $K$-$K$ interaction refers to the element pair 24-26, i.e., in the region of the periodic table for which the fluorescence corrections were originally designed and adapted.

It is also evident that no benefit can be derived from the consideration of data in which the total errors are much larger than the expected fluorescence contribution. For this reason we have excluded the data from the element pair 51-25 in fluorescence 3 and 4, as well as the data #215, in fluorescence 4. In the same fashion, the fluorescence produced in pair 29-30, in the cases of 14-42, and 50-26 can be ignored; several other

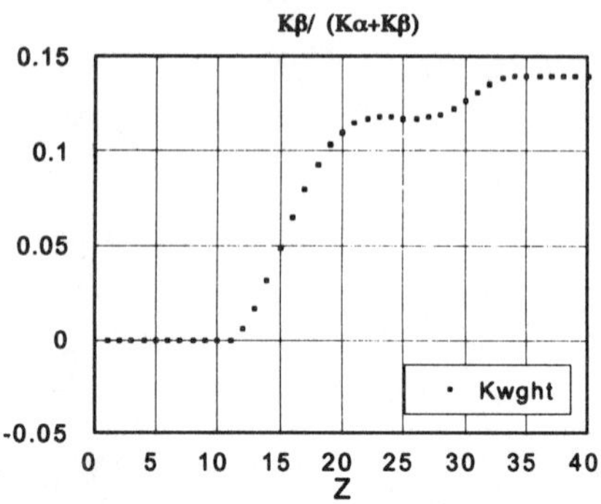



FIGURE 18.8. Ratio of the $K\beta$ intensity to the sum of $K$ intensities.

cases are of marginal significance. Clearly, the experimental evidence on fluorescent excitation of $L$- and $M$-lines is very weak.

I have compared the effects of choice of method and parameters on the calculated fluorescence factor for two binaries: iron-nickel, $Z$: 26-28 (10% iron) and magnesium-aluminum, $Z$:12-13 (10% magnesium). For both sets, the operating voltage was 20 kV and the emergence angle was 40. The factors for absorption and atomic number correction were conventional ZAF and do not affect these results (Table 18.5).

The weight of $K$-lines is not a significant source of error, particularly for low atomic numbers, since the relative intensity of the $K\beta$ lines falls rapidly below atomic number 20, being practically zero for atomic numbers below 12, as shown in Figure 18.8.

For the Mg-Al case, the choice of model for the fluorescent yield, $\omega$, is critical. Experimental values in this region of atomic numbers are scanty, and various models for $\omega$ produce a wide range of values. The model of Burhop[32] is a widely used one; that of Reed[36] is more recent; that of Heinrich (unpublished) was an earlier attempt

TABLE 18.6. Effect of the Model for Fluorescent Yield on the
Measurement of Mg. Aluminum = 10% Mg Alloy; 20 kV;
angle of 40°

| Formula for $w$ | Fluor.factor |
|---|---|
| Wapstra | 1.0318 |
| Burhop | 1.0417 |
| Laberrigue | 1.0435 |
| Reed 1 | 1.0452 |
| Heinrich 2 | 1.0516 |
| Reed 2 | 1.0639 |
| Heinrich 1 | 1.0974 |

TABLE 18.7. Effect of Fluorescent Yield on Experimental Error

| Model | Mean error | S.D. | Root mean sq. |
|---|---|---|---|
| No fluor. correction | +0.0023 | 0.0111 | 0.0110 |
| " #215 del. | − 0.0002 | 0.0065 | 0.0062 |
| Full, Wapstra | +0.0056 | 0.0106 | 0.0116 |
| " #215 del. | +0.0032 | 0.0061 | 0.0067 |
| Full, Heinrich 2 | +0.0076 | 0.0103 | 0.0125 |
| " #215 del. | +0.0053 | 0.0059 | 0.0078 |
| Full, Heinrich 1 | +0.0118 | 0.0095 | 0.0150 |
| " #215 del. | +0.0097 | 0.0054 | 0.0110 |
| Reed 2 | +0.0082 | 0.0100 | 0.0127 |
| " #215 del. | +0.0060 | 0.0057 | 0.0081 |

to fit a curve to newer experimental values, which probably produces too high values for aluminum. The new model of Reed produces negative values of $\omega$ for atomic numbers below 5, a fact that should not be of significance in this context.

The effect of the fluorescent yield model was tested on the group of data on Mg-Al alloys (#203-216 of the binary data list.) One of the data pairs (#215) shows a large error (larger than the fluorescent contribution), which is not influenced by the choice of fluorescent yield. Eliminating this pair substantially improves the statistics.

Surprisingly, the smallest errors are obtained when the fluorescence procedure is omitted; the next smallest set is the full fluorescence, using Wapstra's fluorescent yield, which is the smallest of those tried here. The abbreviated procedure by Reed[36] produces results similar to those with the full procedure and the Heinrich fluorescent yield, which is close to Reed's.

Since these results are derived from a single experimental set, it is not certain that their systematic measurement or calibration errors are not responsible for the residual errors. To the extent that the data can be trusted, they would suggest that all proposed algorithms for fluorescent yield (including the new procedure of Reed's), overestimate the yield for aluminum. The experimental data available are insufficient to clarify this point, and further experimental measurements of the fluorescent yield for elements of atomic number below 26 would be very useful.

### 18.3.3. Cases Involving Two Different Atomic Shells

The cases in which the shell corresponding to the exciting radiation differs from the excited shell are the most uncertain ones in characteristic fluorescence, for the following reasons:

1. The weights of lines and fluorescence yields of levels other than $K$ are poorly known.
2. There are fewer cases of binary data available than in the $K$-$K$ domain.
3. The fluorescence intensities are lower than in the $K$-$K$ case.
4. The atomic number difference of the element pair is such that errors in the atomic number, and sometimes also absorption calculation errors, are dominant in the total error.

It is, therefore, impractical to use the complete fluorescence method; instead, the new approach of Shields[4] is the best choice.


## REFERENCES


1. R. Castaing, Doctoral Thesis, University of Paris (1951).

2. P. J. Statham and J. B. Pawley, in: *Scanning Electron Microscopy—1978* (O. Johari, ed.) SEM Inc., Chicago, p. 469 (1978).

3. J. Small, S. T. Leigh, D. E. Newbury, and R. L. Myklebust, *J. Appl. Phys.* **61**(2), 459 (1987).

4. G. F. Bastin and H. G. M. Heijligers in: *Electron Probe Quantitation* (K. F. J. Heinrich and D. E. Newbury, eds.) Plenum Press, New York, pp. 145, 163 (1991).

5. K. F. J. Heinrich, *Electron Beam X-Ray Microanalysis*, Van Nostrand Reinhold, New York, pp. 350, 405 (1981).

6. H. Margenau and G. M. Murphy, *The Mathematics of Physics and Chemistry*, Second Ed., Van Nostrand, New York, p. 510 (1956).

7. K. J. Heinrich, in: *X-Ray Optics and Microanalysis 1992* (P. B. Kenway, P. J. Duke, G. W. Lorimer, T. Mulvey, I. W. Drummond, G. Love, A. G. Michette, and M. Stedman, eds.) Institute of Physics, Philadelphia, p. 113 (1993).

8. J. Philibert and R. Tixier, in: *Quantitative Electron Probe Microanalysis* (K. F. J. Heinrich, ed.) NBS Special Publication 298, National Bureau of Standards, Washington, DC., p. 13 (1968).

9. J. Henoc, K. F. J. Heinrich, and R. L. Myklebust, NBS Technical Note 769, National Bureau of Standards, Washington, DC (1973).

10. J-L. Pouchou and F. Pichoir, in: *Electron Probe Quantitation* (K. F. J. Heinrich and D. E. Newbury, eds.) Plenum Press, New York, p. 31 (1991).

11. J. D. Brown, in: *Electron Probe Quantitation* (K. F. J. Heinrich and D. E. Newbury, eds.) Plenum Press, New York, p. 77 (1991).

12. R. Packwood, in: *Electron Probe Quantitation* (K. F. J. Heinrich and D. E. Newbury, eds.) Plenum Press, New York, p. 83 (1991).

13. V. D. Scott and G. Love, in: *Electron Probe Quantitation* (K. F. J. Heinrich and D. E. Newbury, eds.) Plenum Press, New York, p. 19 (1991).

14. P. Duncumb, *X-Ray Optics and Microanalysis 1992* (P. B. Kenway, P. J. Duke, G. W. Lorimer, T. Mulvey, I. W. Drummond, G. Love, A. G. Michette, and M. Stedman, eds.) Institute of Physics, Philadelphia, p. 43 (1991).

15. P. Duncumb and S. J. B. Reed, NBS Special Publication 298, National Bureau of Standards, Washington, DC, p. 133 (1968).

16. H. Yakowitz: personal communication (1976).

17. R. L. Myklebust, *J. Physique* **45**, C2 (1984).

18. P. Duncumb, P. K. Shields-Mason and C. da Casa, in: *Proceedings of the 5th International Congress X-Ray Optics and Microanalysis* (G. Möllenstedt and H. K. Gaukler, eds.) Springer, Berlin, p. 146 (1969).

19. K. F. J. Heinrich and H. Yakowitz, *Anal. Chem.* **47**, 2408 (1975).

20. K. F. J. Heinrich, in: *Microbeam Analysis—1985* (J. T. Armstrong, ed.) San Francisco Press, San Francisco, p. 79 (1985).

21. K. F. J. Heinrich, in: *Microbeam Analysis*—1987 (R. H. Geiss, ed.) San Francisco Press, San Francisco, p. 23 (1987).
22. S. J. B. Reed and J. V. B. Long, in: *ICXOM 3* (H. H. Pattee, V. E. Cosslett, and A. Engström, eds.) Academic Press, New York, p. 317 (1963).
23. K. F. J. Heinrich, in: *Microbeam Analysis*—1987 (R. H. Geiss, ed.) San Francisco Press, San Francisco, p. 24 (1987).
24. M. V. Wilson, *Phys. Rev.* **60**, 749 (1941).
25. J. Bloch, *Z. Phys.* **81**, 363 (1933).
26. M. J. Berger and S. M. Seltzer, National Research Council Publication **1133**, National Acad. of Sciences, Washington DC, p. 205 (1964).
27. G. Springer, *Neues Jahrb. Mineral. Monatsh.* **9/10**, 304 (1067).
28. Zeller, cited by J. Ruste and M. Gantois, *J. Phys. D: Appl. Phys.* **8**, 872 (1975).
29. D. C. Joy and S. Luo, *Scanning* **11**, 176 (1989).
30. M. Green and V. E. Cosslett, *Proc. Phys. Soc. London* **78**, 1206 (1961).
31. M. Gryzinski, *Phys. Rev. A* **138**, 336 (1965).
32. E. H. S. Burhop, *J. Phys. Radium* **16**, 625 (1955).
33. W. Bambinek B. Craseman, R. W. Fink, H. U. Freund, H. Mark, C. D. Swift, R. E. Price, and P. V. Rao, *Rev. Mod. Phys.* **44**, 716 (1972).
34. K. F. J. Heinrich, in: *The Electron Microprobe*, (T. D. McKinley, K. F. J. Heinrich, and D. B. Wittry, eds.) J. Wiley & Sons, New York p. 296 (1966).
35. K. F. J. Heinrich, in: *Proceedings of ICXOMII* (J. D. Brown and R. H. Packwood, eds.) London, Ontario, Canada, p. 67 (1986).
36. S. J. B. Reed in: *Microbeam Analysis*–1990 (J. R. Michael and P. Ingram, eds.) San Francisco Press, San Francisco, p. 109 (1990).


APPENDIX 18.1: Table of Binary Data

| Item | Line | Z1 | Z2 | C1 | Eo | Ec | Angle | $k$ exp | Origin |
|---|---|---|---|---|---|---|---|---|---|
| 1 | 1 | 5 | 6 | 0.7820 | 4000 | 188 | 40.0 | 0.7210 | 7 |
| 2 | 1 | 5 | 6 | 0.7981 | 4000 | 188 | 40.0 | 0.7768 | 1 |
| 3 | 1 | 5 | 6 | 0.7981 | 6000 | 188 | 40.0 | 0.7754 | 1 |
| 4 | 1 | 5 | 6 | 0.7981 | 8000 | 188 | 40.0 | 0.7471 | 1 |
| 5 | 1 | 5 | 6 | 0.7981 | 10000 | 188 | 40.0 | 0.7309 | 1 |
| 6 | 1 | 5 | 6 | 0.7981 | 12000 | 188 | 40.0 | 0.7132 | 1 |
| 7 | 1 | 5 | 6 | 0.7981 | 15000 | 188 | 40.0 | 0.6853 | 1 |
| 8 | 1 | 5 | 6 | 0.7981 | 20000 | 188 | 40.0 | 0.6544 | 1 |
| 9 | 1 | 5 | 6 | 0.7981 | 25000 | 188 | 40.0 | 0.6516 | 1 |
| 10 | 1 | 5 | 6 | 0.7981 | 30000 | 188 | 40.0 | 0.6574 | 1 |
| 11 | 1 | 5 | 7 | 0.4348 | 10000 | 188 | 40.0 | 0.2360 | 1 |
| 12 | 1 | 5 | 7 | 0.4348 | 12000 | 188 | 40.0 | 0.2230 | 1 |
| 13 | 1 | 5 | 7 | 0.4348 | 15000 | 188 | 40.0 | 0.2080 | 1 |
| 14 | 1 | 5 | 7 | 0.4348 | 20000 | 188 | 40.0 | 0.1853 | 1 |
| 15 | 1 | 5 | 7 | 0.4348 | 25000 | 188 | 40.0 | 0.1725 | 1 |
| 16 | 1 | 5 | 7 | 0.4348 | 30000 | 188 | 40.0 | 0.1650 | 1 |
| 17 | 1 | 5 | 13 | 0.4449 | 4000 | 188 | 40.0 | 0.1872 | 1 |

| | | | | | | | | | |
|---|---|---|---|---|---|---|---|---|---|
| 18 | 1 | 5 | 13 | 0.4449 | 6000 | 188 | 40.0 | 0.1057 | 1 |
| 19 | 1 | 5 | 13 | 0.4449 | 8000 | 188 | 40.0 | 0.0648 | 1 |
| 20 | 1 | 5 | 13 | 0.4449 | 10000 | 188 | 40.0 | 0.0501 | 1 |
| 21 | 1 | 5 | 13 | 0.4449 | 12000 | 188 | 40.0 | 0.0396 | 1 |
| 22 | 1 | 5 | 13 | 0.4449 | 15000 | 188 | 40.0 | 0.0342 | 1 |
| 23 | 1 | 5 | 13 | 0.8278 | 4000 | 188 | 40.0 | 0.6077 | 1 |
| 24 | 1 | 5 | 13 | 0.8278 | 6000 | 188 | 40.0 | 0.4492 | 1 |
| 25 | 1 | 5 | 13 | 0.8278 | 8000 | 188 | 40.0 | 0.3391 | 1 |
| 26 | 1 | 5 | 13 | 0.8278 | 10000 | 188 | 40.0 | 0.2679 | 1 |
| 27 | 1 | 5 | 13 | 0.8278 | 12000 | 188 | 40.0 | 0.2244 | 1 |
| 28 | 1 | 5 | 13 | 0.8278 | 15000 | 188 | 40.0 | 0.1864 | 1 |
| 29 | 1 | 5 | 13 | 0.8278 | 20000 | 188 | 40.0 | 0.1684 | 1 |
| 30 | 1 | 5 | 13 | 0.8278 | 25000 | 188 | 40.0 | 0.1638 | 1 |
| 31 | 1 | 5 | 13 | 0.8278 | 30000 | 188 | 40.0 | 0.1694 | 1 |
| 32 | 1 | 5 | 14 | 0.5220 | 4000 | 188 | 40.0 | 0.2171 | 1 |
| 33 | 1 | 5 | 14 | 0.5220 | 6000 | 188 | 40.0 | 0.1131 | 1 |
| 34 | 1 | 5 | 14 | 0.5220 | 8000 | 188 | 40.0 | 0.0703 | 1 |
| 35 | 1 | 5 | 14 | 0.5220 | 10000 | 188 | 40.0 | 0.0544 | 1 |
| 36 | 1 | 5 | 14 | 0.5220 | 12000 | 188 | 40.0 | 0.0445 | 1 |
| 37 | 1 | 5 | 14 | 0.5220 | 15000 | 188 | 40.0 | 0.0384 | 1 |
| 38 | 1 | 5 | 14 | 0.6857 | 4000 | 188 | 40.0 | 0.3541 | 1 |
| 39 | 1 | 5 | 14 | 0.6857 | 6000 | 188 | 40.0 | 0.2113 | 1 |
| 40 | 1 | 5 | 14 | 0.6857 | 8000 | 188 | 40.0 | 0.1401 | 1 |
| 41 | 1 | 5 | 14 | 0.6857 | 10000 | 188 | 40.0 | 0.1071 | 1 |
| 42 | 1 | 5 | 14 | 0.6857 | 12000 | 188 | 40.0 | 0.0850 | 1 |
| 43 | 1 | 5 | 14 | 0.6857 | 15000 | 188 | 40.0 | 0.0713 | 1 |
| 44 | 1 | 5 | 22 | 0.1678 | 4000 | 188 | 40.0 | 0.1623 | 1 |
| 45 | 1 | 5 | 22 | 0.1678 | 6000 | 188 | 40.0. | 0.1242 | 1 |
| 46 | 1 | 5 | 22 | 0.1678 | 8000 | 188 | 40.0 | 0.0990 | 1 |
| 47 | 1 | 5 | 22 | 0.1678 | 10000 | 188 | 40.0 | 0.0795 | 1 |
| 48 | 1 | 5 | 22 | 0.1678 | 12000 | 188 | 40.0 | 0.0699 | 1 |
| 49 | 1 | 5 | 22 | 0.1678 | 15000 | 188 | 40.0 | 0.0603 | 1 |
| 50 | 1 | 5 | 22 | 0.1678 | 20000 | 188 | 40.0 | 0.0542 | 1 |
| 51 | 1 | 5 | 22 | 0.1678 | 25000 | 188 | 40.0 | 0.0552 | 1 |
| 52 | 1 | 5 | 22 | 0.1678 | 30000 | 188 | 40.0 | 0.0578 | 1 |
| 53 | 1 | 5 | 22 | 0.3007 | 4000 | 188 | 40.0 | 0.2956 | 1 |
| 54 | 1 | 5 | 22 | 0.3007 | 6000 | 188 | 40.0 | 0.2244 | 1 |
| 55 | 1 | 5 | 22 | 0.3007 | 8000 | 188 | 40.0 | 0.1807 | 1 |
| 56 | 1 | 5 | 22 | 0.3007 | 10000 | 188 | 40.0 | 0.1498 | 1 |
| 57 | 1 | 5 | 22 | 0.3007 | 12000 | 188 | 40.0 | 0.1279 | 1 |
| 58 | 1 | 5 | 22 | 0.3007 | 15000 | 188 | 40.0 | 0.1158 | 1 |
| 59 | 1 | 5 | 22 | 0.3007 | 20000 | 188 | 40.0 | 0.1055 | 1 |
| 60 | 1 | 5 | 22 | 0.3007 | 25000 | 188 | 40.0 | 0.1063 | 1 |
| 61 | 1 | 5 | 22 | 0.2840 | 30000 | 188 | 40.0 | 0.1090 | 1 |
| 62 | 1 | 5 | 23 | 0.2840 | 4000 | 188 | 40.0 | 0.2583 | 1 |
| 63 | 1 | 5 | 23 | 0.2840 | 6000. | 188 | 40.0 | 0.1980 | 1 |

| 64 | 1 | 5 | 23 | 0.2840 | 8000 | 188 | 40.0 | 0.1484 | 1 |
|---|---|---|---|---|---|---|---|---|---|
| 65 | 1 | 5 | 23 | 0.2840 | 10000 | 188 | 40.0 | 0.1235 | 1 |
| 66 | 1 | 5 | 23 | 0.2840 | 12000 | 188 | 40.0 | 0.1058 | 1 |
| 67 | 1 | 5 | 23 | 0.2840 | 15000 | 188 | 40.0 | 0.0865 | 1 |
| 68 | 1 | 5 | 24 | 0.1680 | 4000 | 188 | 40.0 | 0.1387 | 1 |
| 69 | 1 | 5 | 24 | 0.1680 | 6000 | 188 | 40.0 | 0.0991 | 1 |
| 70 | 1 | 5 | 24 | 0.1680 | 8000 | 188 | 40.0. | 0.0747 | 1 |
| 71 | 1 | 5 | 24 | 0.1680 | 10000 | 188 | 40.0 | 0.0611 | 1 |
| 72 | 1 | 5 | 24 | 0.1680 | 12000 | 188 | 40.0 | 0.0512 | 1 |
| 73 | 1 | 5 | 24 | 0.1680 | 15000 | 188 | 40.0 | 0.0444 | 1 |
| 74 | 1 | 5 | 24 | 0.2731 | 4000 | 188 | 40.0 | 0.2301 | 1 |
| 75 | 1 | 5 | 24 | 0.2731 | 6000 | 188 | 40.0 | 0.1682 | 1 |
| 76 | 1 | 5 | 24 | 0.2731 | 8000 | 188 | 40.0 | 0.1268 | 1 |
| 77 | 1 | 5 | 24 | 0.2731 | 10000 | 188 | 40.0 | 0.1037 | 1 |
| 78 | 1 | 5 | 24. | 0.2731 | 12000 | 188 | 40.0 | 0.0891 | 1 |
| 79 | 1 | 5 | 24 | 0.2731 | 15000 | 188 | 40.0 | 0.0767 | 1 |
| 80 | 1 | 5 | 26 | 0.0882 | 4000 | 188 | 40.0 | 0.0611 | 1 |
| 81 | 1 | 5 | 26 | 0.0882 | 8000 | 188 | 40.0 | 0.0290 | 1 |
| 82 | 1 | 5 | 26 | 0.1622 | 15000 | 188 | 40.0 | 0.0292 | 1 |
| 83 | 1 | 5 | 26 | 0.1622 | 4000 | 188 | 40.0 | 0.1131 | 1 |
| 84 | 1 | 5 | 26 | 0.1622 | 8000 | 188 | 40.0 | 0.0533 | 1 |
| 85 | 1 | 5 | 26 | 0.1622 | 10000 | 188 | 40.0 | 0.0414 | 1 |
| 86 | 1 | 5 | 26 | 0.1622 | 12000 | 188 | 40.0 | 0.0345 | 1 |
| 87 | 1 | 5 | 26 | 0.0882 | 6000 | 188 | 52.5 | 0.0409 | 1 |
| 88 | 1 | 5 | 26. | 0.0882 | 10000 | 188 | 52.5 | 0.0226 | 1 |
| 89 | 1 | 5 | 26 | 0.0882 | 12000 | 188 | 52.5 | 0.0190 | 1 |
| 90 | 1 | 5 | 26 | 0.0882 | 15000 | 188 | 52.5 | 0.0169 | 1 |
| 91 | 1 | 5 | 26 | 0.1622 | 6000 | 188 | 52.5 | 0.0757 | 1 |
| 92 | 1 | 5 | 27 | 0.0840 | 4000 | 188 | 40.0 | 0.0492 | 1 |
| 93 | 1 | 5 | 27 | 0.0840 | 6000 | 188 | 40.0 | 0.0301 | 1 |
| 94 | 1 | 5 | 27 | 0.0840 | 8000 | 188 | 40.0 | 0.0212 | 1 |
| 95 | 1 | 5 | 27 | 0.0840 | 10000 | 188 | 40.0 | 0.0167 | 1 |
| 96 | 1 | 5 | 27 | 0.0840 | 12000 | 188 | 40.0 | 0.0137 | 1 |
| 97 | 1 | 5 | 27 | 0.0840 | 15000 | 188 | 40.0 | 0.0117 | 1 |
| 98 | 1 | 5 | 27 | 0.0882 | 5000 | 188 | 40.0 | 0.0446 | 1 |
| 99 | 1 | 5 | 27 | 0.0882 | 10000 | 188 | 40.0 | 0.0210 | 5 |
| 100 | 1 | 5 | 27 | 0.0882 | 15000 | 188 | 40.0 | 0.0160 | 5 |
| 101 | 1 | 5 | 27 | 0.1550 | 4000 | 188 | 40.0 | 0.0981 | 1 |
| 102 | 1 | 5 | 27 | 0.1550 | 6000 | 188 | 40.0 | 0.0619 | 1 |
| 103 | 1 | 5 | 27 | 0.1550 | 8000 | 188 | 40.0 | 0.0423 | 1 |
| 104 | 1 | 5 | 27 | 0.1550 | 10000 | 188 | 40.0 | 0.0333 | 1 |
| 105 | 1 | 5 | 27 | 0.1550 | 12000 | 188 | 40.0 | 0.0282 | 1 |
| 106 | 1 | 5 | 27 | 0.1550 | 15000 | 188 | 40.0 | 0.0239 | 1 |
| 107 | 1 | 5 | 28 | 0.0578 | 4000 | 188 | 40.0 | 0.0281 | 1 |
| 108 | 1 | 5 | 28 | 0.0578 | 6000 | 188 | 40.0 | 0.0164 | 1 |
| 109 | 1 | 5 | 28 | 0.0578 | 8000 | 188 | 40.0 | 0.0116 | 1 |

| | | | | | | | | | |
|---|---|---|---|---|---|---|---|---|---|
| 110 | 1 | 5 | 28 | 0.0578 | 10000 | 188 | 40.0 | 0.0091 | 1 |
| 111 | 1 | 5 | 28 | 0.0578 | 12000 | 188 | 40.0 | 0.0073 | 1 |
| 112 | 1 | 5 | 28 | 0.0578 | 15000 | 188 | 40.0 | 0.0062 | 1 |
| 113 | 1 | 5 | 28 | 0.0843 | 4000 | 188 | 40.0 | 0.0427 | 1 |
| 114 | 1 | 5 | 28 | 0.0843 | 6000 | 188 | 40.0 | 0.0251 | 1 |
| 115 | 1 | 5 | 28 | 0.0843 | 8000 | 188 | 40.0 | 0.0176 | 1 |
| 116 | 1 | 5 | 28 | 0.0843 | 10000 | 188 | 40.0 | 0.0143 | 1 |
| 117 | 1 | 5 | 28 | 0.0843 | 12000 | 188 | 40.0 | 0.0118 | 1 |
| 118 | 1 | 5 | 28 | 0.0843 | 15000 | 188 | 40.0 | 0.0095 | 1 |
| 119 | 1 | 5 | 28 | 0.0882 | 5000 | 188 | 40.0 | 0.0446 | 1 |
| 120 | 1 | 5 | 28 | 0.0882 | 10000 | 188 | 40.0 | 0.0210 | 1 |
| 121 | 1 | 5 | 28 | 0.0882 | 15000 | 188 | 40.0 | 0.0160 | 1 |
| 122 | 1 | 5 | 28 | 0.1440 | 4000 | 188 | 40.0 | 0.0729 | 1 |
| 123 | 1 | 5 | 28 | 0.1440 | 6000 | 188 | 40.0 | 0.0443 | 1 |
| 124 | 1 | 5 | 28 | 0.1440 | 8000 | 188 | 40.0 | 0.0282 | 1 |
| 125 | 1 | 5 | 28 | 0.1440 | 10000 | 188 | 40.0 | 0.0237 | 1 |
| 126 | 1 | 5 | 28 | 0.1440 | 12000 | 188 | 40.0 | 0.0198 | 1 |
| 127 | 1 | 5 | 28 | 0.1440 | 15000 | 188 | 40.0 | 0.0176 | 1 |
| 128 | 1 | 5 | 40 | 0.1789 | 4000 | 188 | 40.0 | 0.2597 | 1 |
| 129 | 1 | 5 | 40 | 0.1789 | 6000 | 188 | 40.0 | 0.2513 | 1 |
| 130 | 1 | 5 | 40 | 0.1789 | 8000 | 188 | 40.0 | 0.2452 | 1 |
| 131 | 1 | 5 | 40 | 0.1789 | 10000 | 188 | 40.0 | 0.2382 | 1 |
| 132 | 1 | 5 | 40 | 0.1789 | 12000 | 188 | 40.0 | 0.2382 | 1 |
| 133 | 1 | 5 | 40 | 0.1789 | 15000 | 188 | 40.0 | 0.2369 | 1 |
| 134 | 1 | 5 | 40 | 0.1789 | 20000 | 188 | 40.0 | 0.2536 | 1 |
| 135 | 1 | 5 | 40 | 0.1789 | 25000 | 188 | 40.0 | 0.2691 | 1 |
| 136 | 1 | 5 | 40 | 0.1789 | 30000 | 188. | 40.0. | 0.2819 | 1 |
| 137 | 1 | 5 | 41 | 0.1042 | 4000 | 188 | 40.0 | 0.1558 | 1 |
| 138 | 1 | 5. | 41 | 0.1042 | 6000 | 188 | 40.0 | 0.1454 | 1 |
| 139 | 1 | 5 | 41 | 0.1042 | 8000 | 188 | 40.0 | 0.1407 | 1 |
| 140 | 1 | 5 | 41 | 0.1042 | 10000 | 188 | 40.0 | 0.1380 | 1 |
| 141 | 1 | 5 | 41 | 0.1042 | 12000 | 188 | 40.0 | 0.1384 | 1 |
| 142 | 1 | 5 | 41 | 0.1042 | 15000 | 188 | 40.0 | 0.1423 | 1 |
| 143 | 1 | 5 | 41 | 0.1715 | 4000 | 188 | 40.0 | 0.2543 | 1 |
| 144 | 1 | 5 | 41 | 0.1715 | 6000. | 188 | 40.0 | 0.2325 | 1 |
| 145 | 1 | 5 | 41 | 0.1715 | 8000 | 188 | 40.0 | 0.2296 | 1 |
| 146 | 1 | 5 | 41 | 0.1715 | 10000 | 188 | 40.0 | 0.2226 | 1 |
| 147 | 1 | 5 | 41 | 0.1715 | 12000 | 188 | 40.0 | 0.2245 | 1 |
| 148 | 1 | 5 | 41 | 0.1715 | 15000 | 188 | 40.0 | 0.2298 | 1 |
| 149 | 1 | 5 | 42 | 0.1013 | 4000 | 188 | 40.0 | 0.1597 | 1 |
| 150 | 1 | 5 | 42 | 0.1013 | 6000 | 188 | 40.0 | 0.1489 | 1 |
| 151 | 1 | 5 | 42 | 0.1013 | 8000 | 188 | 40.0 | 0.1416 | 1 |
| 152 | 1 | 5 | 42 | 0.1013 | 10000 | 188 | 40.0 | 0.1396 | 1 |
| 153 | 1 | 5 | 42 | 0.1013 | 12000 | 188 | 40.0 | 0.1362 | 1 |
| 154 | 1 | 5 | 42 | 0.1013 | 15000 | 188 | 40.0 | 0.1395 | 1 |
| 155 | 1 | 5 | 57 | 0.3183 | 4000 | 188 | 40.0 | 0.4822 | 1 |

| | | | | | | | | | |
|---|---|---|---|---|---|---|---|---|---|
| 156 | 1 | 5 | 57 | 0.3183 | 6000 | 188 | 40.0 | 0.4626 | 1 |
| 157 | 1 | 5 | 57 | 0.3183 | 8000 | 188 | 40.0 | 0.4673 | 1 |
| 158 | 1 | 5 | 57 | 0.3183 | 10000 | 188 | 40.0 | 0.4807 | 1 |
| 159 | 1 | 5 | 57 | 0.3183 | 12000 | 188 | 40.0 | 0.5043 | 1 |
| 160 | 1 | 5 | 57 | 0.3183 | 15000 | 188 | 40.0 | 0.5391 | 1 |
| 161 | 1 | 5 | 57 | 0.3183 | 20000 | 188 | 40.0 | 0.6259 | 1 |
| 162 | 1 | 5 | 57 | 0.3183 | 25000 | 188 | 40.0 | 0.6892 | 1 |
| 163 | 1 | 5 | 57 | 0.3183 | 30000 | 188 | 40.0 | 0.7555 | 1 |
| 164 | 1 | 5 | 60 | 0.3100 | 5000 | 188 | 40.0 | 0.2700 | 5 |
| 165 | 1 | 5 | 60 | 0.3100 | 10000 | 188 | 40.0 | 0.1660 | 5 |
| 166 | 1 | 5 | 60 | 0.3100 | 15000 | 188 | 40.0 | 0.1370 | 5 |
| 167 | 1 | 5 | 73 | 0.0564 | 4000 | 188 | 40.0 | 0.0529 | 1 |
| 168 | 1 | 5 | 73 | 0.0564 | 6000 | 188 | 40.0 | 0.0355 | 1 |
| 169 | 1 | 5 | 73 | 0.0564 | 8000 | 188 | 40.0 | 0.0274 | 1 |
| 170 | 1 | 5 | 73 | 0.0564 | 10000 | 188 | 40.0 | 0.0223 | 1 |
| 171 | 1 | 5 | 73 | 0.0564 | 12000 | 188 | 40.0 | 0.0195 | 1 |
| 172 | 1 | 5 | 73 | 0.0564 | 15000 | 188 | 40.0 | 0.0194 | 1 |
| 173 | 1 | 5 | 73 | 0.0923 | 4000 | 188 | 40.0 | 0.0928 | 1 |
| 174 | 1 | 5 | 73 | 0.0923 | 6000 | 188 | 40.0 | 0.0609 | 1 |
| 175 | 1 | 5 | 73 | 0.0923 | 8000 | 188 | 40.0 | 0.0459 | 1 |
| 176 | 1 | 5 | 73 | 0.0923 | 10000 | 188 | 40.0 | 0.0369 | 1 |
| 177 | 1 | 5 | 73 | 0.0923 | 12000 | 188 | 40.0 | 0.0324 | 1 |
| 178 | 1 | 5 | 73 | 0.0923 | 15000 | 188 | 40.0 | 0.0312 | 1 |
| 179 | 1 | 5 | 74 | 0.0555 | 4000 | 188 | 40.0 | 0.0546 | 1 |
| 180 | 1 | 5 | 74 | 0.0555 | 6000 | 188 | 40.0 | 0.0381 | 1 |
| 181 | 1 | 5 | 74 | 0.0555 | 8000 | 188 | 40.0 | 0.0295 | 1 |
| 182 | 1 | 5 | 74 | 0.0555 | 10000 | 188 | 40.0 | 0.0238 | 1 |
| 183 | 1 | 5 | 74 | 0.0555 | 12000 | 188 | 40.0 | 0.0210 | 1 |
| 184 | 1 | 5 | 74 | 0.0555 | 15000 | 188 | 40.0 | 0.0188 | 1 |
| 185 | 1 | 5 | 92 | 0.1537 | 4000 | 188 | 40.0 | 0.2274 | 1 |
| 186 | 1 | 5 | 92 | 0.1537 | 6000 | 188 | 40.0 | 0.1980 | 1 |
| 187 | 1 | 5 | 92 | 0.1537 | 8000 | 188 | 40.0 | 0.1761 | 1 |
| 188 | 1 | 5 | 92 | 0.1537 | 10000 | 188 | 40.0 | 0.1634 | 1 |
| 189 | 1 | 5 | 92 | 0.1537 | 12000 | 188 | 40.0 | 0.1566 | 1 |
| 190 | 1 | 5 | 92 | 0.1537 | 15000 | 188 | 40.0 | 0.1526 | 1 |
| 191 | 1 | 12 | 8 | 0.6030 | 10000 | 1305 | 20.0 | 0.4990 | 7 |
| 192 | 1 | 12 | 8 | 0.6030 | 25000 | 1305 | 20.0 | 0.2850 | 7 |
| 193 | 1 | 12 | 8 | 0.6031 | 5000 | 1305 | 40.0 | 0.5660 | 6 |
| 194 | 1 | 12 | 8 | 0.6031 | 10000 | 1305 | 40.0 | 0.5270 | 6 |
| 195 | 1 | 12 | 8 | 0.6031 | 15000 | 1305 | 40.0 | 0.4670 | 6 |
| 196 | 1 | 12 | 8 | 0.6031 | 20000 | 1305 | 40.0 | 0.4050 | 6 |
| 197 | 1 | 12 | 8 | 0.6031 | 25000 | 1305 | 40.0 | 0.3470 | 6 |
| 198 | 1 | 12 | 9 | 0.3901 | 5000 | 1305 | 40.0 | 0.3700 | 6 |
| 199 | 1 | 12 | 9 | 0.3901 | 10000 | 1305 | 40.0 | 0.3200 | 6 |
| 200 | 1 | 12 | 9 | 0.3901 | 15000 | 1305 | 40.0 | 0.2570 | 6 |
| 201 | 1 | 12 | 9 | 0.3901 | 20000 | 1305 | 40.0 | 0.2010 | 6 |

| | | | | | | | | | |
|---|---|---|---|---|---|---|---|---|---|
| 202 | 1 | 12 | 9 | 0.3901 | 25000 | 1305 | 40.0 | 0.1590 | 6 |
| 203 | 1 | 12 | 13 | 0.0960 | 15000 | 1305 | 52.5 | 0.1010 | 2 |
| 204 | 1 | 12 | 13 | 0.0960 | 20000 | 1305 | 52.5 | 0.0948 | 2 |
| 205 | 1 | 12 | 13 | 0.3650 | 10000 | 1305 | 52.5 | 0.3730 | 2 |
| 206 | 1 | 12 | 13 | 0.3650 | 15000 | 1305 | 52.5 | 0.3650 | 2 |
| 207 | 1 | 12 | 13 | 0.3650 | 20000 | 1305 | 52.5 | 0.3690 | 2 |
| 208 | 1 | 12 | 13 | 0.4810 | 10000 | 1305 | 52.5 | 0.4760 | 2 |
| 209 | 1 | 12 | 13 | 0.4810 | 15000 | 1305 | 52.5 | 0.4710 | 2 |
| 210 | 1 | 12 | 13 | 0.4810 | 20000 | 1305 | 52.5 | 0.4780 | 2 |
| 211 | 1 | 12 | 13 | 0.5000 | 10000 | 1305 | 52.5 | 0.5060 | 2 |
| 212 | 1 | 12 | 13 | 0.5000 | 15000 | 1305 | 52.5 | 0.4950 | 2 |
| 213 | 1 | 12 | 13 | 0.5000 | 20000 | 1305 | 52.5 | 0.4990 | 2 |
| 214 | 1 | 12 | 13 | 0.7490 | 10000 | 1305 | 52.5 | 0.7440 | 2 |
| 215 | 1 | 12 | 13 | 0.7490 | 15000 | 1305 | 52.5 | 0.7100 | 2 |
| 216 | 1 | 12 | 13 | 0.7490 | 20000 | 1305 | 52.5 | 0.7360 | 2 |
| 217 | 1 | 12 | 39 | 0.4060 | 16500 | 1305 | 17.5 | 0.2740 | 7 |
| 218 | 1 | 13 | 5 | 0.1722 | 4000 | 1559 | 40.0 | 0.1500 | 1 |
| 219 | 1 | 13 | 5 | 0.1722 | 6000 | 1559 | 40.0 | 0.1502 | 1 |
| 220 | 1 | 13 | 5 | 0.1722 | 8000 | 1559 | 40.0 | 0.1503 | 1 |
| 221 | 1 | 13 | 5 | 0.1722 | 10000 | 1559 | 40.0 | 0.1503 | 1 |
| 222 | 1 | 13 | 5 | 0.1722 | 12000 | 1559 | 40.0 | 0.1503 | 1 |
| 223 | 1 | 13 | 5 | 0.1722 | 15000 | 1559 | 40.0 | 0.1504 | 1 |
| 224 | 1 | 13 | 5 | 0.1722 | 20000 | 1559 | 40.0 | 0.1495 | 1 |
| 225 | 1 | 13 | 5 | 0.1722 | 25000 | 1559 | 40.0 | 0.1490 | 1 |
| 226 | 1 | 13 | 5 | 0.1722 | 30000 | 1559 | 40.0 | 0.1455 | 1 |
| 227 | 1 | 13 | 5 | 0.5551 | 4000 | 1559 | 40.0 | 0.5150 | 1 |
| 228 | 1 | 13 | 5 | 0.5551 | 6000 | 1559 | 40.0 | 0.5178 | 1 |
| 229 | 1 | 13 | 5 | 0.5551 | 8000 | 1559 | 40.0 | 0.5188 | 1 |
| 230 | 1 | 13 | 5 | 0.5551 | 10000 | 1559 | 40.0 | 0.5200 | 1 |
| 231 | 1 | 13 | 5 | 0.5551 | 12000 | 1559 | 40.0 | 0.5198 | 1 |
| 232 | 1 | 13 | 5 | 0.5551 | 15000 | 1559 | 40.0 | 0.5170 | 1 |
| 233 | 1 | 13 | 5 | 0.5551 | 20000 | 1559 | 40.0 | 0.5130 | 1 |
| 234 | 1 | 13 | 5 | 0.5551 | 25000 | 1559 | 40.0 | 0.5080 | 1 |
| 235 | 1 | 13 | 5 | 0.5551 | 30000 | 1559 | 40.0 | 0.5030 | 1 |
| 236 | 1 | 13 | 8 | 0.5290 | 10000 | 1559 | 20.0 | 0.4600 | 7 |
| 237 | 1 | 13 | 8 | 0.5290 | 25000 | 1559 | 20.0 | 0.3250 | 7 |
| 238 | 1 | 13 | 8 | 0.5292 | 10000 | 1559 | 40.0 | 0.4880 | 6 |
| 239 | 1 | 13 | 8 | 0.5292 | 15000 | 1559 | 40.0 | 0.4450 | 6 |
| 240 | 1 | 13 | 8 | 0.5292 | 20000 | 1559 | 40.0 | 0.4070 | 6 |
| 241 | 1 | 13 | 8 | 0.5292 | 25000 | 1559 | 40.0 | 0.3770 | 6 |
| 242 | 1 | 13 | 12 | 0.0910 | 10000 | 1559 | 20.0 | 0.0390 | 7 |
| 243 | 1 | 13 | 12 | 0.0910 | 15000 | 1559 | 20.0 | 0.0230 | 7 |
| 244 | 1 | 13 | 12 | 0.0910 | 25000 | 1559 | 20.0 | 0.0120 | 7 |
| 245 | 1 | 13 | 12 | 0.0910 | 30000 | 1559 | 20.0 | 0.0090 | 7 |
| 246 | 1 | 13 | 12 | 0.0910 | 35000 | 1559 | 20.0 | 0.0080 | 7 |
| 247 | 1 | 13 | 12 | 0.6240 | 10000 | 1559 | 20.0 | 0.4600 | 7 |

| | | | | | | | | | |
|---|---|---|---|---|---|---|---|---|---|
| 248 | 1 | 13 | 12 | 0.6240 | 15000 | 1559 | 20.0 | 0.3500 | 7 |
| 249 | 1 | 13 | 12 | 0.6240 | 29000 | 1559 | 29.0 | 0.2120 | 7 |
| 250 | 1 | 13 | 12 | 0.0910 | 10000 | 1559 | 52.5 | 0.0602 | 7 |
| 251 | 1 | 13 | 12. | 0.0910 | 15000 | 1559 | 52.5 | 0.0424 | 7 |
| 252 | 1 | 13 | 12. | 0.0910 | 20000 | 1559 | 52.5 | 0.0302 | 7 |
| 253 | 1 | 13 | 12. | 0.0910 | 25000 | 1559 | 52.5 | 0.0221 | 7 |
| 254 | 1 | 13 | 12 | 0.0910 | 30000 | 1559 | 52.5 | 0.0170 | 7 |
| 255 | 1 | 13 | 12 | 0.1530 | 10000 | 1559 | 52.5 | 0.1110 | 2 |
| 256 | 1 | 13 | 12 | 0.1530 | 15000 | 1559 | 52.5 | 0.0768 | 2 |
| 257 | 1 | 13 | 12 | 0.1530 | 20000 | 1559 | 52.5 | 0.0580 | 2 |
| 258 | 1 | 13 | 12 | 0.2510 | 10000 | 1559 | 52.5 | 0.1840 | 2 |
| 259 | 1 | 13 | 12 | 0.2510 | 15000 | 1559 | 52.5 | 0.1400 | 2 |
| 260 | 1 | 13 | 12 | 0.2510 | 20000 | 1559 | 52.5 | 0.1010 | 2 |
| 261 | 1 | 13 | 12 | 0.5000 | 10000 | 1559 | 52.5 | 0.3870 | 2 |
| 262 | 1 | 13 | 12 | 0.5000 | 15000 | 1559 | 52.5 | 0.3110 | 2 |
| 263 | 1 | 13 | 12 | 0.5000 | 20000 | 1559 | 52.5 | 0.2450 | 2 |
| 264 | 1 | 13 | 12 | 0.5190 | 10000 | 1559 | 52.5 | 0.4200 | 2 |
| 265 | 1 | 13 | 12 | 0.5190 | 15000 | 1559 | 52.5 | 0.3440 | 2 |
| 266 | 1 | 13 | 12 | 0.5190 | 20000 | 1559 | 52.5 | 0.2740 | 2 |
| 267 | 1 | 13 | 12 | 0.6350 | 10000 | 1559 | 52.5 | 0.5250 | 2 |
| 268 | 1 | 13 | 12 | 0.6350 | 15000 | 1559 | 52.5 | 0.4560 | 2 |
| 269 | 1 | 13 | 12 | 0.6350 | 20000 | 1559 | 52.5 | 0.3850 | 2 |
| 270 | 1 | 13 | 12 | 0.9040 | 15000 | 1559 | 52.5 | 0.8220 | 2 |
| 271 | 1 | 13 | 12 | 0.9040 | 20000 | 1559 | 52.5 | 0.7710 | 2 |
| 272 | 1 | 13 | 12 | 0.9790 | 10000 | 1559 | 52.5 | 0.9630 | 2 |
| 273 | 1 | 13 | 12 | 0.9790 | 15000 | 1559 | 52.5 | 0.9580 | 2 |
| 274 | 1 | 13 | 12 | 0.9790 | 20000 | 1559 | 52.5 | 0.9590 | 2 |
| 275 | 1 | 13 | 12 | 0.0910 | 10000 | 1559 | 75.0 | 0.0660 | 7 |
| 276 | 1 | 13 | 12 | 0.0910 | 15000 | 1559 | 75.0 | 0.0473 | 7 |
| 277 | 1 | 13 | 12 | 0.0910 | 20000 | 1559 | 75.0 | 0.0351 | 7 |
| 278 | 1 | 13 | 12 | 0.0910 | 25000 | 1559 | 75.0 | 0.0258 | 7 |
| 279 | 1 | 13 | 12 | 0.0910 | 30000 | 1559 | 75.0 | 0.0190 | 7 |
| 280 | 1 | 13 | 12 | 0.0910 | 35000 | 1559 | 75.0 | 0.0150 | 7 |
| 281 | 1 | 13 | 12 | 0.0910 | 40000 | 1559 | 75.0 | 0.0130 | 7 |
| 282 | 1 | 13 | 22 | 0.6280 | 10000 | 1559 | 17.0 | 0.5110 | 7 |
| 283 | 1 | 13 | 22 | 0.6280 | 15000 | 1559 | 17.0 | 0.4200 | 7 |
| 284 | 1 | 13 | 22 | 0.6280 | 20000 | 1559 | 17.0 | 0.3320 | 7 |
| 285 | 1 | 13 | 22 | 0.6280 | 25000 | 1559 | 17.0 | 0.2670 | 7 |
| 286 | 1 | 13 | 22 | 0.6280 | 30000 | 1559 | 17.0 | 0.2300 | 7 |
| 287 | 1 | 13 | 22 | 0.6280 | 35000 | 1559 | 17.0 | 0.2150 | 7 |
| 288 | 1 | 13 | 22 | 0.3980 | 7000 | 1559 | 40.0 | 0.3853 | 5 |
| 289 | 1 | 13 | 22 | 0.3980 | 10500 | 1559 | 40.0 | 0.3543 | 5 |
| 290 | 1 | 13 | 22 | 0.3980 | 15700 | 1559 | 40.0 | 0.3008 | 5 |
| 291 | 1 | 13 | 22 | 0.3980 | 20800 | 1559 | 40.0 | 0.2585 | 5 |
| 292 | 1 | 13 | 22 | 0.3980 | 26000 | 1559 | 40.0 | 0.2162 | 5 |
| 293 | 1 | 13 | 22 | 0.3980 | 31100 | 1559 | 40.0 | 0.1870 | 5 |

| 294 | 1 | 13 | 22 | 0.3980 | 36400 | 1559 | 40.0 | 0.1579 | 5 |
| 295 | 1 | 13 | 22 | 0.3980 | 40000 | 1559 | 40.0 | 0.1429 | 5 |
| 296 | 1 | 13 | 25 | 0.7460 | 10000 | 1559 | 20.0 | 0.6400 | 7 |
| 297 | 1 | 13 | 25 | 0.7460 | 15000 | 1559 | 20.0 | 0.5340 | 7 |
| 298 | 1 | 13 | 25 | 0.7460 | 29000 | 1559 | 20.0 | 0.3810 | 7 |
| 299 | 1 | 13 | 26 | 0.5920 | 10000 | 1559 | 17.0 | 0.4460 | 7 |
| 300 | 1 | 13 | 26 | 0.5920 | 15000 | 1559 | 17.0 | 0.3350 | 7 |
| 301 | 1 | 13 | 26 | 0.5920 | 20000 | 1559 | 17.0 | 0.2630 | 7 |
| 302 | 1 | 13 | 26 | 0.5920 | 25000 | 1559 | 17.0 | 0.2110 | 7 |
| 303 | 1 | 13 | 26 | 0.5920 | 30000 | 1559 | 17.0 | 0.1820 | 7 |
| 304 | 1 | 13 | 26 | 0.5920 | 35000 | 1559 | 17.0 | 0.1590 | 7 |
| 305 | 1 | 13 | 26 | 0.1000 | 10000 | 1559 | 20.0 | 0.0530 | 7 |
| 306 | 1 | 13 | 26 | 0.1000 | 15000 | 1559 | 20.0 | 0.0350 | 7 |
| 307 | 1 | 13 | 26 | 0.1000 | 20000 | 1559 | 20.0 | 0.0250 | 7 |
| 308 | 1 | 13 | 26 | 0.1000 | 25000 | 1559 | 20.0 | 0.0160 | 7 |
| 309 | 1 | 13 | 26 | 0.1000 | 30000 | 1559 | 20.0 | 0.0150 | 7 |
| 310 | 1 | 13 | 26 | 0.1000 | 35000 | 1559 | 20.0 | 0.0140 | 7 |
| 311 | 1 | 13 | 26 | 0.1000 | 40000 | 1559 | 20.0 | 0.0110 | 7 |
| 312 | 1 | 13 | 26 | 0.6100 | 10000 | 1559 | 20.0 | 0.4590 | 7 |
| 313 | 1 | 13 | 26 | 0.6100 | 15000 | 1559 | 20.0 | 0.3550 | 7 |
| 314 | 1 | 13 | 26 | 0.6100 | 29000 | 1559 | 20.0 | 0.1940 | 7 |
| 315 | 1 | 13 | 26 | 0.5917 | 10000 | 1559 | 40.0 | 0.5540 | 6 |
| 316 | 1 | 13 | 26 | 0.5917 | 15000 | 1559 | 40.0 | 0.4650 | 6 |
| 317 | 1 | 13 | 26 | 0.5917 | 20000 | 1559 | 40.0 | 0.3860 | 6 |
| 318 | 1 | 13 | 26 | 0.1000 | 10000 | 1559 | 52.5 | 0.0765 | 5 |
| 319 | 1 | 13 | 26 | 0.1000 | 15000 | 1559 | 52.5 | 0.0592 | 5 |
| 320 | 1 | 13 | 26 | 0.1000 | 20000 | 1559 | 52.5 | 0.0431 | 5 |
| 321 | 1 | 13 | 26 | 0.1000 | 25000 | 1559 | 52.5 | 0.0335 | 5 |
| 322 | 1 | 13 | 26 | 0.1000 | 30000 | 1559 | 52.5 | 0.0266 | 5 |
| 323 | 1 | 13 | 26 | 0.2410 | 20000 | 1559 | 52.5 | 0.1240 | 7 |
| 324 | 1 | 13 | 26 | 0.2410 | 25000 | 1559 | 52.5 | 0.0980 | 7 |
| 325 | 1 | 13 | 26 | 0.2410 | 30000 | 1559 | 52.5 | 0.0830 | 7 |
| 326 | 1 | 13 | 26 | 0.1000 | 10000 | 1559 | 75.0 | 0.0868 | 5 |
| 327 | 1 | 13 | 26 | 0.1000 | 15000 | 1559 | 75.0 | 0.0692 | 5 |
| 328 | 1 | 13 | 26 | 0.1000 | 02000 | 1559 | 75.0 | 0.0543 | 5 |
| 329 | 1 | 13 | 26 | 0.1000 | 25000 | 1559 | 75.0 | 0.0415 | 5 |
| 330 | 1 | 13 | 26 | 0.1000 | 30000 | 1559 | 75.0 | 0.0337 | 5 |
| 331 | 1 | 13 | 26 | 0.1000 | 35000 | 1559 | 75.0 | 0.0268 | 5 |
| 332 | 1 | 13 | 26 | 0.1000 | 40000 | 1559 | 75.0 | 0.0218 | 5 |
| 333 | 1 | 13 | 27 | 0.6730 | 10000 | 1559 | 20.0 | 0.5410 | 7 |
| 334 | 1 | 13 | 27 | 0.6730 | 15000 | 1559 | 20.0 | 0.4300 | 7 |
| 335 | 1 | 13 | 27 | 0.6730 | 29000 | 1559 | 29.0 | 0.2300 | 7 |
| 336 | 1 | 13 | 28 | 0.4080 | 10000 | 1559 | 17.0 | 0.2470 | 7 |
| 337 | 1 | 13 | 28 | 0.4080 | 15000 | 1559 | 17.0 | 0.1600 | 7 |
| 338 | 1 | 13 | 28 | 0.4080 | 20000 | 1559 | 17.0 | 0.1020 | 7 |
| 339 | 1 | 13 | 28 | 0.4080 | 25000 | 1559 | 17.0 | 0.0800 | 7 |

| | | | | | | | | | |
|---|---|---|---|---|---|---|---|---|---|
| 340 | 1 | 13 | 28 | 0.4080 | 30000 | 1559 | 17.0 | 0.0700 | 7 |
| 341 | 1 | 13 | 28 | 0.4080 | 35000 | 1559 | 17.0 | 0.0650 | 7 |
| 342 | 1 | 13 | 28 | 0.5780 | 10000 | 1559 | 17.0 | 0.3850 | 7 |
| 343 | 1 | 13 | 28 | 0.5780 | 15000 | 1559 | 17.0 | 0.2400 | 7 |
| 344 | 1 | 13 | 28 | 0.5780 | 20000 | 1559 | 17.0 | 0.1750 | 7 |
| 345 | 1 | 13 | 28 | 0.5780 | 25000 | 1559 | 17.0 | 0.1450 | 7 |
| 346 | 1 | 13 | 28 | 0.5780 | 30000 | 1559 | 17.0 | 0.1270 | 7 |
| 347 | 1 | 13 | 28 | 0.5780 | 35000 | 1559 | 17.0 | 0.1150 | 7 |
| 348 | 1 | 13 | 28 | 0.5800 | 10000 | 1559 | 20.0 | 0.4150 | 7 |
| 349 | 1 | 13 | 28 | 0.5800 | 15000 | 1559 | 20.0 | 0.3000 | 7 |
| 350 | 1 | 13 | 28 | 0.5800 | 29000 | 1559 | 20.0 | 0.1440 | 7 |
| 351 | 1 | 13 | 28 | 0.0490 | 5300 | 1559 | 40.0 | 0.0473 | 5 |
| 352 | 1 | 13 | 28 | 0.0490 | 10500 | 1559 | 40.0 | 0.0315 | 5 |
| 353 | 1 | 13 | 28 | 0.0490 | 15900 | 1559 | 40.0 | 0.0210 | 5 |
| 354 | 1 | 13 | 28 | 0.1250 | 5300 | 1559 | 40.0 | 0.1246 | 5 |
| 355 | 1 | 13 | 28 | 0.1250 | 10500 | 1559 | 40.0 | 0.0848 | 5 |
| 356 | 1 | 13 | 28 | 0.1250 | 15900 | 1559 | 40.0 | 0.0585 | 5 |
| 357 | 1 | 13 | 28 | 0.1250 | 21200 | 1559 | 40.0 | 0.0405 | 5 |
| 358 | 1 | 13 | 28 | 0.1250 | 26600 | 1559 | 40.0 | 0.0307 | 5 |
| 359 | 1 | 13 | 28 | 0.1250 | 31900 | 1559 | 40.0 | 0.0233 | 5 |
| 360 | 1 | 13 | 28 | 0.3090 | 5300.0 | 1559 | 40.0 | 0.2957 | 5 |
| 361 | 1 | 13 | 28 | 0.3090 | 10500 | 1559 | 40.0 | 0.2184 | 5 |
| 362 | 1 | 13 | 28 | 0.3090 | 15900 | 1559 | 40.0 | 0.1576 | 5 |
| 363 | 1 | 13 | 28 | 0.3090 | 21200 | 1559 | 40.0 | 0.1141 | 5 |
| 364 | 1 | 13 | 28 | 0.3090 | 26600 | 1559. | 40.0 | 0.0878 | 5 |
| 365 | 1 | 13 | 28 | 0.3090 | 31900 | 1559. | 40.0 | 0.0683 | 5 |
| 366 | 1 | 13 | 28 | 0.3090 | 38200 | 1559. | 40.0 | 0.0570 | 5 |
| 367 | 1 | 13 | 28 | 0.5796 | 5000 | 1559 | 40.0 | 0.6260 | 6 |
| 368 | 1 | 13 | 28 | 0.5796 | 10000 | 1559 | 40.0 | 0.5030 | 6 |
| 369 | 1 | 13 | 28 | 0.5796 | 15000 | 1559 | 40.0 | 0.4000 | 6 |
| 370 | 1 | 13 | 28 | 0.5796 | 20000 | 1559 | 40.0 | 0.3200 | 6 |
| 371 | 1 | 13 | 28 | 0.5796 | 25000 | 1559 | 40.0 | 0.2570 | 6 |
| 372 | 1 | 13 | 29 | 0.4600 | 10000 | 1559 | 20.0 | 0.3100 | 7 |
| 373 | 1 | 13 | 29 | 0.4600 | 15000 | 1559 | 20.0 | 0.2050 | 7 |
| 374 | 1 | 13 | 29 | 0.4600 | 29000 | 1559 | 20.0 | 0.0920 | 7 |
| 375 | 1 | 13 | 29 | 0.4590 | 5000 | 1559 | 40.0 | 0.4910 | 6 |
| 376 | 1 | 13 | 29 | 0.4590 | 10000 | 1559 | 40.0 | 0.3810 | 6 |
| 377 | 1 | 13 | 29 | 0.4590 | 15000 | 1559 | 40.0 | 0.2880 | 6 |
| 378 | 1 | 13 | 29 | 0.4590 | 20000 | 1559 | 40.0 | 0.2200 | 6 |
| 379 | 1 | 13 | 29 | 0.4590 | 25000 | 1559 | 40.0 | 0.1700 | 6 |
| 380 | 1 | 13 | 29 | 0.9445 | 15900 | 1559 | 40.0 | 0.8801 | 5 |
| 381 | 1 | 13 | 29 | 0.9445 | 21200 | 1559 | 40.0 | 0.8441 | 5 |
| 382 | 1 | 13 | 29 | 0.9445 | 31800 | 1559 | 40.0 | 0.7637 | 5 |
| 383 | 1 | 13 | 29 | 0.9445 | 39900 | 1559 | 40.0 | 0.7065 | 5 |
| 384 | 1 | 14 | 5 | 0.3143 | 4000 | 1839 | 40.0 | 0.2900 | 1 |
| 385 | 1 | 14 | 5 | 0.3143 | 6000 | 1839 | 40.0 | 0.2940 | 1 |

| 386 | 1 | 14 | 5 | 0.3143 | 8000 | 1839 | 40.0 | 0.2978 | 1 |
| 387 | 1 | 14 | 5 | 0.3143 | 10000 | 1839 | 40.0 | 0.3002 | 1 |
| 388 | 1 | 14 | 5 | 0.3143 | 12000 | 1839 | 40.0 | 0.3036 | 1 |
| 389 | 1 | 14 | 5 | 0.3143 | 15000 | 1839 | 40.0 | 0.3072 | 1 |
| 390 | 1 | 14 | 5 | 0.3143 | 20000 | 1839 | 40.0 | 0.3120 | 1 |
| 391 | 1 | 14 | 5 | 0.3143 | 25000 | 1839 | 40.0 | 0.3140 | 1 |
| 392 | 1 | 14 | 5 | 0.3143 | 30000 | 1839 | 40.0 | 0.3156 | 1 |
| 393 | 1 | 14 | 5 | 0.4780 | 4000 | 1839 | 40.0 | 0.4465 | 1 |
| 394 | 1 | 14 | 5 | 0.4780 | 6000 | 1839 | 40.0 | 0.4528 | 1 |
| 395 | 1 | 14 | 5 | 0.4780 | 8000 | 1839 | 40.0 | 0.4580 | 1 |
| 396 | 1 | 14 | 5 | 0.4780 | 10000 | 1839 | 40.0 | 0.4622 | 1 |
| 397 | 1 | 14 | 5 | 0.4780 | 12000 | 1839 | 40.0 | 0.4658 | 1 |
| 398 | 1 | 14 | 5 | 0.4780 | 15000 | 1839 | 40.0 | 0.4696 | 1 |
| 399 | 1 | 14 | 5 | 0.4780 | 20000 | 1839 | 40.0 | 0.4738 | 1 |
| 400 | 1 | 14 | 5 | 0.4780 | 25000 | 1839 | 40.0 | 0.4758 | 1 |
| 401 | 1 | 14 | 5 | 0.4780 | 30000 | 1839 | 40.0 | 0.4760 | 1 |
| 402 | 1 | 14 | 6 | 0.7010 | 10000 | 1839 | 20.0 | 0.6800 | 7 |
| 403 | 1 | 14 | 6 | 0.7000 | 4000 | 1839 | 40.0 | 0.7102 | 7 |
| 404 | 1 | 14 | 6 | 0.7000 | 6000 | 1839 | 40.0 | 0.6975 | 7 |
| 405 | 1 | 14 | 6 | 0.7000 | 8000 | 1839 | 40.0 | 0.6854 | 7 |
| 406 | 1 | 14 | 6 | 0.7005 | 5000 | 1839 | 40.0 | 0.6540 | 6 |
| 407 | 1 | 14 | 6 | 0.7005 | 10000 | 1839 | 40.0 | 0.6790 | 6 |
| 408 | 1 | 14 | 6 | 0.7005 | 15000 | 1839 | 40.0 | 0.6750 | 6 |
| 409 | 1 | 14 | 6 | 0.7005 | 20000 | 1839 | 40.0 | 0.6760 | 6 |
| 410 | 1 | 14 | 6 | 0.7005 | 25000 | 1839 | 40.0 | 0.6670 | 6 |
| 411 | 1 | 14 | 6 | 0.7005 | 4000 | 1839 | 40.0 | 0.6800 | 5 |
| 412 | 1 | 14 | 6 | 0.7005 | 6000 | 1839 | 40.0 | 0.6798 | 5 |
| 413 | 1 | 14 | 6 | 0.7005 | 8000 | 1839 | 40.0 | 0.6790 | 5 |
| 414 | 1 | 14 | 6 | 0.7005 | 10000 | 1839 | 40.0 | 0.6787 | 5 |
| 415 | 1 | 14 | 6 | 0.7005 | 12000 | 1839 | 40.0 | 0.6780 | 5 |
| 416 | 1 | 14 | 6 | 0.7005 | 15000 | 1839 | 40.0 | 0.6762 | 5 |
| 417 | 1 | 14 | 6 | 0.7005 | 20000 | 1839 | 40.0 | 0.6730 | 5 |
| 418 | 1 | 14 | 6 | 0.7005 | 25000 | 1839 | 40.0 | 0.6682 | 5 |
| 419 | 1 | 14 | 6 | 0.7005 | 30000 | 1839 | 40.0 | 0.6630 | 5 |
| 420 | 1 | 14 | 8 | 0.4670 | 10000 | 1839 | 20.0 | 0.3890 | 7 |
| 421 | 1 | 14 | 8 | 0.4670 | 25000 | 1839 | 20.0 | 0.3110 | 7 |
| 422 | 1 | 14 | 8 | 0.4675 | 5000 | 1839 | 40.0 | 0.4240 | 6 |
| 423 | 1 | 14 | 8 | 0.4675 | 10000 | 1839 | 40.0 | 0.4190 | 6 |
| 424 | 1 | 14 | 8 | 0.4675 | 15000 | 1839 | 40.0 | 0.4120 | 6 |
| 425 | 1 | 14 | 12 | 0.3660 | 5000 | 1839 | 18.0 | 0.3160 | 7 |
| 426 | 1 | 14 | 26 | 0.0860 | 10000 | 1839 | 20.0 | 0.0660 | 7 |
| 427 | 1 | 14 | 26 | 0.0860 | 15000 | 1839 | 20.0 | 0.0440 | 7 |
| 428 | 1 | 14 | 26 | 0.5917 | 25000 | 1839 | 40.0 | 0.3260 | 6 |
| 429 | 1 | 14 | 26 | 0.0860 | 10000 | 1839 | 52.5 | 0.0829 | 5 |
| 430 | 1 | 14 | 26 | 0.0860 | 15000 | 1839 | 52.5 | 0.0677 | 5 |
| 431 | 1 | 14 | 26 | 0.0860 | 20000 | 1839 | 52.5 | 0.0544 | 5 |

| | | | | | | | | | |
|---|---|---|---|---|---|---|---|---|---|
| 432 | 1 | 14 | 26 | 0.0860 | 25000 | 1839 | 52.5 | 0.0446 | 5 |
| 433 | 1 | 14 | 26 | 0.0860 | 30000 | 1839 | 52.5 | 0.0354 | 5 |
| 434 | 1 | 14 | 26 | 0.0860 | 10000 | 1839 | 75.0 | 0.0846 | 5 |
| 435 | 1 | 14 | 26 | 0.0860 | 15000 | 1839 | 75.0 | 0.0723 | 5 |
| 436 | 1 | 14 | 26 | 0.0860 | 20000 | 1839 | 75.0 | 0.0615 | 5 |
| 437 | 1 | 14 | 26 | 0.0860 | 25000 | 1839 | 75.0 | 0.0517 | 5 |
| 438 | 1 | 14 | 26 | 0.0860 | 30000 | 1839 | 75.0 | 0.0431 | 5 |
| 439 | 1 | 14 | 26 | 0.0860 | 35000 | 1839 | 75.0 | 0.0352 | 5 |
| 440 | 1 | 14 | 26 | 0.0860 | 40000 | 1839 | 75.0 | 0.0277 | 5 |
| 441 | 1 | 14 | 42 | 0.3690 | 5000 | 1839 | 20.0 | 0.3990 | 7 |
| 442 | 1 | 14 | 42 | 0.3690 | 10000 | 1839 | 20.0 | 0.3400 | 7 |
| 443 | 1 | 14 | 42 | 0.3690 | 15000 | 1839 | 20.0 | 0.2840 | 7 |
| 444 | 1 | 14 | 42 | 0.3690 | 20000 | 1839 | 20.0 | 0.2410 | 7 |
| 445 | 1 | 14 | 42 | 0.3690 | 25000 | 1839 | 20.0 | 0.1980 | 7 |
| 446 | 1 | 14 | 42 | 0.3690 | 30000 | 1839 | 20.0 | 0.1760 | 7 |
| 447 | 1 | 14 | 42 | 0.3690 | 35000 | 1839 | 20.0 | 0.1590 | 7 |
| 448 | 1 | 22 | 5 | 0.6993 | 6000. | 4965 | 40.0 | 0.6330 | 1 |
| 449 | 1 | 22 | 5 | 0.6993 | 8000 | 4965 | 40.0 | 0.6400 | 1 |
| 450 | 1 | 22 | 5 | 0.6993 | 10000 | 4965 | 40.0 | 0.6462 | 1 |
| 451 | 1 | 22 | 5 | 0.6993 | 12000 | 4965 | 40.0 | 0.6513 | 1 |
| 452 | 1 | 22 | 5 | 0.6993 | 15000 | 4965 | 40.0 | 0.6578 | 1 |
| 453 | 1 | 22 | 5 | 0.6993 | 20000 | 4965 | 40.0 | 0.6662 | 1 |
| 454 | 1 | 22 | 5 | 0.6993 | 25000 | 4965 | 40.0 | 0.6738 | 1 |
| 455 | 1 | 22 | 5 | 0.6993 | 30000 | 4965 | 40.0 | 0.6782 | 1 |
| 456 | 1 | 22 | 5 | 0.8322 | 6000 | 4965 | 40.0 | 0.7922 | 1 |
| 457 | 1 | 22 | 5 | 0.8322 | 8000 | 4965 | 40.0 | 0.7937 | 1 |
| 458 | 1 | 22 | 5 | 0.8322 | 10000 | 4965 | 40.0 | 0.7942 | 1 |
| 459 | 1 | 22 | 5 | 0.8322 | 12000 | 4965 | 40.0 | 0.7960 | 1 |
| 460 | 1 | 22 | 5 | 0.8322 | 15000 | 4965 | 40.0 | 0.7978 | 1 |
| 461 | 1 | 22 | 5 | 0.8322 | 20000 | 4965 | 40.0 | 0.8012 | 1 |
| 462 | 1 | 22 | 5 | 0.8322 | 25000 | 4965 | 40.0 | 0.8042 | 1 |
| 463 | 1 | 22 | 5 | 0.8322 | 30000 | 4965 | 40.0 | 0.8077 | 1 |
| 464 | 1 | 22 | 6 | 0.8160 | 6000 | 4965 | 40.0 | 0.7832 | 11 |
| 465 | 1 | 22 | 6 | 0.8160 | 8000 | 4965 | 40.0 | 0.7862 | 11 |
| 466 | 1 | 22 | 6 | 0.8160 | 10000 | 4965 | 40.0 | 0.7888 | 11 |
| 467 | 1 | 22 | 6 | 0.8160 | 12000 | 4965 | 40.0 | 0.7910 | 11 |
| 468 | 1 | 22 | 6 | 0.8160 | 15000 | 4965 | 40.0 | 0.7938 | 11 |
| 469 | 1 | 22 | 6 | 0.8160 | 20000 | 4965 | 40.0 | 0.7980 | 11 |
| 470 | 1 | 22 | 6 | 0.8160 | 25000 | 4965 | 40.0 | 0.8012 | 11 |
| 471 | 1 | 22 | 6 | 0.8160 | 30000 | 4965 | 40.0 | 0.8033 | 11 |
| 472 | 1 | 22 | 8 | 0.5990 | 30000 | 4965 | 20.0 | 0.5820 | 7 |
| 473 | 1 | 22 | 8 | 0.5995 | 10000 | 4965 | 40.0 | 0.5500 | 6 |
| 474 | 1 | 22 | 8 | 0.5995 | 15000 | 4965 | 40.0 | 0.5580 | 6 |
| 475 | 1 | 22 | 8 | 0.5995 | 20000 | 4965 | 40.0 | 0.5620 | 6 |
| 476 | 1 | 22 | 13 | 0.3720 | 10000 | 4965 | 17.3 | 0.3200 | 7 |
| 477 | 1 | 22 | 13 | 0.3720 | 15000 | 4965 | 17.3 | 0.3170 | 7 |

| | | | | | | | | | |
|---|---|---|---|---|---|---|---|---|---|
| 478 | 1 | 22 | 13 | 0.3720 | 20000 | 4965 | 17.3 | 0.3070 | 7 |
| 479 | 1 | 22 | 13 | 0.3720 | 25000 | 4965 | 17.3 | 0.2950 | 7 |
| 480 | 1 | 22 | 13 | 0.3720 | 30000 | 4965 | 17.3 | 0.2800 | 7 |
| 481 | 1 | 22 | 13 | 0.3720 | 35000 | 4965 | 17.3 | 0.2650 | 7 |
| 482 | 1 | 22 | 13 | 0.6020 | 15700 | 4965 | 40.0 | 0.5789 | 5 |
| 483 | 1 | 22 | 13 | 0.6020 | 20900 | 4965 | 40.0 | 0.5605 | 5 |
| 484 | 1 | 22 | 13 | 0.6020 | 26100 | 4965 | 40.0 | 0.5580 | 5 |
| 485 | 1 | 22 | 13 | 0.6020 | 41300 | 4965 | 40.0 | 0.5435 | 5 |
| 486 | 1 | 22 | 13 | 0.6020 | 36500 | 4965 | 40.0 | 0.5437 | 5 |
| 487 | 1 | 22 | 13 | 0.6020 | 40000 | 4965 | 40.0 | 0.5380 | 5 |
| 488 | 1 | 22 | 41 | 0.3500 | 25000 | 4965 | 18.0 | 0.2410 | 7 |
| 489 | 1 | 23 | 5 | 0.7160 | 8000 | 5465 | 40.0 | 0.6500 | 1 |
| 490 | 1 | 23 | 5 | 0.7160 | 10000 | 5465 | 40.0 | 0.6602 | 1 |
| 491 | 1 | 23 | 5 | 0.7160 | 12000 | 5465 | 40.0 | 0.6681 | 1 |
| 492 | 1 | 23 | 5 | 0.7160 | 15000 | 5465 | 40.0 | 0.6770 | 1 |
| 493 | 1 | 23 | 5 | 0.7160 | 20000 | 5465 | 40.0 | 0.6878 | 1 |
| 494 | 1 | 23 | 5 | 0.7160 | 25000 | 5465 | 40.0 | 0.6939 | 1 |
| 495 | 1 | 23 | 5 | 0.7160 | 30000 | 5465 | 40.0 | 0.6977 | 1 |
| 496 | 1 | 23 | 5 | 0.7160 | 6000. | 5465 | 40.0 | 0.6378 | 1 |
| 497 | 1 | 23 | 6 | 0.8400 | 6000. | 5465 | 40.0 | 0.7952 | 11 |
| 498 | 1 | 23 | 6 | 0.8400 | 8000. | 5465 | 40.0 | 0.8013 | 11 |
| 499 | 1 | 23 | 6 | 0.8400 | 10000 | 5465 | 40.0 | 0.8065 | 11 |
| 500 | 1 | 23 | 6 | 0.8400 | 12000 | 5465 | 40.0 | 0.8111 | 11 |
| 501 | 1 | 23 | 6 | 0.8400 | 15000 | 5465 | 40.0 | 0.8162 | 11 |
| 502 | 1 | 23 | 6 | 0.8400 | 20000 | 5465 | 40.0 | 0.8220 | 11 |
| 503 | 1 | 23 | 6 | 0.8400 | 25000 | 5465 | 40.0 | 0.8265 | 11 |
| 504 | 1 | 23 | 6 | 0.8400 | 30000 | 5465 | 40.0 | 0.8295 | 11 |
| 505 | 1 | 23 | 14 | 0.4700 | 10000 | 5465 | 20.0 | 0.4310 | 7 |
| 506 | 1 | 23 | 14 | 0.4700 | 15000 | 5465 | 20.0 | 0.4080 | 7 |
| 507 | 1 | 23 | 14 | 0.4700 | 20000 | 5465 | 20.0 | 0.4050 | 7 |
| 508 | 1 | 23 | 14 | 0.4700 | 25000 | 5465 | 20.0 | 0.4140 | 7 |
| 509 | 1 | 23 | 14 | 0.4700 | 30000 | 5465 | 20.0 | 0.3840 | 7 |
| 510 | 1 | 23 | 14 | 0.4700 | 35000 | 5465 | 20.0 | 0.3780 | 7 |
| 511 | 1 | 23 | 14 | 0.4700 | 40000 | 5465 | 20.0 | 0.3580 | 7 |
| 512 | 1 | 23 | 14 | 0.4700 | 10000 | 5465 | 52.5 | 0.4475 | 5 |
| 513 | 1 | 23 | 14 | 0.4700 | 15000 | 5465 | 52.5 | 0.4416 | 5 |
| 514 | 1 | 23 | 14 | 0.4700 | 20000 | 5465 | 52.5 | 0.4342 | 5 |
| 515 | 1 | 23 | 14 | 0.4700 | 25000 | 5465 | 52.5 | 0.4247 | 5 |
| 516 | 1 | 23 | 14 | 0.4700 | 30000 | 5465 | 52.5 | 0.4143 | 5 |
| 517 | 1 | 23 | 14 | 0.4700 | 10000 | 5465 | 75.0 | 0.4371 | 5 |
| 518 | 1 | 23 | 14 | 0.4700 | 15000 | 5465 | 75.0 | 0.4356 | 5 |
| 519 | 1 | 23 | 14 | 0.4700 | 20000 | 5465 | 75.0 | 0.4342 | 5 |
| 520 | 1 | 23 | 14 | 0.4700 | 25000 | 5465 | 75.0 | 0.4327 | 5 |
| 521 | 1 | 23 | 14 | 0.4700 | 30000 | 5465 | 75.0 | 0.4303 | 5 |
| 522 | 1 | 23 | 14 | 0.4700 | 35000 | 5465 | 75.0 | 0.4267 | 5 |
| 523 | 1 | 23 | 14 | 0.4700 | 40000 | 5465 | 75.0 | 0.4234 | 5 |

| | | | | | | | | | |
|---|---|---|---|---|---|---|---|---|---|
| 524 | 1 | 24 | 5 | 0.7269 | 8000 | 5989 | 40.0 | 0.6230 | 1 |
| 525 | 1 | 24 | 5 | 0.7269 | 10000 | 5989 | 40.0 | 0.6374 | 1 |
| 526 | 1 | 24 | 5 | 0.7269 | 12000 | 5989 | 40.0 | 0.6480 | 1 |
| 527 | 1 | 24 | 5 | 0.7269 | 15000 | 5989 | 40.0 | 0.6603 | 1 |
| 528 | 1 | 24 | 5 | 0.7269 | 20000 | 5989 | 40.0 | 0.6738 | 1 |
| 529 | 1 | 24 | 5 | 0.7269 | 25000 | 5989 | 40.0 | 0.6810 | 1 |
| 530 | 1 | 24 | 5 | 0.7269 | 30000 | 5989 | 40.0 | 0.6850 | 1 |
| 531 | 1 | 24 | 5 | 0.8320 | 8000 | 5989 | 40.0 | 0.7785 | 1 |
| 532 | 1 | 24 | 5 | 0.8320 | 10000 | 5989 | 40.0 | 0.7840 | 1 |
| 533 | 1 | 24 | 5 | 0.8320 | 12000 | 5989 | 40.0 | 0.7905 | 1 |
| 534 | 1 | 24 | 5 | 0.8320 | 15000 | 5989 | 40.0 | 0.7981 | 1 |
| 535 | 1 | 24 | 5 | 0.8320 | 20000 | 5989 | 40.0 | 0.8070 | 1 |
| 536 | 1 | 24 | 5 | 0.8320 | 25000 | 5989 | 40.0 | 0.8122 | 1 |
| 537 | 1 | 24 | 5 | 0.8320 | 30000 | 5989 | 40.0 | 0.8156 | 1 |
| 538 | 1 | 24 | 5 | 0.8320 | 15000 | 5989 | 40.0 | 0.7981 | 1 |
| 539 | 1 | 24 | 6 | 0.8666 | 10000 | 5989 | 40.0 | 0.7920 | 6 |
| 540 | 1 | 24 | 6 | 0.8666 | 15000 | 5989 | 40.0 | 0.7980 | 6 |
| 541 | 1 | 24 | 6 | 0.8666 | 8000 | 5989 | 40.0 | 0.8310 | 11 |
| 542 | 1 | 24 | 6 | 0.8666 | 10000 | 5989 | 40.0 | 0.8322 | 11 |
| 543 | 1 | 24 | 6 | 0.8666 | 12000 | 5989 | 40.0 | 0.8358 | 11 |
| 544 | 1 | 24 | 6 | 0.8666 | 15000 | 5989 | 40.0 | 0.8385 | 11 |
| 545 | 1 | 24 | 6 | 0.8666 | 20000 | 5989 | 40.0 | 0.8422 | 11 |
| 546 | 1 | 24 | 6 | 0.8666 | 25000 | 5989 | 40.0 | 0.8455 | 11 |
| 547 | 1 | 24 | 6 | 0.8666 | 30000 | 5989 | 40.0 | 0.8478 | 11 |
| 548 | 1 | 24 | 6 | 0.9090 | 8000 | 5989 | 40.0 | 0.8790 | 11 |
| 549 | 1 | 24 | 6 | 0.9090 | 10000 | 5989 | 40.0 | 0.8819 | 11 |
| 550 | 1 | 24 | 6 | 0.9090 | 12000 | 5989 | 40.0 | 0.8842 | 11 |
| 551 | 1 | 24 | 6 | 0.9090 | 15000 | 5989 | 40.0 | 0.8875 | 11 |
| 552 | 1 | 24 | 6 | 0.9090 | 20000 | 5989 | 40.0 | 0.8918 | 11 |
| 553 | 1 | 24 | 6 | 0.9090 | 25000 | 5989 | 40.0 | 0.8950 | 11 |
| 554 | 1 | 24 | 6 | 0.9090 | 30000 | 5989 | 40.0 | 0.8972 | 11 |
| 555 | 1 | 24 | 6 | 0.9432 | 8000 | 5989 | 40.0 | 0.9252 | 11 |
| 556 | 1 | 24 | 6 | 0.9432 | 10000 | 5989 | 40.0 | 0.9257 | 11 |
| 557 | 1 | 24 | 6 | 0.9432 | 12000 | 5989 | 40.0 | 0.9259 | 11 |
| 558 | 1 | 24 | 6 | 0.9432 | 15000 | 5989 | 40.0 | 0.9261 | 11 |
| 559 | 1 | 24 | 6 | 0.9432 | 20000 | 5989 | 40.0 | 0.9265 | 11 |
| 560 | 1 | 24 | 6 | 0.9432 | 25000 | 5989 | 40.0 | 0.9271 | 11 |
| 561 | 1 | 24 | 6 | 0.9432 | 30000 | 5989 | 40.0 | 0.9278 | 11 |
| 562 | 1 | 24 | 8 | 0.6840 | 20000 | 5989 | 20.0 | 0.6390 | 7 |
| 563 | 1 | 24 | 8 | 0.6840 | 30000 | 5989 | 20.0 | 0.6390 | 7 |
| 564 | 1 | 24 | 8 | 0.6840 | 40000 | 5989 | 20.0 | 0.6640 | 7 |
| 565 | 1 | 24 | 26 | 0.0075 | 15000 | 5989 | 52.5 | 0.0095 | 3 |
| 566 | 1 | 24 | 26 | 0.0075 | 20000 | 5989 | 52.5 | 0.0097 | 3 |
| 567 | 1 | 24 | 26 | 0.0075 | 25000 | 5989 | 52.5 | 0.0099 | 3 |
| 568 | 1 | 24 | 26 | 0.0075 | 30000 | 5989 | 52.5 | 0.0103 | 3 |
| 569 | 1 | 24 | 26 | 0.0075 | 35000 | 5989 | 52.5 | 0.0106 | 3 |

| | | | | | | | | | |
|---|---|---|---|---|---|---|---|---|---|
| 570 | 1 | 24 | 26 | 0.0075 | 40000 | 5989 | 52.5 | 0.0110 | 3 |
| 571 | 1 | 24 | 26 | 0.0075 | 30000 | 5989 | 52.5 | 0.0108 | 4 |
| 572 | 1 | 24 | 26 | 0.0243 | 15000 | 5989 | 52.5 | 0.0308 | 3 |
| 573 | 1 | 24 | 26 | 0.0243 | 20000 | 5989 | 52.5 | 0.0313 | 3 |
| 574 | 1 | 24 | 26 | 0.0243 | 25000 | 5989 | 52.5 | 0.0318 | 3 |
| 575 | 1 | 24 | 26 | 0.0243 | 30000 | 5989 | 52.5 | 0.0323 | 3 |
| 576 | 1 | 24 | 26 | 0.0243 | 35000 | 5989 | 52.5 | 0.0328 | 3 |
| 577 | 1 | 24 | 26 | 0.0243 | 40000 | 5989 | 52.5 | 0.0333 | 3 |
| 578 | 1 | 24 | 26 | 0.0243 | 30000 | 5989 | 52.5 | 0.0337 | 4 |
| 579 | 1 | 24 | 26 | 0.0288 | 15000 | 5989 | 52.5 | 0.0365 | 3 |
| 580 | 1 | 24 | 26 | 0.0288 | 20000 | 5989 | 52.5 | 0.0370 | 3 |
| 581 | 1 | 24 | 26 | 0.0288 | 25000 | 5989 | 52.5 | 0.0376 | 3 |
| 582 | 1 | 24 | 26 | 0.0288 | 30000 | 5989 | 52.5 | 0.0381 | 3 |
| 583 | 1 | 24 | 26 | 0.0288 | 35000 | 5989 | 52.5 | 0.0387 | 3 |
| 584 | 1 | 24 | 26 | 0.0288 | 40000 | 5989 | 52.5 | 0.0392 | 3 |
| 585 | 1 | 24 | 26 | 0.0288 | 10000 | 5989 | 52.5 | 0.0351 | 4 |
| 586 | 1 | 24 | 26 | 0.0288 | 20000 | 5989 | 52.5 | 0.0379 | 4 |
| 587 | 1 | 24 | 26 | 0.0288 | 30000 | 5989 | 52.5 | 0.0396 | 4 |
| 588 | 1 | 24 | 26 | 0.0532 | 15000 | 5989 | 52.5 | 0.0669 | 3 |
| 589 | 1 | 24 | 26 | 0.0532 | 20000 | 5989 | 52.5 | 0.0679 | 3 |
| 590 | 1 | 24 | 26 | 0.0532 | 25000 | 5989 | 52.5 | 0.0688 | 3 |
| 591 | 1 | 24 | 26 | 0.0532 | 30000 | 5989 | 52.5 | 0.0696 | 3 |
| 592 | 1 | 24 | 26 | 0.0532 | 35000 | 5989 | 52.5 | 0.0703 | 3 |
| 593 | 1 | 24 | 26 | 0.0532 | 40000 | 5989 | 52.5 | 0.0709 | 3 |
| 594 | 1 | 24 | 26 | 0.0532 | 10000 | 5989 | 52.5 | 0.0651 | 4 |
| 595 | 1 | 24 | 26 | 0.0532 | 20000 | 5989 | 52.5 | 0.0700 | 4 |
| 596 | 1 | 24 | 26 | 0.0532 | 30000 | 5989 | 52.5 | 0.0719 | 4 |
| 597 | 1 | 24 | 26 | 0.1010 | 15000 | 5989 | 52.5 | 0.1251 | 3 |
| 598 | 1 | 24 | 26 | 0.1010 | 20000 | 5989 | 52.5 | 0.1267 | 3 |
| 599 | 1 | 24 | 26 | 0.1010 | 25000 | 5989 | 52.5 | 0.1281 | 3 |
| 600 | 1 | 24 | 26 | 0.1010 | 30000 | 5989 | 52.5 | 0.1294 | 3 |
| 601 | 1 | 24 | 26 | 0.1010 | 35000 | 5989 | 52.5 | 0.1304 | 3 |
| 602 | 1 | 24 | 26 | 0.1010 | 40000 | 5989 | 52.5 | 0.1313 | 3 |
| 603 | 1 | 24 | 26 | 0.1010 | 10000 | 5989 | 52.5 | 0.1186 | 4 |
| 604 | 1 | 24 | 26 | 0.1010 | 20000 | 5989 | 52.5 | 0.1271 | 4 |
| 605 | 1 | 24 | 26 | 0.1010 | 30000 | 5989 | 52.5 | 0.1314 | 4 |
| 606 | 1 | 24 | 26 | 0.4984 | 15000 | 5989 | 52.5 | 0.5330 | 3 |
| 607 | 1 | 24 | 26 | 0.4984 | 20000 | 5989 | 52.5 | 0.5350 | 3 |
| 608 | 1 | 24 | 26 | 0.4984 | 25000 | 5989 | 52.5 | 0.5368 | 3 |
| 609 | 1 | 24 | 26 | 0.4984 | 30000 | 5989 | 52.5 | 0.5384 | 3 |
| 610 | 1 | 24 | 26 | 0.4984 | 35000 | 5989 | 52.5 | 0.5398 | 3 |
| 611 | 1 | 24 | 26 | 0.4984 | 40000 | 5989 | 52.5 | 0.5410 | 3 |
| 612 | 1 | 24 | 26 | 0.4984 | 15000 | 5989 | 52.5 | 0.5272 | 4 |
| 613 | 1 | 24 | 26 | 0.4984 | 20000 | 5989 | 52.5 | 0.5444 | 4 |
| 614 | 1 | 24 | 26 | 0.4984 | 30000 | 5989 | 52.5 | 0.5436 | 4 |
| 615 | 1 | 24 | 26 | 0.9000 | 15000 | 5989 | 52.5 | 0.9078 | 3 |

| | | | | | | | | | |
|---|---|---|---|---|---|---|---|---|---|
| 616 | 1 | 24 | 26 | 0.9000 | 20000 | 5989 | 52.5 | 0.9082 | 3 |
| 617 | 1 | 24 | 26 | 0.9000 | 25000 | 5989 | 52.5 | 0.9086 | 3 |
| 618 | 1 | 24 | 26 | 0.9000 | 30000 | 5989 | 52.5 | 0.9091 | 3 |
| 619 | 1 | 24 | 26 | 0.9000 | 35000 | 5989 | 52.5 | 0.9095 | 3 |
| 620 | 1 | 24 | 26 | 0.9000 | 40000 | 5989 | 52.5 | 0.9099 | 3 |
| 621 | 1 | 24 | 26 | 0.9608 | 15000 | 5989 | 52.5 | 0.9652 | 3 |
| 622 | 1 | 24 | 26 | 0.9608 | 20000 | 5989 | 52.5 | 0.9653 | 3 |
| 623 | 1 | 24 | 26 | 0.9608 | 25000 | 5989 | 52.5 | 0.9655 | 3 |
| 624 | 1 | 24 | 26 | 0.9608 | 30000 | 5989 | 52.5 | 0.9657 | 3 |
| 625 | 1 | 24 | 26 | 0.9608 | 35000 | 5989 | 52.5 | 0.9659 | 3 |
| 626 | 1 | 24 | 26 | 0.9608 | 40000 | 5989 | 52.5 | 0.9662 | 3 |
| 627 | 1 | 24 | 26 | 0.9703 | 20000 | 5989 | 52.5 | 0.9740 | 3 |
| 628 | 1 | 24 | 26 | 0.9703 | 30000 | 5989 | 52.5 | 0.9742 | 3 |
| 629 | 1 | 24 | 26 | 0.9703 | 40000 | 5989 | 52.5 | 0.9747 | 3 |
| 630 | 1 | 24 | 26 | 0.9792 | 20000 | 5989 | 52.5 | 0.9819 | 3 |
| 631 | 1 | 24 | 26 | 0.9792 | 30000 | 5989 | 52.5 | 0.9821 | 3 |
| 632 | 1 | 24 | 26 | 0.9792 | 40000 | 5989 | 52.5 | 0.9824 | 3 |
| 633 | 1 | 24 | 26 | 0.9900 | 20000 | 5989 | 52.5 | 0.9913 | 3 |
| 634 | 1 | 24 | 26 | 0.9900 | 30000 | 5989 | 52.5 | 0.9916 | 3 |
| 635 | 1 | 24 | 26 | 0.9900 | 40000 | 5989 | 52.5 | 0.9921 | 3 |
| 636 | 1 | 25 | 51 | 0.4740 | 15000 | 6539 | 20.0 | 0.4600 | 7 |
| 637 | 1 | 25 | 51 | 0.4740 | 20000 | 6539 | 20.0 | 0.4210 | 7 |
| 638 | 1 | 25 | 51 | 0.4740 | 25000 | 6539 | 20.0 | 0.3850 | 7 |
| 639 | 1 | 25 | 51 | 0.4740 | 30000 | 6539 | 20.0 | 0.3500 | 7 |
| 640 | 1 | 26 | 5 | 0.8378 | 8000 | 7111 | 40.0 | 0.7823 | 1 |
| 641 | 1 | 26 | 5 | 0.8378 | 10000 | 7111 | 40.0 | 0.7922 | 1 |
| 642 | 1 | 26 | 5 | 0.8378 | 12000 | 7111 | 40.0 | 0.8002 | 1 |
| 643 | 1 | 26 | 5 | 0.8378 | 15000 | 7111 | 40.0 | 0.8098 | 1 |
| 644 | 1 | 26 | 5 | 0.8378 | 20000 | 7111 | 40.0 | 0.8203 | 1 |
| 645 | 1 | 26 | 5 | 0.8378 | 25000 | 7111 | 40.0 | 0.8280 | 1 |
| 646 | 1 | 26 | 5 | 0.8378 | 30000 | 7111 | 40.0 | 0.8330 | 1 |
| 647 | 1 | 26 | 5 | 0.9118 | 8000 | 7111 | 40.0 | 0.8680 | 1 |
| 648 | 1 | 26 | 5 | 0.9118 | 10000 | 7111 | 40.0 | 0.8730 | 1 |
| 649 | 1 | 26 | 5 | 0.9118 | 12000 | 7111 | 40.0 | 0.8764 | 1 |
| 650 | 1 | 26 | 5 | 0.9118 | 15000 | 7111 | 40.0 | 0.8821 | 1 |
| 651 | 1 | 26 | 5 | 0.9118 | 20000 | 7111 | 40.0 | 0.8888 | 1 |
| 652 | 1 | 26 | 5 | 0.9118 | 25000 | 7111 | 40.0 | 0.8940 | 1 |
| 653 | 1 | 26 | 5 | 0.9118 | 30000 | 7111 | 40.0 | 0.8970 | 1 |
| 654 | 1 | 26 | 6 | 0.9333 | 8000 | 7111 | 40.0 | 0.9172 | 11 |
| 655 | 1 | 26 | 6 | 0.9333 | 10000 | 7111 | 40.0 | 0.9194 | 11 |
| 656 | 1 | 26 | 6 | 0.9333 | 12000 | 7111 | 40.0 | 0.9215 | 11 |
| 657 | 1 | 26 | 6 | 0.9333 | 15000 | 7111 | 40.0 | 0.9239 | 11 |
| 658 | 1 | 26 | 6 | 0.9333 | 20000 | 7111 | 40.0 | 0.9270 | 11 |
| 659 | 1 | 26 | 6 | 0.9333 | 25000 | 7111 | 40.0 | 0.9290 | 11 |
| 660 | 1 | 26 | 6 | 0.9333 | 30000 | 7111 | 40.0 | 0.9304 | 11 |
| 661 | 1 | 26 | 8 | 0.6990 | 10000 | 7111 | 20.0 | 0.6240 | 7 |

| | | | | | | | | | |
|---|---|---|---|---|---|---|---|---|---|
| 662 | 1 | 26 | 8 | 0.6990 | 20000 | 7111 | 20.0 | 0.6530 | 7 |
| 663 | 1 | 26 | 8 | 0.6990 | 30000 | 7111 | 20.0 | 0.6530 | 7 |
| 664 | 1 | 26 | 8 | 0.6990 | 40000 | 7111 | 20.0 | 0.6530 | 7 |
| 665 | 1 | 26 | 8 | 0.7000 | 29000 | 7111 | 20.0 | 0.6490 | 7 |
| 666 | 1 | 26 | 8 | 0.6970 | 12000 | 7111 | 40.0 | 0.6420 | 7 |
| 667 | 1 | 26 | 8 | 0.6970 | 15000 | 7111 | 40.0 | 0.6470 | 7 |
| 668 | 1 | 26 | 8 | 0.6970 | 20000 | 7111 | 40.0 | 0.6540 | 7 |
| 669 | 1 | 26 | 8 | 0.6970 | 25000 | 7111 | 40.0 | 0.6610 | 7 |
| 670 | 1 | 26 | 8 | 0.6970 | 29000 | 7111 | 40.0 | 0.6650 | 7 |
| 671 | 1 | 26 | 13 | 0.4080 | 10000 | 7111 | 16.5 | 0.3600 | 7 |
| 672 | 1 | 26 | 13 | 0.4080 | 15000 | 7111 | 16.5 | 0.3620 | 7 |
| 673 | 1 | 26 | 13 | 0.4080 | 20000 | 7111 | 16.5 | 0.3650 | 7 |
| 674 | 1 | 26 | 13 | 0.4080 | 25000 | 7111 | 16.5 | 0.3670 | 7 |
| 675 | 1 | 26 | 13 | 0.4080 | 30000 | 7111 | 16.5 | 0.3700 | 7 |
| 676 | 1 | 26 | 13 | 0.4080 | 35000 | 7111 | 16.5 | 0.3700 | 7 |
| 677 | 1 | 26 | 13 | 0.4083 | 10000 | 7111 | 40.0 | 0.3510 | 6 |
| 678 | 1 | 26 | 13 | 0.4083 | 15000 | 7111 | 40.0 | 0.3670 | 6 |
| 679 | 1 | 26 | 13 | 0.4083 | 20000 | 7111 | 40.0 | 0.3640 | 6 |
| 680 | 1 | 26 | 13 | 0.4083 | 25000 | 7111 | 40.0 | 0.3670 | 6 |
| 681 | 1 | 26 | 13 | 0.4083 | 30000 | 7111 | 40.0 | 0.3670 | 6 |
| 682 | 1 | 26 | 13 | 0.4083 | 35000 | 7111 | 40.0 | 0.3660 | 6 |
| 683 | 1 | 26 | 13 | 0.4083 | 40000 | 7111 | 40.0 | 0.3690 | 6 |
| 684 | 1 | 26 | 13 | 0.7590 | 20000 | 7111 | 52.5 | 0.7360 | 7 |
| 685 | 1 | 26 | 13 | 0.7590 | 25000 | 7111 | 52.5 | 0.7420 | 7 |
| 686 | 1 | 26 | 13 | 0.7590 | 30000 | 7111 | 52.5 | 0.7480 | 7 |
| 687 | 1 | 26 | 16 | 0.4660 | 10000 | 7111 | 75.0 | 0.4060 | 7 |
| 688 | 1 | 26 | 16 | 0.4660 | 12000 | 7111 | 75.0 | 0.4210 | 7 |
| 689 | 1 | 26 | 16 | 0.4660 | 15000 | 7111 | 75.0 | 0.4250 | 7 |
| 690 | 1 | 26 | 16 | 0.4660 | 20000 | 7111 | 75.0 | 0.4250 | 7 |
| 691 | 1 | 26 | 16 | 0.4660 | 25000 | 7111 | 75.0 | 0.4220 | 7 |
| 692 | 1 | 26 | 16 | 0.4660 | 30000 | 7111 | 75.0 | 0.4190 | 7 |
| 693 | 1 | 26 | 24 | 0.9020 | 10000 | 7111 | 18.0 | 0.8870 | 7 |
| 694 | 1 | 26 | 24 | 0.9020 | 20000 | 7111 | 18.0 | 0.8700 | 7 |
| 695 | 1 | 26 | 24 | 0.0109 | 15000 | 7111 | 52.5 | 0.0113 | 3 |
| 696 | 1 | 26 | 24 | 0.0109 | 20000 | 7111 | 52.5 | 0.0105 | 3 |
| 697 | 1 | 26 | 24 | 0.0109 | 25000 | 7111 | 52.5 | 0.0096 | 3 |
| 698 | 1 | 26 | 24 | 0.0109 | 30000 | 7111 | 52.5 | 0.0084 | 3 |
| 699 | 1 | 26 | 24 | 0.0109 | 35000 | 7111 | 52.5 | 0.0071 | 3 |
| 700 | 1 | 26 | 24 | 0.0109 | 40000 | 7111 | 52.5 | 0.0056 | 3 |
| 701 | 1 | 26 | 24 | 0.0220 | 15000 | 7111 | 52.5 | 0.0202 | 3 |
| 702 | 1 | 26 | 24 | 0.0220 | 20000 | 7111 | 52.5 | 0.0193 | 3 |
| 703 | 1 | 26 | 24 | 0.0220 | 25000 | 7111 | 52.5 | 0.0180 | 3 |
| 704 | 1 | 26 | 24 | 0.0220 | 30000 | 7111 | 52.5 | 0.0164 | 3 |
| 705 | 1 | 26 | 24 | 0.0220 | 35000 | 7111 | 52.5 | 0.0144 | 3 |
| 706 | 1 | 26 | 24 | 0.0220 | 40000 | 7111 | 52.5 | 0.0120 | 3 |
| 707 | 1 | 26 | 24 | 0.0308 | 15000 | 7111 | 52.5 | 0.0274 | 3 |

| | | | | | | | | | |
|---|---|---|---|---|---|---|---|---|---|
| 708 | 1 | 26 | 24 | 0.0308 | 20000 | 7111 | 52.5 | 0.0263 | 3 |
| 709 | 1 | 26 | 24 | 0.0308 | 25000 | 7111 | 52.5 | 0.0248 | 3 |
| 710 | 1 | 26 | 24 | 0.0308 | 30000 | 7111 | 52.5 | 0.0227 | 3 |
| 711 | 1 | 26 | 24 | 0.0308 | 35000 | 7111 | 52.5 | 0.0202 | 3 |
| 712 | 1 | 26 | 24 | 0.0308 | 40000 | 7111 | 52.5 | 0.0171 | 3 |
| 713 | 1 | 26 | 24 | 0.0406 | 15000 | 7111 | 52.5 | 0.0353 | 3 |
| 714 | 1 | 26 | 24 | 0.0406 | 20000 | 7111 | 52.5 | 0.0342 | 3 |
| 715 | 1 | 26 | 24 | 0.0406 | 25000 | 7111 | 52.5 | 0.0323 | 3 |
| 716 | 1 | 26 | 24 | 0.0406 | 30000 | 7111 | 52.5 | 0.0298 | 3 |
| 717 | 1 | 26 | 24 | 0.0406 | 35000 | 7111 | 52.5 | 0.0267 | 3 |
| 718 | 1 | 26 | 24 | 0.0406 | 40000 | 7111 | 52.5 | 0.0229 | 3 |
| 719 | 1 | 26 | 24 | 0.1003 | 15000 | 7111 | 52.5 | 0.0849 | 3 |
| 720 | 1 | 26 | 24 | 0.1003 | 20000 | 7111 | 52.5 | 0.0829 | 3 |
| 721 | 1 | 26 | 24 | 0.1003 | 25000 | 7111 | 52.5 | 0.0794 | 3 |
| 722 | 1 | 26 | 24 | 0.1003 | 30000 | 7111 | 52.5 | 0.0744 | 3 |
| 723 | 1 | 26 | 24 | 0.1003 | 35000 | 7111 | 52.5 | 0.0679 | 3 |
| 724 | 1 | 26 | 24 | 0.1003 | 40000 | 7111 | 52.5 | 0.0599 | 3 |
| 725 | 1 | 26 | 24 | 0.4991 | 15000 | 7111 | 52.5 | 0.4537 | 3 |
| 726 | 1 | 26 | 24 | 0.4991 | 20000 | 7111 | 52.5 | 0.4493 | 3 |
| 727 | 1 | 26 | 24 | 0.4991 | 25000 | 7111 | 52.5 | 0.4403 | 3 |
| 728 | 1 | 26 | 24 | 0.4991 | 30000 | 7111 | 52.5 | 0.4267 | 3 |
| 729 | 1 | 26 | 24 | 0.4991 | 35000 | 7111 | 52.5 | 0.4086 | 3 |
| 730 | 1 | 26 | 24 | 0.4991 | 40000 | 7111 | 52.5 | 0.3858 | 3 |
| 731 | 1 | 26 | 24 | 0.5016 | 10000 | 7111 | 52.5 | 0.4811 | 4 |
| 732 | 1 | 26 | 24 | 0.5016 | 20000 | 7111 | 52.5 | 0.4535 | 4 |
| 733 | 1 | 26 | 24 | 0.5016 | 30000 | 7111 | 52.5 | 0.4343 | 4 |
| 734 | 1 | 26 | 24 | 0.8984 | 15000 | 7111 | 52.5 | 0.8894 | 3 |
| 735 | 1 | 26 | 24 | 0.8984 | 20000 | 7111 | 52.5 | 0.8871 | 3 |
| 736 | 1 | 26 | 24 | 0.8984 | 25000 | 7111 | 52.5 | 0.8821 | 3 |
| 737 | 1 | 26 | 24 | 0.8984 | 30000 | 7111 | 52.5 | 0.8741 | 3 |
| 738 | 1 | 26 | 24 | 0.8984 | 35000 | 7111 | 52.5 | 0.8634 | 3 |
| 739 | 1 | 26 | 24 | 0.8984 | 40000 | 7111 | 52.5 | 0.8498 | 3 |
| 740 | 1 | 26 | 24 | 0.8990 | 10000 | 7111 | 52.5 | 0.8883 | 4 |
| 741 | 1 | 26 | 24 | 0.8990 | 20000 | 7111 | 52.5 | 0.8842 | 4 |
| 742 | 1 | 26 | 24 | 0.8990 | 30000 | 7111 | 52.5 | 0.8751 | 4 |
| 743 | 1 | 26 | 24 | 0.9455 | 20000 | 7111 | 52.5 | 0.9435 | 3 |
| 744 | 1 | 26 | 24 | 0.9455 | 30000 | 7111 | 52.5 | 0.9332 | 3 |
| 745 | 1 | 26 | 24 | 0.9455 | 40000 | 7111 | 52.5 | 0.9136 | 3 |
| 746 | 1 | 26 | 24 | 0.9468 | 10000 | 7111 | 52.5 | 0.9286 | 4 |
| 747 | 1 | 26 | 24 | 0.9468 | 20000 | 7111 | 52.5 | 0.9405 | 4 |
| 748 | 1 | 26 | 24 | 0.9468 | 30000 | 7111 | 52.5 | 0.9321 | 4 |
| 749 | 1 | 26 | 24 | 0.9686 | 20000 | 7111 | 52.5 | 0.9715 | 3 |
| 750 | 1 | 26 | 24 | 0.9686 | 30000 | 7111 | 52.5 | 0.9626 | 3 |
| 751 | 1 | 26 | 24 | 0.9686 | 40000 | 7111 | 52.5 | 0.9455 | 3 |
| 752 | 1 | 26 | 24 | 0.9738 | 20000 | 7111 | 52.5 | 0.9778 | 3 |
| 753 | 1 | 26 | 24 | 0.9738 | 30000 | 7111 | 52.5 | 0.9693 | 3 |

| | | | | | | | | | |
|---|---|---|---|---|---|---|---|---|---|
| 754 | 1 | 26 | 24 | 0.9738 | 40000 | 7111 | 52.5 | 0.9528 | 3 |
| 755 | 1 | 26 | 24 | 0.9893 | 20000 | 7111 | 52.5 | 0.9968 | 3 |
| 756 | 1 | 26 | 24 | 0.9893 | 30000 | 7111 | 52.5 | 0.9892 | 3 |
| 757 | 1 | 26 | 24 | 0.9893 | 40000 | 7111 | 52.5 | 0.9746 | 3 |
| 758 | 1 | 26 | 50 | 0.1905 | 20000 | 7111 | 40.0 | 0.1859 | 7 |
| 759 | 1 | 26 | 50 | 0.3200 | 20000 | 7111 | 40.0 | 0.3145 | 7 |
| 760 | 1 | 27 | 5 | 0.8450 | 10000 | 7709 | 40.0 | 0.7823 | 1 |
| 761 | 1 | 27 | 5 | 0.8450 | 12000 | 7709 | 40.0 | 0.7939 | 1 |
| 762 | 1 | 27 | 5 | 0.8450 | 15000 | 7709 | 40.0 | 0.8058 | 1 |
| 763 | 1 | 27 | 5 | 0.8450 | 20000 | 7709 | 40.0 | 0.8180 | 1 |
| 764 | 1 | 27 | 5 | 0.8450 | 25000 | 7709 | 40.0 | 0.8244 | 1 |
| 765 | 1 | 27 | 5 | 0.8450 | 30000 | 7709 | 40.0 | 0.8278 | 1 |
| 766 | 1 | 27 | 5 | 0.9160 | 10000 | 7709 | 40.0 | 0.8660 | 1 |
| 767 | 1 | 27 | 5 | 0.9160 | 12000 | 7709 | 40.0 | 0.8832 | 1 |
| 768 | 1 | 27 | 5 | 0.9160 | 15000 | 7709 | 40.0 | 0.8980 | 1 |
| 769 | 1 | 27 | 5 | 0.9160 | 20000 | 7709 | 40.0 | 0.9101 | 1 |
| 770 | 1 | 27 | 5 | 0.9160 | 25000 | 7709 | 40.0 | 0.9120 | 1 |
| 771 | 1 | 27 | 5 | 0.9160 | 30000 | 7709 | 40.0 | 0.9078 | 1 |
| 772 | 1 | 28 | 5 | 0.8560 | 10000 | 8331 | 40.0 | 0.8083 | 1 |
| 773 | 1 | 28 | 5 | 0.8560 | 12000 | 8331 | 40.0 | 0.8180 | 1 |
| 774 | 1 | 28 | 5 | 0.8560 | 15000 | 8331 | 40.0 | 0.8283 | 1 |
| 775 | 1 | 28 | 5 | 0.8560 | 20000 | 8331 | 40.0 | 0.8388 | 1 |
| 776 | 1 | 28 | 5 | 0.8560 | 25000 | 8331 | 40.0 | 0.8438 | 1 |
| 777 | 1 | 28 | 5 | 0.8560 | 30000 | 8331 | 40.0 | 0.8450 | 1 |
| 778 | 1 | 28 | 5 | 0.9157 | 10000 | 8331 | 40.0 | 0.8858 | 1 |
| 779 | 1 | 28 | 5 | 0.9157 | 12000 | 8331 | 40.0 | 0.8899 | 1 |
| 780 | 1 | 28 | 5 | 0.9157 | 15000 | 8331 | 40.0 | 0.8950 | 1 |
| 781 | 1 | 28 | 5 | 0.9157 | 20000 | 8331 | 40.0 | 0.9005 | 1 |
| 782 | 1 | 28 | 5 | 0.9157 | 25000 | 8331 | 40.0 | 0.9040 | 1 |
| 783 | 1 | 28 | 5 | 0.9157 | 30000 | 8331 | 40.0 | 0.9060 | 1 |
| 784 | 1 | 28 | 5 | 0.9422 | 10000 | 8331 | 40.0 | 0.9238 | 1 |
| 785 | 1 | 28 | 5 | 0.9422 | 12000 | 8331 | 40.0 | 0.9262 | 1 |
| 786 | 1 | 28 | 5 | 0.9422 | 15000 | 8331 | 40.0 | 0.9303 | 1 |
| 787 | 1 | 28 | 5 | 0.9422 | 20000 | 8331 | 40.0 | 0.9358 | 1 |
| 788 | 1 | 28 | 5 | 0.9422 | 25000 | 8331 | 40.0 | 0.9380 | 1 |
| 789 | 1 | 28 | 5 | 0.9422 | 30000 | 8331 | 40.0 | 0.9382 | 1 |
| 790 | 1 | 28 | 8 | 0.7860 | 10000 | 8331 | 40.0 | 0.7170 | 6 |
| 791 | 1 | 28 | 8 | 0.7860 | 15000 | 8331 | 40.0 | 0.7540 | 6 |
| 792 | 1 | 28 | 8 | 0.7860 | 20000 | 8331 | 40.0 | 0.7600 | 6 |
| 793 | 1 | 28 | 8 | 0.7860 | 25000 | 8331 | 40.0 | 0.7640 | 6 |
| 794 | 1 | 28 | 13 | 0.4210 | 10000 | 8331 | 16.2 | 0.3600 | 7 |
| 795 | 1 | 28 | 13 | 0.4210 | 15000 | 8331 | 16.2 | 0.3700 | 7 |
| 796 | 1 | 28 | 13 | 0.4210 | 20000 | 8331 | 16.2 | 0.3800 | 7 |
| 797 | 1 | 28 | 13 | 0.4210 | 25000 | 8331 | 16.2 | 0.3850 | 7 |
| 798 | 1 | 28 | 13 | 0.4210 | 30000 | 8331 | 16.2 | 0.3850 | 7 |
| 799 | 1 | 28 | 13 | 0.4210 | 35000 | 8331 | 16.2 | 0.3850 | 7 |

| | | | | | | | | | |
|---|---|---|---|---|---|---|---|---|---|
| 800 | 1 | 28 | 13 | 0.5920 | 10000 | 8331 | 16.2 | 0.5250 | 7 |
| 801 | 1 | 28 | 13 | 0.5920 | 15000 | 8331 | 16.2 | 0.5400 | 7 |
| 802 | 1 | 28 | 13 | 0.5920 | 20000 | 8331 | 16.2 | 0.5550 | 7 |
| 803 | 1 | 28 | 13 | 0.5920 | 25000 | 8331 | 16.2 | 0.5600 | 7 |
| 804 | 1 | 28 | 13 | 0.5920 | 30000 | 8331 | 16.2 | 0.5600 | 7 |
| 805 | 1 | 28 | 13 | 0.5920 | 35000 | 8331 | 16.2 | 0.5600 | 7 |
| 806 | 1 | 28 | 13 | 0.4204 | 10000 | 8331 | 40.0 | 0.3650 | 6 |
| 807 | 1 | 28 | 13 | 0.4204 | 15000 | 8331 | 40.0 | 0.3960 | 6 |
| 808 | 1 | 28 | 13 | 0.4204 | 20000 | 8331 | 40.0 | 0.3940 | 6 |
| 809 | 1 | 28 | 13 | 0.4204 | 25000 | 8331 | 40.0 | 0.4010 | 6 |
| 810 | 1 | 28 | 13 | 0.4204 | 30000 | 8331 | 40.0 | 0.4020 | 6 |
| 811 | 1 | 28 | 13 | 0.4204 | 35000 | 8331 | 40.0 | 0.4070 | 6 |
| 812 | 1 | 28 | 23 | 0.2000 | 30000 | 8331 | 30.0 | 0.1640 | 7 |
| 813 | 1 | 28 | 26 | 0.0510 | 20000 | 8331 | 18.0 | 0.0400 | 7 |
| 814 | 1 | 28 | 26 | 0.1000 | 20000 | 8331 | 18.0 | 0.0790 | 7 |
| 815 | 1 | 28 | 26 | 0.1980 | 20000 | 8331 | 18.0 | 0.1610 | 7 |
| 816 | 1 | 28 | 26 | 0.5630 | 20000 | 8331 | 18.0 | 0.5070 | 7 |
| 817 | 1 | 28 | 26 | 0.6300 | 20000 | 8331 | 18.0 | 0.5680 | 7 |
| 818 | 1 | 28 | 26 | 0.8950 | 20000 | 8331 | 18.0 | 0.8690 | 7 |
| 819 | 1 | 28 | 26 | 0.0980 | 10000 | 8331 | 20.0 | 0.0980 | 7 |
| 820 | 1 | 28 | 26 | 0.0980 | 15000 | 8331 | 20.0 | 0.0800 | 7 |
| 821 | 1 | 28 | 26 | 0.0980 | 20000 | 8331 | 20.0 | 0.0800 | 7 |
| 822 | 1 | 28 | 26 | 0.0980 | 25000 | 8331 | 20.0 | 0.0740 | 7 |
| 823 | 1 | 28 | 26 | 0.0980 | 30000 | 8331 | 20.0 | 0.0660 | 7 |
| 824 | 1 | 28 | 26 | 0.0980 | 35000 | 8331 | 20.0 | 0.0570 | 7 |
| 825 | 1 | 28 | 26 | 0.0980 | 40000 | 8331 | 20.0 | 0.0480 | 7 |
| 826 | 1 | 28 | 26 | 0.0980 | 10000 | 8331 | 75.0 | 0.0973 | 5 |
| 827 | 1 | 28 | 26 | 0.0980 | 15000 | 8331 | 75.0 | 0.0916 | 5 |
| 828 | 1 | 28 | 26 | 0.0980 | 20000 | 8331 | 75.0 | 0.0886 | 5 |
| 829 | 1 | 28 | 26 | 0.0980 | 25000 | 8331 | 75.0 | 0.0858 | 5 |
| 830 | 1 | 28 | 26 | 0.0980 | 30000 | 8331 | 75.0 | 0.0835 | 5 |
| 831 | 1 | 28 | 26 | 0.0980 | 35000 | 8331 | 75.0 | 0.0776 | 5 |
| 832 | 1 | 28 | 26 | 0.0980 | 40000 | 8331 | 75.0 | 0.0746 | 5 |
| 833 | 2 | 28 | 29 | 0.4350 | 10200 | 8331 | 40.0 | 0.4473 | 5 |
| 834 | 2 | 28 | 29 | 0.4350 | 12500 | 8331 | 40.0 | 0.4509 | 5 |
| 835 | 2 | 28 | 29 | 0.4350 | 15300 | 8331 | 40.0 | 0.4504 | 5 |
| 836 | 2 | 28 | 29 | 0.4350 | 20500 | 8331 | 40.0 | 0.4387 | 5 |
| 837 | 2 | 28 | 29 | 0.4350 | 25600 | 8331 | 40.0 | 0.4454 | 5 |
| 838 | 2 | 28 | 29 | 0.4350 | 30700 | 8331 | 40.0 | 0.4457 | 5 |
| 839 | 2 | 28 | 29 | 0.4350 | 40000 | 8331 | 40.0 | 0.4460 | 5 |
| 840 | 1 | 28 | 47 | 0.2000 | 30000 | 8331 | 30.0 | 0.1730 | 7 |
| 841 | 1 | 28 | 50 | 0.4259 | 20000 | 8331 | 40.0 | 0.4587 | 7 |
| 842 | 1 | 28 | 78 | 0.0650 | 30000 | 8331 | 15.5 | 0.0670 | 7 |
| 843 | 1 | 28 | 78 | 0.1640 | 30000 | 8331 | 15.5 | 0.1660 | 7 |
| 844 | 1 | 28 | 78 | 0.2970 | 30000 | 8331 | 15.5 | 0.3050 | 7 |
| 845 | 1 | 28 | 78 | 0.5510 | 30000 | 8331 | 15.5 | 0.5600 | 7 |

| | | | | | | | | | |
|---|---|---|---|---|---|---|---|---|---|
| 846 | 1 | 29 | 13 | 0.5360 | 28500 | 8980 | 16.0 | 0.4960 | 5 |
| 847 | 1 | 29 | 13 | 0.0555 | 15900 | 8980 | 40.0 | 0.0469 | 5 |
| 848 | 1 | 29 | 13 | 0.0555 | 21200 | 8980 | 40.0 | 0.0479 | 5 |
| 849 | 1 | 29 | 13 | 0.0555 | 31800 | 8980 | 40.0 | 0.0487 | 5 |
| 850 | 1 | 29 | 13 | 0.0555 | 39300 | 8980 | 40.0 | 0.0483 | 5 |
| 851 | 1 | 29 | 13 | 0.5410 | 10000 | 8980 | 40.0 | 0.4560 | 6 |
| 852 | 1 | 29 | 13 | 0.5410 | 15000 | 8980 | 40.0 | 0.4910 | 6 |
| 853 | 1 | 29 | 13 | 0.5410 | 20000 | 8980 | 40.0 | 0.5010 | 6 |
| 854 | 1 | 29 | 13 | 0.5410 | 25000 | 8980 | 40.0 | 0.5020 | 6 |
| 855 | 1 | 29 | 13 | 0.5410 | 30000 | 8980 | 40.0 | 0.5100 | 6 |
| 856 | 1 | 29 | 13 | 0.5410 | 35000 | 8980 | 40.0 | 0.5110 | 6 |
| 857 | 3 | 29 | 30 | 0.5550 | 15900 | 931 | 40.0 | 0.0515 | 5 |
| 858 | 3 | 29 | 30 | 0.5550 | 21200 | 931 | 40.0 | 0.0523 | 5 |
| 859 | 3 | 29 | 30 | 0.5550 | 31800 | 931 | 40.0 | 0.0514 | 5 |
| 860 | 3 | 29 | 30 | 0.5550 | 39900 | 931 | 40.0 | 0.0509 | 5 |
| 861 | 1 | 29 | 30 | 5.6500 | 10100 | 8980 | 40.0 | 0.5580 | 5 |
| 862 | 1 | 29 | 28 | 0.5650 | 12500 | 8980 | 40.0 | 0.5547 | 5 |
| 863 | 1 | 29 | 28 | 0.5650 | 15300 | 8980 | 40.0 | 0.5556 | 5 |
| 864 | 1 | 29 | 28 | 0.5650 | 20400 | 8980 | 40.0 | 0.5574 | 5 |
| 865 | 1 | 29 | 28 | 0.5650 | 25600 | 8980 | 40.0 | 0.5610 | 5 |
| 866 | 1 | 29 | 28 | 0.5650 | 30700 | 8980 | 40.0 | 0.5587 | 5 |
| 867 | 1 | 29 | 28 | 0.5650 | 40000 | 8980 | 40.0 | 0.5557 | 5 |
| 868 | 2 | 29 | 28 | 0.5650 | 10200 | 8980 | 40.0 | 0.5494 | 5 |
| 869 | 2 | 29 | 28 | 0.5650 | 12500 | 8980 | 40.0 | 0.5418 | 5 |
| 870 | 2 | 29 | 28 | 0.5650 | 15300 | 8980 | 40.0 | 0.5334 | 5 |
| 871 | 2 | 29 | 28 | 0.5650 | 20500 | 8980 | 40.0 | 0.5308 | 5 |
| 872 | 2 | 29 | 28 | 0.5650 | 25600 | 8980 | 40.0 | 0.5143 | 5 |
| 873 | 2 | 29 | 28 | 0.5650 | 30700 | 8980 | 40.0 | 0.5014 | 5 |
| 874 | 2 | 29 | 28 | 0.5650 | 40000 | 8980 | 40.0 | 0.4770 | 5 |
| 875 | 1 | 29 | 30 | 0.7290 | 25000 | 8980 | 18.0 | 0.7300 | 7 |
| 876 | 1 | 29 | 79 | 0.2010 | 30000 | 8980 | 15.5 | 0.1940 | 5 |
| 877 | 1 | 29 | 79 | 0.3990 | 30000 | 8980 | 15.5 | 0.3860 | 7 |
| 878 | 1 | 29 | 79 | 0.5980 | 30000 | 8980 | 15.5 | 0.5850 | 7 |
| 879 | 1 | 29 | 79 | 0.7940 | 30000 | 8980 | 15.5 | 0.7890 | 6 |
| 880 | 1 | 29 | 79 | 0.7940 | 30000 | 8980 | 15.5 | 0.7890 | 7 |
| 881 | 1 | 29 | 79 | 0.4000 | 16800 | 8980 | 16.0 | 0.4560 | 7 |
| 882 | 1 | 29 | 79 | 0.4000 | 22500 | 8980 | 16.0 | 0.4260 | 7 |
| 883 | 1 | 29 | 79 | 0.4730 | 28500 | 8980 | 16.0 | 0.4700 | 7 |
| 884 | 1 | 29 | 79 | 0.7620 | 28500 | 8980 | 16.0 | 0.7580 | 7 |
| 885 | 1 | 29 | 79 | 0.8000 | 16000 | 8980 | 16.0 | 0.7980 | 6 |
| 886 | 1 | 29 | 79 | 0.8000 | 16800 | 8980 | 16.0 | 0.7980 | 7 |
| 887 | 1 | 29 | 79 | 0.4000 | 28200 | 8980 | 17.1 | 0.3980 | 7 |
| 888 | 1 | 29 | 79 | 0.4000 | 39300 | 8980 | 17.1 | 0.3620 | 7 |
| 889 | 1 | 29 | 79 | 0.1990 | 25000 | 8980 | 18.0 | 0.2130 | 7 |
| 890 | 1 | 29 | 79 | 0.3960 | 25000 | 8980 | 18.0 | 0.4170 | 7 |
| 891 | 1 | 29 | 79 | 0.5990 | 25000 | 8980 | 18.0 | 0.6200 | 7 |

| | | | | | | | | | |
|---|---|---|---|---|---|---|---|---|---|
| 892 | 1 | 29 | 79 | 0.7900 | 25000 | 8980 | 18.0 | 0.8070 | 7 |
| 893 | 1 | 29 | 79 | 0.1980 | 12000 | 8980 | 40.0 | 0.2640 | 5 |
| 894 | 1 | 29 | 79 | 0.1980 | 14000 | 8980 | 40.0 | 0.2570 | 5 |
| 895 | 1 | 29 | 79 | 0.1980 | 16000 | 8980 | 40.0 | 0.2520 | 5 |
| 896 | 1 | 29 | 79 | 0.1980 | 20000 | 8980 | 40.0 | 0.2450 | 5 |
| 897 | 1 | 29 | 79 | 0.1980 | 25000 | 8980 | 40.0 | 0.2350 | 5 |
| 898 | 1 | 29 | 79 | 0.3960 | 16000 | 8980 | 40.0 | 0.4720 | 1 |
| 899 | 1 | 29 | 79 | 0.3960 | 12000 | 8980 | 40.0 | 0.4940 | 5 |
| 900 | 1 | 29 | 79 | 0.3960 | 14000 | 8980 | 40.0 | 0.4760 | 5 |
| 901 | 1 | 29 | 79 | 0.3960 | 20000 | 8980 | 40.0 | 0.4630 | 5 |
| 902 | 1 | 29 | 79 | 0.3960 | 25000 | 8980 | 40.0 | 0.4470 | 5 |
| 903 | 1 | 29 | 79 | 0.4920 | 13000 | 8980 | 40.0 | 0.5850 | 5 |
| 904 | 1 | 29 | 79 | 0.4920 | 15700 | 8980 | 40.0 | 0.5697 | 5 |
| 905 | 1 | 29 | 79 | 0.4920 | 20900 | 8980 | 40.0 | 0.5534 | 5 |
| 906 | 1 | 29 | 79 | 0.4920 | 26100 | 8980 | 40.0 | 0.5403 | 5 |
| 907 | 1 | 29 | 79 | 0.4920 | 31300 | 8980 | 40.0 | 0.5298 | 5 |
| 908 | 1 | 29 | 79 | 0.4920 | 39500 | 8980 | 40.0 | 0.5160 | 5 |
| 910 | 1 | 29 | 79 | 0.5990 | 12000 | 8980 | 40.0 | 0.6940 | 5 |
| 911 | 1 | 29 | 79 | 0.5990 | 14000 | 8980 | 40.0 | 0.6750 | 5 |
| 912 | 1 | 29 | 79 | 0.5990 | 16000 | 8980 | 40.0 | 0.6700 | 5 |
| 913 | 1 | 29 | 79 | 0.5990 | 20000 | 8980 | 40.0 | 0.6550 | 5 |
| 914 | 1 | 29 | 79 | 0.5990 | 25000 | 8980 | 40.0 | 0.6430 | 5 |
| 915 | 1 | 29 | 79 | 0.6040 | 14000 | 8980 | 40.0 | 0.5110 | 5 |
| 916 | 1 | 29 | 79 | 0.7000 | 13000 | 8980 | 40.0 | 0.7758 | 5 |
| 917 | 1 | 29 | 79 | 0.7000 | 15700 | 8980 | 40.0 | 0.7583 | 5 |
| 918 | 1 | 29 | 79 | 0.7000 | 20900 | 8980 | 40.0 | 0.7468 | 5 |
| 919 | 1 | 29 | 79 | 0.7000 | 26100 | 8980 | 40.0 | 0.7387 | 5 |
| 920 | 1 | 29 | 79 | 0.7000 | 31300 | 8980 | 40.0 | 0.7270 | 5 |
| 921 | 1 | 29 | 79 | 0.7000 | 36500 | 8980 | 40.0 | 0.7223 | 5 |
| 922 | 1 | 29 | 79 | 0.7000 | 38500 | 8980 | 40.0 | 0.7161 | 5 |
| 923 | 1 | 29 | 79 | 0.7990 | 12000 | 8980 | 40.0 | 0.8530 | 5 |
| 924 | 1 | 29 | 79 | 0.7990 | 14000 | 8980 | 40.0 | 0.8450 | 5 |
| 925 | 1 | 29 | 79 | 0.7990 | 16000 | 8980 | 40.0 | 0.8450 | 5 |
| 926 | 1 | 29 | 79 | 0.7990 | 20000 | 8980 | 40.0 | 0.8350 | 5 |
| 927 | 1 | 29 | 79 | 0.7990 | 25000 | 8980 | 40.0 | 0.8270 | 5 |
| 928 | 1 | 29 | 79 | 0.1983 | 12000 | 8980 | 52.5 | 0.2730 | 4 |
| 929 | 1 | 29 | 79 | 0.1983 | 15000 | 8980 | 52.5 | 0.2540 | 4 |
| 930 | 1 | 29 | 79 | 0.1983 | 20000 | 8980 | 52.5 | 0.2470 | 4 |
| 931 | 1 | 29 | 79 | 0.1983 | 25000 | 8980 | 52.5 | 0.2400 | 4 |
| 932 | 1 | 29 | 79 | 0.1983 | 30000 | 8980 | 52.5 | 0.2350 | 4 |
| 933 | 1 | 29 | 79 | 0.1983 | 40000 | 8980 | 52.5 | 0.2180 | 4 |
| 934 | 1 | 29 | 79 | 0.1983 | 48500 | 8980 | 52.5 | 0.2090 | 4 |
| 935 | 1 | 29 | 79 | 0.3964 | 12000 | 8980 | 52.5 | 0.5000 | 4 |
| 936 | 1 | 29 | 79 | 0.3964 | 15000 | 8980 | 52.5 | 0.4800 | 4 |
| 937 | 1 | 29 | 79 | 0.3964 | 20000 | 8980 | 52.5 | 0.4640 | 4 |
| 938 | 1 | 29 | 79 | 0.3964 | 25000 | 8980 | 52.5 | 0.4430 | 4 |

| 939 | 1 | 29 | 79 | 0.3964 | 30000 | 8980 | 52.5 | 0.4530 | 4 |
|-----|---|----|----|--------|-------|------|------|--------|---|
| 940 | 1 | 29 | 79 | 0.3964 | 40000 | 8980 | 52.5 | 0.4160 | 4 |
| 941 | 1 | 29 | 79 | 0.3964 | 48500 | 8980 | 52.5 | 0.4140 | 4 |
| 942 | 1 | 29 | 79 | 0.5992 | 12000 | 8980 | 52.5 | 0.6980 | 4 |
| 943 | 1 | 29 | 79 | 0.5992 | 15000 | 8980 | 52.5 | 0.6760 | 4 |
| 944 | 1 | 29 | 79 | 0.5992 | 20000 | 8980 | 52.5 | 0.6630 | 4 |
| 945 | 1 | 29 | 79 | 0.5992 | 25000 | 8980 | 52.5 | 0.6510 | 4 |
| 946 | 1 | 29 | 79 | 0.5992 | 30000 | 8980 | 52.5 | 0.6440 | 4 |
| 947 | 1 | 29 | 79 | 0.5992 | 40000 | 8980 | 52.5 | 0.6160 | 4 |
| 948 | 1 | 29 | 79 | 0.5992 | 48500 | 8980 | 52.5 | 0.6040 | 4 |
| 949 | 1 | 29 | 79 | 0.7985 | 12000 | 8980 | 52.5 | 0.8690 | 4 |
| 950 | 1 | 29 | 79 | 0.7985 | 15000 | 8980 | 52.5 | 0.8510 | 4 |
| 951 | 1 | 29 | 79 | 0.7985 | 20000 | 8980 | 52.5 | 0.8410 | 4 |
| 952 | 1 | 29 | 79 | 0.7985 | 25000 | 8980 | 52.5 | 0.8340 | 4 |
| 953 | 1 | 29 | 79 | 0.7985 | 30000 | 8980 | 52.5 | 0.8260 | 4 |
| 954 | 1 | 29 | 79 | 0.7985 | 40000 | 8980 | 52.5 | 0.8150 | 4 |
| 955 | 1 | 29 | 79 | 0.7985 | 48500 | 8980 | 52.5 | 0.8000 | 4 |
| 956 | 3 | 29 | 79 | 0.1983 | 5000 | 931 | 52.5 | 0.2100 | 4 |
| 957 | 3 | 29 | 79 | 0.1983 | 7500 | 931 | 52.5 | 0.2030 | 4 |
| 958 | 3 | 29 | 79 | 0.1983 | 10000 | 931 | 52.5 | 0.1660 | 4 |
| 959 | 3 | 29 | 79 | 0.1983 | 12000 | 931 | 52.5 | 0.1550 | 4 |
| 960 | 3 | 29 | 79 | 0.1983 | 15000 | 931 | 52.5 | 0.1360 | 4 |
| 961 | 3 | 29 | 79 | 0.1983 | 20000 | 931 | 52.5 | 0.1090 | 4 |
| 962 | 3 | 29 | 79 | 0.1983 | 25000 | 931 | 52.5 | 0.0970 | 4 |
| 963 | 3 | 29 | 79 | 0.1983 | 30000 | 931 | 52.5 | 0.0870 | 4 |
| 964 | 3 | 29 | 79 | 0.1983 | 40000 | 931 | 52.5 | 0.0670 | 4 |
| 965 | 3 | 29 | 79 | 0.1983 | 48500 | 931 | 52.5 | 0.0690 | 4 |
| 966 | 3 | 29 | 79 | 0.3964 | 5000 | 931 | 52.5 | 0.4170 | 4 |
| 967 | 3 | 29 | 79 | 0.3964 | 7500 | 931 | 52.5 | 0.4040 | 4 |
| 968 | 3 | 29 | 79 | 0.3964 | 10000 | 931 | 52.5 | 0.3540 | 4 |
| 969 | 3 | 29 | 79 | 0.3964 | 12000 | 931 | 52.5 | 0.3250 | 4 |
| 970 | 3 | 29 | 79 | 0.3964 | 15000 | 931 | 52.5 | 0.2950 | 4 |
| 971 | 3 | 29 | 79 | 0.3964 | 20000 | 931 | 52.5 | 0.2470 | 4 |
| 972 | 3 | 29 | 79 | 0.3964 | 25000 | 931 | 52.5 | 0.2190 | 4 |
| 973 | 3 | 29 | 79 | 0.3964 | 30000 | 931 | 52.5 | 0.2000 | 4 |
| 974 | 3 | 29 | 79 | 0.3964 | 40000 | 931 | 52.5 | 0.1660 | 4 |
| 975 | 3 | 29 | 79 | 0.3964 | 48500 | 931 | 52.5 | 0.1610 | 4 |
| 976 | 3 | 29 | 79 | 0.5992 | 5000 | 931 | 52.5 | 0.6190 | 4 |
| 977 | 3 | 29 | 79 | 0.5992 | 7500 | 931 | 52.5 | 0.6040 | 4 |
| 978 | 3 | 29 | 79 | 0.5992 | 10000 | 931 | 52.5 | 0.5570 | 4 |
| 979 | 3 | 29 | 79 | 0.5992 | 12000 | 931 | 52.5 | 0.5240 | 4 |
| 980 | 3 | 29 | 79 | 0.5992 | 15000 | 931 | 52.5 | 0.4840 | 4 |
| 981 | 3 | 29 | 79 | 0.5992 | 20000 | 931 | 52.5 | 0.4280 | 4 |
| 982 | 3 | 29 | 79 | 0.5992 | 25000 | 931 | 52.5 | 0.3870 | 4 |
| 983 | 3 | 29 | 79 | 0.5992 | 30000 | 931 | 52.5 | 0.3600 | 4 |
| 984 | 3 | 29 | 79 | 0.5992 | 40000 | 931 | 52.5 | 0.3130 | 4 |

| | | | | | | | | | |
|---|---|---|---|---|---|---|---|---|---|
| 985 | 3 | 29 | 79 | 0.5992 | 48500 | 931 | 52.5 | 0.3040 | 4 |
| 986 | 3 | 29 | 79 | 0.7985 | 5000 | 931 | 52.5 | 0.8150 | 4 |
| 987 | 3 | 29 | 79 | 0.7985 | 7500 | 931 | 52.5 | 0.8090 | 4 |
| 988 | 3 | 29 | 79 | 0.7985 | 10000 | 931 | 52.5 | 0.7640 | 4 |
| 989 | 3 | 29 | 79 | 0.7985 | 12000 | 931 | 52.5 | 0.7470 | 4 |
| 990 | 3 | 29 | 79 | 0.7985 | 15000 | 931 | 52.5 | 0.7160 | 4 |
| 991 | 3 | 29 | 79 | 0.7985 | 20000 | 931 | 52.5 | 0.6720 | 4 |
| 992 | 3 | 29 | 79 | 0.7985 | 25000 | 931 | 52.5 | 0.6320 | 4 |
| 993 | 3 | 29 | 79 | 0.7985 | 30000 | 931 | 52.5 | 0.6020 | 4 |
| 994 | 3 | 29 | 79 | 0.7985 | 40000 | 931 | 52.5 | 0.5500 | 4 |
| 995 | 3 | 29 | 79 | 0.7985 | 48500 | 931 | 52.5 | 0.5440 | 4 |
| 996 | 1 | 30 | 16 | 0.6710 | 15000 | 9669 | 75.0 | 0.6200 | 7 |
| 997 | 1 | 30 | 16 | 0.6710 | 20000 | 9669 | 75.0 | 0.6200 | 7 |
| 998 | 1 | 30 | 16 | 0.6710 | 25000 | 9669 | 75.0 | 0.6260 | 7 |
| 999 | 1 | 30 | 16 | 0.6710 | 30000 | 9669 | 75.0 | 0.6280 | 7 |
| 1000 | 1 | 30 | 29 | 0.2710 | 25000 | 9669 | 18.0 | 0.2730 | 7 |
| 1001 | 1 | 33 | 31 | 0.5180 | 15000 | 11866 | 20.0 | 0.4920 | 7 |
| 1002 | 1 | 33 | 31 | 0.5180 | 20000 | 11866 | 20.0 | 0.4770 | 7 |
| 1003 | 1 | 33 | 31 | 0.5180 | 25000 | 11866 | 20.0 | 0.4650 | 7 |
| 1004 | 1 | 33 | 31 | 0.5180 | 30000 | 11866 | 20.0 | 0.4460 | 7 |
| 1005 | 1 | 33 | 31 | 0.5180 | 35000 | 11866 | 20.0 | 0.4280 | 7 |
| 1006 | 1 | 33 | 49 | 0.3950 | 15000 | 11866 | 20.0 | 0.4180 | 7 |
| 1007 | 1 | 33 | 49 | 0.3950 | 20000 | 11866 | 20.0 | 0.4050 | 7 |
| 1008 | 1 | 33 | 49 | 0.3950 | 25000 | 11866 | 20.0 | 0.3940 | 7 |
| 1009 | 1 | 33 | 49 | 0.3950 | 30000 | 11866 | 20.0 | 0.3850 | 7 |
| 1010 | 1 | 33 | 49 | 0.3950 | 35000 | 11866 | 20.0 | 0.3770 | 7 |
| 1011 | 3 | 40 | 5 | 0.8211 | 4000 | 2222 | 40.0 | 0.7540 | 1 |
| 1012 | 3 | 40 | 5 | 0.8211 | 6000 | 2222 | 40.0 | 0.7662 | 1 |
| 1013 | 3 | 40 | 5 | 0.8211 | 8000 | 2222 | 40.0 | 0.7770 | 1 |
| 1014 | 3 | 40 | 5 | 0.8211 | 10000 | 2222 | 40.0 | 0.7855 | 1 |
| 1015 | 3 | 40 | 5 | 0.8211 | 12000 | 2222 | 40.0 | 0.7930 | 1 |
| 1016 | 3 | 40 | 5 | 0.8211 | 20000 | 2222 | 40.0 | 0.8140 | 1 |
| 1017 | 3 | 40 | 5 | 0.8211 | 25000 | 2222 | 40.0 | 0.8225 | 1 |
| 1018 | 3 | 40 | 5 | 0.8211 | 30000 | 2222 | 40.0 | 0.8290 | 1 |
| 1019 | 3 | 40 | 5 | 0.8211 | 15000 | 2222 | 40.0 | 0.8024 | 5 |
| 1020 | 3 | 40 | 6 | 0.9145 | 4000 | 2222 | 40.0 | 0.8981 | 11 |
| 1021 | 3 | 40 | 6 | 0.9145 | 6000 | 2222 | 40.0 | 0.9011 | 11 |
| 1022 | 3 | 40 | 6 | 0.9145 | 8000 | 2222 | 40.0 | 0.9033 | 11 |
| 1023 | 3 | 40 | 6 | 0.9145 | 10000 | 2222 | 40.0 | 0.9055 | 11 |
| 1024 | 3 | 40 | 6 | 0.9145 | 12000 | 2222 | 40.0 | 0.9070 | 11 |
| 1025 | 3 | 40 | 6 | 0.9145 | 15000 | 2222 | 40.0 | 0.9088 | 11 |
| 1026 | 3 | 40 | 6 | 0.9145 | 20000 | 2222 | 40.0 | 0.9095 | 11 |
| 1027 | 3 | 40 | 6 | 0.9145 | 25000 | 2222 | 40.0 | 0.9100 | 11 |
| 1028 | 3 | 40 | 6 | 0.9145 | 30000 | 2222 | 40.0 | 0.9110 | 11 |
| 1029 | 1 | 40 | 8 | 0.9450 | 28500 | 17998 | 15.0 | 0.9200 | 6 |
| 1030 | 1 | 40 | 8 | 0.9450 | 28500 | 17998 | 15.0 | 0.9200 | 7 |

| | | | | | | | | | |
|---|---|---|---|---|---|---|---|---|---|
| 1031 | 3 | 40 | 8 | 0.7400 | 10000 | 2222 | 20.0 | 0.6430 | 7 |
| 1032 | 3 | 40 | 8 | 0.7400 | 20000 | 2222 | 20.0 | 0.6490 | 7 |
| 1033 | 3 | 40 | 8 | 0.7400 | 10000 | 2222 | 40.0 | 0.6820 | 6 |
| 1034 | 3 | 40 | 8 | 0.7400 | 15000 | 2222 | 40.0 | 0.6990 | 6 |
| 1035 | 1 | 40 | 13 | 0.5300 | 28500 | 17998 | 15.0 | 0.4720 | 6 |
| 1036 | 1 | 40 | 13 | 0.5300 | 28500 | 17998 | 15.0 | 0.4720 | 7 |
| 1037 | 1 | 40 | 14 | 0.6200 | 28500 | 17998 | 15.0 | 0.5440 | 6 |
| 1038 | 1 | 40 | 14 | 0.6200 | 28500 | 17998 | 15.0 | 0.5440 | 7 |
| 1039 | 1 | 40 | 14 | 0.7960 | 28500 | 17998 | 15.0 | 0.7480 | 6 |
| 1040 | 1 | 40 | 14 | 0.7960 | 28500 | 17998 | 15.0 | 0.7480 | 7 |
| 1041 | 3 | 41 | 5 | 0.8958 | 6000 | 2371 | 30.0 | 0.8680 | 1 |
| 1042 | 3 | 41 | 5 | 0.8285 | 4000 | 2371 | 40.0 | 0.7502 | 1 |
| 1043 | 3 | 41 | 5 | 0.8285 | 6000 | 2371 | 40.0 | 0.7640 | 1 |
| 1044 | 3 | 41 | 5 | 0.8285 | 8000 | 2371 | 40.0 | 0.7760 | 1 |
| 1045 | 3 | 41 | 5 | 0.8285 | 10000 | 2371 | 40.0 | 0.7862 | 1 |
| 1046 | 3 | 41 | 5 | 0.8285 | 12000 | 2371 | 40.0 | 0.7950 | 1 |
| 1047 | 3 | 41 | 5 | 0.8285 | 15000 | 2371 | 40.0 | 0.8060 | 1 |
| 1048 | 3 | 41 | 5 | 0.8285 | 20000 | 2371 | 40.0 | 0.8210 | 1 |
| 1049 | 3 | 41 | 5 | 0.8285 | 25000 | 2371 | 40.0 | 0.8338 | 1 |
| 1050 | 3 | 41 | 5 | 0.8285 | 30000 | 2371 | 40.0 | 0.8442 | 1 |
| 1051 | 3 | 41 | 5 | 0.8958 | 4000 | 2371 | 40.0 | 0.8588 | 1 |
| 1052 | 3 | 41 | 5 | 0.8958 | 8000 | 2371 | 40.0 | 0.8758 | 1 |
| 1053 | 3 | 41 | 5 | 0.8958 | 10000 | 2371 | 40.0 | 0.8820 | 1 |
| 1054 | 3 | 41 | 5 | 0.8958 | 12000 | 2371 | 40.0 | 0.8878 | 1 |
| 1055 | 3 | 41 | 5 | 0.8958 | 15000 | 2371 | 40.0 | 0.8942 | 1 |
| 1056 | 3 | 41 | 5 | 0.8958 | 20000 | 2371 | 40.0 | 0.9036 | 1 |
| 1057 | 3 | 41 | 5 | 0.8958 | 25000 | 2371 | 40.0 | 0.9100 | 1 |
| 1058 | 3 | 41 | 5 | 0.8958 | 30000 | 2371 | 40.0 | 0.9157 | 1 |
| 1059 | 3 | 41 | 6 | 0.8860 | 7000 | 2371 | 18.0 | 0.8680 | 7 |
| 1060 | 3 | 41 | 6 | 0.8850 | 10000 | 2371 | 40.0 | 0.8760 | 7 |
| 1061 | 3 | 41 | 6 | 0.8860 | 4000 | 2371 | 40.0 | 0.8845 | 7 |
| 1062 | 3 | 41 | 6 | 0.8860 | 6000 | 2371 | 40.0 | 0.8823 | 7 |
| 1063 | 3 | 41 | 6 | 0.8860 | 8000 | 2371 | 40.0 | 0.8864 | 7 |
| 1064 | 3 | 41 | 6 | 0.9145 | 4000 | 2371 | 40.0 | 0.8750 | 11 |
| 1065 | 3 | 41 | 6 | 0.9145 | 6000 | 2371 | 40.0 | 0.8822 | 11 |
| 1066 | 3 | 41 | 6 | 0.9145 | 8000 | 2371 | 40.0 | 0.8880 | 11 |
| 1067 | 3 | 41 | 6 | 0.9145 | 10000 | 2371 | 40.0 | 0.8930 | 11 |
| 1068 | 3 | 41 | 6 | 0.9145 | 12000 | 2371 | 40.0 | 0.8970 | 11 |
| 1069 | 3 | 41 | 6 | 0.9145 | 15000 | 2371 | 40.0 | 0.9022 | 11 |
| 1070 | 3 | 41 | 6 | 0.9145 | 20000 | 2371 | 40.0 | 0.9088 | 11 |
| 1071 | 3 | 41 | 6 | 0.9145 | 25000 | 2371 | 40.0 | 0.9135 | 11 |
| 1072 | 3 | 41 | 6 | 0.9145 | 30000 | 2371 | 40.0 | 0.9175 | 11 |
| 1073 | 3 | 41 | 22 | 0.6500 | 25000 | 2371 | 18.0 | 0.5990 | 7 |
| 1074 | 3 | 41 | 23 | 0.0800 | 10000 | 2371 | 20.0 | 0.0700 | 7 |
| 1075 | 3 | 41 | 23 | 0.0800 | 15000 | 2371 | 20.0 | 0.0680 | 7 |
| 1076 | 3 | 41 | 23 | 0.0800 | 10000 | 2371 | 52.5 | 0.0757 | 5 |

| | | | | | | | | | |
|---|---|---|---|---|---|---|---|---|---|
| 1077 | 3 | 41 | 23 | 0.0800 | 15000 | 2371 | 52.5 | 0.0761 | 5 |
| 1078 | 3 | 41 | 23 | 0.0800 | 20000 | 2371 | 52.5 | 0.0730 | 5 |
| 1079 | 3 | 41 | 23 | 0.0800 | 25000 | 2371 | 52.5 | 0.0698 | 5 |
| 1080 | 3 | 41 | 23 | 0.0800 | 30000 | 2371 | 52.5 | 0.0657 | 5 |
| 1081 | 3 | 41 | 23 | 0.0800 | 10000 | 2371 | 75.0 | 0.0730 | 5 |
| 1082 | 3 | 41 | 23 | 0.0800 | 15000 | 2371 | 75.0 | 0.0740 | 5 |
| 1083 | 3 | 41 | 23 | 0.0800 | 20000 | 2371 | 75.0 | 0.0739 | 5 |
| 1084 | 3 | 41 | 23 | 0.0800 | 25000 | 2371 | 75.0 | 0.0730 | 5 |
| 1085 | 3 | 41 | 23 | 0.0800 | 30000 | 2371 | 75.0 | 0.0722 | 5 |
| 1086 | 3 | 41 | 23 | 0.0800 | 35000 | 2371 | 75.0 | 0.0707 | 5 |
| 1087 | 3 | 41 | 23 | 0.0800 | 40000 | 2371 | 75.0 | 0.0684 | 5 |
| 1088 | 3 | 41 | 23 | 0.0800 | 10000 | 2371 | 75.0 | 0.0730 | 7 |
| 1089 | 3 | 41 | 92 | 0.0132 | 10000 | 2371 | 40.0 | 0.0144 | 9 |
| 1090 | 3 | 41 | 92 | 0.0132 | 15000 | 2371 | 40.0 | 0.0119 | 9 |
| 1091 | 3 | 41 | 92 | 0.0132 | 20000 | 2371 | 40.0 | 0.0108 | 9 |
| 1092 | 3 | 41 | 92 | 0.0132 | 25000 | 2371 | 40.0 | 0.0097 | 9 |
| 1093 | 3 | 41 | 92 | 0.0132 | 30000 | 2371 | 40.0 | 0.0086 | 9 |
| 1094 | 3 | 41 | 92 | 0.0132 | 35000 | 2371 | 40.0 | 0.0082 | 9 |
| 1095 | 3 | 41 | 92 | 0.0132 | 40000 | 2371 | 40.0 | 0.0077 | 9 |
| 1096 | 3 | 41 | 92 | 0.0189 | 10000 | 2371 | 40.0 | 0.0205 | 9 |
| 1097 | 3 | 41 | 92 | 0.0189 | 15000 | 2371 | 40.0 | 0.0172 | 9 |
| 1098 | 3 | 41 | 92 | 0.0189 | 20000 | 2371 | 40.0 | 0.0156 | 9 |
| 1099 | 3 | 41 | 92 | 0.0189 | 25000 | 2371 | 40.0 | 0.0137 | 9 |
| 1100 | 3 | 41 | 92 | 0.0189 | 30000 | 2371 | 40.0 | 0.0128 | 9 |
| 1101 | 3 | 41 | 92 | 0.0189 | 35000 | 2371 | 40.0 | 0.0118 | 9 |
| 1102 | 3 | 41 | 92 | 0.0189 | 40000 | 2371 | 40.0 | 0.0112 | 9 |
| 1103 | 3 | 41 | 92 | 0.0406 | 10000 | 2371 | 40.0 | 0.0440 | 9 |
| 1104 | 3 | 41 | 92 | 0.0406 | 15000 | 2371 | 40.0 | 0.0362 | 9 |
| 1105 | 3 | 41 | 92 | 0.0406 | 20000 | 2371 | 40.0 | 0.0334 | 9 |
| 1106 | 3 | 41 | 92 | 0.0406 | 25000 | 2371 | 40.0 | 0.0301 | 9 |
| 1107 | 3 | 41 | 92 | 0.0406 | 30000 | 2371 | 40.0 | 0.0271 | 9 |
| 1108 | 3 | 41 | 92 | 0.0406 | 35000 | 2371 | 40.0 | 0.0253 | 9 |
| 1109 | 3 | 41 | 92 | 0.0406 | 40000 | 2371 | 40.0 | 0.0235 | 9 |
| 1110 | 3 | 41 | 92 | 0.0587 | 10000 | 2371 | 40.0 | 0.0629 | 9 |
| 1111 | 3 | 41 | 92 | 0.0587 | 15000 | 2371 | 40.0 | 0.0527 | 9 |
| 1112 | 3 | 41 | 92 | 0.0587 | 20000 | 2371 | 40.0 | 0.0479 | 9 |
| 1113 | 3 | 41 | 92 | 0.0587 | 25000 | 2371 | 40.0 | 0.0433 | 9 |
| 1114 | 3 | 41 | 92 | 0.0587 | 30000 | 2371 | 40.0 | 0.0396 | 9 |
| 1115 | 3 | 41 | 92 | 0.0587 | 35000 | 2371 | 40.0 | 0.0370 | 9 |
| 1116 | 3 | 41 | 92 | 0.0587 | 40000 | 2371 | 40.0 | 0.0340 | 9 |
| 1117 | 3 | 41 | 92 | 0.0741 | 10000 | 2371 | 40.0 | 0.0817 | 9 |
| 1118 | 3 | 41 | 92 | 0.0741 | 15000 | 2371 | 40.0 | 0.0661 | 9 |
| 1119 | 3 | 41 | 92 | 0.0741 | 20000 | 2371 | 40.0 | 0.0611 | 9 |
| 1120 | 3 | 41 | 92 | 0.0741 | 25000 | 2371 | 40.0 | 0.0549 | 9 |
| 1121 | 3 | 41 | 92 | 0.0741 | 30000 | 2371 | 40.0 | 0.0501 | 9 |
| 1122 | 3 | 41 | 92 | 0.0741 | 35000 | 2371 | 40.0 | 0.0468 | 9 |

| 1123 | 3 | 41 | 92 | 0.0741 | 40000 | 2371 | 40.0 | 0.0434 | 9 |
|---|---|---|---|---|---|---|---|---|---|
| 1124 | 3 | 41 | 92 | 0.1286 | 10000 | 2371 | 40.0 | 0.1417 | 9 |
| 1125 | 3 | 41 | 92 | 0.1286 | 15000 | 2371 | 40.0 | 0.1158 | 9 |
| 1126 | 3 | 41 | 92 | 0.1286 | 20000 | 2371 | 40.0 | 0.1066 | 9 |
| 1127 | 3 | 41 | 92 | 0.1286 | 25000 | 2371 | 40.0 | 0.0965 | 9 |
| 1128 | 3 | 41 | 92 | 0.1286 | 30000 | 2371 | 40.0 | 0.0884 | 9 |
| 1129 | 3 | 41 | 92 | 0.1286 | 35000 | 2371 | 40.0 | 0.0831 | 9 |
| 1130 | 3 | 41 | 92 | 0.1286 | 40000 | 2371 | 40.0 | 0.0762 | 9 |
| 1131 | 3 | 42 | 5 | 0.8987 | 4000 | 2520 | 40.0 | 0.8401 | 1 |
| 1132 | 3 | 42 | 5 | 0.8987 | 6000 | 2520 | 40.0 | 0.8500 | 1 |
| 1133 | 3 | 42 | 5 | 0.8987 | 8000 | 2520 | 40.0 | 0.8580 | 1 |
| 1134 | 3 | 42 | 5 | 0.8987 | 10000 | 2520 | 40.0 | 0.8660 | 1 |
| 1135 | 3 | 42 | 5 | 0.8987 | 12000 | 2520 | 40.0 | 0.8731 | 1 |
| 1136 | 3 | 42 | 5 | 0.8987 | 15000 | 2520 | 40.0 | 0.8828 | 1 |
| 1137 | 3 | 42 | 5 | 0.8987 | 20000 | 2520 | 40.0 | 0.8978 | 1 |
| 1138 | 3 | 42 | 5 | 0.8987 | 25000 | 2520 | 40.0 | 0.9100 | 1 |
| 1139 | 3 | 42 | 5 | 0.8987 | 30000 | 2520 | 40.0 | 0.9202 | 1 |
| 1140 | 3 | 42 | 6 | 0.9442 | 4000 | 2520 | 40.0 | 0.9060 | 1 |
| 1141 | 3 | 42 | 6 | 0.9442 | 6000 | 2520 | 40.0 | 0.9128 | 1 |
| 1142 | 3 | 42 | 6 | 0.9442 | 8000 | 2520 | 40.0 | 0.9182 | 1 |
| 1143 | 3 | 42 | 6 | 0.9442 | 10000 | 2520 | 40.0 | 0.9225 | 1 |
| 1144 | 3 | 42 | 6 | 0.9442 | 12000 | 2520 | 40.0 | 0.9264 | 1 |
| 1145 | 3 | 42 | 6 | 0.9442 | 15000 | 2520 | 40.0 | 0.9312 | 1 |
| 1146 | 3 | 42 | 6 | 0.9442 | 20000 | 2520 | 40.0 | 0.9373 | 1 |
| 1147 | 3 | 42 | 6 | 0.9442 | 25000 | 2520 | 40.0 | 0.9418 | 1 |
| 1148 | 3 | 42 | 6 | 0.9442 | 30000 | 2520 | 40.0 | 0.9458 | 11 |
| 1149 | 3 | 42 | 8 | 0.6700 | 5000 | 2520 | 40.0 | 0.5390 | 6 |
| 1150 | 3 | 42 | 8 | 0.6700 | 10000 | 2520 | 40.0 | 0.5810 | 6 |
| 1151 | 3 | 42 | 8 | 0.6700 | 15000 | 2520 | 40.0 | 0.5990 | 6 |
| 1152 | 1 | 42 | 14 | 0.6310 | 30000 | 20000 | 20.0 | 0.5600 | 7 |
| 1153 | 1 | 42 | 14 | 0.6310 | 35000 | 20000 | 20.0 | 0.5670 | 7 |
| 1154 | 3 | 42 | 14 | 0.6310 | 10000 | 2520 | 20.0 | 0.4850 | 7 |
| 1155 | 3 | 42 | 14 | 0.6310 | 15000 | 2520 | 20.0 | 0.4540 | 7 |
| 1156 | 3 | 42 | 14 | 0.6310 | 20000 | 2520 | 20.0 | 0.4130 | 7 |
| 1157 | 3 | 42 | 14 | 0.6310 | 25000 | 2520 | 20.0 | 0.3770 | 7 |
| 1158 | 3 | 42 | 14 | 0.6310 | 30000 | 2520 | 20.0 | 0.3540 | 7 |
| 1159 | 3 | 42 | 74 | 0.2150 | 5000 | 2520 | 18.0 | 0.2300 | 7 |
| 1160 | 3 | 42 | 74 | 0.2000 | 10000 | 2520 | 52.5 | 0.2120 | 7 |
| 1161 | 3 | 42 | 74 | 0.2000 | 15000 | 2520 | 52.5 | 0.1640 | 7 |
| 1162 | 3 | 42 | 74 | 0.2000 | 20000 | 2520 | 52.5 | 0.1430 | 7 |
| 1163 | 3 | 44 | 73 | 0.1930 | 30000 | 2838 | 15.5 | 0.0670 | 7 |
| 1164 | 3 | 44 | 73 | 0.3580 | 30000 | 2838 | 15.5 | 0.1400 | 7 |
| 1165 | 3 | 44 | 73 | 0.4560 | 30000 | 2838 | 15.5 | 0.1970 | 7 |
| 1166 | 3 | 44 | 73 | 0.5660 | 30000 | 2838 | 15.5 | 0.2820 | 7 |
| 1167 | 3 | 44 | 73 | 0.8350 | 30000 | 2838 | 15.5 | 0.6140 | 7 |
| 1168 | 3 | 45 | 74 | 0.0240 | 30000 | 3004 | 15.5 | 0.0080 | 7 |

| | | | | | | | | | |
|---|---|---|---|---|---|---|---|---|---|
| 1169 | 3 | 45 | 74 | 0.4100 | 30000 | 3004 | 15.5 | 0.1850 | 7 |
| 1170 | 3 | 45 | 74 | 0.5640 | 30000 | 3004 | 15.5 | 0.2930 | 7 |
| 1171 | 3 | 45 | 74 | 0.6880 | 30000 | 3004 | 15.5 | 0.4250 | 7 |
| 1172 | 3 | 45 | 74 | 0.9110 | 30000 | 3004 | 15.5 | 0.7760 | 7 |
| 1173 | 3 | 46 | 79 | 0.3590 | 5200 | 3173 | 40.0 | 0.3921 | 5 |
| 1174 | 3 | 46 | 79 | 0.3590 | 10300 | 3173 | 40.0 | 0.3342 | 5 |
| 1175 | 3 | 46 | 79 | 0.3590 | 15500 | 3173 | 40.0 | 0.2876 | 5 |
| 1176 | 3 | 46 | 79 | 0.3590 | 20700 | 3173 | 40.0 | 0.2445 | 5 |
| 1177 | 3 | 46 | 79 | 0.3590 | 25800 | 3173 | 40.0 | 0.2154 | 5 |
| 1178 | 3 | 46 | 79 | 0.3590 | 31000 | 3173 | 40.0 | 0.1898 | 5 |
| 1179 | 3 | 46 | 79 | 0.3590 | 36200 | 3173 | 40.0 | 0.1742 | 5 |
| 1180 | 3 | 47 | 79 | 0.1995 | 10000 | 3351 | 40.0 | 0.1853 | 8 |
| 1181 | 3 | 47 | 79 | 0.1995 | 15000 | 3351 | 40.0 | 0.1576 | 8 |
| 1182 | 3 | 47 | 79 | 0.1995 | 20000 | 3351 | 40.0 | 0.1343 | 8 |
| 1183 | 3 | 47 | 79 | 0.1995 | 25000 | 3351 | 40.0 | 0.1153 | 8 |
| 1184 | 3 | 47 | 79 | 0.1995 | 30000 | 3351 | 40.0 | 0.1006 | 8 |
| 1185 | 3 | 47 | 79 | 0.3995 | 10000 | 3351 | 40.0 | 0.3783 | 8 |
| 1186 | 3 | 47 | 79 | 0.3995 | 15000 | 3351 | 40.0 | 0.3349 | 8 |
| 1187 | 3 | 47 | 79 | 0.3995 | 20000 | 3351 | 40.0 | 0.2959 | 8 |
| 1188 | 3 | 47 | 79 | 0.3995 | 25000 | 3351 | 40.0 | 0.2606 | 8 |
| 1189 | 3 | 47 | 79 | 0.3995 | 30000 | 3351 | 40.0 | 0.2323 | 8 |
| 1190 | 3 | 47 | 79 | 0.5997 | 10000 | 3351 | 40.0 | 0.5803 | 8 |
| 1191 | 3 | 47 | 79 | 0.5997 | 15000 | 3351 | 40.0 | 0.5327 | 8 |
| 1192 | 3 | 47 | 79 | 0.5997 | 20000 | 3351 | 40.0 | 0.4875 | 8 |
| 1193 | 3 | 47 | 79 | 0.5997 | 25000 | 3351 | 40.0 | 0.4458 | 8 |
| 1194 | 3 | 47 | 79 | 0.5997 | 30000 | 3351 | 40.0 | 0.4074 | 8 |
| 1195 | 3 | 47 | 79 | 0.7757 | 10000 | 3351 | 40.0 | 0.7627 | 8 |
| 1196 | 3 | 47 | 79 | 0.7757 | 15000 | 3351 | 40.0 | 0.7248 | 8 |
| 1197 | 3 | 47 | 79 | 0.7757 | 20000 | 3351 | 40.0 | 0.6869 | 8 |
| 1198 | 3 | 47 | 79 | 0.7757 | 25000 | 3351 | 40.0 | 0.6510 | 8 |
| 1199 | 3 | 47 | 79 | 0.7757 | 30000 | 3351 | 40.0 | 0.6161 | 8 |
| 1200 | 3 | 47 | 79 | 0.1996 | 5000 | 3351 | 52.5 | 0.2230 | 4 |
| 1201 | 3 | 47 | 79 | 0.1996 | 10000 | 3351 | 52.5 | 0.1910 | 4 |
| 1202 | 3 | 47 | 79 | 0.1996 | 20000 | 3351 | 52.5 | 0.1440 | 4 |
| 1203 | 3 | 47 | 79 | 0.1996 | 30000 | 3351 | 52.5 | 0.1090 | 4 |
| 1204 | 3 | 47 | 79 | 0.1996 | 40000 | 3351 | 52.5 | 0.0887 | 4 |
| 1205 | 3 | 47 | 79 | 0.1996 | 48500 | 3351 | 52.5 | 0.0795 | 4 |
| 1206 | 3 | 47 | 79 | 0.1996 | 5000 | 3351 | 52.5 | 0.2230 | 6 |
| 1207 | 3 | 47 | 79 | 0.3992 | 5000 | 3351 | 52.5 | 0.4360 | 4 |
| 1208 | 3 | 47 | 79 | 0.3992 | 10000 | 3351 | 52.5 | 0.3860 | 4 |
| 1209 | 3 | 47 | 79 | 0.3992 | 20000 | 3351 | 52.5 | 0.3090 | 4 |
| 1210 | 3 | 47 | 79 | 0.3992 | 30000 | 3351 | 52.5 | 0.2450 | 4 |
| 1211 | 3 | 47 | 79 | 0.3992 | 40000 | 3351 | 52.5 | 0.2080 | 4 |
| 1212 | 3 | 47 | 79 | 0.3992 | 48500 | 3351 | 52.5 | 0.1810 | 4 |
| 1213 | 3 | 47 | 79 | 0.3992 | 5000 | 3351 | 52.5 | 0.4360 | 6 |
| 1214 | 3 | 47 | 79 | 0.5993 | 5000 | 3351 | 52.5 | 0.6240 | 4 |

| 1215 | 3 | 47 | 79 | 0.5993 | 10000 | 3351 | 52.5 | 0.5840 | 4 |
| 1216 | 3 | 47 | 79 | 0.5993 | 20000 | 3351 | 52.5 | 0.5040 | 4 |
| 1217 | 3 | 47 | 79 | 0.5993 | 30000 | 3351 | 52.5 | 0.4260 | 4 |
| 1218 | 3 | 47 | 79 | 0.5993 | 40000 | 3351 | 52.5 | 0.3750 | 4 |
| 1219 | 3 | 47 | 79 | 0.5993 | 48500 | 3351 | 52.5 | 0.3390 | 4 |
| 1220 | 3 | 47 | 79 | 0.7758 | 5000 | 3351 | 52.5 | 0.8070 | 4 |
| 1221 | 3 | 47 | 79 | 0.7758 | 10000 | 3351 | 52.5 | 0.7640 | 4 |
| 1222 | 3 | 47 | 79 | 0.7758 | 20000 | 3351 | 52.5 | 0.7060 | 4 |
| 1223 | 3 | 47 | 79 | 0.7758 | 30000 | 3351 | 52.5 | 0.6360 | 4 |
| 1224 | 3 | 47 | 79 | 0.7758 | 40000 | 3351 | 52.5 | 0.5860 | 4 |
| 1225 | 3 | 47 | 79 | 0.7758 | 48500 | 3351 | 52.5 | 0.5420 | 4 |
| 1226 | 4 | 47 | 79 | 0.1996 | 15000 | 3524 | 52.5 | 0.1670 | 4 |
| 1227 | 4 | 47 | 79 | 0.1996 | 20000 | 3524 | 52.5 | 0.1460 | 4 |
| 1228 | 4 | 47 | 79 | 0.1996 | 30000 | 3524 | 52.5 | 0.1120 | 4 |
| 1229 | 4 | 47 | 79 | 0.1996 | 40000 | 3524 | 52.5 | 0.0898 | 4 |
| 1230 | 4 | 47 | 79 | 0.1996 | 48500 | 3524 | 52.5 | 0.0720 | 4 |
| 1231 | 4 | 47 | 79 | 0.3992 | 30000 | 3524 | 52.5 | 0.2530 | 1 |
| 1232 | 4 | 47 | 79 | 0.3992 | 15000 | 3524 | 52.5 | 0.3490 | 4 |
| 1233 | 4 | 47 | 79 | 0.3992 | 20000 | 3524 | 52.5 | 0.3120 | 4 |
| 1234 | 4 | 47 | 79 | 0.3992 | 40000 | 3524 | 52.5 | 0.2090 | 4 |
| 1235 | 4 | 47 | 79 | 0.3992 | 48500 | 3524 | 52.5 | 0.1750 | 4 |
| 1236 | 4 | 47 | 79 | 0.5993 | 15000 | 3524 | 52.5 | 0.5480 | 4 |
| 1237 | 4 | 47 | 79 | 0.5993 | 20000 | 3524 | 52.5 | 0.5060 | 4 |
| 1238 | 4 | 47 | 79 | 0.5993 | 30000 | 3524 | 52.5 | 0.4360 | 4 |
| 1239 | 4 | 47 | 79 | 0.5993 | 40000 | 3524 | 52.5 | 0.3680 | 4 |
| 1240 | 4 | 47 | 79 | 0.5993 | 48500 | 3524 | 52.5 | 0.3400 | 4 |
| 1241 | 4 | 47 | 79 | 0.7758 | 15000 | 3524 | 52.5 | 0.7420 | 4 |
| 1242 | 4 | 47 | 79 | 0.7758 | 20000 | 3524 | 52.5 | 0.7000 | 4 |
| 1243 | 4 | 47 | 79 | 0.7758 | 30000 | 3524 | 52.5 | 0.6450 | 4 |
| 1244 | 4 | 47 | 79 | 0.7758 | 40000 | 3524 | 52.5 | 0.5740 | 4 |
| 1245 | 4 | 47 | 79 | 0.7758 | 48500 | 3524 | 52.5 | 0.5420 | 4 |
| 1246 | 3 | 48 | 12 | 0.5500 | 30000 | 3538 | 52.5 | 0.4540 | 7 |
| 1247 | 3 | 48 | 12 | 0.7200 | 30000 | 3538 | 52.5 | 0.6390 | 7 |
| 1248 | 3 | 48 | 16 | 0.7780 | 10000 | 3538 | 20.0 | 0.6710 | 7 |
| 1249 | 3 | 48 | 16 | 0.7780 | 20000 | 3538 | 20.0 | 0.6650 | 7 |
| 1250 | 3 | 48 | 16 | 0.7780 | 30000 | 3538 | 20.0 | 0.6430 | 7 |
| 1251 | 3 | 49 | 15 | 0.7870 | 10000 | 3730 | 20.0 | 0.7180 | 7 |
| 1252 | 3 | 49 | 15 | 0.7870 | 15000 | 3730 | 20.0 | 0.7170 | 7 |
| 1253 | 3 | 49 | 15 | 0.7870 | 20000 | 3730 | 20.0 | 0.7090 | 7 |
| 1254 | 3 | 49 | 15 | 0.7870 | 25000 | 3730 | 20.0 | 0.6910 | 7 |
| 1255 | 3 | 49 | 15 | 0.7870 | 30000 | 3730 | 20.0 | 0.6770 | 7 |
| 1256 | 3 | 49 | 15 | 0.7870 | 35000 | 3730 | 20.0 | 0.6740 | 7 |
| 1257 | 3 | 49 | 33 | 0.6050 | 10000 | 3730 | 20.0 | 0.5690 | 7 |
| 1258 | 3 | 49 | 33 | 0.6050 | 15000 | 3730 | 20.0 | 0.5630 | 7 |
| 1259 | 3 | 49 | 33 | 0.6050 | 20000 | 3730 | 20.0 | 0.5390 | 7 |
| 1260 | 3 | 49 | 33 | 0.6050 | 25000 | 3730 | 20.0 | 0.5210 | 7 |

| | | | | | | | | | |
|---|---|---|---|---|---|---|---|---|---|
| 1261 | 3 | 49 | 33 | 0.6050 | 30000 | 3730 | 20.0 | 0.4980 | 7 |
| 1262 | 3 | 49 | 33 | 0.6050 | 35000 | 3730 | 20.0 | 0.5000 | 7 |
| 1263 | 3 | 49 | 51 | 0.4850 | 10000 | 3730 | 20.0 | 0.4920 | 7 |
| 1264 | 3 | 49 | 51 | 0.4850 | 20000 | 3730 | 20.0 | 0.4990 | 7 |
| 1265 | 3 | 49 | 51 | 0.4850 | 30000 | 3730 | 20.0 | 0.4940 | 7 |
| 1266 | 3 | 50 | 26 | 0.6800 | 20000 | 3929 | 40.0 | 0.6586 | 7 |
| 1267 | 3 | 50 | 26 | 0.8095 | 20000 | 3929 | 40.0 | 0.7872 | 7 |
| 1268 | 3 | 50 | 28 | 0.5741 | 20000 | 3929 | 40.0 | 0.5162 | 7 |
| 1269 | 3 | 51 | 25 | 0.5260 | 10000 | 4132 | 20.0 | 0.4980 | 7 |
| 1270 | 3 | 51 | 25 | 0.5260 | 15000 | 4132 | 20.0 | 0.5300 | 7 |
| 1271 | 3 | 51 | 25 | 0.5260 | 20000 | 4132 | 20.0 | 0.5340 | 7 |
| 1272 | 3 | 51 | 25 | 0.5260 | 25000 | 4132 | 20.0 | 0.5510 | 7 |
| 1273 | 3 | 51 | 25 | 0.5260 | 30000 | 4132 | 20.0 | 0.5740 | 7 |
| 1274 | 3 | 51 | 25 | 0.5260 | 35000 | 4132 | 20.0 | 0.5790 | 7 |
| 1275 | 3 | 51 | 31 | 0.6360 | 10000 | 4132 | 20.0 | 0.5990 | 7 |
| 1276 | 3 | 51 | 31 | 0.6360 | 15000 | 4132 | 20.0 | 0.5980 | 7 |
| 1277 | 3 | 51 | 31 | 0.6360 | 20000 | 4132 | 20.0 | 0.5940 | 7 |
| 1278 | 3 | 51 | 31 | 0.6360 | 25000 | 4132 | 20.0 | 0.5800 | 7 |
| 1279 | 3 | 51 | 31 | 0.6360 | 30000 | 4132 | 20.0 | 0.5710 | 7 |
| 1280 | 3 | 51 | 31 | 0.6360 | 35000 | 4132 | 20.0 | 0.5710 | 7 |
| 1281 | 3 | 51 | 49 | 0.5150 | 10000 | 4132 | 20.0 | 0.5300 | 7 |
| 1282 | 3 | 51 | 49 | 0.5150 | 20000 | 4132 | 20.0 | 0.5410 | 7 |
| 1283 | 3 | 51 | 49 | 0.5150 | 30000 | 4132 | 20.0 | 0.5430 | 7 |
| 1284 | 3 | 51 | 83 | 0.0277 | 10000 | 4132 | 52.5 | 0.0277 | 5 |
| 1285 | 3 | 51 | 83 | 0.0277 | 15000 | 4132 | 52.5 | 0.0240 | 5 |
| 1286 | 3 | 51 | 83 | 0.0277 | 20000 | 4132 | 52.5 | 0.0210 | 5 |
| 1287 | 3 | 51 | 83 | 0.0502 | 10000 | 4132 | 52.5 | 0.0502 | 5 |
| 1288 | 3 | 51 | 83 | 0.0502 | 15000 | 4132 | 52.5 | 0.0437 | 5 |
| 1289 | 3 | 51 | 83 | 0.0502 | 20000 | 4132 | 52.5 | 0.0382 | 5 |
| 1290 | 3 | 51 | 83 | 0.0502 | 25000 | 4132 | 52.5 | 0.0340 | 5 |
| 1291 | 3 | 51 | 83 | 0.0502 | 30000 | 4132 | 52.5 | 0.0306 | 5 |
| 1292 | 3 | 51 | 83 | 0.1058 | 10000 | 4132 | 52.5 | 0.1058 | 5 |
| 1293 | 3 | 51 | 83 | 0.1058 | 15000 | 4132 | 52.5 | 0.0928 | 5 |
| 1294 | 3 | 51 | 83 | 0.1058 | 20000 | 4132 | 52.5 | 0.0817 | 5 |
| 1295 | 3 | 51 | 83 | 0.1058 | 25000 | 4132 | 52.5 | 0.0731 | 5 |
| 1296 | 3 | 51 | 83 | 0.1058 | 30000 | 4132 | 52.5 | 0.0659 | 5 |
| 1297 | 3 | 51 | 83 | 0.1664 | 10000 | 4132 | 52.5 | 0.1664 | 5 |
| 1298 | 3 | 51 | 83 | 0.1664 | 15000 | 4132 | 52.5 | 0.1471 | 5 |
| 1299 | 3 | 51 | 83 | 0.1664 | 20000 | 4132 | 52.5 | 0.1305 | 5 |
| 1300 | 3 | 51 | 83 | 0.1664 | 25000 | 4132 | 52.5 | 0.1174 | 5 |
| 1301 | 3 | 51 | 83 | 0.1664 | 30000 | 4132 | 52.5 | 0.1064 | 5 |
| 1302 | 3 | 51 | 83 | 0.3130 | 10000 | 4132 | 52.5 | 0.3130 | 5 |
| 1303 | 3 | 51 | 83 | 0.3130 | 15000 | 4132 | 52.5 | 0.2825 | 5 |
| 1304 | 3 | 51 | 83 | 0.3130 | 20000 | 4132 | 52.5 | 0.2552 | 5 |
| 1305 | 3 | 51 | 83 | 0.3130 | 25000 | 4132 | 52.5 | 0.2330 | 5 |
| 1306 | 3 | 51 | 83 | 0.3130 | 30000 | 4132 | 52.5 | 0.2136 | 5 |

| | | | | | | | | |
|---|---|---|---|---|---|---|---|---|
| 1307 | 3 | 51 | 83 | 0.4602 | 10000 | 4132 | 52.5 | 0.4602 | 5 |
| 1308 | 3 | 51 | 83 | 0.4602 | 15000 | 4132 | 52.5 | 0.4242 | 5 |
| 1309 | 3 | 51 | 83 | 0.4602 | 20000 | 4132 | 52.5 | 0.3906 | 5 |
| 1310 | 3 | 51 | 83 | 0.4602 | 25000 | 4132 | 52.5 | 0.3624 | 5 |
| 1311 | 3 | 51 | 83 | 0.4602 | 30000 | 4132 | 52.5 | 0.3370 | 5 |
| 1312 | 3 | 51 | 83 | 0.5344 | 10000 | 4132 | 52.5 | 0.5344 | 5 |
| 1313 | 3 | 51 | 83 | 0.5344 | 15000 | 4132 | 52.5 | 0.4980 | 5 |
| 1314 | 3 | 51 | 83 | 0.5344 | 20000 | 4132 | 52.5 | 0.4632 | 5 |
| 1315 | 3 | 51 | 83 | 0.5344 | 25000 | 4132 | 52.5 | 0.4335 | 5 |
| 1316 | 3 | 51 | 83 | 0.5344 | 30000 | 4132 | 52.5 | 0.4063 | 5 |
| 1317 | 3 | 51 | 83 | 0.6255 | 10000 | 4132 | 52.5 | 0.6255 | 5 |
| 1318 | 3 | 51 | 83 | 0.6255 | 15000 | 4132 | 52.5 | 0.5908 | 5 |
| 1319 | 3 | 51 | 83 | 0.6255 | 20000 | 4132 | 52.5 | 0.5567 | 5 |
| 1320 | 3 | 51 | 83 | 0.6255 | 25000 | 4132 | 52.5 | 0.5268 | 5 |
| 1321 | 3 | 51 | 83 | 0.6255 | 30000 | 4132 | 52.5 | 0.4990 | 5 |
| 1322 | 3 | 73 | 5 | 0.9077 | 12000 | 9881 | 40.0 | 0.8230 | 1 |
| 1323 | 3 | 73 | 5 | 0.9077 | 15000 | 9881 | 40.0 | 0.8365 | 1 |
| 1324 | 3 | 73 | 5 | 0.9077 | 20000 | 9881 | 40.0 | 0.8525 | 1 |
| 1325 | 3 | 73 | 5 | 0.9077 | 25000 | 9881 | 40.0 | 0.8622 | 1 |
| 1326 | 3 | 73 | 5 | 0.9077 | 30000 | 9881 | 40.0 | 0.8690 | 1 |
| 1327 | 3 | 73 | 5 | 0.9436 | 12000 | 9881 | 40.0 | 0.9158 | 1 |
| 1328 | 3 | 73 | 5 | 0.9436 | 15000 | 9881 | 40.0 | 0.9193 | 1 |
| 1329 | 3 | 73 | 5 | 0.9436 | 20000 | 9881 | 40.0 | 0.9230 | 1 |
| 1330 | 3 | 73 | 5 | 0.9436 | 25000 | 9881 | 40.0 | 0.9258 | 1 |
| 1331 | 3 | 73 | 5 | 0.9436 | 30000 | 9881 | 40.0 | 0.9260 | 1 |
| 1332 | 8 | 73 | 5 | 0.9077 | 4000 | 1735 | 40.0 | 0.8478 | 1 |
| 1333 | 8 | 73 | 5 | 0.9077 | 6000 | 1735 | 40.0 | 0.8560 | 1 |
| 1334 | 8 | 73 | 5 | 0.9077 | 8000 | 1735 | 40.0 | 0.8625 | 1 |
| 1335 | 8 | 73 | 5 | 0.9077 | 10000 | 1735 | 40.0 | 0.8690 | 1 |
| 1336 | 8 | 73 | 5 | 0.9077 | 12000 | 1735 | 40.0 | 0.8740 | 1 |
| 1337 | 8 | 73 | 5 | 0.9077 | 15000 | 1735 | 40.0 | 0.8803 | 1 |
| 1338 | 8 | 73 | 5 | 0.9077 | 20000 | 1735 | 40.0 | 0.8895 | 1 |
| 1339 | 8 | 73 | 5 | 0.9077 | 25000 | 1735 | 40.0 | 0.8960 | 1 |
| 1340 | 8 | 73 | 5 | 0.9077 | 30000 | 1735 | 40.0 | 0.9002 | 1 |
| 1341 | 8 | 73 | 5 | 0.9436 | 4000 | 1735 | 40.0 | 0.9160 | 1 |
| 1342 | 8 | 73 | 5 | 0.9436 | 6000 | 1735 | 40.0 | 0.9208 | 1 |
| 1343 | 8 | 73 | 5 | 0.9436 | 8000 | 1735 | 40.0 | 0.9258 | 1 |
| 1344 | 8 | 73 | 5 | 0.9436 | 10000 | 1735 | 40.0 | 0.9295 | 1 |
| 1345 | 8 | 73 | 5 | 0.9436 | 12000 | 1735 | 40.0 | 0.9324 | 1 |
| 1346 | 8 | 73 | 5 | 0.9436 | 15000 | 1735 | 40.0 | 0.9365 | 1 |
| 1347 | 8 | 73 | 5 | 0.9436 | 20000 | 1735 | 40.0 | 0.9430 | 1 |
| 1348 | 8 | 73 | 5 | 0.9436 | 25000 | 1735 | 40.0 | 0.9462 | 1 |
| 1349 | 8 | 73 | 5 | 0.9436 | 30000 | 1735 | 40.0 | 0.9482 | 1 |
| 1350 | 3 | 73 | 6 | 0.9380 | 20000 | 9881 | 18.0 | 0.9120 | 7 |
| 1351 | 3 | 73 | 6 | 0.9400 | 12000 | 9881 | 40.0 | 0.8820 | 11 |
| 1352 | 3 | 73 | 6 | 0.9400 | 15000 | 9881 | 40.0 | 0.8886 | 11 |

| | | | | | | | | | |
|---|---|---|---|---|---|---|---|---|---|
| 1353 | 3 | 73 | 6 | 0.9400 | 20000 | 9881 | 40.0 | 0.8980 | 11 |
| 1354 | 3 | 73 | 6 | 0.9400 | 25000 | 9881 | 40.0 | 0.9060 | 11 |
| 1355 | 3 | 73 | 6 | 0.9400 | 30000 | 9881 | 40.0 | 0.9130 | 11 |
| 1356 | 3 | 73 | 6 | 0.9700 | 12000 | 9881 | 40.0 | 0.9348 | 11 |
| 1357 | 3 | 73 | 6 | 0.9700 | 15000 | 9881 | 40.0 | 0.9406 | 11 |
| 1358 | 3 | 73 | 6 | 0.9700 | 20000 | 9881 | 40.0 | 0.9453 | 11 |
| 1359 | 3 | 73 | 6 | 0.9700 | 25000 | 9881 | 40.0 | 0.9503 | 11 |
| 1360 | 3 | 73 | 6 | 0.9700 | 30000 | 9881 | 40.0 | 0.9543 | 11 |
| 1361 | 8 | 73 | 6 | 0.9400 | 4000 | 1735 | 40.0 | 0.8765 | 11 |
| 1362 | 8 | 73 | 6 | 0.9400 | 6000 | 1735 | 40.0 | 0.8815 | 11 |
| 1363 | 8 | 73 | 6 | 0.9400 | 8000 | 1735 | 40.0 | 0.8862 | 11 |
| 1364 | 8 | 73 | 6 | 0.9400 | 10000 | 1735 | 40.0 | 0.8905 | 11 |
| 1365 | 8 | 73 | 6 | 0.9400 | 12000 | 1735 | 40.0 | 0.8943 | 11 |
| 1366 | 8 | 73 | 6 | 0.9400 | 15000 | 1735 | 40.0 | 0.9000 | 11 |
| 1367 | 8 | 73 | 6 | 0.9400 | 20000 | 1735 | 40.0 | 0.9088 | 11 |
| 1368 | 8 | 73 | 6 | 0.9400 | 25000 | 1735 | 40.0 | 0.9162 | 11 |
| 1369 | 8 | 73 | 6 | 0.9400 | 30000 | 1735 | 40.0 | 0.9220 | 11 |
| 1370 | 8 | 73 | 6 | 0.9700 | 4000 | 1735 | 40.0 | 0.9342 | 11 |
| 1371 | 8 | 73 | 6 | 0.9700 | 6000 | 1735 | 40.0 | 0.9390 | 11 |
| 1372 | 8 | 73 | 6 | 0.9700 | 8000 | 1735 | 40.0 | 0.9428 | 11 |
| 1373 | 8 | 73 | 6 | 0.9700 | 10000 | 1735 | 40.0 | 0.9462 | 11 |
| 1374 | 8 | 73 | 6 | 0.9700 | 12000 | 1735 | 40.0 | 0.9490 | 11 |
| 1375 | 8 | 73 | 6 | 0.9700 | 15000 | 1735 | 40.0 | 0.9540 | 11 |
| 1376 | 8 | 73 | 6 | 0.9700 | 20000 | 1735 | 40.0 | 0.9602 | 11 |
| 1377 | 8 | 73 | 6 | 0.9700 | 25000 | 1735 | 40.0 | 0.9660 | 11 |
| 1378 | 8 | 73 | 6 | 0.9700 | 30000 | 1735 | 40.0 | 0.9702 | 11 |
| 1379 | 9 | 73 | 6 | 0.9380 | 7000 | 1793 | 18.0 | 0.9140 | 7 |
| 1380 | 9 | 73 | 6 | 0.9370 | 4000 | 1793 | 40.0 | 0.8770 | 7 |
| 1381 | 9 | 73 | 6 | 0.9380 | 6000 | 1793 | 40.0 | 0.8933 | 7 |
| 1382 | 9 | 73 | 6 | 0.9380 | 8000 | 1793 | 40.0 | 0.8908 | 7 |
| 1383 | 9 | 73 | 6 | 0.9380 | 10000 | 1793 | 40.0 | 0.9020 | 7 |
| 1384 | 3 | 73 | 44 | 0.1650 | 30000 | 9881 | 15.5 | 0.1600 | 7 |
| 1385 | 3 | 73 | 44 | 0.3090 | 30000 | 9881 | 15.5 | 0.2890 | 7 |
| 1386 | 3 | 73 | 44 | 0.4340 | 30000 | 9881 | 15.5 | 0.4140 | 7 |
| 1387 | 3 | 73 | 44 | 0.5440 | 30000 | 9881 | 15.5 | 0.5320 | 7 |
| 1388 | 3 | 73 | 44 | 0.6420 | 30000 | 9881 | 15.5 | 0.6350 | 7 |
| 1389 | 3 | 73 | 44 | 0.8070 | 30000 | 9881 | 15.5 | 0.8000 | 7 |
| 1390 | 3 | 73 | 44 | 0.9410 | 30000 | 9881 | 15.5 | 0.9600 | 7 |
| 1391 | 3 | 74 | 5 | 0.9445 | 12000 | 10206 | 40.0 | 0.8918 | 1 |
| 1392 | 3 | 74 | 5 | 0.9445 | 15000 | 10206 | 40.0 | 0.8990 | 1 |
| 1393 | 3 | 74 | 5 | 0.9445 | 20000 | 10206 | 40.0 | 0.9103 | 1 |
| 1394 | 3 | 74 | 5 | 0.9445 | 25000 | 10206 | 40.0 | 0.9198 | 1 |
| 1395 | 3 | 74 | 5 | 0.9445 | 30000 | 10206 | 40.0 | 0.9264 | 1 |
| 1396 | 8 | 74 | 5 | 0.9445 | 4000 | 1809 | 40.0 | 0.8850 | 1 |
| 1397 | 8 | 74 | 5 | 0.9445 | 6000 | 1809 | 40.0 | 0.8950 | 1 |
| 1398 | 8 | 74 | 5 | 0.9445 | 8000 | 1809 | 40.0 | 0.9038 | 1 |

| | | | | | | | | | |
|---|---|---|---|---|---|---|---|---|---|
| 1399 | 8 | 74 | 5 | 0.9445 | 10000 | 1809 | 40.0 | 0.9105 | 1 |
| 1400 | 8 | 74 | 5 | 0.9445 | 12000 | 1809 | 40.0 | 0.9178 | 1 |
| 1401 | 8 | 74 | 5 | 0.9445 | 15000 | 1809 | 40.0 | 0.9258 | 1 |
| 1402 | 8 | 74 | 5 | 0.9445 | 20000 | 1809 | 40.0 | 0.9375 | 1 |
| 1403 | 8 | 74 | 5 | 0.9445 | 25000 | 1809 | 40.0 | 0.9463 | 1 |
| 1404 | 8 | 74 | 5 | 0.9445 | 30000 | 1809 | 40.0 | 0.9530 | 1 |
| 1405 | 3 | 74 | 6 | 0.9387 | 15000 | 10206 | 40.0 | 0.8800 | 6 |
| 1406 | 3 | 74 | 6 | 0.9387 | 20000 | 10206 | 40.0 | 0.8910 | 6 |
| 1407 | 3 | 74 | 6 | 0.9387 | 30000 | 10206 | 40.0 | 0.9080 | 6 |
| 1408 | 3 | 74 | 6 | 0.9387 | 35000 | 10206 | 40.0 | 0.9110 | 6 |
| 1409 | 3 | 74 | 6 | 0.9387 | 12000 | 10206 | 40.0 | 0.8558 | 11 |
| 1410 | 3 | 74 | 6 | 0.9387 | 15000 | 10206 | 40.0 | 0.8782 | 11 |
| 1411 | 3 | 74 | 6 | 0.9387 | 20000 | 10206 | 40.0 | 0.8975 | 11 |
| 1412 | 3 | 74 | 6 | 0.9387 | 25000 | 10206 | 40.0 | 0.9093 | 11 |
| 1413 | 3 | 74 | 6 | 0.9387 | 30000 | 10206 | 40.0 | 0.9178 | 11 |
| 1414 | 3 | 74 | 6 | 0.9700 | 12000 | 10206 | 40.0 | 0.9153 | 11 |
| 1415 | 3 | 74 | 6 | 0.9700 | 15000 | 10206 | 40.0 | 0.9290 | 11 |
| 1416 | 3 | 74 | 6 | 0.9700 | 20000 | 10206 | 40.0 | 0.9432 | 11 |
| 1417 | 3 | 74 | 6 | 0.9700 | 25000 | 10206 | 40.0 | 0.9522 | 11 |
| 1418 | 3 | 74 | 6 | 0.9700 | 30000 | 10206 | 40.0 | 0.9580 | 11 |
| 1419 | 8 | 74 | 6 | 0.9387 | 4000 | 1809 | 40.0 | 0.8560 | 11 |
| 1420 | 8 | 74 | 6 | 0.9387 | 6000 | 1809 | 40.0 | 0.8712 | 11 |
| 1421 | 8 | 74 | 6 | 0.9387 | 8000 | 1809 | 40.0 | 0.8823 | 11 |
| 1422 | 8 | 74 | 6 | 0.9387 | 10000 | 1809 | 40.0 | 0.8910 | 11 |
| 1423 | 8 | 74 | 6 | 0.9387 | 12000 | 1809 | 40.0 | 0.8987 | 11 |
| 1424 | 8 | 74 | 6 | 0.9387 | 15000 | 1809 | 40.0 | 0.9081 | 11 |
| 1425 | 8 | 74 | 6 | 0.9387 | 20000 | 1809 | 40.0 | 0.9212 | 11 |
| 1426 | 8 | 74 | 6 | 0.9387 | 25000 | 1809 | 40.0 | 0.9328 | 11 |
| 1427 | 8 | 74 | 6 | 0.9387 | 30000 | 1809 | 40.0 | 0.9428 | 11 |
| 1428 | 8 | 74 | 6 | 0.9700 | 4000 | 1809 | 40.0 | 0.8865 | 11 |
| 1429 | 8 | 74 | 6 | 0.9700 | 6000 | 1809 | 40.0 | 0.9060 | 11 |
| 1430 | 8 | 74 | 6 | 0.9700 | 8000 | 1809 | 40.0 | 0.9178 | 11 |
| 1431 | 8 | 74 | 6 | 0.9700 | 10000 | 1809 | 40.0 | 0.9272 | 11 |
| 1432 | 8 | 74 | 6 | 0.9700 | 12000 | 1809 | 40.0 | 0.9350 | 11 |
| 1433 | 8 | 74 | 6 | 0.9700 | 15000 | 1809 | 40.0 | 0.9443 | 11 |
| 1434 | 8 | 74 | 6 | 0.9700 | 20000 | 1809 | 40.0 | 0.9568 | 11 |
| 1435 | 8 | 74 | 6 | 0.9700 | 25000 | 1809 | 40.0 | 0.9668 | 11 |
| 1436 | 8 | 74 | 6 | 0.9700 | 30000 | 1809 | 40.0 | 0.9756 | 11 |
| 1437 | 3 | 74 | 42 | 0.7850 | 30000 | 10206 | 18.0 | 0.7640 | 7 |
| 1438 | 3 | 74 | 42 | 0.8000 | 15000 | 10206 | 52.5 | 0.7420 | 7 |
| 1439 | 3 | 74 | 42 | 0.8000 | 20000 | 10206 | 52.5 | 0.7720 | 7 |
| 1440 | 8 | 74 | 42 | 0.8000 | 10000 | 1809 | 52.5 | 0.7260 | 7 |
| 1441 | 9 | 74 | 42 | 0.7850 | 5000 | 1872 | 18.0 | 0.7640 | 7 |
| 1442 | 3 | 74 | 44 | 0.4380 | 30000 | 10206 | 15.5 | 0.4440 | 7 |
| 1443 | 3 | 74 | 44 | 0.6450 | 30000 | 10206 | 15.5 | 0.6550 | 7 |
| 1444 | 3 | 74 | 44 | 0.8090 | 30000 | 10206 | 15.5 | 0.8250 | 7 |

| | | | | | | | | | |
|---|---|---|---|---|---|---|---|---|---|
| 1445 | 3 | 74 | 44 | 0.9420 | 30000 | 10206 | 15.5 | 0.9650 | 7 |
| 1446 | 3 | 78 | 28 | 0.4490 | 30000 | 11564 | 15.5 | 0.3450 | 7 |
| 1447 | 3 | 78 | 28 | 0.7030 | 30000 | 11564 | 15.5 | 0.5800 | 7 |
| 1448 | 3 | 78 | 28 | 0.8360 | 30000 | 11564 | 15.5 | 0.7750 | 7 |
| 1449 | 3 | 78 | 28 | 0.9350 | 30000 | 11564 | 15.5 | 0.8840 | 7 |
| 1450 | 3 | 79 | 25 | 0.7660 | 28500 | 11919 | 17.3 | 0.6830 | 7 |
| 1451 | 3 | 79 | 29 | 0.2060 | 30000 | 11919 | 15.5 | 0.1520 | 7 |
| 1452 | 3 | 79 | 29 | 0.4020 | 30000 | 11919 | 15.5 | 0.3240 | 7 |
| 1453 | 3 | 79 | 29 | 0.6010 | 30000 | 11919 | 15.5 | 0.5050 | 7 |
| 1454 | 3 | 79 | 29 | 0.7990 | 30000 | 11919 | 15.5 | 0.7300 | 7 |
| 1455 | 3 | 79 | 29 | 0.0250 | 28200 | 11919 | 17.3 | 0.0160 | 7 |
| 1456 | 3 | 79 | 29 | 0.2000 | 16800 | 11919 | 17.3 | 0.1470 | 7 |
| 1457 | 3 | 79 | 29 | 0.2000 | 19800 | 11919 | 17.3 | 0.1450 | 7 |
| 1458 | 3 | 79 | 29 | 0.2000 | 22500 | 11919 | 17.3 | 0.1470 | 7 |
| 1459 | 3 | 79 | 29 | 0.2000 | 28200 | 11919 | 17.3 | 0.1460 | 7 |
| 1460 | 3 | 79 | 29 | 0.2000 | 33900 | 11919 | 17.3 | 0.1400 | 7 |
| 1461 | 3 | 79 | 29 | 0.2380 | 28500 | 11919 | 17.3 | 0.1740 | 7 |
| 1462 | 3 | 79 | 29 | 0.4000 | 28200 | 11919 | 17.3 | 0.3100 | 5 |
| 1463 | 3 | 79 | 29 | 0.5270 | 28500 | 11919 | 17.3 | 0.4330 | 5 |
| 1464 | 3 | 79 | 29 | 0.6000 | 16800 | 11919 | 17.3 | 0.4980 | 7 |
| 1465 | 3 | 79 | 29 | 0.6000 | 19800 | 11919 | 17.3 | 0.5000 | 7 |
| 1466 | 3 | 79 | 29 | 0.6000 | 22500 | 11919 | 17.3 | 0.5140 | 7 |
| 1467 | 3 | 79 | 29 | 0.6000 | 28200 | 11919 | 17.3 | 0.5100 | 7 |
| 1468 | 3 | 79 | 29 | 0.6000 | 33900 | 11919 | 17.3 | 0.5020 | 7 |
| 1469 | 3 | 79 | 29 | 0.8000 | 28200 | 11919 | 17.3 | 0.7370 | 5 |
| 1470 | 3 | 79 | 29 | 0.2010 | 25000 | 11919 | 18.0 | 0.1490 | 7 |
| 1471 | 3 | 79 | 29 | 0.4010 | 25000 | 11919 | 18.0 | 0.3140 | 7 |
| 1472 | 3 | 79 | 29 | 0.6040 | 25000 | 11919 | 18.0 | 0.5130 | 7 |
| 1473 | 3 | 79 | 29 | 0.8010 | 25000 | 11919 | 18.0 | 0.7410 | 7 |
| 1474 | 3 | 79 | 29 | 0.2010 | 16000 | 11919 | 40.0 | 0.1470 | 5 |
| 1475 | 3 | 79 | 29 | 0.2010 | 20000 | 11919 | 40.0 | 0.1530 | 5 |
| 1476 | 3 | 79 | 29 | 0.2010 | 25000 | 11919 | 40.0 | 0.1550 | 5 |
| 1477 | 3 | 79 | 29 | 0.3000 | 13000 | 11919 | 40.0 | 0.2319 | 5 |
| 1478 | 3 | 79 | 29 | 0.3000 | 15700 | 11919 | 40.0 | 0.2278 | 5 |
| 1479 | 3 | 79 | 29 | 0.3000 | 20900 | 11919 | 40.0 | 0.2365 | 5 |
| 1480 | 3 | 79 | 29 | 0.3000 | 26100 | 11919 | 40.0 | 0.2402 | 5 |
| 1481 | 3 | 79 | 29 | 0.3000 | 31300 | 11919 | 40.0 | 0.2413 | 5 |
| 1482 | 3 | 79 | 29 | 0.3000 | 36300 | 11919 | 40.0 | 0.2381 | 5 |
| 1483 | 3 | 79 | 29 | 0.3000 | 39500 | 11919 | 40.0 | 0.2377 | 5 |
| 1484 | 3 | 79 | 29 | 0.4010 | 16000 | 11919 | 40.0 | 0.3230 | 5 |
| 1485 | 3 | 79 | 29 | 0.4010 | 20000 | 11919 | 40.0 | 0.3340 | 5 |
| 1486 | 3 | 79 | 29 | 0.4010 | 25000 | 11919 | 40.0 | 0.3380 | 5 |
| 1487 | 3 | 79 | 29 | 0.5080 | 13000 | 11919 | 40.0 | 0.4174 | 5 |
| 1488 | 3 | 79 | 29 | 0.5080 | 15700 | 11919 | 40.0 | 0.4142 | 5 |
| 1489 | 3 | 79 | 29 | 0.5080 | 20900 | 11919 | 40.0 | 0.4285 | 5 |
| 1490 | 3 | 79 | 29 | 0.5080 | 26100 | 11919 | 40.0 | 0.4348 | 5 |

| | | | | | | | | | |
|---|---|---|---|---|---|---|---|---|---|
| 1491 | 3 | 79 | 29 | 0.5080 | 31300 | 11919 | 40.0 | 0.4337 | 5 |
| 1492 | 3 | 79 | 29 | 0.5080 | 36300 | 11919 | 40.0 | 0.4282 | 5 |
| 1493 | 3 | 79 | 29 | 0.5080 | 39500 | 11919 | 40.0 | 0.4279 | 5 |
| 1494 | 3 | 79 | 29 | 0.6040 | 16000 | 11919 | 40.0 | 0.5180 | 5 |
| 1495 | 3 | 79 | 29 | 0.6040 | 20000 | 11919 | 40.0 | 0.5300 | 5 |
| 1496 | 3 | 79 | 29 | 0.6040 | 25000 | 11919 | 40.0 | 0.5370 | 5 |
| 1497 | 3 | 79 | 29 | 0.8020 | 16000 | 11919 | 40.0 | 0.7510 | 5 |
| 1498 | 3 | 79 | 29 | 0.8020 | 20000 | 11919 | 40.0 | 0.7610 | 5 |
| 1499 | 3 | 79 | 29 | 0.8020 | 25000 | 11919 | 40.0 | 0.7620 | 5 |
| 1500 | 3 | 79 | 29 | 0.2012 | 15000 | 11919 | 52.5 | 0.1450 | 4 |
| 1501 | 3 | 79 | 29 | 0.2012 | 20000 | 11919 | 52.5 | 0.1580 | 4 |
| 1502 | 3 | 79 | 29 | 0.2012 | 25000 | 11919 | 52.5 | 0.1540 | 4 |
| 1503 | 3 | 79 | 29 | 0.2012 | 30000 | 11919 | 52.5 | 0.1570 | 4 |
| 1504 | 3 | 79 | 29 | 0.2012 | 39500 | 11919 | 52.5 | 0.1640 | 4 |
| 1505 | 3 | 79 | 29 | 0.2012 | 48500 | 11919 | 52.5 | 0.1520 | 4 |
| 1506 | 3 | 79 | 29 | 0.4010 | 15000 | 11919 | 52.5 | 0.3130 | 4 |
| 1507 | 3 | 79 | 29 | 0.4010 | 20000 | 11919 | 52.5 | 0.3330 | 4 |
| 1508 | 3 | 79 | 29 | 0.4010 | 25000 | 11919 | 52.5 | 0.3310 | 4 |
| 1509 | 3 | 79 | 29 | 0.4010 | 30000 | 11919 | 52.5 | 0.3310 | 4 |
| 1510 | 3 | 79 | 29 | 0.4010 | 40000 | 11919 | 52.5 | 0.3260 | 4 |
| 1511 | 3 | 79 | 29 | 0.4010 | 48500 | 11919 | 52.5 | 0.3500 | 4 |
| 1512 | 3 | 79 | 29 | 0.6036 | 15000 | 11919 | 52.5 | 0.5110 | 4 |
| 1513 | 3 | 79 | 29 | 0.6036 | 20000 | 11919 | 52.5 | 0.5310 | 4 |
| 1514 | 3 | 79 | 29 | 0.6036 | 25000 | 11919 | 52.5 | 0.5290 | 4 |
| 1515 | 3 | 79 | 29 | 0.6036 | 30000 | 11919 | 52.5 | 0.5330 | 4 |
| 1516 | 3 | 79 | 29 | 0.6036 | 40000 | 11919 | 52.5 | 0.5520 | 4 |
| 1517 | 3 | 79 | 29 | 0.6036 | 48500 | 11919 | 52.5 | 0.5210 | 4 |
| 1518 | 3 | 79 | 29 | 0.8015 | 20000 | 11919 | 52.5 | 0.7520 | 4 |
| 1519 | 3 | 79 | 29 | 0.8015 | 25000 | 11919 | 52.5 | 0.7450 | 4 |
| 1520 | 3 | 79 | 29 | 0.8015 | 30000 | 11919 | 52.5 | 0.7540 | 4 |
| 1521 | 3 | 79 | 29 | 0.8015 | 40000 | 11919 | 52.5 | 0.7580 | 4 |
| 1522 | 3 | 79 | 29 | 0.8015 | 48500 | 11919 | 52.5 | 0.7510 | 4 |
| 1523 | 4 | 79 | 29 | 0.2012 | 20000 | 13734 | 25.2 | 0.1620 | 4 |
| 1524 | 4 | 79 | 29 | 0.2012 | 30000 | 13734 | 52.5 | 0.1680 | 4 |
| 1525 | 4 | 79 | 29 | 0.2012 | 40000 | 13734 | 52.5 | 0.1720 | 4 |
| 1526 | 4 | 79 | 29 | 0.2012 | 48500 | 13734 | 52.5 | 0.1670 | 4 |
| 1527 | 4 | 79 | 29 | 0.4010 | 20000 | 13734 | 52.5 | 0.3410 | 4 |
| 1528 | 4 | 79 | 29 | 0.4010 | 30000 | 13734 | 52.5 | 0.3550 | 4 |
| 1529 | 4 | 79 | 29 | 0.4010 | 40000 | 13734 | 52.5 | 0.3580 | 4 |
| 1530 | 4 | 79 | 29 | 0.4010 | 48500 | 13734 | 52.5 | 0.3500 | 4 |
| 1531 | 4 | 79 | 29 | 0.6036 | 20000 | 13734 | 52.5 | 0.5420 | 4 |
| 1532 | 4 | 79 | 29 | 0.6036 | 30000 | 13734 | 52.5 | 0.5490 | 4 |
| 1533 | 4 | 79 | 29 | 0.6036 | 40000 | 13734 | 52.5 | 0.5410 | 4 |
| 1534 | 4 | 79 | 29 | 0.6036 | 48500 | 13734 | 52.5 | 0.5450 | 4 |
| 1535 | 4 | 79 | 29 | 0.8015 | 20000 | 13734 | 52.5 | 0.7660 | 4 |
| 1536 | 4 | 79 | 29 | 0.8015 | 30000 | 13734 | 52.5 | 0.7750 | 4 |

| | | | | | | | | | |
|---|---|---|---|---|---|---|---|---|---|
| 1537 | 4 | 79 | 29 | 0.8015 | 40000 | 13734 | 52.5 | 0.7490 | 4 |
| 1538 | 4 | 79 | 29 | 0.8015 | 48500 | 13734 | 52.5 | 0.7600 | 4 |
| 1539 | 8 | 79 | 29 | 0.2010 | 12000 | 2206 | 40.0 | 0.1470 | 5 |
| 1540 | 8 | 79 | 29 | 0.2010 | 14000 | 2206 | 40.0 | 0.1430 | 5 |
| 1541 | 8 | 79 | 29 | 0.3000 | 5200 | 2206 | 40.0 | 0.2317 | 5 |
| 1542 | 8 | 79 | 29 | 0.3000 | 10400 | 2206 | 40.0 | 0.2317 | 5 |
| 1543 | 8 | 79 | 29 | 0.3000 | 15700 | 2206 | 40.0 | 0.2186 | 5 |
| 1544 | 8 | 79 | 29 | 0.3000 | 26100 | 2206 | 40.0 | 0.1886 | 5 |
| 1545 | 8 | 79 | 29 | 0.3000 | 31500 | 2206 | 40.0 | 0.1773 | 5 |
| 1546 | 8 | 79 | 29 | 0.3000 | 36600 | 2206 | 40.0 | 0.1717 | 5 |
| 1547 | 8 | 79 | 29 | 0.3000 | 39700 | 2206 | 40.0 | 0.1689 | 5 |
| 1548 | 8 | 79 | 29 | 0.4010 | 12000 | 2206 | 40.0 | 0.3160 | 5 |
| 1549 | 8 | 79 | 29 | 0.4010 | 12000 | 2206 | 40.0 | 0.3220 | 5 |
| 1550 | 8 | 79 | 29 | 0.4010 | 14000 | 2206 | 40.0 | 0.3140 | 5 |
| 1551 | 8 | 79 | 29 | 0.5080 | 5200 | 2206 | 40.0 | 0.4259 | 5 |
| 1552 | 8 | 79 | 29 | 0.5080 | 10400 | 2206 | 40.0 | 0.4203 | 5 |
| 1553 | 8 | 79 | 29 | 0.5080 | 15700 | 2206 | 40.0 | 0.4034 | 5 |
| 1554 | 8 | 79 | 29 | 0.5080 | 20800 | 2206 | 40.0 | 0.3865 | 5 |
| 1555 | 8 | 79 | 29 | 0.5080 | 26100 | 2206 | 40.0 | 0.3668 | 5 |
| 1556 | 8 | 79 | 29 | 0.5080 | 31500 | 2206 | 40.0 | 0.3499 | 5 |
| 1557 | 8 | 79 | 29 | 0.5080 | 36600 | 2206 | 40.0 | 0.3405 | 5 |
| 1558 | 8 | 79 | 29 | 0.5080 | 39700 | 2206 | 40.0 | 0.3349 | 5 |
| 1559 | 8 | 79 | 29 | 0.6040 | 14000 | 2206 | 40.0 | 0.5110 | 1 |
| 1560 | 8 | 79 | 29 | 0.6040 | 10000 | 2206 | 40.0 | 0.5230 | 5 |
| 1561 | 8 | 79 | 29 | 0.6040 | 12000 | 2206 | 40.0 | 0.5140 | 5 |
| 1562 | 8 | 79 | 29 | 0.8020 | 10000 | 2206 | 40.0 | 0.7560 | 5 |
| 1563 | 8 | 79 | 29 | 0.8020 | 12000 | 2206 | 40.0 | 0.7450 | 5 |
| 1564 | 8 | 79 | 29 | 0.8020 | 14000 | 2206 | 40.0 | 0.7420 | 5 |
| 1565 | 8 | 79 | 29 | 0.2012 | 5000 | 2206 | 52.5 | 0.1570 | 4 |
| 1566 | 8 | 79 | 29 | 0.2012 | 10000 | 2206 | 52.5 | 0.1540 | 4 |
| 1567 | 8 | 79 | 29 | 0.2012 | 20000 | 2206 | 52.5 | 0.1400 | 4 |
| 1568 | 8 | 79 | 29 | 0.2012 | 30000 | 2206 | 52.5 | 0.1220 | 4 |
| 1569 | 8 | 79 | 29 | 0.2012 | 40000 | 2206 | 52.5 | 0.1100 | 4 |
| 1570 | 8 | 79 | 29 | 0.4010 | 5000 | 2206 | 52.5 | 0.3340 | 4 |
| 1571 | 8 | 79 | 29 | 0.4010 | 10000 | 2206 | 52.5 | 0.3320 | 4 |
| 1572 | 8 | 79 | 29 | 0.4010 | 20000 | 2206 | 52.5 | 0.3040 | 4 |
| 1573 | 8 | 79 | 29 | 0.4010 | 30000 | 2206 | 52.5 | 0.2750 | 4 |
| 1574 | 8 | 79 | 29 | 0.4010 | 40000 | 2206 | 52.5 | 0.2530 | 4 |
| 1575 | 8 | 79 | 29 | 0.6036 | 5000 | 2206 | 52.5 | 0.5330 | 4 |
| 1576 | 8 | 79 | 29 | 0.6036 | 10000 | 2206 | 52.5 | 0.5310 | 4 |
| 1577 | 8 | 79 | 29 | 0.6036 | 20000 | 2206 | 52.5 | 0.5000 | 4 |
| 1578 | 8 | 79 | 29 | 0.6036 | 30000 | 2206 | 52.5 | 0.4620 | 4 |
| 1579 | 8 | 79 | 29 | 0.6036 | 40000 | 2206 | 52.5 | 0.4280 | 4 |
| 1580 | 8 | 79 | 29 | 0.8015 | 5000 | 2206 | 52.5 | 0.7650 | 4 |
| 1581 | 8 | 79 | 29 | 0.8015 | 10000 | 2206 | 52.5 | 0.7490 | 4 |
| 1582 | 8 | 79 | 29 | 0.8015 | 20000 | 2206 | 52.5 | 0.7280 | 4 |

| | | | | | | | | | |
|---|---|---|---|---|---|---|---|---|---|
| 1583 | 8 | 79 | 29 | 0.8015 | 30000 | 2206 | 52.5 | 0.7040 | 4 |
| 1584 | 8 | 79 | 29 | 0.8015 | 40000 | 2206 | 52.5 | 0.6750 | 4 |
| 1585 | 8 | 79 | 29 | 0.3000 | 20800 | 2206 | 40.0 | 0.2054 | 5 |
| 1586 | 3 | 79 | 31 | 0.7390 | 28500 | 11919 | 17.3 | 0.7180 | 6 |
| 1587 | 3 | 79 | 31 | 0.7390 | 28500 | 11919 | 17.3 | 0.7180 | 7 |
| 1588 | 8 | 79 | 46 | 0.6410 | 5200 | 2206 | 40.0 | 0.5786 | 5 |
| 1589 | 8 | 79 | 46 | 0.6410 | 10300 | 2206 | 40.0 | 0.5936 | 5 |
| 1590 | 8 | 79 | 46 | 0.6410 | 15500 | 2206 | 40.0 | 0.6005 | 5 |
| 1591 | 8 | 79 | 46 | 0.6410 | 20700 | 2206 | 40.0 | 0.6033 | 5 |
| 1592 | 8 | 79 | 46 | 0.6410 | 25800 | 2206 | 40.0 | 0.5962 | 5 |
| 1593 | 8 | 79 | 46 | 0.6410 | 31000 | 2206 | 40.0 | 0.5928 | 5 |
| 1594 | 8 | 79 | 46 | 0.6410 | 36200 | 2206 | 40.0 | 0.5883 | 5 |
| 1595 | 3 | 79 | 47 | 0.1990 | 30000 | 11919 | 15.5 | 0.1860 | 6 |
| 1596 | 3 | 79 | 47 | 0.1990 | 30000 | 11919 | 15.5 | 0.1860 | 7 |
| 1597 | 3 | 79 | 47 | 0.4050 | 30000 | 11919 | 15.5 | 0.3990 | 6 |
| 1598 | 3 | 79 | 47 | 0.4050 | 30000 | 11919 | 15.5 | 0.3990 | 7 |
| 1599 | 3 | 79 | 47 | 0.5940 | 30000 | 11919 | 15.5 | 0.5930 | 6 |
| 1600 | 3 | 79 | 47 | 0.5940 | 30000 | 11919 | 15.5 | 0.5930 | 7 |
| 1601 | 3 | 79 | 47 | 0.8040 | 30000 | 11919 | 15.5 | 0.7800 | 6 |
| 1602 | 3 | 79 | 47 | 0.8040 | 30000 | 11919 | 15.5 | 0.7800 | 7 |
| 1603 | 3 | 79 | 47 | 0.5000 | 19400 | 11919 | 17.3 | 0.4500 | 7 |
| 1604 | 3 | 79 | 47 | 0.5000 | 30000 | 11919 | 17.3 | 0.4700 | 7 |
| 1605 | 3 | 79 | 47 | 0.2240 | 20000 | 11919 | 18.0 | 0.1990 | 7 |
| 1606 | 3 | 79 | 47 | 0.4000 | 20000 | 11919 | 18.0 | 0.3590 | 7 |
| 1607 | 3 | 79 | 47 | 0.6010 | 20000 | 11919 | 18.0 | 0.5590 | 7 |
| 1608 | 3 | 79 | 47 | 0.8000 | 20000 | 11919 | 18.0 | 0.7670 | 7 |
| 1609 | 3 | 79 | 47 | 0.2243 | 15000 | 11919 | 40.0 | 0.2081 | 8 |
| 1610 | 3 | 79 | 47 | 0.2243 | 20000 | 11919 | 40.0 | 0.2072 | 8 |
| 1611 | 3 | 79 | 47 | 0.2243 | 25000 | 11919 | 40.0 | 0.2101 | 8 |
| 1612 | 3 | 79 | 47 | 0.2243 | 30000 | 11919 | 40.0 | 0.2092 | 8 |
| 1613 | 3 | 79 | 47 | 0.4003 | 15000 | 11919 | 40.0 | 0.3658 | 8 |
| 1614 | 3 | 79 | 47 | 0.4003 | 20000 | 11919 | 40.0 | 0.3755 | 8 |
| 1615 | 3 | 79 | 47 | 0.4003 | 25000 | 11919 | 40.0 | 0.3759 | 8 |
| 1616 | 3 | 79 | 47 | 0.4003 | 30000 | 11919 | 40.0 | 0.3791 | 8 |
| 1617 | 3 | 79 | 47 | 0.6005 | 15000 | 11919 | 40.0 | 0.5859 | 8 |
| 1618 | 3 | 79 | 47 | 0.6005 | 20000 | 11919 | 40.0 | 0.5800 | 8 |
| 1619 | 3 | 79 | 47 | 0.6005 | 25000 | 11919 | 40.0 | 0.5808 | 8 |
| 1620 | 3 | 79 | 47 | 0.6005 | 30000 | 11919 | 40.0 | 0.5806 | 8 |
| 1621 | 3 | 79 | 47 | 0.8005 | 15000 | 11919 | 40.0 | 0.8251 | 8 |
| 1622 | 3 | 79 | 47 | 0.8005 | 20000 | 11919 | 40.0 | 0.7989 | 8 |
| 1623 | 3 | 79 | 47 | 0.8005 | 25000 | 11919 | 40.0 | 0.7939 | 8 |
| 1624 | 3 | 79 | 47 | 0.8005 | 30000 | 11919 | 40.0 | 0.7942 | 8 |
| 1625 | 3 | 79 | 47 | 0.2243 | 15000 | 11919 | 52.5 | 0.2030 | 4 |
| 1626 | 3 | 79 | 47 | 0.2243 | 20000 | 11919 | 52.5 | 0.2030 | 4 |
| 1627 | 3 | 79 | 47 | 0.2243 | 30000 | 11919 | 52.5 | 0.2110 | 4 |
| 1628 | 3 | 79 | 47 | 0.2243 | 40000 | 11919 | 52.5 | 0.2160 | 4 |

| | | | | | | | | | |
|---|---|---|---|---|---|---|---|---|---|
| 1629 | 3 | 79 | 47 | 0.2243 | 48500 | 11919 | 52.5 | 0.2140 | 4 |
| 1630 | 3 | 79 | 47 | 0.4003 | 15000 | 11919 | 52.5 | 0.3680 | 4 |
| 1631 | 3 | 79 | 47 | 0.4003 | 20000 | 11919 | 52.5 | 0.3660 | 4 |
| 1632 | 3 | 79 | 47 | 0.4003 | 30000 | 11919 | 52.5 | 0.3760 | 4 |
| 1633 | 3 | 79 | 47 | 0.4003 | 40000 | 11919 | 52.5 | 0.3840 | 4 |
| 1634 | 3 | 79 | 47 | 0.4003 | 48500 | 11919 | 52.5 | 0.3850 | 4 |
| 1635 | 3 | 79 | 47 | 0.6005 | 15000 | 11919 | 52.5 | 0.5660 | 4 |
| 1636 | 3 | 79 | 47 | 0.6005 | 20000 | 11919 | 52.5 | 0.5730 | 4 |
| 1637 | 3 | 79 | 47 | 0.6005 | 30000 | 11919 | 52.5 | 0.5800 | 4 |
| 1638 | 3 | 79 | 47 | 0.6005 | 40000 | 11919 | 52.5 | 0.5840 | 4 |
| 1639 | 3 | 79 | 47 | 0.6005 | 48500 | 11919 | 52.5 | 0.5840 | 4 |
| 1640 | 3 | 79 | 47 | 0.8005 | 15000 | 11919 | 52.5 | 0.7890 | 4 |
| 1641 | 3 | 79 | 47 | 0.8005 | 20000 | 11919 | 52.5 | 0.7830 | 4 |
| 1642 | 3 | 79 | 47 | 0.8005 | 30000 | 11919 | 52.5 | 0.7830 | 4 |
| 1643 | 3 | 79 | 47 | 0.8005 | 40000 | 11919 | 52.5 | 0.7900 | 4 |
| 1644 | 3 | 79 | 47 | 0.8005 | 48500 | 11919 | 52.5 | 0.7880 | 4 |
| 1645 | 8 | 79 | 47 | 0.2243 | 10000 | 2206 | 40.0 | 0.2028 | 8 |
| 1646 | 8 | 79 | 47 | 0.2243 | 15000 | 2206 | 40.0 | 0.2000 | 8 |
| 1647 | 8 | 79 | 47 | 0.2243 | 20000 | 2206 | 40.0 | 0.1964 | 8 |
| 1648 | 8 | 79 | 47 | 0.2243 | 25000 | 2206 | 40.0 | 0.1921 | 8 |
| 1649 | 8 | 79 | 47 | 0.2243 | 30000 | 2206 | 40.0 | 0.1885 | 8 |
| 1650 | 8 | 79 | 47 | 0.4003 | 10000 | 2206 | 40.0 | 0.3680 | 8 |
| 1651 | 8 | 79 | 47 | 0.4003 | 15000 | 2206 | 40.0 | 0.3660 | 8 |
| 1652 | 8 | 79 | 47 | 0.4003 | 20000 | 2206 | 40.0 | 0.3602 | 8 |
| 1653 | 8 | 79 | 47 | 0.4003 | 25000 | 2206 | 40.0 | 0.3541 | 8 |
| 1654 | 8 | 79 | 47 | 0.4003 | 30000 | 2206 | 40.0 | 0.3492 | 8 |
| 1655 | 8 | 79 | 47 | 0.6005 | 10000 | 2206 | 40.0 | 0.5770 | 8 |
| 1656 | 8 | 79 | 47 | 0.6005 | 15000 | 2206 | 40.0 | 0.5676 | 8 |
| 1657 | 8 | 79 | 47 | 0.6005 | 20000 | 2206 | 40.0 | 0.5626 | 8 |
| 1658 | 8 | 79 | 47 | 0.6005 | 25000 | 2206 | 40.0 | 0.5528 | 8 |
| 1659 | 8 | 79 | 47 | 0.6005 | 30000 | 2206 | 40.0 | 0.5497 | 8 |
| 1660 | 8 | 79 | 47 | 0.8005 | 10000 | 2206 | 40.0 | 0.7963 | 8 |
| 1661 | 8 | 79 | 47 | 0.8005 | 15000 | 2206 | 40.0 | 0.7873 | 8 |
| 1662 | 8 | 79 | 47 | 0.8005 | 20000 | 2206 | 40.0 | 0.7794 | 8 |
| 1663 | 8 | 79 | 47 | 0.8005 | 25000 | 2206 | 40.0 | 0.7724 | 8 |
| 1664 | 8 | 79 | 47 | 0.8005 | 30000 | 2206 | 40.0 | 0.7711 | 8 |
| 1665 | 8 | 79 | 47 | 0.2243 | 5000 | 2206 | 52.5 | 0.1980 | 4 |
| 1666 | 8 | 79 | 47 | 0.2243 | 10000 | 2206 | 52.5 | 0.2010 | 4 |
| 1667 | 8 | 79 | 47 | 0.2243 | 20000 | 2206 | 52.5 | 0.1970 | 4 |
| 1668 | 8 | 79 | 47 | 0.2243 | 30000 | 2206 | 52.5 | 0.1920 | 4 |
| 1669 | 8 | 79 | 47 | 0.2243 | 40000 | 2206 | 52.5 | 0.1850 | 4 |
| 1670 | 8 | 79 | 47 | 0.2243 | 48500 | 2206 | 52.5 | 0.1810 | 4 |
| 1671 | 8 | 79 | 47 | 0.4003 | 5000 | 2206 | 52.5 | 0.3650 | 4 |
| 1672 | 8 | 79 | 47 | 0.4003 | 10000 | 2206 | 52.5 | 0.3620 | 4 |
| 1673 | 8 | 79 | 47 | 0.4003 | 20000 | 2206 | 52.5 | 0.3640 | 4 |
| 1674 | 8 | 79 | 47 | 0.4003 | 30000 | 2206 | 52.5 | 0.3510 | 4 |

| | | | | | | | | | |
|---|---|---|---|---|---|---|---|---|---|
| 1675 | 8 | 79 | 47 | 0.4003 | 40000 | 2206 | 52.5 | 0.3440 | 4 |
| 1676 | 8 | 79 | 47 | 0.4003 | 48500 | 2206 | 52.5 | 0.3370 | 4 |
| 1677 | 8 | 79 | 47 | 0.6005 | 5000 | 2206 | 52.5 | 0.5660 | 4 |
| 1678 | 8 | 79 | 47 | 0.6005 | 10000 | 2206 | 52.5 | 0.5590 | 4 |
| 1679 | 8 | 79 | 47 | 0.6005 | 20000 | 2206 | 52.5 | 0.5510 | 4 |
| 1680 | 8 | 79 | 47 | 0.6005 | 30000 | 2206 | 52.5 | 0.5470 | 4 |
| 1681 | 8 | 79 | 47 | 0.6005 | 40000 | 2206 | 52.5 | 0.5410 | 4 |
| 1682 | 8 | 79 | 47 | 0.6005 | 48500 | 2206 | 52.5 | 0.5370 | 4 |
| 1683 | 8 | 79 | 47 | 0.8005 | 5000 | 2206 | 52.5 | 0.7890 | 4 |
| 1684 | 8 | 79 | 47 | 0.8005 | 10000 | 2206 | 52.5 | 0.7710 | 4 |
| 1685 | 8 | 79 | 47 | 0.8005 | 20000 | 2206 | 52.5 | 0.7640 | 4 |
| 1686 | 8 | 79 | 47 | 0.8005 | 30000 | 2206 | 52.5 | 0.7660 | 4 |
| 1687 | 8 | 79 | 47 | 0.8005 | 40000 | 2206 | 52.5 | 0.7610 | 4 |
| 1688 | 8 | 79 | 47 | 0.8005 | 48500 | 2206 | 52.5 | 0.7550 | 4 |
| 1689 | 3 | 82 | 16 | 0.8660 | 20000 | 13035 | 75.0 | 0.8070 | 7 |
| 1690 | 3 | 82 | 16 | 0.8660 | 25000 | 13035 | 75.0 | 0.8090 | 7 |
| 1691 | 3 | 82 | 16 | 0.8660 | 30000 | 13035 | 75.0 | 0.8130 | 7 |
| 1692 | 3 | 82 | 16 | 0.8660 | 35000 | 13035 | 75.0 | 0.8180 | 7 |
| 1693 | 3 | 82 | 16 | 0.8660 | 40000 | 13035 | 75.0 | 0.8260 | 7 |
| 1694 | 3 | 92 | 6 | 0.9520 | 30000 | 17166 | 15.5 | 0.9240 | 7 |
| 1695 | 3 | 92 | 6 | 0.9080 | 31500 | 17166 | 15.6 | 0.8660 | 7 |
| 1696 | 3 | 92 | 6 | 0.9520 | 31500 | 17166 | 15.6 | 0.9200 | 7 |
| 1697 | 9 | 92 | 6 | 0.9080 | 10000 | 3728 | 52.5 | 0.8550 | 7 |
| 1698 | 9 | 92 | 6 | 0.9080 | 15000 | 3728 | 52.5 | 0.8680 | 7 |
| 1699 | 9 | 92 | 6 | 0.9080 | 20000 | 3728 | 52.5 | 0.8780 | 7 |
| 1700 | 9 | 92 | 6 | 0.9080 | 25000 | 3728 | 52.5 | 0.8890 | 7 |
| 1701 | 9 | 92 | 6 | 0.9080 | 30000 | 3728 | 52.5 | 0.8980 | 7 |
| 1702 | 9 | 92 | 6 | 0.9080 | 35000 | 3728 | 52.5 | 0.9070 | 7 |
| 1703 | 9 | 92 | 6 | 0.9084 | 10000 | 3728 | 52.5 | 0.8493 | 5 |
| 1704 | 9 | 92 | 6 | 0.9084 | 15000 | 3728 | 52.5 | 0.8588 | 5 |
| 1705 | 9 | 92 | 6 | 0.9084 | 20000 | 3728 | 52.5 | 0.8648 | 5 |
| 1706 | 9 | 92 | 6 | 0.9084 | 25000 | 3728 | 52.5 | 0.8709 | 5 |
| 1707 | 9 | 92 | 6 | 0.9084 | 30000 | 3728 | 52.5 | 0.8747 | 5 |
| 1708 | 9 | 92 | 6 | 0.9084 | 35000 | 3728 | 52.5 | 0.8746 | 5 |
| 1709 | 9 | 92 | 7 | 0.9440 | 10000 | 3728 | 52.5 | 0.9090 | 7 |
| 1710 | 9 | 92 | 7 | 0.9440 | 15000 | 3728 | 52.5 | 0.9180 | 7 |
| 1711 | 9 | 92 | 7 | 0.9440 | 20000 | 3728 | 52.5 | 0.9260 | 7 |
| 1712 | 9 | 92 | 7 | 0.9440 | 25000 | 3728 | 52.5 | 0.9330 | 7 |
| 1713 | 9 | 92 | 7 | 0.9440 | 30000 | 3728 | 52.5 | 0.9380 | 7 |
| 1714 | 9 | 92 | 7 | 0.9440 | 35000 | 3728 | 52.5 | 0.9430 | 7 |
| 1715 | 9 | 92 | 7 | 0.9444 | 10000 | 3728 | 52.5 | 0.9061 | 5 |
| 1716 | 9 | 92 | 7 | 0.9444 | 15000 | 3728 | 52.5 | 0.9129 | 5 |
| 1717 | 9 | 92 | 7 | 0.9444 | 20000 | 3728 | 52.5 | 0.9179 | 5 |
| 1718 | 9 | 92 | 7 | 0.9444 | 25000 | 3728 | 52.5 | 0.9218 | 5 |
| 1719 | 9 | 92 | 7 | 0.9444 | 30000 | 3728 | 52.5 | 0.9243 | 5 |
| 1720 | 9 | 92 | 7 | 0.9444 | 35000 | 3728 | 52.5 | 0.9259 | 5 |

| 1721 | 3 | 92 | 8 | 0.8820 | 28500 | 17166 | 15.2 | 0.8370 | 7 |
|---|---|---|---|---|---|---|---|---|---|
| 1722 | 3 | 92 | 8 | 0.8810 | 30000 | 17166 | 15.5 | 0.8260 | 7 |
| 1723 | 3 | 92 | 13 | 0.8150 | 28500 | 17166 | 15.2 | 0.7250 | 7 |
| 1724 | 3 | 92 | 13 | 0.6880 | 31500 | 17166 | 15.6 | 0.5780 | 7 |
| 1725 | 3 | 92 | 13 | 0.7460 | 31500 | 17166 | 15.6 | 0.6440 | 7 |
| 1726 | 8 | 92 | 13 | 0.8150 | 11000 | 3552 | 15.4 | 0.7280 | 7 |
| 1727 | 9 | 92 | 14 | 0.9620 | 10000 | 3728 | 52.5 | 0.9390 | 7 |
| 1728 | 9 | 92 | 14 | 0.9620 | 15000 | 3728 | 52.5 | 0.9430 | 7 |
| 1729 | 9 | 92 | 14 | 0.9620 | 20000 | 3728 | 52.5 | 0.9460 | 7 |
| 1730 | 9 | 92 | 14 | 0.9620 | 25000 | 3728 | 52.5 | 0.9480 | 7 |
| 1731 | 9 | 92 | 14 | 0.9620 | 30000 | 3728 | 52.5 | 0.9490 | 7 |
| 1732 | 9 | 92 | 14 | 0.9620 | 35000 | 3728 | 52.5 | 0.9500 | 7 |
| 1733 | 9 | 92 | 14 | 0.9622 | 10000 | 3728 | 52.5 | 0.9359 | 5 |
| 1734 | 9 | 92 | 14 | 0.9622 | 15000 | 3728 | 52.5 | 0.9433 | 5 |
| 1735 | 9 | 92 | 14 | 0.9622 | 20000 | 3728 | 52.5 | 0.9471 | 5 |
| 1736 | 9 | 92 | 14 | 0.9622 | 25000 | 3728 | 52.5 | 0.9493 | 5 |
| 1737 | 9 | 92 | 14 | 0.9622 | 30000 | 3728 | 52.5 | 0.9507 | 5 |
| 1738 | 9 | 92 | 14 | 0.9622 | 35000 | 3728 | 52.5 | 0.9521 | 5 |
| 1739 | 9 | 92 | 15 | 0.8840 | 10000 | 3728 | 52.5 | 0.8220 | 7 |
| 1740 | 9 | 92 | 15 | 0.8840 | 15000 | 3728 | 52.5 | 0.8310 | 7 |
| 1741 | 9 | 92 | 15 | 0.8840 | 20000 | 3728 | 52.5 | 0.8380 | 7 |
| 1742 | 9 | 92 | 15 | 0.8840 | 25000 | 3728 | 52.5 | 0.8430 | 7 |
| 1743 | 9 | 92 | 15 | 0.8840 | 30000 | 3728 | 52.5 | 0.8450 | 7 |
| 1744 | 9 | 92 | 15 | 0.8840 | 35000 | 3728 | 52.5 | 0.8470 | 7 |
| 1745 | 9 | 92 | 15 | 0.8849 | 10000 | 3728 | 52.5 | 0.8242 | 5 |
| 1746 | 9 | 92 | 15 | 0.8849 | 15000 | 3728 | 52.5 | 0.8349 | 5 |
| 1747 | 9 | 92 | 15 | 0.8849 | 20000 | 3728 | 52.5 | 0.8432 | 5 |
| 1748 | 9 | 92 | 15 | 0.8849 | 25000 | 3728 | 52.5 | 0.8506 | 5 |
| 1749 | 9 | 92 | 15 | 0.8849 | 30000 | 3728 | 52.5 | 0.8544 | 5 |
| 1750 | 9 | 92 | 15 | 0.8849 | 35000 | 3728 | 52.5 | 0.8592 | 5 |
| 1751 | 9 | 92 | 16 | 0.8810 | 25000 | 3728 | 52.5 | 0.8290 | 5 |
| 1752 | 9 | 92 | 16 | 0.8810 | 10000 | 3728 | 52.5 | 0.8140 | 7 |
| 1753 | 9 | 92 | 16 | 0.8810 | 15000 | 3728 | 52.5 | 0.8210 | 7 |
| 1754 | 9 | 92 | 16 | 0.8810 | 20000 | 3728 | 52.5 | 0.8270 | 7 |
| 1755 | 9 | 92 | 16 | 0.8810 | 30000 | 3728 | 52.5 | 0.8290 | 7 |
| 1756 | 9 | 92 | 16 | 0.8810 | 35000 | 3728 | 52.5 | 0.8270 | 7 |
| 1757 | 9 | 92 | 16 | 0.8813 | 10000 | 3728 | 52.5 | 0.8191 | 5 |
| 1758 | 9 | 92 | 16 | 0.8813 | 15000 | 3728 | 52.5 | 0.8291 | 5 |
| 1759 | 9 | 92 | 16 | 0.8813 | 20000 | 3728 | 52.5 | 0.8394 | 5 |
| 1760 | 9 | 92 | 16 | 0.8813 | 25000 | 3728 | 52.5 | 0.8452 | 5 |
| 1761 | 9 | 92 | 16 | 0.8813 | 30000 | 3728 | 52.5 | 0.8500 | 5 |
| 1762 | 9 | 92 | 16 | 0.8813 | 35000 | 3728 | 52.5 | 0.8530 | 5 |
| 1763 | 3 | 92 | 22 | 0.5540 | 31500 | 17166 | 15.6 | 0.4740 | 7 |
| 1764 | 3 | 92 | 22 | 0.8330 | 31500 | 17166 | 15.6 | 0.7790 | 7 |
| 1765 | 3 | 92 | 22 | 0.9520 | 31500 | 17166 | 15.6 | 0.9370 | 7 |
| 1766 | 3 | 92 | 26 | 0.6810 | 31500 | 17166 | 15.6 | 0.5860 | 7 |

| | | | | | | | | | |
|---|---|---|---|---|---|---|---|---|---|
| 1767 | 9 | 92 | 26 | 0.9620 | 10000 | 3728 | 52.5 | 0.9440 | 7 |
| 1768 | 9 | 92 | 26 | 0.9620 | 15000 | 3728 | 52.5 | 0.9490 | 7 |
| 1769 | 9 | 92 | 26 | 0.9620 | 20000 | 3728 | 52.5 | 0.9530 | 7 |
| 1770 | 9 | 92 | 26 | 0.9620 | 25000 | 3728 | 52.5 | 0.9560 | 7 |
| 1771 | 9 | 92 | 26 | 0.9620 | 30000 | 3728 | 52.5 | 0.9580 | 7 |
| 1772 | 9 | 92 | 26 | 0.9620 | 35000 | 3728 | 52.5 | 0.9600 | 7 |
| 1773 | 9 | 92 | 26 | 0.9624 | 10000 | 3728 | 52.5 | 0.9436 | 5 |
| 1774 | 9 | 92 | 26 | 0.9624 | 15000 | 3728 | 52.5 | 0.9478 | 5 |
| 1775 | 9 | 92 | 26 | 0.9624 | 20000 | 3728 | 52.5 | 0.9506 | 5 |
| 1776 | 9 | 92 | 26 | 0.9624 | 25000 | 3728 | 52.5 | 0.9524 | 5 |
| 1777 | 9 | 92 | 26 | 0.9624 | 30000 | 3728 | 52.5 | 0.9538 | 5 |
| 1778 | 9 | 92 | 26 | 0.9624 | 35000 | 3728 | 52.5 | 0.9549 | 5 |
| 1779 | 3 | 92 | 28 | 0.4480 | 31500 | 17166 | 15.6 | 0.3530 | 7 |
| 1780 | 3 | 92 | 28 | 0.6700 | 31500 | 17166 | 15.6 | 0.5800 | 7 |
| 1781 | 3 | 92 | 29 | 0.4280 | 31500 | 17166 | 15.6 | 0.3430 | 7 |
| 1782 | 3 | 92 | 32 | 0.5220 | 31500 | 17166 | 15.6 | 0.4360 | 7 |
| 1783 | 8 | 92 | 41 | 0.8714 | 10000 | 3552 | 40.0 | 0.8161 | 9 |
| 1784 | 8 | 92 | 41 | 0.8714 | 15000 | 3552 | 40.0 | 0.8075 | 9 |
| 1785 | 8 | 92 | 41 | 0.8714 | 20000 | 3552 | 40.0 | 0.8027 | 9 |
| 1786 | 8 | 92 | 41 | 0.8714 | 25000 | 3552 | 40.0 | 0.7926 | 9 |
| 1787 | 8 | 92 | 41 | 0.8714 | 30000 | 3552 | 40.0 | 0.7791 | 9 |
| 1788 | 8 | 92 | 41 | 0.8714 | 35000 | 3552 | 40.0 | 0.7632 | 9 |
| 1789 | 8 | 92 | 41 | 0.8714 | 40000 | 3552 | 40.0 | 0.7555 | 9 |
| 1790 | 8 | 92 | 41 | 0.9259 | 10000 | 3552 | 40.0 | 0.9026 | 9 |
| 1791 | 8 | 92 | 41 | 0.9259 | 15000 | 3552 | 40.0 | 0.8796 | 9 |
| 1792 | 8 | 92 | 41 | 0.9259 | 20000 | 3552 | 40.0 | 0.8800 | 9 |
| 1793 | 8 | 92 | 41 | 0.9259 | 25000 | 3552 | 40.0 | 0.8792 | 9 |
| 1794 | 8 | 92 | 41 | 0.9259 | 30000 | 3552 | 40.0 | 0.8654 | 9 |
| 1795 | 8 | 92 | 41 | 0.9259 | 35000 | 3552 | 40.0 | 0.8562 | 9 |
| 1796 | 8 | 92 | 41 | 0.9259 | 40000 | 3552 | 40.0 | 0.8489 | 9 |
| 1797 | 8 | 92 | 41 | 0.9413 | 10000 | 3552 | 40.0 | 0.9026 | 9 |
| 1798 | 8 | 92 | 41 | 0.9413 | 15000 | 3552 | 40.0 | 0.9083 | 9 |
| 1799 | 8 | 92 | 41 | 0.9413 | 20000 | 3552 | 40.0 | 0.9066 | 9 |
| 1800 | 8 | 92 | 41 | 0.9413 | 25000 | 3552 | 40.0 | 0.9052 | 9 |
| 1801 | 8 | 92 | 41 | 0.9413 | 30000 | 3552 | 40.0 | 0.8964 | 9 |
| 1802 | 8 | 92 | 41 | 0.9413 | 35000 | 3552 | 40.0 | 0.8888 | 9 |
| 1803 | 8 | 92 | 41 | 0.9413 | 40000 | 3552 | 40.0 | 0.8793 | 9 |
| 1804 | 8 | 92 | 41 | 0.9594 | 10000 | 3552 | 40.0 | 0.9308 | 9 |
| 1805 | 8 | 92 | 41 | 0.9594 | 15000 | 3552 | 40.0 | 0.9319 | 9 |
| 1806 | 8 | 92 | 41 | 0.9594 | 20000 | 3552 | 40.0 | 0.9327 | 9 |
| 1807 | 8 | 92 | 41 | 0.9594 | 25000 | 3552 | 40.0 | 0.9340 | 9 |
| 1808 | 8 | 92 | 41 | 0.9594 | 30000 | 3552 | 40.0 | 0.9218 | 9 |
| 1809 | 8 | 92 | 41 | 0.9594 | 35000 | 3552 | 40.0 | 0.9255 | 9 |
| 1810 | 8 | 92 | 41 | 0.9594 | 40000 | 3552 | 40.0 | 0.9203 | 9 |
| 1811 | 8 | 92 | 41 | 0.9811 | 10000 | 3552 | 40.0 | 0.9675 | 9 |
| 1812 | 8 | 92 | 41 | 0.9811 | 15000 | 3552 | 40.0 | 0.9686 | 9 |

| | | | | | | | | | |
|---|---|---|---|---|---|---|---|---|---|
| 1813 | 8 | 92 | 41 | 0.9811 | 20000 | 3552 | 40.0 | 0.9669 | 9 |
| 1814 | 8 | 92 | 41 | 0.9811 | 25000 | 3552 | 40.0 | 0.9739 | 9 |
| 1815 | 8 | 92 | 41 | 0.9811 | 35000 | 3552 | 40.0 | 0.9680 | 9 |
| 1816 | 8 | 92 | 41 | 0.9811 | 40000 | 3552 | 40.0 | 0.9663 | 9 |
| 1817 | 8 | 92 | 41 | 0.9868 | 10000 | 3552 | 40.0 | 0.9760 | 9 |
| 1818 | 8 | 92 | 41 | 0.9868 | 15000 | 3552 | 40.0 | 0.9750 | 9 |
| 1819 | 8 | 92 | 41 | 0.9868 | 20000 | 3552 | 40.0 | 0.9747 | 9 |
| 1820 | 8 | 92 | 41 | 0.9868 | 25000 | 3552 | 40.0 | 0.9824 | 9 |
| 1821 | 8 | 92 | 41 | 0.9868 | 30000 | 3552 | 40.0 | 0.9691 | 9 |
| 1822 | 8 | 92 | 41 | 0.9868 | 35000 | 3552 | 40.0 | 0.9736 | 9 |
| 1823 | 8 | 92 | 41 | 0.9868 | 40000 | 3552 | 40.0 | 0.9709 | 9 |
| 1824 | 3 | 92 | 50 | 0.4010 | 31500 | 17166 | 15.6 | 0.3710 | 7 |
| 1825 | 8 | 92 | 51 | 0.9811 | 30000 | 3552 | 40.0 | 0.9691 | 9 |
| 1826 | 3 | 92 | 77 | 0.3810 | 31500 | 17166 | 15.6 | 0.3130 | 7 |

The sources of data used in the compilation are as follows:

1. Bastin's measurements of borides: G. F. Bastin and H. J. M. Heijligers, Report ISBN 90–6819–006–7 CIP, University for Physical Chemistry, Eindhoven, Netherlands.
2. Goldstein *et al.* measurements: J. I. Goldstein, F. J. Majeske, and H. Yakowitz, *Adv. X-Ray Anal.* **10**, 431 (1967).
3. J. Colby and D. K. Conley's measurements: J. Colby and D. K. Conley, in: *X-Ray Optics and Microanalysis*, (Castaing *et al.*, eds.), Hermann, Paris, p. 263 (1966).
4. NBS 1 measurements: K. F. J. Heinrich *et al.*, NBS Special Publication 260–28 (1971).
5. Pouchou's compilation: J.-L. Pouchou, in: *Electron Probe Quantitation* (K. F. J. Heinrich and D. E. Newbury, eds.) Plenum, New York, p. 64 (1991).
6. Sewell *et al.* compilation: D. A. Sewell, G. Love and V. D. Scott, *J. Phys. D: Appl. Phys.* **18**, 1245 (1985).
7. Bastin Data Basis: personal communication (1990).
8. NBS 2 measurements: R. Marinenko (NBS) and Hiroyoshi Konuma (Jap. Nat. Res. Inst. of Police Science, Tokyo), 1986 (unpublished).
9. P. F. Hlava's (Sandia) measurements: personal communication. See also: A. D. Romig Jr. *et al.*, in: *Microbeam Analysis—1987* (R. H. Geiss, ed.) San Francisco Press, San Francisco, p. 15 (1987)
10. Schreiber's measurements: T. P. Schreiber and R. A. Waldo, in: *Proceedings of ICXOM11*, (J. D. Brown and R. H. Packwood, eds.), U. London, Ontario, Canada, p.265 (1986).
11. Bastin compilation 2: personal communication (1990).

For the names of the authors of measurements in sources 5–7, the respective compilations should be consulted.